*Praise for*

# Solar Energy International

Quite possibly the best 10 days of instruction I've received on any topic!
The enthusiasm and technical expertise of the SEI staff is encouraging in
light of the problems of the world's energy demands. Thanks!

— Participant, PV Design & Installation, 2002

The course was exactly what I had hoped for. It covered all of the
key areas and built on them in each successive chapter. Well planned
and increasingly challenging. Well worth the cost and I am looking forward
to applying the knowledge to my own design. The contacts and links
were exceptional and will be used over and over again.

— Alan Greszler, Participant, PV Design Online workshop, 2003

This class was perfect and exactly what I needed
in order to get started with a career in the PV/RE industry.

— Participant, Advanced PV, 2002

I feel totally inspired to go home and apply what I have learned.
The instructors were a dynamic team who worked together very
well ... very complimentary. I would come back for sure!!!

— Participant, Women's PV, 2002

I will definitely return to SEI and suggest to anyone
who is interested in renewables to attend an SEI workshop.
Thank you so much for providing my first real step
towards living a path with a heart.

— Participant, Women's PV, 2002

This was a great introduction to solar energy.
I got much more out of it than I was expecting.

— Participant, PV Design & Installation, 2003

Could not ask for a better introduction, overview, and practical
experience for PV system design and installation – Thanks SEI!

— Fred Sharkey, Participant, PV Design & Installation, 2003

The instructors were extraordinary. I have never
experienced greater commitment, patience, patience, patience,
educational skills, technical skills, and care from anyone.
They're high on my "admired persons" list.

— Brian Burke, Participant, PV Design & Installation, 2003

## *Praise continued*

Another incredible workshop – I am so happy with the
education I have received and with the connections I have made.
Incredible people at an incredible organization!
— Participant, SEI workshops, 2002

This class concludes a full curriculum of the most intense and
informative and well-directed courses I could imagine having
the good fortune to participate in. Thank you very much!
— Jon Crowley, Participant, SEI workshop, 2003

# PHOTOVOLTAICS
## Design and Installation Manual

# PHOTOVOLTAICS
## Design and Installation Manual

Renewable Energy
Education for a
Sustainable Future

SOLAR ENERGY INTERNATIONAL

NEW SOCIETY PUBLISHERS

**Cataloguing in Publication Data:**
A catalog record for this publication is available from the National Library of Canada.

Cover design by Melody Warford. Photos by Solar Energy International staff.

Printed in Canada.
Revised and updated edition May 2007. Fifth printing April 2010.
Paperback ISBN: 978-0-86571-520-2

Inquiries regarding requests to reprint all or part of *Photovoltaics: Design and Installation Manual* should be addressed to New Society Publishers at the address below.

This book is a copublication of Solar Energy International and New Society Publishers.
To order directly from the publishers, please contact the publishers, below.

NEW SOCIETY PUBLISHERS
P.O. Box 189,
Gabriola Island, BC
V0R 1X0, Canada
1-800-567-6772 / www.newsociety.com

or

SOLAR ENERGY INTERNATIONAL
P.O. Box 715
Carbondale, CO 81623-0715
970-963-8855 / www. solarenergy.org

New Society Publishers' mission is to publish books that contribute in fundamental ways to building an ecologically sustainable and just society, and to do so with the least possible impact on the environment, in a manner that models this vision. We are committed to doing this not just through education, but through action. We are acting on our commitment to the world's remaining ancient forests by phasing out our paper supply from ancient forests worldwide. This book is one step towards ending global deforestation and climate change. It is printed on acid-free paper that is **100% old growth forest-free** (100% post-consumer recycled), processed chlorine free, and printed with vegetable based, low VOC inks. For further information, or to browse our full list of books and purchase securely, visit our website at www.newsociety.com

NEW SOCIETY PUBLISHERS                                                                 www.newsociety.com

Dedicated to the two billion people on this planet without access to electricity, in the hope that many will get electricity from solar energy, and to all the pioneers currently generating their own sustainable power with photovoltaics.

# Contents

List of Figures . . . . . . . . . . . . . . . . . . . . . . . . . . . . . . . . . . . . . . . . xv

List of Tables and Worksheets . . . . . . . . . . . . . . . . . . . . . . . . . . . . xvii

Acknowledgments and Disclaimer . . . . . . . . . . . . . . . . . . . . . . . . . xviii

Preface . . . . . . . . . . . . . . . . . . . . . . . . . . . . . . . . . . . . . . . . . . . . . xix

**Chapter 1: An Overview of Photovoltaics**

1.1  The Development of Photovoltaics  . . . . . . . . . . . . . . . . . . . 2
1.2  Current and Emerging Opportunities  . . . . . . . . . . . . . . . . . 2
1.3  Advantages of Photovoltaic Technology  . . . . . . . . . . . . . . . 3
1.4  Disadvantages of Photovoltaic Technology  . . . . . . . . . . . . . 3
1.5  Environmental, Health, and Safety Issues  . . . . . . . . . . . . . 3
1.6  Photovoltaic System Components  . . . . . . . . . . . . . . . . . . . 4
1.7  Photovoltaic System Types  . . . . . . . . . . . . . . . . . . . . . . . 4

**Chapter 2: Photovoltaic Electric Principles**

2.1  Terminology . . . . . . . . . . . . . . . . . . . . . . . . . . . . . . . . . . 10
2.2  Matching Appliances to the System . . . . . . . . . . . . . . . . . 11
2.3  Electrical Circuits . . . . . . . . . . . . . . . . . . . . . . . . . . . . . . . 11
2.4  Series and Parallel Circuits in Power Sources . . . . . . . . . . . 12
2.5  Series and Parallel Circuits in Electric Loads . . . . . . . . . . . 15
2.6  Series and Parallel Wiring Exercises . . . . . . . . . . . . . . . . . . 16
2.7  Solutions to Wiring Exercises . . . . . . . . . . . . . . . . . . . . . . 23

**Chapter 3: The Solar Resource**

3.1  Solar Radiation Fundamentals . . . . . . . . . . . . . . . . . . . . . . 30
3.2  Gathering Site Data . . . . . . . . . . . . . . . . . . . . . . . . . . . . . . 35
3.3  Completing the Solar Site Analysis . . . . . . . . . . . . . . . . . . . 36

**Chapter 4: Electric Load Analysis**

4.1  Using Energy Efficiently . . . . . . . . . . . . . . . . . . . . . . . . . . 38
4.2  Electrical Load Requirements . . . . . . . . . . . . . . . . . . . . . . . 38
4.3  Refrigeration . . . . . . . . . . . . . . . . . . . . . . . . . . . . . . . . . . . 39
4.4  Lighting . . . . . . . . . . . . . . . . . . . . . . . . . . . . . . . . . . . . . . . 41
4.5  Considerations for Calculating Load Estimates . . . . . . . . . . 43
4.6  Calculating Load Estimates . . . . . . . . . . . . . . . . . . . . . . . . 43

**Chapter 5: Photovoltaic Modules**

5.1    Photovoltaic Principles . . . . . . . . . . . . . . . . . . . . . . . . . . . . . . . . . 50

5.2    Characteristics of Modules . . . . . . . . . . . . . . . . . . . . . . . . . . . . . 51

5.3    Module Performance . . . . . . . . . . . . . . . . . . . . . . . . . . . . . . . . . . 52

5.4    Factors of Module Performance . . . . . . . . . . . . . . . . . . . . . . . . 54

**Chapter 6: Batteries**

6.1    Battery Types and Operation . . . . . . . . . . . . . . . . . . . . . . . . . . 60

6.2    Battery Specifications . . . . . . . . . . . . . . . . . . . . . . . . . . . . . . . . 62

6.3    Battery Safety . . . . . . . . . . . . . . . . . . . . . . . . . . . . . . . . . . . . . . 68

6.4    Battery Wiring Configuration . . . . . . . . . . . . . . . . . . . . . . . . . 69

6.5    Battery Sizing Exercise . . . . . . . . . . . . . . . . . . . . . . . . . . . . . . . 71

**Chapter 7: PV Controllers**

7.1    Controller Types . . . . . . . . . . . . . . . . . . . . . . . . . . . . . . . . . . . . 74

7.2    Controller Features . . . . . . . . . . . . . . . . . . . . . . . . . . . . . . . . . . 75

7.3    Specifying a Controller . . . . . . . . . . . . . . . . . . . . . . . . . . . . . . . 78

7.4    Controller Sizing Exercise . . . . . . . . . . . . . . . . . . . . . . . . . . . . 79

**Chapter 8: Inverters**

8.1    Inverter Operating Principles . . . . . . . . . . . . . . . . . . . . . . . . . 84

8.2    Inverter Types . . . . . . . . . . . . . . . . . . . . . . . . . . . . . . . . . . . . . . 84

8.3    Inverter Features . . . . . . . . . . . . . . . . . . . . . . . . . . . . . . . . . . . . 85

8.4    Batteryless Grid-tied Inverters . . . . . . . . . . . . . . . . . . . . . . . . 86

8.5    Grid-tied with Battery Back-up Inverters . . . . . . . . . . . . . . . 87

8.6    Stand-alone Inverters . . . . . . . . . . . . . . . . . . . . . . . . . . . . . . . . 88

8.7    AC Coupled Systems . . . . . . . . . . . . . . . . . . . . . . . . . . . . . . . . 89

8.8    Stand-alone Inverter Sizing Exercise . . . . . . . . . . . . . . . . . . . 90

**Chapter 9: Photovoltaic System Wiring**

9.1    Introduction . . . . . . . . . . . . . . . . . . . . . . . . . . . . . . . . . . . . . . . . 92

9.2    Wire Size . . . . . . . . . . . . . . . . . . . . . . . . . . . . . . . . . . . . . . . . . . 94

9.3    System Wire Sizing Exercise . . . . . . . . . . . . . . . . . . . . . . . . . 100

9.4    Overcurrent Protection . . . . . . . . . . . . . . . . . . . . . . . . . . . . . 107

9.5    Overcurrent Protection Sizing Exercise . . . . . . . . . . . . . . . . 107

9.6    Disconnects . . . . . . . . . . . . . . . . . . . . . . . . . . . . . . . . . . . . . . . 108

9.7    Grounding . . . . . . . . . . . . . . . . . . . . . . . . . . . . . . . . . . . . . . . . 108

9.8    Surge Suppression . . . . . . . . . . . . . . . . . . . . . . . . . . . . . . . . . 113

**Chapter 10: Sizing Stand-alone PV Systems**

10.1 Introduction to Sizing PV Systems . . . . . . . . . . . . . . . . . . . . . 116
10.2 Design Penalties . . . . . . . . . . . . . . . . . . . . . . . . . . . . . . . . 116
10.3 Sizing Worksheet Explanation . . . . . . . . . . . . . . . . . . . . . . 117
10.4 Sample System Exercise . . . . . . . . . . . . . . . . . . . . . . . . . . 119
10.5 Hybrid Systems with Generators . . . . . . . . . . . . . . . . . . . . 124

**Chapter 11: Grid-tied PV Systems**

11.1 Introduction . . . . . . . . . . . . . . . . . . . . . . . . . . . . . . . . . . 126
11.2 Grid-tied System Types and Advantages . . . . . . . . . . . . . . . 126
11.3 System Sizing and Economics . . . . . . . . . . . . . . . . . . . . . . 130
11.4 Obtaining an Interconnection Agreement . . . . . . . . . . . . . . 131
11.5 Net Metering . . . . . . . . . . . . . . . . . . . . . . . . . . . . . . . . . 131
11.6 Sample System Exercise . . . . . . . . . . . . . . . . . . . . . . . . . . 133

**Chapter 12: Mounting Photovoltaic Modules**

12.1 Mounting System Types . . . . . . . . . . . . . . . . . . . . . . . . . . 140
12.2 Building Integrated Photovoltaics . . . . . . . . . . . . . . . . . . . . 143

**Chapter 13: PV Applications for the Developing World**

13.1 The Need for Reliable Electricity . . . . . . . . . . . . . . . . . . . . 146
13.2 Lighting . . . . . . . . . . . . . . . . . . . . . . . . . . . . . . . . . . . . . 146
13.3 Television and Radio . . . . . . . . . . . . . . . . . . . . . . . . . . . . 147
13.4 Health Care and Refrigeration . . . . . . . . . . . . . . . . . . . . . . 147
13.5 Micro-Enterprises . . . . . . . . . . . . . . . . . . . . . . . . . . . . . . 147
13.6 Water Pumping . . . . . . . . . . . . . . . . . . . . . . . . . . . . . . . . 148
13.7 Determining Solar Access with a Sun Chart . . . . . . . . . . . . . 152
13.8 Sample Installation Materials . . . . . . . . . . . . . . . . . . . . . . . 154

**Chapter 14: System Installation**

14.1 Site Evaluation . . . . . . . . . . . . . . . . . . . . . . . . . . . . . . . . 158
14.2 Photovoltaic Array Installation . . . . . . . . . . . . . . . . . . . . . . 158
14.3 Battery Installation . . . . . . . . . . . . . . . . . . . . . . . . . . . . . . 158
14.4 Controller and Inverter Installation . . . . . . . . . . . . . . . . . . . 159
14.5 Photovoltaic System Wire Installation . . . . . . . . . . . . . . . . . 160
14.6 PV System Installations Final Checklist . . . . . . . . . . . . . . . . 162

**Chapter 15: Maintenance and Troubleshooting**

15.1 Materials and Tools List . . . . . . . . . . . . . . . . . . . . . . . . . 168
15.2 Maintaining PV Components . . . . . . . . . . . . . . . . . . . . . 168
15.3 Troubleshooting Common System Faults . . . . . . . . . . . . . 170
15.4 Troubleshooting Wiring Problems Using a Multimeter . . . . . . 170
15.5 Troubleshooting Specific Problems . . . . . . . . . . . . . . . . . 173

**Chapter 16: Installation Safety**

16.1 Introduction . . . . . . . . . . . . . . . . . . . . . . . . . . . . . . 180
16.2 Basic Safety . . . . . . . . . . . . . . . . . . . . . . . . . . . . . . . 180
16.3 Safely Testing High Voltage . . . . . . . . . . . . . . . . . . . . . 182
16.4 Hazards . . . . . . . . . . . . . . . . . . . . . . . . . . . . . . . . . 182
16.5 Safety Equipment . . . . . . . . . . . . . . . . . . . . . . . . . . . 184
16.6 Site Safety . . . . . . . . . . . . . . . . . . . . . . . . . . . . . . . . 185
16.7 First Aid . . . . . . . . . . . . . . . . . . . . . . . . . . . . . . . . . 186

**Appendix A: Glossary** . . . . . . . . . . . . . . . . . . . . . . . . . . . **191**

**Appendix B: Solar Data** . . . . . . . . . . . . . . . . . . . . . . . . . . **201**

**Appendix C: Sun Charts** . . . . . . . . . . . . . . . . . . . . . . . . . **275**

**Appendix D: System Sizing Worksheets** . . . . . . . . . . . . . . . . **283**

**Resource Guide** . . . . . . . . . . . . . . . . . . . . . . . . . . . . . . . **305**

**Index** . . . . . . . . . . . . . . . . . . . . . . . . . . . . . . . . . . . . . **321**

# List of Figures

Figure 1-1: Day Use System . . . . . . . . . . . . . . . . . . . . . . . . . . . . . . . 5

Figure 1-2: DC System with Batteries . . . . . . . . . . . . . . . . . . . . . . . . 5

Figure 1-3: System with DC and AC Loads . . . . . . . . . . . . . . . . . . . . 6

Figure 1-4: Hybrid System . . . . . . . . . . . . . . . . . . . . . . . . . . . . . . . . . 6

Figure 1-5: Grid-tied System without Batteries . . . . . . . . . . . . . . . . . 7

Figure 1-6: Grid-tied System with Batteries . . . . . . . . . . . . . . . . . . . 8

Figure 2-1: Electrical Circuits . . . . . . . . . . . . . . . . . . . . . . . . . . . . . 11

Figure 2-2: PV Modules in Series . . . . . . . . . . . . . . . . . . . . . . . . . . 12

Figure 2-3: PV Modules in Parallel . . . . . . . . . . . . . . . . . . . . . . . . . 12

Figure 2-4: PV Modules in Series and Parallel . . . . . . . . . . . . . . . . . 13

Figure 2-5: Batteries in Series and Parallel . . . . . . . . . . . . . . . . . . . 13

Figure 2-6: High Voltage Array . . . . . . . . . . . . . . . . . . . . . . . . . . . . 14

Figure 2-7: Loads in Series . . . . . . . . . . . . . . . . . . . . . . . . . . . . . . . 15

Figure 2-8: Loads in Parallel . . . . . . . . . . . . . . . . . . . . . . . . . . . . . . 15

Figure 3-1: The Sun's Path Throughout the Year — Northern Latitudes . 30

Figure 3-2: Magnetic Declination in the United States . . . . . . . . . . . . 31

Figure 3-3: World Magnetic Declination Chart . . . . . . . . . . . . . . . . . 32

Figure 3-4: Azimuth and Altitude for all Northern Latitudes . . . . . . . . 32

Figure 3-5: Effect of Tilt Angle . . . . . . . . . . . . . . . . . . . . . . . . . . . . 33

Figure 3-6: Effects of Array Tilt on Energy Production . . . . . . . . . . . . 34

Figure 3-7: The Solar Window . . . . . . . . . . . . . . . . . . . . . . . . . . . . . 35

Figure 4-1: Watts Consumed by Common Phantom Loads . . . . . . . . . . 39

Figure 5-1: Photovoltaic Terminology . . . . . . . . . . . . . . . . . . . . . . . . 50

Figure 5-2: Module (Brand X) I-V Curve (12VDC nominal) . . . . . . . . 52

Figure 5-3: Specification Label . . . . . . . . . . . . . . . . . . . . . . . . . . . . . 53

Figure 5-4: Effect of Insolation on Module Performance . . . . . . . . . . . 54

Figure 5-5: Effect of Cell Temperature on Module Performance . . . . . . . 55

Figure 5-6: Effect of Shading on Module Performance . . . . . . . . . . . . . . . 57

Figure 6-1: Cut-away of a Standard Lead Acid Battery Cell . . . . . . . . . . 60

Figure 6-2: Number of Battery Cycles in Relation to Depth of Discharge . . 65

Figure 6-3: Effects of Temperature on Battery Capacity . . . . . . . . . . . . . 66

Figure 6-4: 12 Volt Battery Configurations . . . . . . . . . . . . . . . . . . . . . 69

Figure 6-5: 24-Volt Battery Configurations . . . . . . . . . . . . . . . . . . . . . 70

Figure 6-6: 48-Volt Battery Configurations . . . . . . . . . . . . . . . . . . . . . 71

Figure 7-1: Affects of Battery on Module Power Production . . . . . . . . . . 76

Figure 7-2: Controller Step Down Feature . . . . . . . . . . . . . . . . . . . . . . 77

Figure 8-1: Common Waveforms Produced by Inverters . . . . . . . . . . . . 85

Figure 8-2: Efficiency of a 4000 Watt Inverter . . . . . . . . . . . . . . . . . . . 86

Figure 9-1: AC and DC Load Schematic . . . . . . . . . . . . . . . . . . . . . . . 106

Figure 9-2: Grounding . . . . . . . . . . . . . . . . . . . . . . . . . . . . . . . . . . . 110

Figure 9-3: Equipment Grounding Schematic for Stand-alone System . . . 111

Figure 9-4: Equipment Grounding Schematic for Grid-tied System . . . 112

Figure 9-5: System and Equipment Grounding Schematic . . . . . . . . . . . 114

Figure 10-1: Sample System Exercise Schematic . . . . . . . . . . . . . . . . . . 123

Figure 11-1: Grid-tied System without Battery Back-up . . . . . . . . . . . . 128

Figure 11-2: Grid-tied with Battery Back-up System . . . . . . . . . . . . . . 129

Figure 11-3: Uninterruptible Power Supply System . . . . . . . . . . . . . . . 130

Figure 11-4: Net Metering: Grid-tied System . . . . . . . . . . . . . . . . . . . . 132

Figure 11-5: Sample System Exercise Schematic . . . . . . . . . . . . . . . . . . 137

Figure 12-1: Basic Mounting Strategies . . . . . . . . . . . . . . . . . . . . . . . 140

Figure 12-2: PV Facade - Sawtooth Design . . . . . . . . . . . . . . . . . . . . . 144

Figure 13-1: Pump Curves . . . . . . . . . . . . . . . . . . . . . . . . . . . . . . . . 149

Figure 13-2: Visualizing the Four Sides of the Solar Window . . . . . . . . 153

# List of Tables and Worksheets

Load Estimation Worksheet . . . . . . . . . . . . . . . . . . . . . . . . . . . . . . . . . . . 44

Table 4-1: Typical Wattage Requirements for Common Appliances . . . . . 46

Load Estimation Worksheet . . . . . . . . . . . . . . . . . . . . . . . . . . . . . . . . . . . 47

Table 5-1: Effects of shading on module power . . . . . . . . . . . . . . . . . . . 56

Table 6-1: Voltage Set Points for Lead-Acid Batteries in a 12V System . . 63

Table 6-2: Effect of Discharge Rate on Battery Capacity . . . . . . . . . . . . 64

Table 6-3: Battery Temperature Multiplier at Various Temperatures . . . . . 66

Table 6-4: Liquid Electrolyte Freeze Points, Specific Gravity, and Voltage  67

Table 6-5: Battery Sizing Worksheet  . . . . . . . . . . . . . . . . . . . . . . . . . . 71

Table 6-6: Answers to the Battery Sizing Exercise  . . . . . . . . . . . . . . . . 72

Table 7-1: Controller Sizing Worksheet . . . . . . . . . . . . . . . . . . . . . . . . 79

Table 7-2: Answers to the Controller Sizing Exercise . . . . . . . . . . . . . . 79

Stand-alone Array & MPPT Controller Sizing Worksheet  . . . . . . . . 80-82

Table 8-1: Inverter Sizing Worksheet . . . . . . . . . . . . . . . . . . . . . . . . . . 90

Table 8-2: Answers to the Inverter Sizing Exercise . . . . . . . . . . . . . . . . 90

Table 9-1: Wire Types  . . . . . . . . . . . . . . . . . . . . . . . . . . . . . . . . . . . . 92

Table 9-2: Color Coding of Wires . . . . . . . . . . . . . . . . . . . . . . . . . . . . 93

Table 9-3: Cable Types . . . . . . . . . . . . . . . . . . . . . . . . . . . . . . . . . . . . 94

Table 9-4: Ampacity of Copper Wire . . . . . . . . . . . . . . . . . . . . . . . . . . 95

Table 9-5: Length (feet) of 12V Copper Wire for 2% Voltage Drop  . . . . 97

Table 9-6: Length (feet) of 24V Copper Wire for 2% Voltage Drop  . . . . 98

Table 9-7: Length (feet) of 48V Copper Wire for 2% Voltage Drop  . . . . 99

Table 9-8: Voltage Drop Index Chart . . . . . . . . . . . . . . . . . . . . . . . . . 103

System Wire Sizing Worksheet . . . . . . . . . . . . . . . . . . . . . . . . . 104-105

Answers to the Stand-Alone Sizing Exercise . . . . . . . . . . . . . . . . 121-122

Answers to the Grid-tied Sizing Exercise . . . . . . . . . . . . . . . . . . . 134-136

Table 13-1: Vertical Angle from Horizon to Four Sides of Solar Window 154

Table 13-2: Sample Installation Materials . . . . . . . . . . . . . . . . . . . . . . 156

## Acknowledgments

Solar Energy International extends a great big thank you to: Steve McCarney, Ken Olson, and Johnny Weiss, who began developing PV training materials while teaching together in the '80s; sincere gratitude goes to the dynamic SEI teachers, Laurie Stone, Ed Eaton, Justine Sanchez, and Carol Weis, who provided invaluable contributions to the fact-finding, researching and written composition of this book. Additional thanks go to the current administrative staff of Sandy Pickard, Kathy Fontaine, Jeff Tobe,d and Kevin Lundy. SEI would like to thank technical proofreaders Dan Rauch- SEI intern, Kevin Ulrich and the late Dan Dean of Solar Flare Institute, Paul Owens of San Juan Community College, and Jay Peltz of Peltz Power. We also wish to give thanks to John Wiles and the Southwest Technology Development Institute, National Renewable Energy Laboratory, Sandia National Laboratories, Juan Livingstone, Mark Colby, Ian Woofenden- SEI's NW coordinator, the USGS, and the National Fire Protection Association, for supporting this project through technical contributions. Also, for book production help we thank our co-publisher New Society Publishing, Harlan Feder, Marianne Ackerman, Gregory Keith, Rachel Tomich and Melody Warford. We give a special thanks to Richard and Karen Perez and to the *Home Power* magazine crew for their enduring inspiration, generosity and dedication to hands-on renewable energy. Finally, we would like to warmly thank Mark Fitzgerald of the Institute for Sustainable Power for his strategic advice and guidance that has been instrumental to the development of this manual.

For this revised edition (07), Solar Energy International would like to sincerely thank Carol Weis, Jeff Tobe, Laurie Stone, Justine Sanchez, Laura Walters, Skye-Laurel Riggs, and Khanti Munro for their invaluable contributions in enhancing and updating this book. We would also like to thank the technical proof readers Kris Sutton, Christopher Freitas, Darren Emmons, Jeffrey Philpott, Bill Brooks, John Wiles, boB Gudgel, and Robin Gudgel. Additional thanks to SEI staff, Sandy Pickard, Kathy Fontaine, Kevin Lundy, Matthew Harris, Rachel Connor, Soozie Lindbloom, Kathy Swartz, and Dave Straumfjord. Finally we would like to express thanks for the production of this book to New Society Publishing, Marianne Ackerman, and SEI Intern Jason Albert for updating the textbook cover.

## Disclaimer

Reference herein to any specific commercial product, process or service by trade name, trademark, manufacturer, or otherwise, does not necessarily constitute or imply the endorsement, recommendation or preference by the authors or publishers. The authors and publishers do not maintain that their methods and recommendations are exclusive to others for the design of photovoltaic power systems. Furthermore, no local, state, or federal code regulations have been referenced in this manual. Performances of photovoltaic power systems may differ depending upon the nature of the particular application, the specific circumstances relevant to the application, and the quality of the installation. Neither the authors nor the publisher assume any legal liability for systems installed using any information, apparatus, product or process disclosed within this publication, or for the performance of systems installed using the information contained in this publication.

## Safety Disclaimer

We encourage all users of this manual to work safely with photovoltaic systems. Use sound judgment with all photovoltaic equipment, assembly and installation practices on and off the worksite. Always consult a photovoltaic professional, local, and/or state electrical authority. US users must follow National Electrical Code (NEC) practices. This book is not intended as a do-it-yourself manual. Solar Energy International disclaims responsibility for any injury, damage, or other loss suffered related to any information presented in this manual. The reader of this manual acknowledges the inherently dangerous

# Preface

## Preface to the Original Edition ('04)

The evolution of this manual has been a great adventure. It has grown organically, been rewritten, updated and revised many times by many solar professionals. In the early 80's, Steve McCarney, Ken Olson and Johnny Weiss originated a PV textbook for the Solar Training Program we were developing and team-teaching at Colorado Mountain College. Over the decade, with assistance from Luke Elliot and Dr. Mark McCray, different versions served as a teaching tool in conjunction with our vocational training. Later, as Appropriate Technology Associates, Solar Technology Institute, and as Sustainable Technologies International, the manual became an important training aid. Since 1991, Solar Energy International (SEI), a 501(c)(3) nonprofit educational organization, has revised editions as part of its ongoing PV training programs.

Now, the current SEI instructional team is proud to present Photovoltaics: Design and Installation Manual. This major new effort includes new chapters on System Wiring, Utility Interconnected Systems, and Building Integrated PV. All chapters, the Appendixes, and the glossary, have been significantly updated and expanded. The "textbook" format is performance-based and generally presents a non-product-specific approach.

It is SEI's hope that this new manual will make a contribution to the growing understanding of PV system design and installation by providing a thorough basis for classroom education, laboratory training and hands-on field installation work.

May the PV adventure continue!

Johnny Weiss
Executive Director, SEI
April 2004


## Preface to the Revised Edition ('07)

Now is an exciting time for photovoltaics and for renewable energy. The entire solar energy market has been growing at 30% annually and will soon surpass $50 billion per year. Keeping our PV educational curriculum current is an ongoing challenge for the SEI Instructional Staff. This revised edition includes information about the latest PV industry innovations of both maximum power point tracking controllers and grid-tied inverters. We have also updated sections on specifying and sizing equipment and have included new information on high voltage PV modules and arrays. Today's environmental awareness, policy directives, peak oil crisis, technological advances, and common sense all create growing demand for clean energy. PV plays a significant part in the world's sustainable energy future. On with the PV adventure!

Johnny Weiss
Executive Director, SEI
April 2007

# Chapter 1
# An Overview of Photovoltaics

## Contents:

1.1  The Development of Photovoltaics . . . . . . . . . . . . . . . . . . . . . . . . 2
1.2  Current and Emerging Opportunities . . . . . . . . . . . . . . . . . . . . . 2
1.3  Advantages of Photovoltaic Technology . . . . . . . . . . . . . . . . . . . 3
1.4  Disadvantages of Photovoltaic Technology . . . . . . . . . . . . . . . . . 3
1.5  Environmental, Health, and Safety Issues . . . . . . . . . . . . . . . . . 3
1.6  Photovoltaic System Components . . . . . . . . . . . . . . . . . . . . . . . 4
1.7  Photovoltaic System Types . . . . . . . . . . . . . . . . . . . . . . . . . . . . 4

# 1.1 The Development of Photovoltaics

Photovoltaic systems are solar energy systems that produce electricity directly from sunlight. Photovoltaic (PV) systems produce clean, reliable energy without consuming fossil fuels and can be used in a wide variety of applications. A common application of PV technology is providing power for watches and radios. On a larger scale, many utilities have recently installed large photovoltaic arrays to provide consumers with solar-generated electricity, or as back-up systems for critical equipment.

Research into photovoltaic technology began over one hundred years ago. In 1873, British scientist Willoughby Smith noticed that selenium was sensitive to light. Smith concluded that selenium's ability to conduct electricity increased in direct proportion to the degree of its exposure to light. This observation of the photovoltaic effect led many scientists to experiment with this relatively uncommon element with the hope of using the material to create electricity. In 1880, Charles Fritts developed the first selenium-based solar electric cell. The cell produced electricity without consuming any material substance, and without generating heat.

Broader acceptance of photovoltaics as a power source didn't occur until 1905, when Albert Einstein offered his explanation of the photoelectric effect. Einstein's theories led to a greater understanding of the physical process of generating electricity from sunlight. Scientists continued limited research on the selenium solar cell through the 1930's, despite its low efficiency and high production costs.

In the early 1950s, Bell Laboratories began a search for a dependable way to power remote communication systems. Bell scientists discovered that silicon, the second most abundant element on earth, was sensitive to light and, when treated with certain impurities, generated a substantial voltage. By 1954, Bell developed a silicon-based cell that achieved six percent efficiency.

The first non-laboratory use of photovoltaic technology was to power a telephone repeater station in rural Georgia in the late 1950s. National Aeronautics and Space Administration (NASA) scientists, seeking a lightweight, rugged and reliable energy source suitable for outer space, installed a PV system consisting of 108 cells on the United States' first satellite, Vanguard I. By the early 1960s, PV systems were being installed on most satellites and spacecraft.

Today, solar modules supply electricity to more than 1 million homes worldwide, producing thousands of jobs and creating sustainable economic opportunities. In 2005, world solar photovoltaic market installations totaled 1,460 megawatts and over $7 billion in the global market (2005 Solar Buzz Inc. Report). The applications include communications, refrigeration for health care, crop irrigation, water purification, lighting, cathodic protection, environmental monitoring, marine and air navigation, utility power, and other residential and commercial applications. The intense interest generated by current photovoltaic applications provides promise for this rapidly developing technology.

# 1.2 Current and Emerging Opportunities

Conventional fuel sources have created a myriad of environmental problems, such as global warming, acid rain, smog, water pollution, rapidly filling waste disposal sites, destruction of habitat from fuel spills, and the loss of natural resources. Photovoltaic systems do not pose these environmental consequences. Today, the majority of PV modules use silicon as their major component. The silicon cells manufactured from one ton of sand can produce as much electricity as burning 500,000 tons of coal.

Photovoltaic technology also creates jobs. In 2005 the solar industry directly employed at least 55,000 people in jobs such as research, manufacturing, development, and installation. According to the 2005 Solar Buzz Inc. Report, the PV installation market grew 34% worldwide in 2005. The field is growing and there is a need for qualified installation professionals.

Economists have predicted that photovoltaics will be the most rapidly growing form of commercial energy after 2030, with sales exceeding $100 billion. In fact, the use of solar and renewable energy is expected to double by the year 2010, which would create more than 350,000 new jobs. It is no surprise that this clean, reliable source of electric power is regarded as the future of energy production.

# 1.3 Advantages of Photovoltaic Technology

Photovoltaic systems offer substantial advantages over conventional power sources:

**Reliability:** Even in harsh conditions, photovoltaic systems have proven their reliability. PV arrays prevent costly power failures in situations where continuous operation is critical.

**Durability:** Most modules are guaranteed from the manufacturer to produce power for 25 years, and will keep producing well beyond that time frame.

**Low maintenance cost:** Transporting materials and personnel to remote areas for equipment maintenance or service work is expensive. Since PV systems require only periodic inspection and occasional maintenance, these costs are usually less than with conventionally fueled systems.

**No fuel cost:** Since no fuel source is required, there are no costs associated with purchasing, storing, or transporting fuel.

**Reduced sound pollution:** Photovoltaic systems operate silently and with minimal movement.

**Photovoltaic modularity:** Modules may be added incrementally to a photovoltaic system to increase available power.

**Safety:** PV systems do not require the use of combustible fuels and are safe when properly designed and installed.

**Independence:** Many residential PV users cite energy independence from utilities as their primary motivation for adopting the new technology.

**Electrical grid decentralization:** Small-scale decentralized power stations reduce the possibility of outages on the electric grid.

**High altitude performance:** Increased insolation at high altitudes makes using photovoltaics advantageous, since power output is optimized. In contrast, a diesel generator at higher altitudes must be de-rated because of losses in efficiency and power output.

# 1.4 Disadvantages of Photovoltaic Technology

Photovoltaics have some disadvantages when compared to conventional power systems:

**Initial cost:** Each PV installation must be evaluated from an economic perspective and compared to existing alternatives. As the initial cost of PV systems decreases and the cost of conventional fuel sources increases, these systems will become more economically competitive.

**Variability of available solar radiation:** Weather can greatly affect the power output of any solar-based energy system. Variations in climate or site conditions require modifications in system design.

**Energy storage:** Some PV systems use batteries for storing energy, increasing the size, cost, and complexity of a system.

**Efficiency improvements:** A cost-effective use of photovoltaics requires a high-efficiency approach to energy consumption. This often dictates replacing inefficient appliances.

**Education:** PV systems present a new and unfamiliar technology; few people understand their value and feasibility. This lack of information slows market and technological growth.

# 1.5 Environmental, Health, and Safety Issues

Electricity produced from photovoltaics is more environmentally benign than conventional sources of energy production. However, there are environmental, safety, and health issues associated with manufacturing, using, and disposing of photovoltaic equipment.

The manufacturing of electronic equipment is energy intensive. On the other hand, photovoltaic modules produce more electricity in their lifetime than it takes to produce them. An energy break-even point is usually achieved after approximately one to three years.

As with any manufacturing process, producing photovoltaic modules often poses environmental and health hazards. Workers may be exposed to toxic and potentially explosive gases, such as phosphine, diborane, hydrogen deselenide, and cadmium compounds. Manufacturers have made steps to minimize environmental and worker hazards by implementing carefully designed industrial processes and monitoring systems.

Safety for installation technicians is also a concern. Only qualified personnel, using equipment that complies with national safety standards, should install photovoltaic systems.

The disposal of photovoltaic system components poses a moderate environmental hazard. Most solar modules have an expected useful life of at least 25 years. Most of the components can be recycled or reused (for example, glass and plastic encasements, and aluminum frames), but semiconductor recycling is just starting to be addressed in the industry.

# 1.6 Photovoltaic System Components

Photovoltaic systems are built from several important components:

**Photovoltaic cell**: Thin squares, discs, or films of semiconductor material that generate voltage and current when exposed to sunlight.

**Module**: A configuration of PV cells laminated between a clear superstrate (glazing) and an encapsulating substrate.

**Panel**: One or more modules fastened together (often used interchangeably with "module").

**Array**: One or more panels wired together for a specific voltage and fastened to a mounting structure.

**Charge controller**: Equipment that regulates battery voltage.

**Battery**: A device that chemically stores direct current (DC) electrical energy.

**Inverter**: An electrical device that changes direct current to alternating current (AC).

**DC loads**: Appliances, motors, and equipment powered by direct current.

**AC loads**: Appliances, motors, and equipment powered by alternating current.

# 1.7 Photovoltaic System Types

Photovoltaic systems can be configured in many ways. For example, many residential systems use battery storage to power appliances during the night. In contrast, water pumping systems often operate only during the day and require no storage device. A large commercial system would likely have an inverter to power AC appliances, whereas a system in a small cabin would likely power only DC appliances and wouldn't need an inverter. Some systems are linked to the utility grid, while others operate independently.

## Integrated Photovoltaic Battery-charging Systems

These systems incorporate all their components, including the application, in a single package. This arrangement may be economical when it compliments or replaces a disposable battery system. Small appliances, complete with a rechargeable battery and integrated PV battery-chargers, are a common example. Solar lanterns and photovoltaic chargers for radio batteries have worldwide market potential. Kits for photovoltaic flashlights, clocks, and radios may eventually replace similar units that use expensive, wasteful, disposable batteries.

## Day Use Systems

The simplest and least expensive photovoltaic systems are designed for day use only. These systems consist of modules wired directly to a DC appliance, with no storage device. See Figure 1-1. When the sun shines on the modules, the appliance consumes the electricity they generate. Higher insolation (sunshine) levels result in increased power output and greater load capacity.

Examples of day use systems include:

- Remote water pumping for a storage tank
- Operation of fans, blowers, or circulators to distribute thermal energy for solar water heating systems or ventilation systems

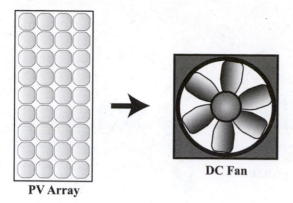

**Figure 1-1**
**DAY USE SYSTEM**

## Direct Current Systems with Storage Batteries

To operate loads at night or during cloudy weather, PV systems must include a means of storing electrical energy. Batteries are the most common solution. System loads can be powered from the batteries during the day or night, continuously or intermittently, regardless of weather. In addition, a battery bank has the capacity to supply high-surge currents for a brief period, giving the system the ability to start large motors or to perform other difficult tasks. A simple DC system that uses batteries is illustrated in Figure 1-2. This system's basic components include a PV module, charge controller, storage batteries, and appliances (the system's electrical load).

A battery bank can range from small flashlight-size batteries to dozens of heavy-duty industrial batteries. Deep-cycle batteries are designed to withstand being deeply discharged and then fully recharged when the sun shines. (Conventional automobile batteries are not well suited for use in photovoltaic systems and will have short effective lives.) The size and configuration of the battery bank depends on the operating voltage of the system and the amount of nighttime usage. In addition, local weather conditions must be considered in sizing a battery bank. The number of modules must be chosen to adequately recharge the batteries during the day.

Batteries must not be allowed to discharge too deeply or be overcharged — either situation will damage them severely. A charge controller will prevent the battery from overcharging by automatically disconnecting the module from the battery bank when it is fully loaded. Some charge controllers also prevent batteries from reaching dangerously low charge levels by stopping the supply of power to the DC load. Providing charge control is critical to maintaining battery performance in all but the simplest of PV systems.

## Direct Current Systems Powering Alternating Current Loads

Photovoltaic modules produce DC electrical power, but many common appliances require AC power. Direct current systems that power AC loads must use an inverter to convert DC electricity into AC. Inverters provide convenience and flexibility in a photovoltaic system, but add complexity and cost. Because AC appliances are mass-produced, they are generally offered in a wider selection, at lower cost, and with higher reliability than DC appliances. High quality inverters are commercially available in a wide range of capacities. See Figure 1-3.

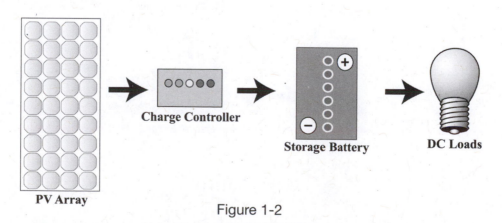

**Figure 1-2**
**DC SYSTEM WITH BATTERIES**

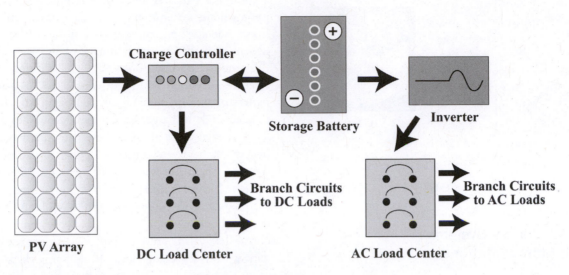

Figure 1-3

**SYSTEM WITH DC AND AC LOADS**

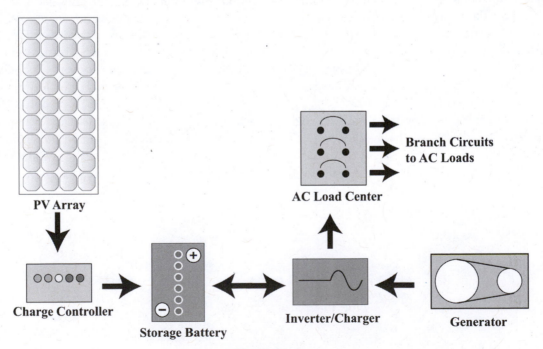

Figure 1-4

**HYBRID SYSTEM**

## Hybrid Systems

Most people do not run their entire load solely off their PV system. The majority of systems use a hybrid approach by integrating another power source. The most common form of hybrid system incorporates a gas or diesel-powered engine generator, which can greatly reduce the initial cost. See Figure 1-4. Meeting the full load with a PV system means the array and batteries need to support the load under worst-case weather conditions. This also means the battery bank must be large enough to power large loads, such as washing machines, dryers, and large tools. A generator can provide the extra power needed during cloudy weather and during periods of heavier than normal electrical use, and can also be charging the batteries at the same time. A hybrid system provides increased reliability because there are two independent charging systems at work.

Another hybrid approach is a PV system integrated with a wind turbine. Adding a wind turbine makes sense in locations where the wind blows when the sun doesn't shine. In this case, consecutive days of cloudy weather are not a problem, so long as the wind turbine is spinning. For even greater reliability and flexibility, a generator can be included in a PV/Wind system. A PV/Wind/Generator system has all of the advantages of a PV/Generator system, with the added benefit of a third charging source for the batteries.

## Grid-tied Systems

Photovoltaic systems that are connected to the utility grid (utility-connected, grid-tie, or line-tie systems) do not need battery storage in their design because the utility grid acts as a power reserve. Instead of storing surplus energy that is not used during the day, the homeowner sells the excess energy to a local utility through a specially designed inverter. When homeowners need more electricity than the photovoltaic system produces, they can draw power from the utility grid. See Figure 1-5.

If the utility grid goes down, the inverter automatically shuts off and will not feed solar-generated electricity back into the grid. This ensures the safety of linepersons working on the grid. Because utility-connected systems use the grid for storage these systems will not have power if the utility grid goes down. For that reason, some of these systems are also equipped with battery storage to provide power in the event of power loss from the utility grid. See Figure 1-6.

The Public Utilities Regulatory Policies Act (PURPA) of 1978 requires electric utilities to purchase power from qualified, small power producing system owners. The utilities must pay the small power producers based on their "avoided costs," or costs the utility does not have to pay to generate that power themselves. Additional terms

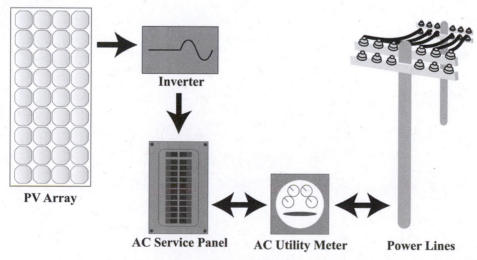

Figure 1-5

**GRID-TIED SYSTEM WITHOUT BATTERIES**

and conditions for these purchases are set by state utility commissions and vary from state to state. While this law allows homeowners in areas with utility power to purchase photovoltaic systems and sell their excess power to an electric utility, people contemplating doing so should remember that this is rarely a profitable venture at the present time.

Some utility companies offer "net metering" to their customers, where a single meter spins in either direction depending upon whether the utility is providing power to the customer or the customer is producing excess power. The customer or independent power producer pays or collects the net value on the meter. Net metering is very desirable to the independent power producer because he/she can sell power at the same retail rate that the utility charges its customers.

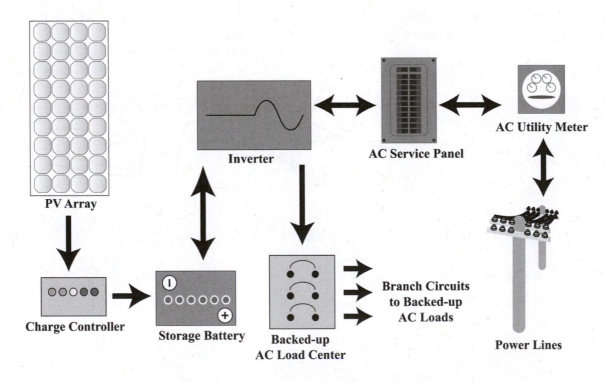

Figure 1-6

**GRID-TIED SYSTEM WITH BATTERIES**

# Chapter 2
# Photovoltaic Electric Principles

**Contents:**

2.1   Terminology . . . . . . . . . . . . . . . . . . . . . . . . . . . 10

2.2   Matching Appliances to the System . . . . . . . . . . . . . . . . . . 11

2.3   Electrical Circuits . . . . . . . . . . . . . . . . . . . . . . . . . 11

2.4   Series and Parallel Circuits in Power Sources . . . . . . . . . . . . . 12

2.5   Series and Parallel Circuits in Electric Loads . . . . . . . . . . . . . 15

2.6   Series and Parallel Wiring Exercises . . . . . . . . . . . . . . . . . . 16

2.7   Solutions to Wiring Exercises . . . . . . . . . . . . . . . . . . . . . 23

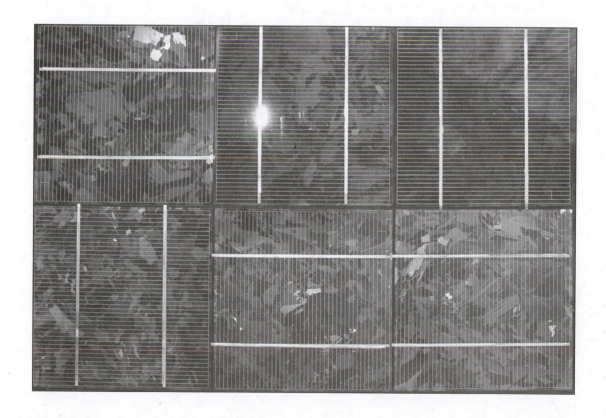

# 2.1 Terminology

**Electricity** is the flow of electrons through a circuit. The force or pressure of moving electrons in a circuit is measured as voltage. The flow rate of electrons is measured as amperage. The power of a system is measured as watts.

A **volt** is the unit of force (electrical pressure) that causes electrons to flow through a wire. Volts are abbreviated V, or expressed by the symbol E. Electrical pressure is sometimes referred to as the electromotive force (EMF). Some common voltages used in light-duty electrical systems include 12v, 24v, 48v. Most homes use 120v and 240v systems.

An **ampere** or **amp** is the unit of electrical current flowing through a wire. Amps are abbreviated A or expressed by the symbol I (for intensity of current). Just as pipe is sized by the rate of water passing through it, a wire is sized according to the rate of electrons (amps) flowing through it. One amp of current flowing for one hour is referred to as an **amp-hour** (Ah). The term amp-hour is commonly used when describing battery capacity. (For more information on wire-sizing methods, refer to Chapter 9.)

A **watt** is a unit of electrical power equivalent to a current of one ampere under a pressure of one volt. Watts indicate the *rate* at which an appliance uses electrical energy or the rate at which electrical energy is produced. Since consumers need to gauge how much electricity they use, the **watt-hour**, an electrical unit of energy, is an important measurement. An appliance that consumes electrical energy at a rate of one watt for one hour will have consumed a quantity of electricity equal to one watt-hour.

To calculate watt-hours, there are two things you'll need to know:

- An appliance's rated watts.
- The estimated duration of time the appliance will be operated.

The term watt-hours probably sounds familiar, since utility companies bill their customers for the number of kilowatt-hours consumed. **Kilowatt-hours** of electricity are equal to 1,000 watt-hours and are abbreviated kWh.

## Types of Current

There are two types of electrical current. **Alternating current** (AC) is electric current in which direction of flow reverses at frequent, regular intervals. This type of current is produced by alternators. In an alternator, a magnetic field causes electrons to flow first in one direction, then in the reverse direction. Electric utility companies supply alternating current.

**Direct current** (DC) is electric current that flows in one direction. Direct Current is the type of current produced by PV modules and stored in batteries.

## Equations:

Power = Watts (W) = Volts (V) x Amps (A)

1,000W = 1 Kilowatt (kW)

Energy = Watt-hours (Wh) = Watts x Hours

1,000Wh = 1 Kilowatt-hour (kWh)

Amp-hours (Ah) = Amps x Hours

---

**Problem:** How much electrical energy is consumed if a 100-watt light bulb is used for 10 hours?

**Solution:** 100 watts x 10 hours = 1,000 watt-hours (or 1 kilowatt hour).

---

# 2.2 Matching Appliances to the System

Photovoltaic system designers adapt their systems by using the manufacturers' power ratings for appliances, in conjunction with a careful estimation of how long each appliance will be used. You can find an appliance's electrical rating and power requirements on the nameplate. To use the information on an appliance's nameplate and correctly match the electrical supply to the appliance's requirements, you must understand the terms discussed in this chapter, including watts, amps, volts, alternating current, and direct current.

When choosing an appliance for a PV system, there are two important rules that must be observed:

- The voltage of an appliance must match the voltage supplied to it. The power source, such as a battery, generator, or photovoltaic module, determines the voltage supplied.

- An appliance must be compatible with the type of current (AC or DC) that is supplied to it.

# 2.3 Electrical Circuits

An **electrical circuit** is the continuous path of electron flow from a voltage source, such as a battery or photovoltaic module, through a conductor (wire)

to a load and back to the source. A simple electrical circuit is shown in Figure 2-1 as a schematic and a diagram. This example shows a single voltage source, a 12-volt battery, wired to a single load, a 12-volt, 24-watt light bulb, with a switch to turn the light on and off.

The switch controls the continuity of current flow. If the switch is turned off (an **open circuit**), the wire between the source and the load is disconnected, and the light will be off. If the switch is turned on (a **closed circuit**), the wire between source and load is connected, and the light will shine. Relay switching devices are often used as controls to open or close a circuit. Relays are rated by voltage, type of current (AC or DC), and whether the circuit is normally open or closed.

An electrical system can be compared to a water pumping system. A pump lifts two gallons of water per minute from a lower tank to an upper tank, increasing its height and pressure by 12 feet, the distance between the two tanks. The pressure created by the 12-foot height of the upper tank is like the 12-volt electrical pressure in the battery. The water falls at two gallons per minute from the upper tank and turns a water wheel, losing its height and pressure as it returns to the lower tank. The falling of water at two gallons per minute to turn the water wheel is like the two amp flow of electrons that powers the light and returns to the battery.

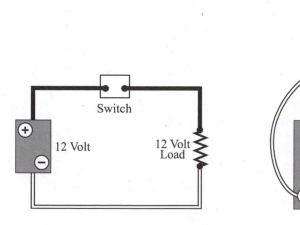

Figure 2-1

**ELECTRICAL CIRCUITS**

# 2.4 Series and Parallel Circuits in Power Sources

Photovoltaic modules and batteries are a system's building blocks. While each module or battery has a rated voltage or amperage, they can also be wired together to obtain a desired system voltage.

## Series Circuits

**Series** wiring connections are made at the positive (+) end of one module to the negative (-) end of another module. When loads or power sources are connected in series, the voltage increases. Series wiring does not increase the amperage produced. Figure 2-2 shows two modules wired in series resulting in 24V and 3A.

Series circuits can also be illustrated with flashlight batteries. Flashlight batteries are often connected in series to increase the voltage and power a higher voltage lamp than one battery only could power alone.

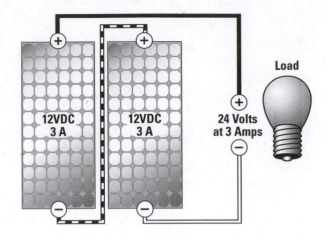

Figure 2-2

**PV MODULES IN *SERIES***

---

**Problem:** When four 1.5V DC batteries are connected in series, what is the resulting voltage?

**Solution:** 6 volts

---

## Parallel Circuits

**Parallel** wiring connections are made from the positive (+) to positive (+) terminals and negative (-) to negative (-) terminals between modules. When loads or sources are wired in parallel, currents are additive and voltage is equal through all parts of the circuit. To increase the amperage of a system, the voltage sources must be wired in parallel. Figure 2-3 shows PV modules wired in parallel to get a 12V, 6-amp system. Notice that parallel wiring increases the current produced and does not increase voltage.

Batteries are also often connected in parallel to increase the total amp-hours, which increases the storage capacity and prolongs the operating time.

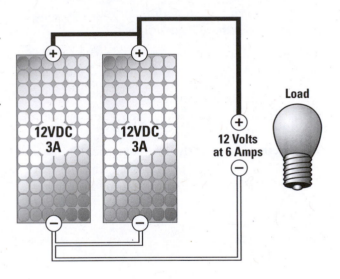

Figure 2-3

**PV MODULES IN *PARALLEL***

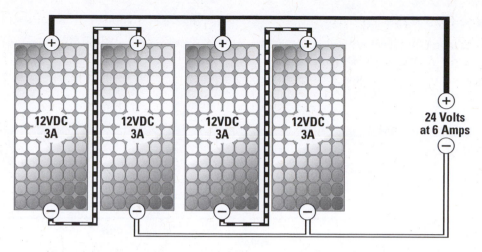

Figure 2-4

PV MODULES IN SERIES *AND* PARALLEL

## Series and Parallel Circuits

Systems may use a mix of series and parallel wiring to obtain required voltages and amperages. In Figure 2-4, four 3-amp, 12V DC modules are wired in series and parallel. Strings of two modules are wired in series, increasing the voltage to 24V. Each of these strings is wired in parallel to the circuit, increasing the amperage to 6 amps. The result is a 6-amp, 24V DC system.

## Batteries in Series and Parallel

The advantages of a parallel circuit can be illustrated by observing how long a flashlight will operate before the batteries fully discharge. To make the flashlight last twice as long, battery storage would have to be doubled.

In Figure 2-5, a series string of four batteries has been added in parallel to another string of four batteries to increase storage (amp-hours). The new string of batteries is wired in parallel, which increases the available amp-hours, thereby adding additional storage capacity and increasing the usage time. The second string could not be added in series because the total voltage would be 12 volts, which is not compatible with the six-volt lamp.

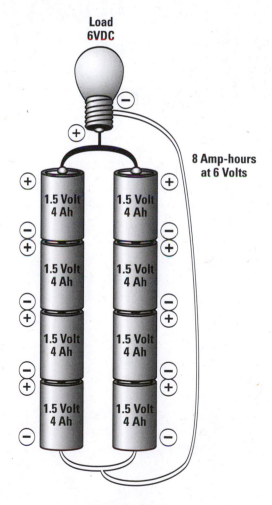

Figure 2-5

BATTERIES IN SERIES *AND* PARALLEL

## High Voltage PV Arrays

So far in this chapter, we have only discussed input voltages up to 24V nominal. Today, most batteryless grid-tied inverters on the market require a high voltage DC input. This input window is generally in the range of 75 to 600VDC. Because of the inverter's high voltage input requirements, PV modules must be wired together in series in order to sufficiently increase the voltage. See Figure 2-6 for an example of a high voltage system.

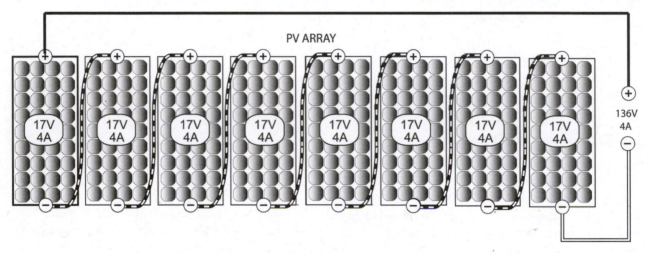

Figure 2-6

**HIGH VOLTAGE ARRAY**

# 2.5 Series and Parallel Circuits in Electric Loads

Like the photovoltaic modules and flashlight batteries described in the previous sections, loads wired in series, parallel, or series/parallel configurations act similarly.

## Loads in Series

Loads wired in series result in a voltage drop that is additive. The total voltage drop is equal to the sum of all the loads in the circuit. Current is equal through all loads in the circuit. Figure 2-7 shows two 6V light bulbs wired in series and supplied by a 12V battery. The voltage drop caused by each bulb is 6V; thus the total voltage drop is 12V, which is equal to the 12V pressure in the battery.

The two lights in Figure 2-7 are wired in series and are jointly controlled. If one light burns out, the circuit will be open, and all loads in the circuit will lose power. For this reason it is not recommended to wire loads in series.

## Loads in Parallel

What happens when loads are wired in parallel? Remember how the batteries added to the flashlight in Figure 2-5 did not increase the voltage supplied to the lamp? As loads are added in parallel, the voltage drop for each remains equal to the source voltage. Current drawn from the source is increased with each load added in parallel. See Figure 2-8.

Electrical circuits are commonly wired with all the loads in parallel for the following two reasons:

- Each load can be controlled individually.
- Adding more loads does not affect the operating voltage of any other load.

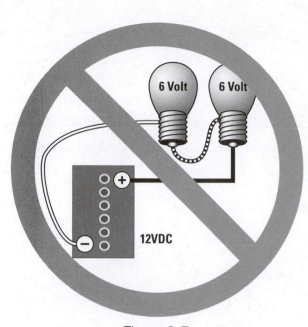

Figure 2-7

**LOADS IN *SERIES***

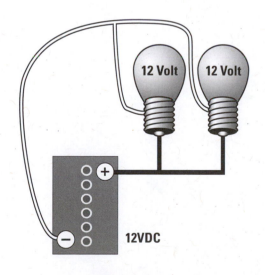

Figure 2-8

**LOADS IN *PARALLEL***

# 2.6 Series and Parallel Wiring Exercises

Use the following worksheets to practice series and parallel wiring for 12-, 24-, and 48-volt systems. Enter your answers in the blanks on each page or copy these pages for future practice. Draw lines to make your connections. The answers are listed in Section 2.7.

## Instructions:

1. Connect the photovoltaic modules (array) either in series or parallel or series/ parallel to get the desired system voltage.

2. Calculate total module output for volts and amps.

3. Connect the array to a charge controller.

4. Connect batteries either in series or parallel to get the desired system voltage.

5. Calculate total battery bank voltage and amp-hour capacity.

6. Connect the battery bank to the charge controller.

**Exercise A: DESIGN A 12V SYSTEM WITH FOUR 12V PV MODULES**

**PV Array**

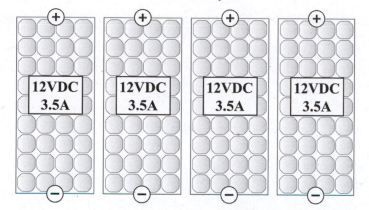

Total Volts = _____

Total Amps = _____

Charge Controller

Total Volts = _____

Total Amp-Hours = _____

6VDC 350Ah   6VDC 350Ah   6VDC 350Ah   6VDC 350Ah

**Battery Storage**

**Exercise B: DESIGN A 24V SYSTEM WITH FOUR 12V PV MODULES**

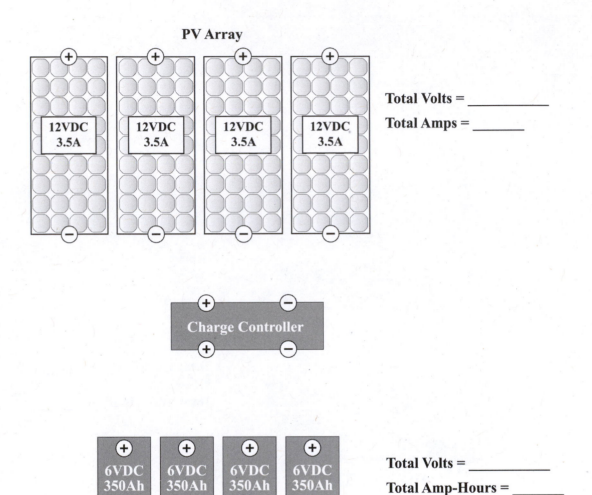

PV Array

Total Volts = _____

Total Amps = _____

Charge Controller

Total Volts = _____

Total Amp-Hours = _____

Battery Storage

**Exercise C: DESIGN A 48V SYSTEM WITH EIGHT 12V PV MODULES**

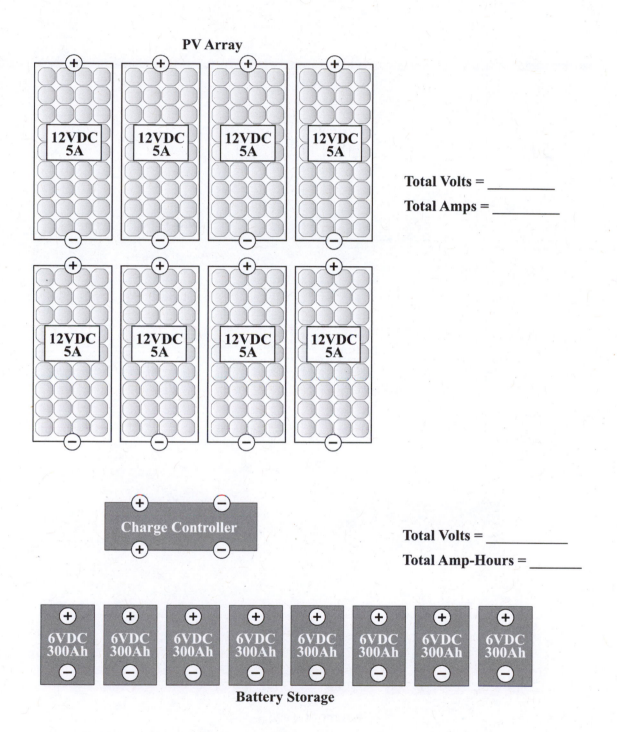

PV Array

Total Volts = _____

Total Amps = _____

Charge Controller

Total Volts = _____

Total Amp-Hours = _____

Battery Storage

## Exercise D: DESIGN A 48V SYSTEM WITH EIGHT 24V PV MODULES

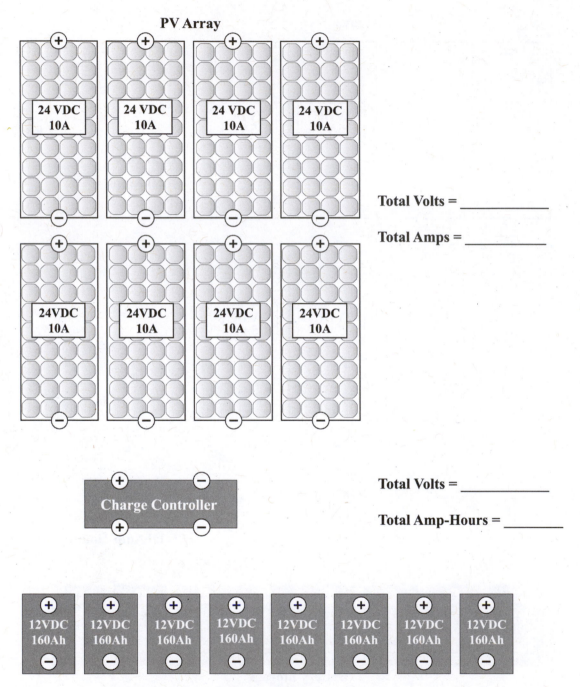

PV Array

Total Volts = _____

Total Amps = _____

Total Volts = _____

Total Amp-Hours = _____

Charge Controller

Battery Storage

**Exercise E: DESIGN A 48V SYSTEM WITH SIXTEEN 12V PV MODULES**

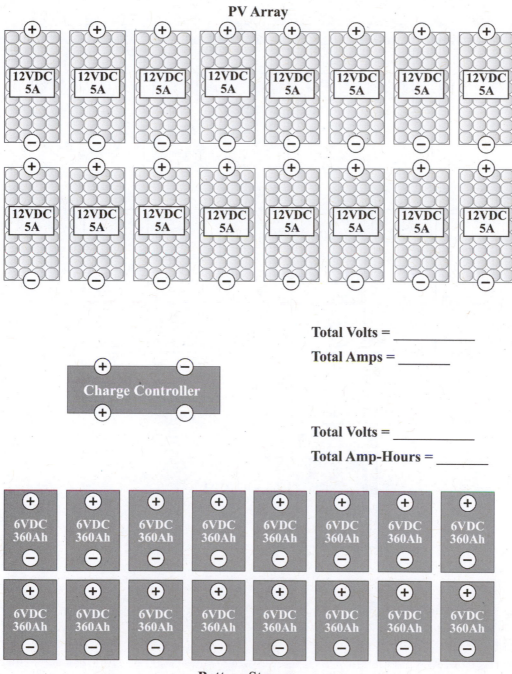

PV Array

Total Volts = _____

Total Amps = _____

Total Volts = _____

Total Amp-Hours = _____

Charge Controller

Battery Storage

## Exercise F: DESIGN A GRID-TIED SYSTEM WITH 2 SERIES-STRINGS USING SIXTEEN 34V PV MODULES

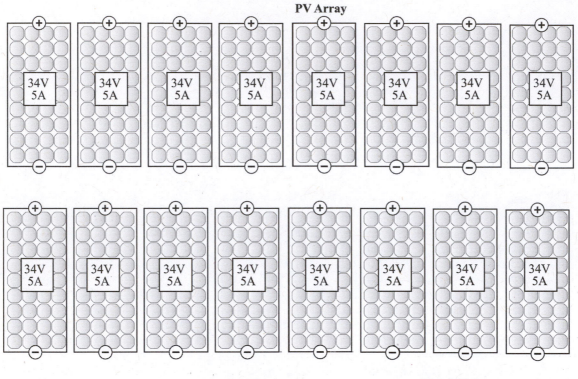

PV Array

**Total Volts = _____**

**Total Amps = _____**

**DC Combiner Box**

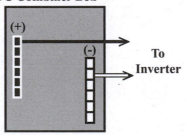

(+)

(−)

**To Inverter**

# 2.7 Solutions to Wiring Exercises

**Answer A: 12V SYSTEM WITH FOUR 12V PV MODULES**

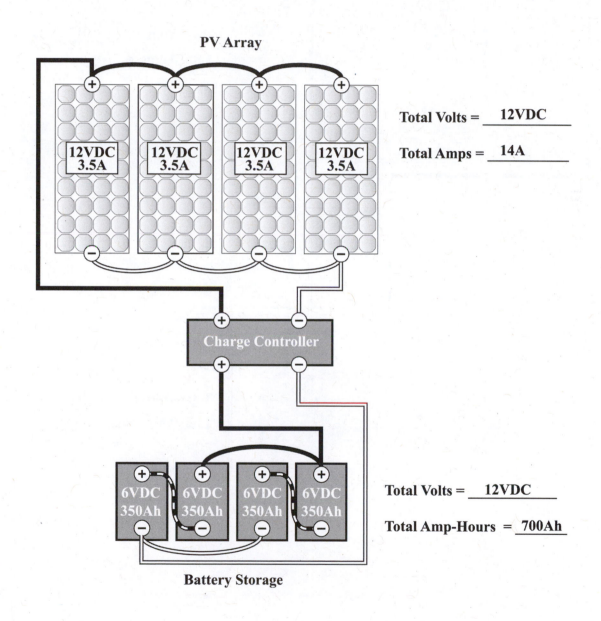

Total Volts = ___12VDC___

Total Amps = ___14A___

Total Volts = ___12VDC___

Total Amp-Hours = ___700Ah___

## Answer B: 24V SYSTEM WITH FOUR 12V PV MODULES

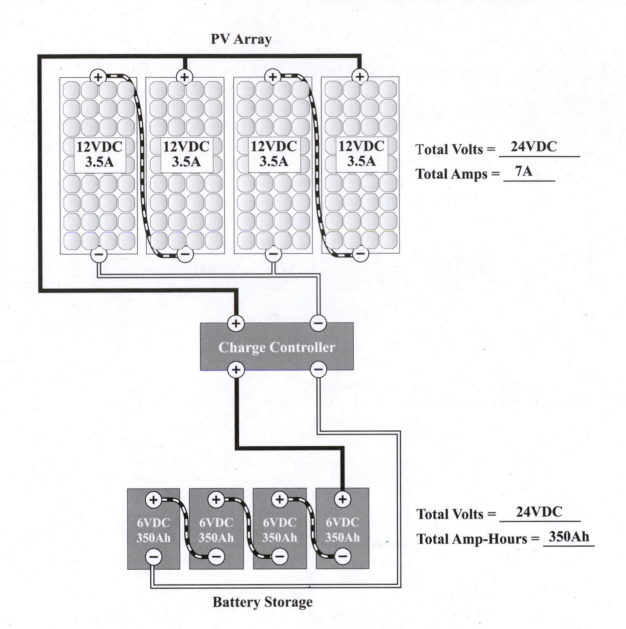

**PV Array**

Total Volts = <u>24VDC</u>

Total Amps = <u>7A</u>

**Charge Controller**

Total Volts = <u>24VDC</u>

Total Amp-Hours = <u>350Ah</u>

**Battery Storage**

**Answer C: 48V SYSTEM WITH EIGHT 12V PV MODULES**

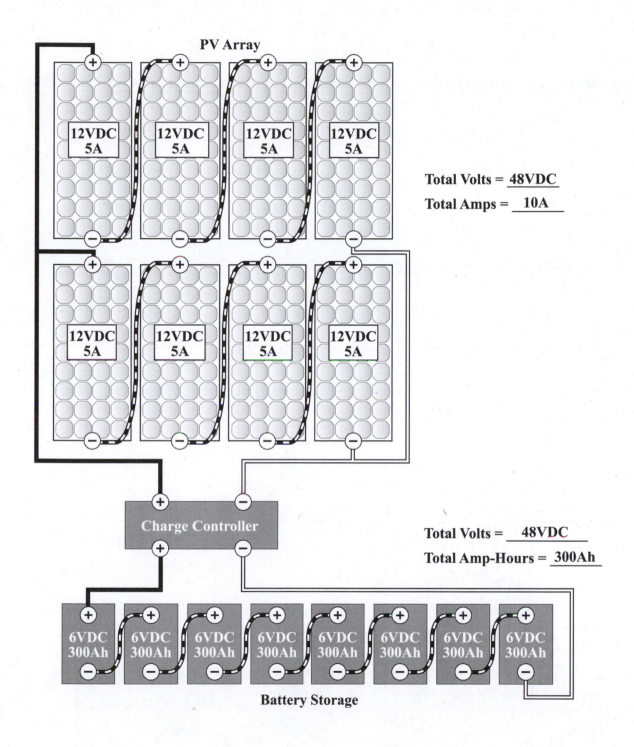

PV Array

12VDC 5A · 12VDC 5A · 12VDC 5A · 12VDC 5A

12VDC 5A · 12VDC 5A · 12VDC 5A · 12VDC 5A

Total Volts = <u>48VDC</u>

Total Amps = <u>10A</u>

**Charge Controller**

Total Volts = <u>48VDC</u>

Total Amp-Hours = <u>300Ah</u>

6VDC 300Ah · 6VDC 300Ah · 6VDC 300Ah · 6VDC 300Ah · 6VDC 300Ah · 6VDC 300Ah · 6VDC 300Ah · 6VDC 300Ah

**Battery Storage**

**Answer D: 48V SYSTEM WITH EIGHT 24V PV MODULES**

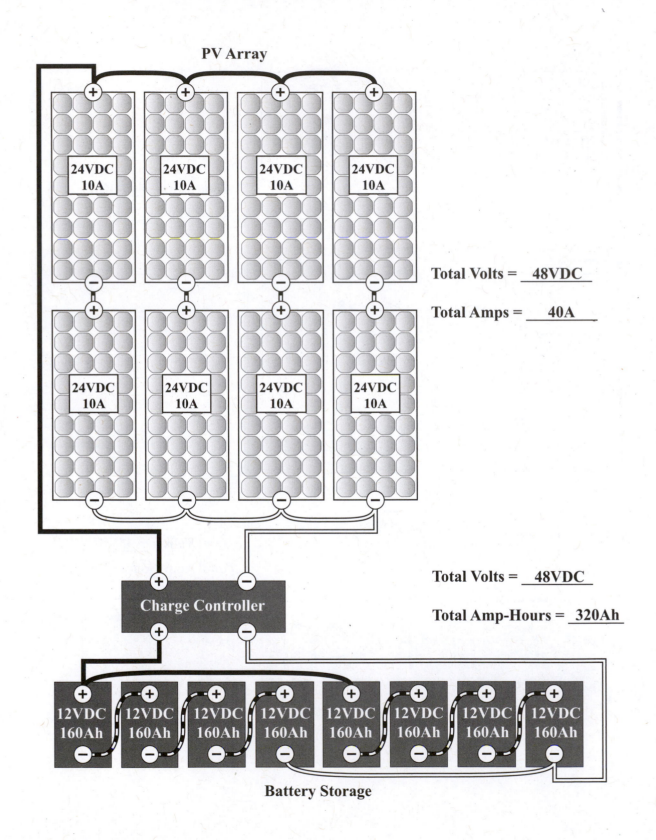

PV Array

Total Volts = __48VDC__

Total Amps = __40A__

Total Volts = __48VDC__

Total Amp-Hours = __320Ah__

Charge Controller

Battery Storage

**Answer E: 48V SYSTEM WITH SIXTEEN 12V PV MODULES**

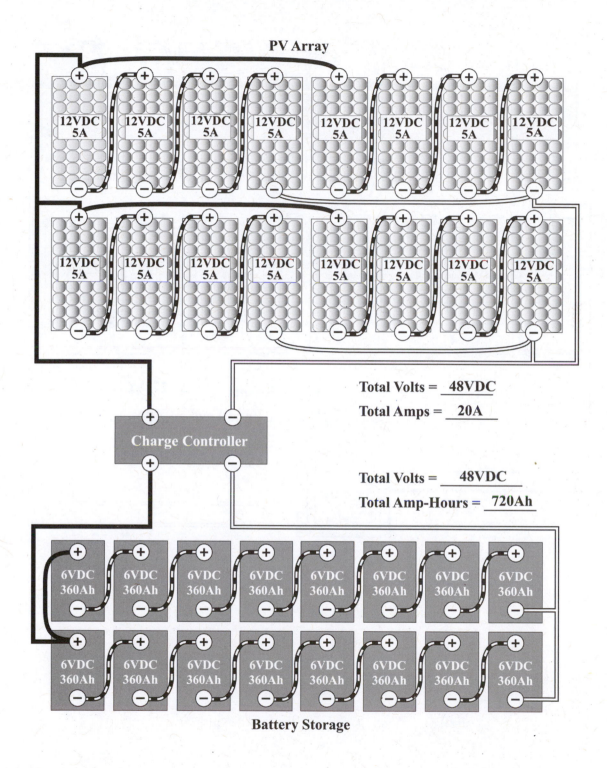

**PV Array**

Total Volts = __48VDC__

Total Amps = __20A__

Total Volts = __48VDC__

Total Amp-Hours = __720Ah__

**Charge Controller**

**Battery Storage**

## Answer F: GRID-TIED SYSTEM WITH 2 SERIES-STRINGS USING SIXTEEN 34V PV MODULES

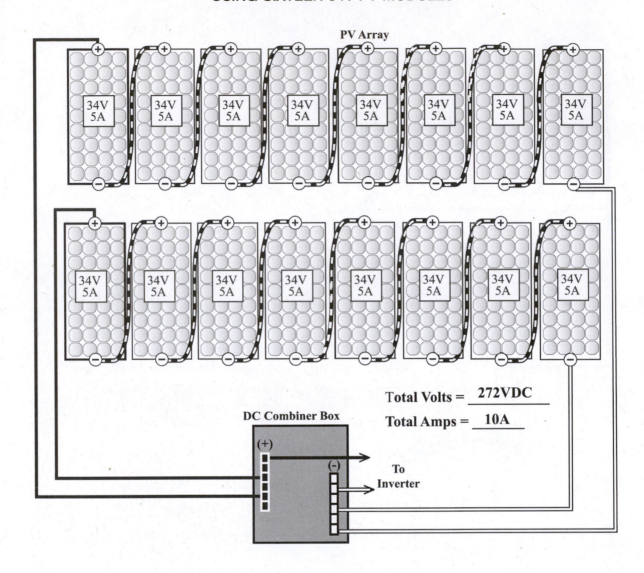

Total Volts = __272VDC__

Total Amps = __10A__

Section 2.7

# Chapter 3
# The Solar Resource

**Contents:**

3.1 Solar Radiation Fundamentals . . . . . . . . . . . . . . . . . . . . . . . . . .30

3.2 Gathering Site Data . . . . . . . . . . . . . . . . . . . . . . . . . . . . . . . .35

3.3 Completing the Solar Site Analysis . . . . . . . . . . . . . . . . . . . . .36

# 3.1 Solar Radiation Fundamentals

The term for solar radiation striking a surface at a particular time and place is **insolation**. When insolation is described as power, it is expressed as a number of watts per square meter and usually presented as an average daily value for each month. On a clear day, the total insolation striking the earth is about 1,000 watts per square meter. However, many factors determine how much sunlight will be available at a given site, including atmospheric conditions, the earth's position in relation to the sun, and obstructions at the site.

> **Note:** Although this book expresses insolation as kilowatt-hours per square meter, it can also be described in BTUs, Joules (J), or Langleys. The following equivalences can be used:
> $1 kWh/m2 = 317.1 BTU/ft2 = 3.6MJ/m2 = 1 Langley/85.93 = 1$ peak sun hour

Solar radiation received at the earth's surface is subject to variations caused by atmospheric attenuation. The primary causes of this phenomenon are the following:

- Air molecules, water vapor, and dust in the atmosphere scattering light.

- Ozone, water vapor, and carbon dioxide in the atmosphere absorbing light.

**Peak sun hours** are the number of hours per day when the solar insolation equals 1,000 w/m². For example, 5 peak sun hours = 5 kWh/m², where the energy received during total daylight hours equals the energy received if the sun shines for 5 hours at 1,000 w/m². See Appendix B, Solar Insolation Data, for information on a specific location.

The earth's distance from the sun and the earth's tilt also affect the amount of available solar energy. The earth's northern latitudes are tilted towards the sun from June to August, which brings summer to the northern hemisphere. The longer summer days and the more favorable tilt of the earth's axis create significantly more available energy on a summer day than on a winter day.

In the northern hemisphere, where the sun is predominantly in the southern sky, solar collectors or photovoltaic modules should point towards the southern sky to collect solar energy. Designers should optimize solar collection by positioning the array to take full advantage of the maximum amount of sunlight available at a particular location. Fortunately, the sun's path across the sky is predictable.

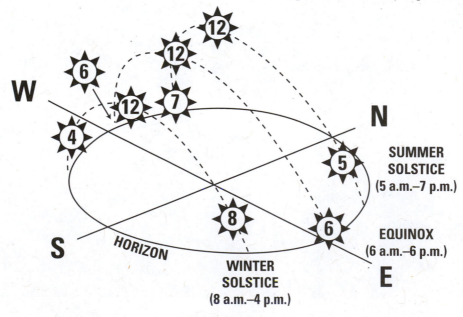

Figure 3-1

**THE SUN'S PATH THROUGHOUT THE YEAR—NORTHERN LATITUDES**

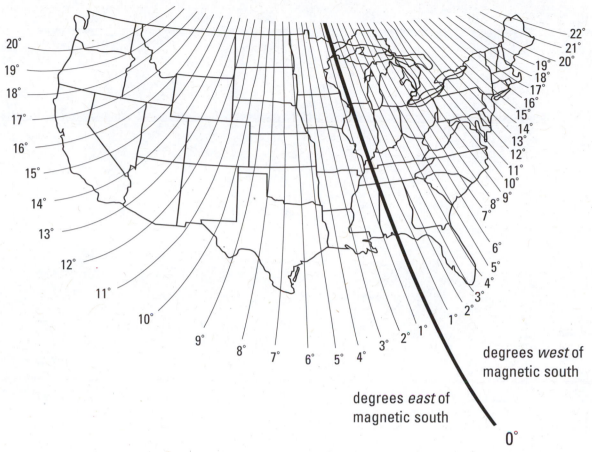

Figure 3-2

**MAGNETIC DECLINATION IN THE UNITED STATES**

The site's latitude (the distance north or south of the earth's equator) determines whether the sun appears to travel in the northern or southern sky. For example, Denver, Colorado, is located at approximately 40 degrees north latitude, and the sun appears to move across the southern sky. At midday, the sun is exactly true south.

Once a day, the earth rotates on its axis, which is tilted approximately 23.5 degrees from vertical. The sun appears to rise and set at different points on the horizon throughout the year because of this tilt. On the fall and spring equinoxes (September 21 and March 21) the sun appears to rise exactly due east of south and appears to set exactly due west of south. During the winter months, the sun appears to rise south of true east and set south of true west; in the summer months, it appears to rise north of true east and set north of true west. Figure 3-1 illustrates the solar position at different times of day and year.

## Orientation

The sun's apparent location east and west of true south is called **azimuth**, which is measured in degrees east or west of true south. See Figure 3-4. Since there are 360 degrees in a circle and 24 hours in a day, the sun appears to move 15 degrees in azimuth each hour (360 degrees divided by 24 hours). Magnetic south or south on a compass is not the same as true south. A compass aligns with the earth's magnetic field, which is not necessarily aligned with the earth's rotational axis. The deviation of magnetic south from true south is called magnetic declination. Refer to a map or ask a local surveyor for your location's **magnetic declination**. Figures 3-2 and 3-3 provide approximate magnetic declination for the United States and the world respectively. These maps are sufficient for our purposes.

Daily performance will be optimized if fixed mounted collectors are faced true south or 0 degrees

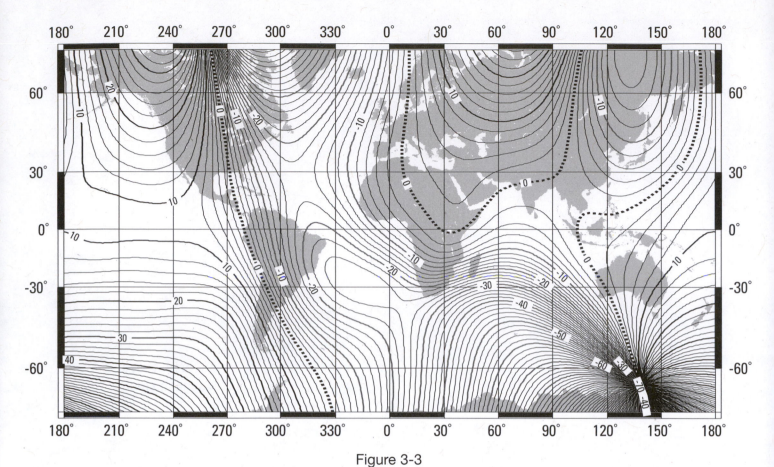

Figure 3-3

**WORLD MAGNETIC DECLINATION CHART**

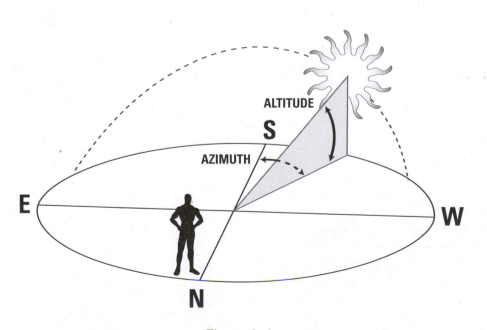

Figure 3-4

**AZIMUTH AND ALTITUDE FOR ALL NORTHERN LATITUDES**

azimuth, which is the best generic orientation for locations in the Northern hemisphere. An array that deviates 30 degrees from true south will collect a lesser percentage of the sun's available energy on an average daily basis.

---

A site in Montana has a magnetic declination of 20 degrees east, meaning that true south is 20 degrees east of magnetic south. On a compass with the north needle at 360 degrees, true south is in the direction indicated by 160 degrees.

Local climate characteristics should be carefully evaluated and taken into consideration. For example, you can compensate for early morning fog by adjusting the photovoltaic array west of south to gain additional late afternoon insolation.

---

## Tilt Angle

The sun's height above the horizon is called **altitude**, which is measured in degrees above the horizon. See Figure 3-4. When the sun appears to be just rising or just setting, its altitude is 0 degrees. When the sun is true south in the sky at 0 degrees azimuth, it will be at its highest altitude for that day. This time is called **solar noon**.

A location's latitude determines how high the sun appears above the horizon at solar noon throughout the year. As a result of the earth's orbit around the sun with a tilted axis, the sun is at different altitudes above the horizon at solar noon throughout the year.

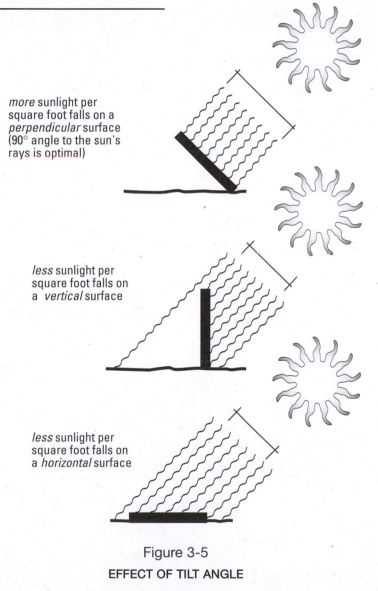

*more* sunlight per square foot falls on a *perpendicular* surface (90° angle to the sun's rays is optimal)

*less* sunlight per square foot falls on a *vertical* surface

*less* sunlight per square foot falls on a *horizontal* surface

Figure 3-5

**EFFECT OF TILT ANGLE**

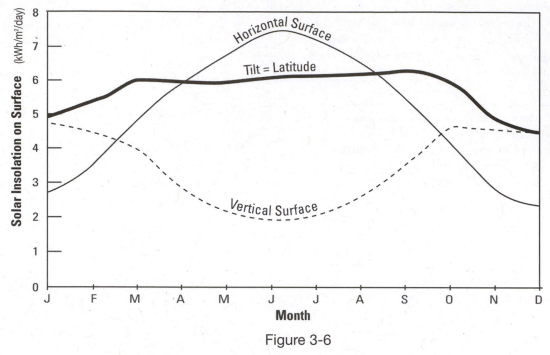

Figure 3-6

**EFFECTS OF ARRAY TILT ON ENERGY PRODUCTION**

Seasonal changes of the sun's altitude must be considered to optimize a system's performance. The following outlines the optimum tilt angle of a photovoltaic array for different seasonal loads.

**Year-round loads:** Tilt angle equals latitude

**Winter loads:** Tilt angle equals latitude plus 15°

**Summer loads:** Tilt angle equals latitude minus 15°

Photovoltaic arrays work best when the sun's rays shine perpendicular (90 degrees) to the cells. See Figure 3-5. When the cells are directly facing the sun in both azimuth and altitude, the angle of incidence is "normal." Figure 3-6 illustrates the effect of this tilt angle on available monthly insolation.

**Note:** Adjusting the tilt angle of the PV array seasonally can increase power production significantly for year-round loads.

Since grid-tied systems have the utility as a back-up source of power, there is more flexibility to siting the array than for stand-alone systems. For a grid-tied array, if the optimal tilt angle is not feasible, the system will produce a percentage of the total available energy, as the utility will provide the balance. Since grid-tied systems are offsetting average annual energy use, (instead of powering loads directly), designers may choose to flush mount an array on a roof for aesthetic reasons even if it is not the optimal tilt and orientation angle. A good resource for calculating annual energy production for grid-tied PV systems is NREL's PV Watts website. This program allows the designer to input the tilt and orientation angle and calculates corresponding energy production values.

# 3.2 Gathering Site Data

Correctly siting stand-alone systems is more critical than for utility interactive systems. The first step in designing a stand-alone system is to determine what time of the year will have the largest loads and then to select a month that you will use to design the system. You will also need to gather solar insolation data for the sizing calculations.

## Determining Design Month

Insolation data is most often presented as an average daily value for each month. When sizing a system, it is important to use the correct month. If the load is constant throughout the year, the **design month** will be the month with the lowest insolation. The array should then be installed with a tilt angle that yields the highest value of insolation during that month. This ensures that the system is designed to meet the load and keep the battery fully charged in the worst month for the average year.

If the load is variable for each month, you should use the load estimation worksheet in Appendix D to calculate the design current for each month. The design current is the average daily load for the month divided by the monthly insolation. The month corresponding to the largest design current should be used as the design month.

## Gathering Insolation Data

Many locations around the world have years of weather records that will provide average data sufficient for designing PV systems. Appendix B contains average daily insolation data for major cities worldwide, including values for different tilt angles and tracking options. This type of data may also be available from meteorological stations, universities, government ministries, Internet, or other information depositories.

If no long-term data exists for your specific site, the availability and amount of sunshine must be estimated. Even though local solar conditions may vary significantly from place to place, particularly in mountainous areas, you can estimate local weather by studying the variation in average data from several cities located around the proposed site.

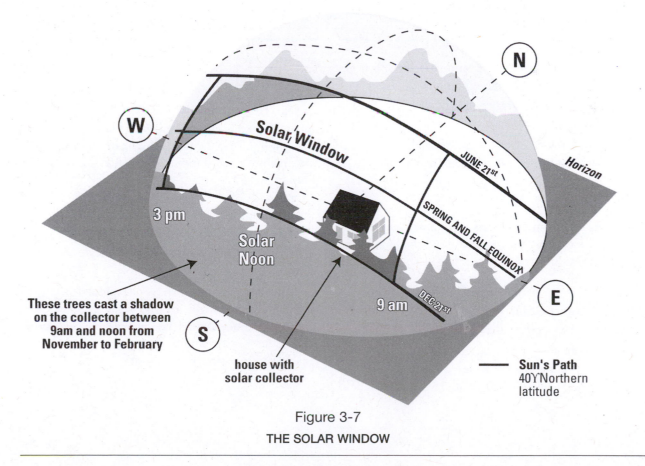

Figure 3-7

**THE SOLAR WINDOW**

# 3.3 Completing the Solar Site Analysis

With the site data and an understanding of how the array should be oriented, you are now ready to evaluate the site and locate where the array will be installed.

## Identifying Shading Obstacles

Shading critically affects a photovoltaic array's performance. Even a small amount of shade on a PV panel can reduce the panel's performance significantly. For this reason, minimizing shading is much more important in PV system design than in solar thermal system design. Carefully determining solar access or a shade-free location is fundamental to cost-effective photovoltaic performance.

Unwanted shading can occur from trees, vegetation, structures, other arrays, poles, and wires. As a general rule, an array should be free of shade from 9:00 AM until 3:00 PM. This optimum collection timeframe is called the **solar window** (see Figure 3-7). Shading is often a greater problem during winter months when the sun's altitude is low and shadows are longer. For locations in the northern hemisphere, December 21 should be used for worst-case shadow calculations.

When a site is selected, be sure that the following parameters are met and tasks completed:

- Be sure that the array is not shaded from 9 a.m. to 3 p.m. on any day.

- Be sure that the array is not shaded in any month of the year during the solar window.

- Identify the obstacles that shade the array between 9 a.m. and 3 p.m.

- Make recommendations to eliminate any shading, move the array to avoid shading, or increase the array size to offset losses due to shading.

## Determining Solar Access

You should carefully examine a site to identify any possible shading on your solar panels. This is accomplished by getting a clear picture of the sun's path across the horizon from east to west. One method for finding a site with good year-round exposure is through long-term observation. Unfortunately, this is not always possible or practical.

The only type of "time" that truly makes sense for living organisms (and solar energy collection) is "solar time." Solar time signifies that at noon the sun should be at its peak in the sky for the day. Thus, when we refer to solar noon, we mean maximum solar altitude. The invention of time zones marked a human attempt to link all the site-specific solar times throughout the world. Time zones are merely approximations. The logic behind time zones is as follows: since the earth rotates at a rate of 15 degrees per hour, a one hour time change exists every 15 degrees of longitude. Divide 360 degrees (a full circle) by 15 degrees and the result is 24 hours; therefore, a time zone for every hour of day.

Another way to evaluate a site is with a **sun chart**. Since solar altitude is of utmost importance to the designer, special charts are available for specific latitudes. Appendix C shows sun charts for altitudes located between 28 degrees and 56 degrees north latitude. Sun charts are also available from state energy offices or local solar suppliers.

If a site is partially shaded, you can use a sun chart to determine the amount of available sunlight. These charts will help you locate the position of the sun in the sky at any time of the year and help you determine if your solar collectors will be shaded from direct beam radiation during critical times of the day or year. Find the appropriate chart for your latitude in Appendix C. The curved solid lines represent the sun's path during the day. The top and bottom lines are drawn for the summer and winter solstices. The lines between the solstices represent the 21st day of each month of the year. The dotted lines represent the solar time of day.

**Note:** When charting the sun's path, remember that the winter sun is low in the sky.

Solar professionals have developed tools to provide quick insight to the solar window at a specific location. Commercially available solar siting devices have sun charts built directly into them and enable you to easily evaluate a site.

# Chapter 4
# Electric Load Analysis

**Contents:**

4.1   Using Energy Efficiently . . . . . . . . . . . . . . . . . . . . . . . . . . . . . . . 38
4.2   Electrical Load Requirements . . . . . . . . . . . . . . . . . . . . . . . . . . . 38
4.3   Refrigeration . . . . . . . . . . . . . . . . . . . . . . . . . . . . . . . . . . . . . . . 39
4.4   Lighting . . . . . . . . . . . . . . . . . . . . . . . . . . . . . . . . . . . . . . . . . . . 41
4.5   Considerations for Calculating Load Estimates . . . . . . . . . . . . . . 43
4.6   Calculating Load Estimates . . . . . . . . . . . . . . . . . . . . . . . . . . . . . 43

# 4.1 Using Energy Efficiently

Devices that operate using electrical power are often referred to as loads. They are usually the largest single influence on the size and cost of a photovoltaic system. A photovoltaic system designer can minimize a PV system's cost by efficiently using the energy available. A designer should thoroughly analyze the energy requirements to identify energy conserving opportunities. For example, many common household appliances use electric resistance to perform their function. Resistive heating is used in electric ovens, clothes dryers, water heaters, and space heating systems. As a rule, powering these loads can be too cost prohibitive for residential photovoltaics systems, and you should find an alternative way to heat. A water heater element may be rated around 2,500 watts and operate 25 percent of the time. Water heating and space heating loads can be better accomplished through other means including solar thermal heating, gas, propane, and wood. This method of supplying thermal loads from thermal sources is a more efficient use of energy, and is an example of **load shifting.**

Although small convenience items, such as toasters and hair dryers, require significant amounts of instantaneous power (watts), they can still be powered by photovoltaic systems. Because these loads are typically not used for a prolonged period of time, their overall energy consumption can be low.

Designers should also suggest using more efficient appliances within a photovoltaic system. For example, incandescent lamps can be replaced with compact fluorescent lamps that provide equal illumination and use about one quarter the power.

To reduce the cost of a PV system, avoid the following appliances:

- Electric space heaters.
- Electric water heaters.
- Electric clothes dryers.
- Electric ranges.
- Inefficient refrigerators.

In addition to load shifting and increased equipment efficiency, designers can lessen the need for additional electric power by implementing the following practices:

- Living without unnecessary items.
- Reducing use of appliances when appropriate.
- Designing a system without an inverter and using only direct current loads, if appropriate, to avoid inverter efficiency losses (usually only considered for small systems like RVs and cabins).
- Doing tasks during sunlight hours to maximize battery efficiency (for stand-alone systems).

Designers should involve system owners or operators in the design process and system installation. Their increased awareness will also reduce the need for electric power, since they will be more likely to use their electric resource more prudently.

# 4.2 Electrical Load Requirements

Manufacturers' literature and equipment nameplates often list the watts required for a load. When the watts required by a given load are not listed, you will usually find volts and amps listed instead. You can calculate the watts required by a load by multiplying volts times amperes.

Table 4-1, at the end of this chapter, lists typical wattages for many common household loads. This chart is only for reference, actual load specifications should be used when designing a system.

## Cycling Loads

Most loads consume power continuously when they are switched on. However, some loads will automatically turn themselves on when they are connected to a power source. A **duty cycle** is the percentage of time an appliance that is "on" is actually drawing power. Good examples of such appliances are refrigerators and freezers. A refrigerator may operate 50 to 60 percent of the time, depending on its efficiency.

In addition, appliances that create or use heat usually cycle on and off. For example, electric blankets, irons, and cooking appliances. Thermostats control these types of appliances.

## Phantom Loads

Many electrical loads draw power even when they are turned "off." When estimating the energy use of a home, **phantom loads** must also be taken into account. Phantom loads are small loads that constantly draw power. For example, instant-on television sets and appliances with digital clocks, such as microwave ovens, VCRs, any item with a remote control, and some personal computers. Other phantom loads are appliances with a "wall-cube," such as AC telephone machines, battery chargers, and dust busters. Phantom loads may seem negligible, but they are using power 24 hrs a day, 7 days a week. It is recommended to keep phantom loads to a minimum by either unplugging them, or placing them on a switched circuit or a power strip. Figure 4-1 displays some common phantom loads and the watts they consume.

## Estimating Surge Requirements

When estimating an electrical load, **surge loads** are another factor. These are appliances with motors that draw more current when they start than when they are operating. For example, a power saw that uses 900 watts continuously might use up to 3,000 watts to start the motor. Consult the manufacturer or measure the load with an ammeter to determine surge requirements of specific loads. As a rough rule of thumb, minimum surge requirements may be calculated by multiplying the required watts of a load by three. (For specific surge requirements see manufacturer's equipment specifications.)

$$\text{Surge requirements} \approx \text{Operating watts} \times 3$$

# 4.3 Refrigeration

Commonly, the refrigerator is one of the single largest energy loads in a residential home. Though there are other ways of storing food, people in the United States rarely live without this appliance. Choosing an efficient refrigerator will not only reduce your monthly utility bill, but it will also significantly lower the upfront cost of a PV system. This is often a good place to start. There are many energy efficient models on the market today. Consider appliances that are Energy Star rated by the Department of Energy.

> **Note:** There is a wide range of energy consumption within Energy Star ratings. The American Council for an Energy Efficient Economy (ACEEE) also publishes a useful buying guide for some of the most efficient appliances commercially available today. See the Reference Guide at the end of this textbook for more information.

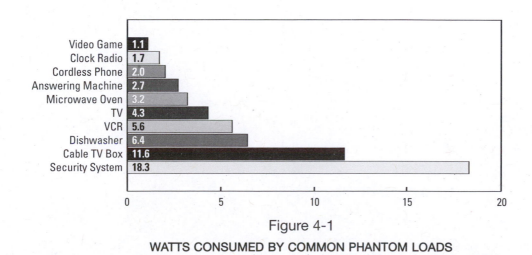

Figure 4-1

**WATTS CONSUMED BY COMMON PHANTOM LOADS**

## Refrigeration Options

In stand-alone PV systems, buying an energy efficient refrigerator is extremely important. There are four basic options that are commercially available:

- Propane powered.
- Kerosene powered.
- AC powered.
- DC powered (12, 24, or 48V).

**Propane powered:** Propane refrigerators have been used in remote locations in the United States for over 70 years. Modern propane refrigerators are widely available and used throughout the recreational vehicle and marine industry. They are more dependable and require less maintenance than the earlier models. Modern propane refrigerators can be a safe, cost-effective solution for remote refrigeration applications. However, while the initial cost of these refrigerators makes them an attractive purchase for many users, they do have some limitations. They require periodic refueling, making them less "independent" than refrigerators powered by a photovoltaic system. You should consider a refrigerator's life-cycle cost when deciding which unit to purchase.

**AC powered refrigerators:** Alternating current refrigerators are normally used in homes with utility-supplied electricity. They require an inverter if they are to be in a stand-alone system. The availability of efficient AC refrigerators combined with the high efficiency of today's inverters, makes them a common choice.

**DC powered refrigerators:** Direct current refrigerators have a higher initial cost than alternating current units, but direct current units operate directly off battery power and do not require an inverter. DC powered refrigerators were more common in the U.S. when inverters were less efficient.

## Design Criteria

When deciding which mechanical refrigerator to purchase, you should choose a unit with the following specifications:

- Superior insulation with a high R value.
- Small current draw when operating.
- Compartmentalized storage/separate freezer.
- Efficient compressor waste heat removal.
- Avoid side by side models.
- Compressor located on top.
- Energy Star rated.

There are many ways to design a refrigerator into a kitchen to allow more efficient operation:

- Avoid direct sunlight.
- Avoid placing adjacent to heating appliances (ie. stove or dishwasher).
- Place on an interior wall.
- Insure adequate air flow around the unit.

## Refrigeration Load Estimating

Refrigeration loads can be accurately estimated from the manufacturer's specifications and a thorough knowledge of use patterns. The manufacturer should be able to supply voltage, amperage draw, and actual on-time under given conditions. Refrigerator on-time is seasonal because the load is affected by ambient temperature. Routine maintenance, such as cleaning or defrosting, can reduce unnecessary on-time and improve performance.

# 4.4 Lighting

Lighting is an essential element in most lifestyles, and is generally provided by electricity or natural day lighting. In some cases, kerosene or propane lamps are used. These lamps can be an appropriate option because they effectively reduce the electrical load requirements of a photovoltaic system.

The following section describes lighting controls, three common lamp types, and their applications. This information will help you determine the type of lamp to use for specific applications. There are many other types of lighting on the market today. If you are still unsure of the right choice, expert advice is available from local utilities, electrical equipment distributors and manufacturers, electrical contractors, consulting engineers, and state and federal energy offices.

## Lamp Efficiency

When selecting a lamp type, efficiency is an important consideration, but it should not be the only criteria used. In most cases, a more efficient light source can be substituted for a less efficient source with little or no loss in visibility or color rendition. The total annual cost savings help to decrease the size of the photovoltaic system. Lamp efficiency is measured in lumens per watt. Lumens are a measure of the light output from the lamp. If a lamp produces more lumens from each watt of electrical energy input, it is more efficient.

## Lighting Controls

Lighting control and operation is an important concern when sizing a photovoltaic system since a lower load will reduce the size of the photovoltaic system, and therefore, reduced cost. Lighting controls include the following:

**Manual switches:** These controls, typified by a wall switch or pull-chain mounted directly on a fixture, are the least expensive and most commonly used controls. Switching each light separately with a manual switch offers the greatest potential for minimizing energy use, but this method is effective only if people consistently use the switches. Switches must be conveniently located and easy to use. For example, stairway lights should be switched at the top and bottom of the stairs using three-way switches. Remember that the standard wall mounted light switches commonly used with 120V alternating current systems are usually inappropriate for use with 12V direct current lights and systems due to higher currents and arcing in a 12V system. Hence, if using DC lighting, DC rated switches are needed.

**Timers:** These controls can be set to automatically turn lights on or off or to limit the time a light will be on. You should consider safety when using timers, for example lights should not turn off without warning the occupants first. Timers can require a small amount of additional power for their own operation.

**Photocells:** Security or safety lighting can be controlled by photocells, which are devices that sense light levels. Photocells sense a loss of daylight at dusk and activate the light, and conversely, they sense daylight at dawn and turn the light off. Photocells are more dependable than manual switching and more accurate than timers. There is a wide selection of 120-volt alternating current photocells on the market. Twelve-volt direct current photocells are less common.

**Sensors:** Sensors are used when precise control is desired. Sensors activate lights when they detect motion or infrared heat.

---

**Problem:** A 12-volt direct current, 13-watt fluorescent light is controlled by a photocell. The average daily on-time is 12 hours per day. How many watt-hours are consumed by the light on an average day?

**Solution:** Multiply 13 watts by 12 hours, resulting in 156 watt-hours.

---

## Lamp Types

The three lamps discussed in this section are:

- Incandescent.
- Fluorescent / compact-fluorescent.
- Light Emitting Diodes (LED).

**Incandescent lamps:** Incandescent lamps are the most commonly used even though they have the poorest efficiency or lowest lumens per watt ratings. Their main benefits are low initial cost, attractive color, cold weather operation, and simple installation. In typical incandescent lamps, electricity is conducted through a filament that resists the flow of electricity, heats up, and glows. Incandescent lamps use the familiar "Edison base" bulb and require no special equipment or ballasts to modify the characteristics of the power supplied to the fixture. Incandescent lamps are available in many wattages, both in 120-volt alternating current and 12-volt direct current.

**Compact fluorescent lamps:** Given the superior efficiency and longevity of compact fluorescent lamps (CFL) compared to incandescent, it's not surprising that they have become the most utilized type of lamp in homes incorporating renewable energy systems. CFL's take multiple forms such as 'mini-tubes' or a shape similar to incandescent bulbs. CFL's not only fit in existing incandescent lamp holders, but they also use about a quarter of the energy compared to an incandescent bulb.

Compact fluorescent lamps employ ballasts to regulate the flow of power through the lamp. These bulbs generally work in either alternating or direct current systems, though the ballast must match the system's nominal voltage and current type. A large selection of AC and DC ballasts with a wide range of wattages are commercially available. When operating, an electric arc is drawn along the length of the tube. The ultraviolet light produced by the arc activates a phosphorescent coating on the inside of the tube wall, causing light to be emitted.

Today's energy efficient ballasts and the wide variety of bulbs available eliminate many of the previous drawbacks associated with fluorescent lighting. Some of the earlier drawbacks include: higher startup consumption, poor color rendition, and incompatibility with dimmer and 3-way switches. Although modern compact fluorescent bulbs are generally more expensive than incandescents, they no longer suffer from these prior limitations and can save the homeowner a considerable amount of energy. In addition, they experience significantly longer lives than incandescents. One existing drawback that remains, is that in cold temperatures, it can take a few minutes for the bulb to fully light up. CFLs also contain a small amount of mercury and need to be properly disposed. In conclusion, compact fluorescent lamps can last longer, require no extra installation costs, and reduce electricity bills.

**LED lamps:** Though still a developing technology for residential applications, the LED (light emitting diode) is worth mentioning. Common applications today are traffic lights, car brake lights, and flash lights. In a light emitting diode, the creation of light happens much more efficiently at the molecular level. As previously mentioned, a diode is an electronic device which limits the direction that electrons may flow in an electronic circuit. An LED is a special type of diode that has been optimized to release energy in the form of light instead of as heat as in a traditional diode. An unbreakable, crystal clear solid resin encases each LED and makes it nearly indestructible, contributing to the long life of LEDs, which typically last 5 to 10 years of constant use and draw as little as 1/10th to 1/20th the current of an incandescent light bulb producing equivalent lumens. The main drawback of LED lighting is the quality of the light tends to be a bit too bright and glaring. These lights are great for short term space lighting, though not pleasing to read by for extended periods of time. Currently research is being done to reduce cost, and to improve light quality. Worldwide, LEDs are starting to be used for low wattage PV lighting systems.

# 4.5 Considerations for Calculating Load Estimates

Load estimations can be difficult to calculate due to the number of variables. Consider the following suggestions when estimating electrical loads:

- Use the correct load estimates.

- Account for duty cycle when sizing equipment that cycles on and off.

- Use manufacturer's literature when available, instead of generalized sources, such as the tables in this chapter.

- Recheck your numbers, but keep in mind this is only an estimate.

- Consider all energy conservation, equipment efficiency, and load shifting opportunities.

- Remember that future loads may vary because of the following:

  - People change

  - Seasons change

  - Efficiency of loads may decrease with age

  - Appliances fail and will be replaced

  - People forget to turn appliances off

  - Loads may be added

- Remember that not all appliances available in alternating current are available in direct current. Make sure you can get the equipment you need and want for the system you are designing.

# 4.6 Calculating Load Estimates

You can calculate the average daily electrical energy use in watt-hours as well as the total connected watts using the Load Estimation Worksheet found at the end of this chapter. First, list the desired electric loads. If possible, obtain the quantity of each load and the electrical specifications, including alternating current and direct current voltages, amperage, and wattage of each load. Then list the average hours per day and days per week that the load will be used. Depending on the information you have available, you can either calculate or estimate these figures. Be sure to allow for seasonal variations in the use of the load and solar insolation.

Many electric loads that a designer would like to include may be cost prohibitive to power with a photovoltaic system. Usually a mix of load shifting and increased equipment efficiency can significantly reduce the cost of photovoltaic systems. A system designer will usually change the electric load estimate several times before the final system is efficiently sized.

The following examples are load-estimating exercises. Using the Load Estimation Worksheet at the end of this chapter, calculate the average daily load in watt-hours for each example.

## Load Calculation Exercises

**Problem One:** A retired couple has sold their home in the city and now live in a recently-purchased recreational vehicle. They follow the seasons, living in the desert near Apache Junction, Arizona in the winter and in the mountains near Lake Tahoe, California in the summer. They treasure their new lifestyle's peace and quiet and have resisted purchasing a noisy diesel generator. Their RV is equipped with propane fueled appliances, including their refrigerator, range, water heater, and space heater. They use kerosene lamps for reading but would use the single 50-watt (12V direct current, 4.17 amps) incandescent reading light in the RV more often if they had a steady power source. They would like to read for four hours each night, if power were available. Using the Load Estimation Worksheet (below), estimate their electrical load and propose a solution.

**Problem Two:** A homeowner in a remote mountainous area near Colorado Springs, Colorado wants to power the television and refrigerator in his cabin with a photovoltaic system. His refrigerator is a 17-cubic-foot model, rated at 500 watts (120V alternating current, 4.17 amps). He has timed its operation and says it runs 30 minutes each hour (50 percent of the time). The cabin owner's small television is rated at 20 watts (12V direct current, 1.67 amps). At first, he admits to watching only one hour of television per day, but upon further questioning, he realizes his first estimate did not account for an additional hour each day of news, weather, and sports. Estimate the homeowner's load using the load estimation worksheet below.

### Load Estimation Worksheet

| Individual Loads | Qty **X** Volts **X** Amps = | Watts AC DC | **X** Use hrs/day | **X** Use days/wk | ÷ 7 days | = Watt Hours AC DC |
|---|---|---|---|---|---|---|
| | | | | | 7 | |
| | | | | | 7 | |
| | | | | | 7 | |

AC Total Connected Watts: _____     AC Average Daily Load: _____

DC Total Connected Watts: _____     DC Average Daily Load: _____

### Load Estimation Worksheet

| Individual Loads | Qty **X** Volts **X** Amps = | Watts AC DC | **X** Use hrs/day | **X** Use days/wk | ÷ 7 days | = Watt Hours AC DC |
|---|---|---|---|---|---|---|
| | | | | | 7 | |
| | | | | | 7 | |
| | | | | | 7 | |

AC Total Connected Watts: _____     AC Average Daily Load: _____

DC Total Connected Watts: _____     DC Average Daily Load: _____

**Solution One:**

12V X 4.17amps = 50 watts X 4 hours per day
X 7 days per week ÷ 7 days per week
= 200 watt-hours per day DC

The electrical load could be reduced without reducing the available lighting by using a 13-watt (12V direct current, 1.08 amps) fluorescent lamp and ballast. Recalculate the estimate of this couple's electrical load, noting the effects of this substitution.

**Solution Two:**

Refrigerator: 500 watts AC X 12 hours per day
X 7 days per week ÷ 7 days per week
= 6,000 watt-hours per day AC

TV: 20 watts DC X 2 hours per day
= 40 watt-hours per day DC

A responsible designer should suggest alternate, more cost-effective equipment for operating the homeowner's refrigerator. The owner should consider switching to a propane or kerosene refrigerator to eliminate his largest electrical load. If these units are not available or cannot be used for some reason, the owner should consider purchasing a more efficient direct current refrigerator.

## Table 4-1
## Typical Wattage Requirements for Common Appliances

*These ratings are only estimates

### General household:

Air conditioner (room) . . 1,000
Air conditioner (central) . 3,500
Alarm/security system . . . . . . . 3
Blow dryer . . . . . . . . . . . 1,000
Ceiling fan . . . . . . . . . . 10-50
Vacuum (upright) . . . . . . . 800
Clock radio . . . . . . . . . . . . . 2
Clothes washer . . . . . . . . 1,450
Dryer (electric) . . . . . . . . 4,000
Dryer (gas) . . . . . . . . . . . 300
Electric blanket . . . . . . . . 200
Electric clock . . . . . . . . . . . 2
Furnace fan . . . . . . . . . . . 500
Garage door opener . . . . . . 350
Heater (portable) . . . . . . . 1,500
Iron (electric) . . . . . . . . . 1,200
Radio/phone transmit . . 40-150
Sewing machine . . . . . . . . 100
Table fan . . . . . . . . . . . 10-25

### Refrigeration:

Energy Star fridge/freezer . . 110
    16 ft³ (10 hrs/day)
Refrigerator/freezer . . . . . . 475
    16 ft³ (13 hrs/day)
Sun Frost refrigerator . . . . . 112
    16 ft³ (7 hrs/day)
Vestfrost fridge/freezer . . . . . 60
    10.5 ft³
Standard freezer . . . . . . . . 440
    14 ft³ (15 hrs/day)
Sun frost freezer . . . . . . . . 112
    19 ft³ (10 hrs/day)

### Kitchen appliances:

Blender . . . . . . . . . . . . . 300
Can opener (electric) . . . . . 100
Coffee grinder . . . . . . . . . 100
Coffee maker . . . . . . . . . . 800
Dishwasher . . . . . . . . . . 1,500
Exhaust fans (3) . . . . . . . . 144
Food dehydrator . . . . . . . . 600
Food processor . . . . . . . . . 500
Microwave (.5 ft3) . . . . . . . 750
Microwave (.8 to 1.5 ft3) . . 1,400
Mixer . . . . . . . . . . . . . . 120
Popcorn popper . . . . . . . . 250
Range (large burner) . . . . 2,100
Range (small burner) . . . . 1,250
Trash compactor . . . . . . . 1,500
Toaster . . . . . . . . . 800-1,500

### Lighting:

Incandescent (100W) . . . . . 100
Incandescent light (60W) . . . 60
Compact fluorescent . . . . . . 16
    (60W equivalent)
Incandescent (40W) . . . . . . 40
Compact fluorescent . . . . . . 11
    (40W equivalent)

### Water Pumping:

AC jet pump (¼hp) . . . . . . . 500
    165 gal per day, 20 ft. well
DC pump for house . . . . . . 60
    pressure system (1-2 hrs/day)
DC submersible pump . . . . . 50
    (6 hrs/day)

### Entertainment:

CB radio . . . . . . . . . . . . . . 10
CD player . . . . . . . . . . . . . 35
Cell phone . . . . . . . . . . . . 24
Cell phone charger . . . . . . 6-20
Radio telephone . . . . . . . . . 10
Satellite system (12 ft dish) . . 45
Stereo . . . . . . . . . . . . . 25-50
TV (19-inch color) . . . . . . . . 60
TV (25-inch color) . . . . . . . 130
TV (32-inch color) . . . . . . . 300
VCR . . . . . . . . . . . . . . 20-50
DVD . . . . . . . . . . . . . . . . 11

### Tools:

Band saw (14") . . . . . . . . 1,100
Chain saw (12") . . . . . . . 1,100
Circular saw (7 ¼") . . . . . 1,400
Disc sander (9") . . . . . . . . 1,200
Drill (¼") . . . . . . . . . . . . 300
Drill (½") . . . . . . . . . . . . . 600
Drill (1") . . . . . . . . . . . . 1,000
Electric mower . . . . . . . . 1,500
Weed eater . . . . . . . . . . . 500

### Office:

Computer (desktop) . . . 80-450
Computer (laptop) . . . . 20-140
Printer (ink jet) . . . . . . . . 50-75
Printer (laser) . . . . . . 600-1,200
Fax (stand-by) . . . . . . . . 15-45
Fax (printing) . . . . . . . 120-350

## Load Estimation Worksheet

| Individual Loads | Qty | X Volts | X Amps | = Watts AC | Watts DC | X Use hrs/day | X Use days/wk | ÷ 7 days | = Watt Hours AC | Watt Hours DC |
|---|---|---|---|---|---|---|---|---|---|---|
| | | | | | | | | 7 | | |
| | | | | | | | | 7 | | |
| | | | | | | | | 7 | | |
| | | | | | | | | 7 | | |
| | | | | | | | | 7 | | |
| | | | | | | | | 7 | | |
| | | | | | | | | 7 | | |
| | | | | | | | | 7 | | |
| | | | | | | | | 7 | | |
| | | | | | | | | 7 | | |
| | | | | | | | | 7 | | |
| | | | | | | | | 7 | | |
| | | | | | | | | 7 | | |
| | | | | | | | | 7 | | |
| | | | | | | | | 7 | | |
| | | | | | | | | 7 | | |
| | | | | | | | | 7 | | |
| | | | | | | | | 7 | | |
| | | | | | | | | 7 | | |
| | | | | | | | | 7 | | |
| | | | | | | | | 7 | | |
| | | | | | | | | 7 | | |
| | | | | | | | | 7 | | |
| | | | | | | | | 7 | | |
| | | | | | | | | 7 | | |
| | | | | | | | | 7 | | |
| | | | | | | | | 7 | | |
| | | | | | | | | 7 | | |
| | | | | | | | | 7 | | |

AC Total Connected Watts: _____      AC Average Daily Load: _____

DC Total Connected Watts: _____      DC Average Daily Load: _____

# Chapter 5
# Photovoltaic Modules

**Contents:**

5.1  Photovoltaic Principles . . . . . . . . . . . . . . . . . . . . . . . . . . . . . . . .50

5.2  Characteristics of Modules . . . . . . . . . . . . . . . . . . . . . . . . . . . . .51

5.3  Module Performance . . . . . . . . . . . . . . . . . . . . . . . . . . . . . . . . . .52

5.4  Factors of Module Performance . . . . . . . . . . . . . . . . . . . . . . . . .54

# 5.1 Photovoltaic Principles

Photovoltaic modules and arrays have proven to be a reliable source of electrical energy, but they must be properly designed as a reliable system to be effective. This chapter discusses the basic physical characteristics of photovoltaic modules and explains how some climate and site-specific factors will affect their performance. System designers and users should be aware of these factors when choosing panels and designing photovoltaic systems.

The basic unit of a photovoltaic system is the **photovoltaic cell**. Cells are electrical devices about 1/100th of an inch thick that convert sunlight into direct current electricity through the photovoltaic effect. They do not consume fuel and have a life span of at least 25 years. PV cells have the potential to produce a significant amount of our electric energy.

A **module** is an assembly of photovoltaic cells wired in series or series/parallel to produce a desired voltage and current. Like small batteries, when PV cells are wired in series, the voltage is additive while the current remains constant. Most cells produce approximately one-half of a volt. Therefore, a 36-cell module will typically have an operating voltage of 18 volts under standard test conditions (STC), and a nominal voltage of 12 volts. The current output of the module is dictated by the amount of surface area and the cell efficiency of an individual cell in the module.

The PV cells are encapsulated within the module framework to protect them from weather and other environmental factors. Modules are available in a variety of sizes and shapes. Typically, they are flat rectangular panels that produce anywhere from 5 watts to around 300 watts. The terms "module" and "panel" are often used interchangeably, though more accurately a **panel** is one or more modules wired together. An **array** is a group of modules or panels wired together to produce a desired voltage and fastened to a mounting structure. See Figure 5-1.

A module or array can convert about 10 percent of the available solar radiation to usable electrical energy. For example, at solar noon on a clear day, an array may receive 1000 watts of solar radiation per square meter. This would result in approximately 100 watts of peak power per square meter of array.

Because the sun's position in the sky changes throughout the day and year, the array will receive varying amounts of sunlight. Since the array's power output is directly related to the amount of light it receives and its temperature, an array rarely produces the maximum power possible. For example, an array in Albuquerque, New Mexico (35 degrees NL), which faces due south and is tilted at a 35-degree angle, receives 6.4 full-sun hours each day (averaged over a typical year). If the module is rated at 10 watts per square foot under full sun, 1 square foot will produce 64 Watt-hours of energy each day.

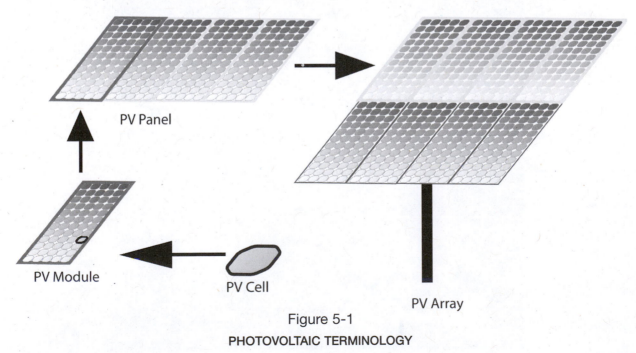

PV Panel

PV Module

PV Cell

PV Array

Figure 5-1

**PHOTOVOLTAIC TERMINOLOGY**

## The Photovoltaic Reaction

Photovoltaic cells do not need moving parts to create electrical energy from the sun's energy. When sunlight strikes a cell, electrons are excited and generate an electric voltage and current that is carried through wires within the cell to an electrical circuit.

We will describe the manufacturing process for single crystalline cells in order to help you understand the photovoltaic reaction. To manufacture single crystalline PV cells, silicon, one of the earth's most abundant elements, is purified and grown into a crystalline structure. Silicon, in its pure form, is a semi-conductor, meaning its electrical properties fall between those of an insulator and a conductor and make it a relatively poor conductor of electricity. By adding special impurities to the silicon through a process known as "doping," silicon's natural properties are modified to better facilitate electric flow. The impurities diffused in the silicon, boron and phosphorous, create a permanent imbalance in the molecular charge, thereby enhancing the silicon's ability to carry electrons.

Once the silicon is grown into a cylinder-shaped crystalline mass, it is sliced into wafers. The wafers are then doped with either boron or phosphorous. When boron, which has an electron deficiency, is diffused into a wafer of silicon, it creates a positively charged material (**p-type material**). When phosphorous, which has an excess of electrons, is diffused into the silicon, it creates a negatively charged material (**n-type material**). A crystalline solar cell is a wafer doped on one side with boron (+) and on the other side with phosphorous (-). The region created between the positive and negative layers is called the **p-n junction**.

When sunlight strikes a cell, it "knocks" loosely held electrons from the negative layer. These excited electrons are attracted to the positively charged boron layer, creating static electrical charge. The loose electrons build an electrical pressure at the p-n junction and begin to flow through the metal contacts built into the cell. All the contacts in a cell join together into a wire that connects the front of one cell to the back of another cell in the module. This electrical circuit enables the electrons to flow through p-n junctions of each cell, building voltage with each cell wired in series.

The voltage increase occurring at the p-n junction of each cell has an electromotive force of approximately one-half volt. The cell voltage is independent of a cell's size, although, current is affected by cell area and sunlight intensity. The larger a cell's area, the greater current it can produce.

Each photovoltaic module manufacturer uses specific designs and construction methods for fabricating wafers of silicon into a module. Once the wafers are formed, they are embedded with metal contacts to sweep electrons into the electrical circuit. The cells are covered with an anti-reflective coating to enhance sunlight absorption. The individual cells are then placed on a backing and wired together to achieve the desired voltage and current. This configuration of cells is framed and encapsulated to create a structural framework and to protect it from environmental factors.

# 5.2 Characteristics of Modules

The following components of a photovoltaic module differentiate the various types of modules:

- Cell material
- Glazing material
- Electrical connections

The primary difference between modules is the cell material. The most common cell material is crystalline silicon. The crystalline material can be grown as a single crystal (mono- or single-crystalline), cast into an ingot of multiple crystals (poly- or multi-crystalline), or deposited as a thin film (amorphous silicon). The two types of crystalline silicon cells perform similarly, although single crystalline cells are slightly more efficient than poly-crystalline due to the poly-crystalline inter-grain boundaries within the cell. Thin film or amorphous silicon, which may also be deposited on a substrate or superstrate, is inexpensive to manufacture but is only about half as efficient as crystalline silicon cells.

Some solar cells, called multi-junction cells, stack several layers of p-type and n-type material. Each layer captures part of the sunlight passing through the cell, to boost overall cell efficiency. There are also advancements being made in **concentrator modules** which incorporate small solar cells surrounded by reflective material that concentrates sunlight onto the cell. Other types of solar cells on the market use different elements other than silicon such as copper indium selenium (CIS) and cadmium telluride.

Each module has a positive and negative terminal, but how these terminals are accessed can vary depending on the module. Some modules have junction boxes with knock-out holes for conduit to enter. These junction boxes can be opened for access and include screw down terminals inside the box where the wires are connected. Alternatively, many modules today come pre-wired with a positive and negative conductor already attached to the back of the module with inaccessible junction boxes. These conductors have a male and female connector attached to the wire ends to expedite the installation process. They are commonly referred to as 'quick-connects.'

# 5.3 Module Performance

The total electrical power output (wattage) of a photovoltaic module is equal to its operating voltage multiplied by its operating current. Photovoltaic modules may produce current over a wide range of voltages. This is unlike voltage sources such as batteries, which produce current at a relatively constant voltage.

The output characteristics of any given module are characterized by a performance curve, called an **I-V curve**, that shows the relationship between current and voltage output. Figure 5-2 shows a typical I-V curve. Voltage (V) is plotted along the horizontal axis. Current (I) is plotted along the vertical axis. Most I-V curves are given for the standard test conditions (STC) of 1000 watts per square meter irradiance (often referred to as one peak sun hour) and 25°C (77°F) cell temperature and 1.5 ATM spectral conditions. It should be noted that STC represent the optimal conditions as a consistent means for measuring — rarely are these conditions recreated in outside environments. The I-V curve contains three significant points:

- Maximum Power Point (both Vmp and Imp)
- Open Circuit Voltage (Voc)
- Short Circuit Current (Isc)

## Maximum Power Point (Vmp & Imp)

This point, labeled **Vmp** and **Imp**, is the operating point at which the maximum output will be produced by the module at operating conditions indicated for that curve. In other words, the Vmp and Imp of the module can be measured when the system is under load at 25°C cell temperature and 1000 Watts per square meter. The voltage at the maximum power point can be determined by extending a vertical line from the curve downward to read a value on the horizontal voltage scale. The example in Figure 5-2 displays a voltage of approximately 17 volts at the maximum power (Vmp).

The current at the maximum power point can be determined by extending a horizontal line from the curve to the left to read a value on the vertical current scale (Imp). The example in Figure 5-2 displays a current of approximately 2.5 amps at the maximum power.

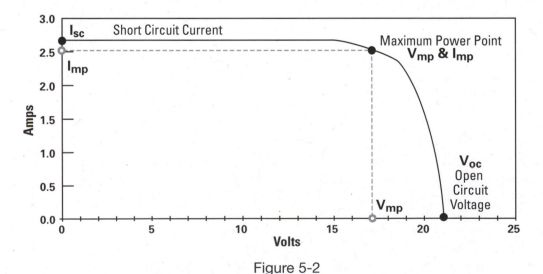

Figure 5-2

**MODULE (*BRAND X*) I-V CURVE (12VDC NOMINAL)**

The wattage at the maximum power point is determined by multiplying the voltage at maximum power by the current at maximum power. In Figure 5-2, the maximum Wattage at STC would be approximately 43 watts. This power is represented by the rectangle under the curve.

The power output decreases as the voltage drops. Current and power output of most modules drops off as the voltage increases beyond the maximum power point.

## Open Circuit Voltage (Voc)

This point, labeled **Voc**, is the maximum potential voltage achieved when no current is being drawn from the module. Since no current is flowing, the module experiences maximum electrical pressure. The example in Figure 5-2 displays an open circuit voltage of approximately 21 volts. The power output at Voc is zero watts.

Open Circuit Voltage can be measured in the field in several common circumstances. When buying a module, it is recommended to test the voltage to see if it matches the manufacture's specifications. When testing voltage with a digital multimeter from the positive to the negative terminal, an open circuit is created by the meter which allows Voc to be measured. It is also common to see a module operating at Voc early in the morning and late in the evening. Intensity of sunlight and cell temperature, as they affect voltage and current, will be discussed later this chapter.

See Section 15.5 and Chapter 16 for directions on safely testing voltage on modules.

## Short Circuit Current (Isc)

This point, labeled **Isc**, is the maximum current output that can be reached by the module under the conditions of a circuit with no resistance or a short circuit. The example in Figure 5-2 displays a current of approximately 2.65 amps. The power output at Isc is zero watts.

When first purchasing a module, it is recommended to test the short circuit current to see if it matches the specification sheet. The short circuit current can be measured only when making a direct short across the positive and negative terminals of a module. Creating a direct short across more than one module at a time (or a module with voltage greater than 24V nominal) is not recommended and can be extremely dangerous. All Isc measurements should be taken when the module is not connected to other components in the system.

> **Note:** When testing modules with 'quick-connects' it is recommended to use test leads to avoid leaving carbon deposits (which cause high resistance) on the module's leads.

Before testing amperage with a digital multimeter, check to ensure the module's Isc does not exceed the meter's DC amperage rating and always use the appropriate personal protective equipment. Intensity of sunlight will affect the amperage reading and is discussed in the next section. See Chapter 16 for directions on reading amperage on high voltage modules or arrays.

See Section 15.5 and Chapter 16 for directions on safely testing current on modules.

## Specification Label

All of the values found on the I-V curve are used to create a specification label for each module. All modules are rated under standard test conditions, thereby allowing their values to be compared. The specification label can be found on the back side of the module or through the manufacturer. Figure 5-3 shows an example specification label. Notice how it correlates to Figure 5-2.

---

### Module Brand X

Electrical Ratings at 1,000W/m$^2$
AM 1.5, Cell Temp. 25°C

| | |
|---|---|
| Max Power: | 43W |
| Voc: | 21.4V |
| Vmp: | 17.3V |
| Isc: | 2.65A |
| Imp: | 2.5A |

Figure 5-3

**SPECIFICATION LABEL**

---

## Safety Procedures

Reference the following materials for procedures when working on or near live parts:

- National Fire Protection Association (NFPA) 70E: *Standard for Electrical Safety in the Workplace.*

- Occupational Safety and Health Administration (OSHA): *Standards for the Construction Industry*, Article 1926.400.

- Occupational Safety and Health Administration (OSHA): *Standards for General Industry*, Article 1910.300.

# 5.4 Factors of Module Performance

Five major factors affect the performance output of photovoltaic modules:

- Cell Material.
- Load Resistance.
- Sunlight Intensity.
- Cell Temperature.
- Shading.

## Load Resistance

When directly coupling to a PV module, the load or battery determines the voltage at which the module will operate. For example, in a nominal 12-volt battery system, the battery voltage is usually between 11.5 and 15 volts. For the batteries to charge, the modules must operate at a slightly higher voltage than the battery bank voltage.

When working with directly coupled systems, designers should ensure that the PV system operates at voltages close to the maximum power point of the array. If a load's resistance is well matched to a module's I-V curve, the module will operate at or near the maximum power point, resulting in the highest possible efficiency. As the load's resistance increases, the module will operate at voltages higher than the maximum power point, causing efficiency and power output to decrease. Efficiency also decreases as the voltage drops below the maximum power point.

This relationship between the load and photovoltaic array is particularly significant when an inductive load, such as a pump or motor, is powered directly by the array. A control device that tracks the maximum power point may be used to continuously match voltage and current operating requirements of the load to the photovoltaic array for maximum efficiency.

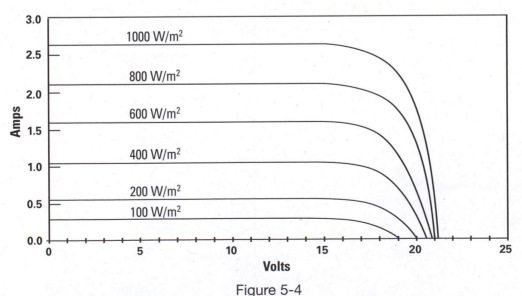

Figure 5-4

**EFFECT OF INSOLATION ON MODULE PERFORMANCE (12VDC NOMINAL)**

## Intensity of Sunlight

A module's current output is proportional to the intensity of solar radiation to which it is exposed. More intense sunlight will result in greater module output. As illustrated in Figure 5-4, as the sunlight level drops, the shape of the I-V curve remains the same, but it shifts downward indicating lower current and power output. However, voltage is not changed appreciably by variations in sunlight intensity.

## Cell Temperature

As the cell temperature rises above the standard operating temperature of 25°C, the module operates less efficiently and the voltage decreases. Figure 5-5 illustrates that as cell temperature rises above 25°C (cell temp, not ambient air temp), the shape of the I-V curve remains the same, but it shifts to the left at higher cell temperatures indicating lower voltage and power output. Heat, in this case, may be thought of as electrical resistance to the flow of electrons. Effective current output may also be decreased if the maximum power point of a module or array drops below the operating voltage of the load.

Airflow around all sides of the modules is critical to remove heat build-up which causes high cell temperatures. A mounting scheme that provides for adequate airflow, such as a stand-off roof mount, a ground mount, or a pole mount, can help maintain lower cell temperatures. Some modules are designed to offset high temperatures by having a better temperature coefficient. Other modules use a greater number of cells in series to offset the lower voltage caused by high temperatures.

Remember, a module's wattage rating is based on Standard Test Conditions in which the cell temperature is 25°C. This is an unrealistic cell temperature for most systems since PV cells are directly in the sun, are dark colored, and are often times mounted on a hot roof. A designer must account for this when sizing a system. Even though a module may be rated at 100 watts, if the system is installed in a hot climate, it is unlikely to produce 100 watts. For this reason, another more realistic rating system has been created called PTC (PV USA Test Conditions). **PTC ratings** still use 1,000 Watts per square meter as one of the parameters, but it also utilizes a more realistic value for temperature. PTC uses 20°C ambient (rather than cell) temperature when rating modules. On average, the wattage ratings will be approximately 88% of the rated STC watts. For example, a 100W STC rated module, will have a PTC rating around 88W.

When calculating the effects of temperature, every module has its own temperature coefficient. An average temperature coefficient for a module is as follows:

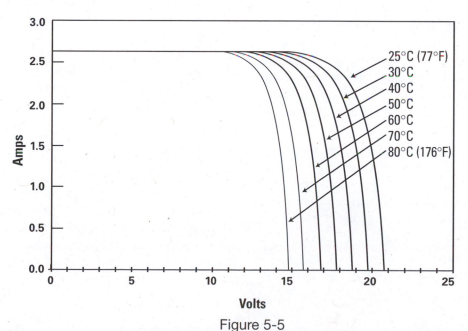

Figure 5-5

**EFFECT OF CELL TEMPERATURE ON MODULE PERFORMANCE (12VDC NOMINAL)**

**Table 5-1**
**Effects of shading on module power**

| Percent of One Cell Shaded | Percent of Module Power Loss |
| --- | --- |
| 0 % | 0% |
| 25 % | 25 % |
| 50 % | 50 % |
| 75 % | 66 % |
| 100 % | 75 % |
| 3 cells shaded | 93 % |

For every degree that cell temperature increases or decreases above or below 25°C, the module voltage shall be adjusted by +/- 0.5%. In other words, as the PV cell temperature increases above 25°C, the module's voltage decreases by approximately 0.5% per degree C. The opposite is true for cold temperatures. See manufacturer's specifications for an individual module's temperature coefficient.

## Shading

Even partial shading of photovoltaic modules will result in a dramatic output reduction. Some modules are more affected by shading than others. Figure 5-6 and Table 5-1 show the extreme effect of shading on one cell of a single crystalline module that has no internal bypass diodes. In Figure 5-6, one completely shaded cell reduces this module's output by as much as 75%. Some modules are less affected by shading than this example.

Note: Remember that at a minimum, the array should not be shaded from 9 a.m. to 3 p.m. If there is shading during this period, more modules will be needed to produce adequate power. The installer is responsible for selecting an appropriate site.

Locating shading obstacles at the site is an extremely important part of a site evaluation. An entire system's performance can be diminished and a client's investment undermined by underestimating the effects of shading, even partial shading. Most manufacturers today use bypass diodes in the module to reduce the effects of shading by allowing current to bypass shaded or failed cells. A bypass diode is a semiconductor device that allows electric current to only flow in one direction, and prevents current from flowing into shaded areas. Even with diodes, modules can be severely impacted by shading.

A common scenario causing shading effects in the field, can be created by mounting modules at a different tilt and orientation angles on a roof. Having an array on both an east and west roof, for example, will cause shading on different parts of the array throughout the day as the sun changes position in the sky. Under these conditions, the wiring configuration of the modules (and series strings) becomes critical to the maximum power output. The MPPT feature of grid-tied inverters (and charge controllers) can only track the peak power of a single I-V curve. For this reason, all modules wired together in series must have the same tilt and orientation angle for optimum power production. If it is necessary to use different roof pitches, series strings (with the same tilt and orientation) can then be wired in parallel. Under no conditions should modules be wired in series from different roof pitches and orientations.

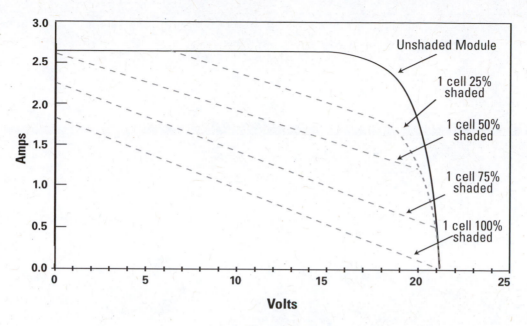

Figure 5-6

**EFFECT OF SHADING ON MODULE PERFORMANCE (12VDC NOMINAL)**

# Chapter 6
# Batteries

**Contents:**

6.1   Battery Types and Operation . . . . . . . . . . . . . . . . . . . . . . . . . . . . . .60

6.2   Battery Specifications . . . . . . . . . . . . . . . . . . . . . . . . . . . . . . . . . . .62

6.3   Battery Safety . . . . . . . . . . . . . . . . . . . . . . . . . . . . . . . . . . . . . . . . .68

6.4   Battery Wiring Configuration . . . . . . . . . . . . . . . . . . . . . . . . . . . . .69

6.5   Battery Sizing Exercise . . . . . . . . . . . . . . . . . . . . . . . . . . . . . . . . . .71

# 6.1 Battery Types and Operation

Batteries store direct current electrical energy in chemical form for later use. In a photovoltaic system, the energy is used at night and during periods of cloudy weather. Batteries also serve as a portable power source for appliances, such as flashlights and radios. Since a photovoltaic system's power output varies throughout any given day, a battery storage system can provide a relatively constant source of power when the PV system is producing minimal power during periods of reduced insolation. Batteries can even power the loads when the PV array is disconnected for repair and maintenance. Batteries can also provide the necessary amounts of surge power required to start some motors.

Batteries are not one hundred percent efficient. Some energy is lost as heat and in chemical reactions, during charging and discharging. Therefore, additional photovoltaic modules must be added to a system to compensate for battery loss.

Remember, there are several types of day-use systems that do not require batteries. A water pumping system can be designed to pump during the day to a storage tank. If water is needed at night, the water can be gravity fed from a storage tank located higher than the faucet or point of use. Some applications, such as greenhouse ventilation fans, proportionally require more electricity as the intensity of the sun increases. The fans are only needed and can only operate when the sun shines, an ideal use for PV-supplied power.

Utility grid-connected photovoltaic systems do not necessarily require batteries, though they can be used as an emergency back-up power source. Some inefficiency issues exist when using batteries in a grid-connected system. See Chapter 11.

Many battery types and sizes are available. Smaller sizes, commonly used in flashlights or portable radios, are available in the "disposable" (primary) or rechargeable (secondary) options. Rechargeable nickel cadmium (NiCad) batteries are commonly used for large standby loads, such as industrial applications, daily loads in cold climates, and small portable appliances. These batteries may be recharged using a solar or AC battery charger. Manufacturers of NiCad batteries claim that NiCads will last through more charge/discharge cycles than lead-acid batteries. A battery is charging when energy is being put in and discharging when energy is being taken out. A cycle is considered one charge-discharge sequence, which often occurs over a period of one day in residential photovoltaic systems.

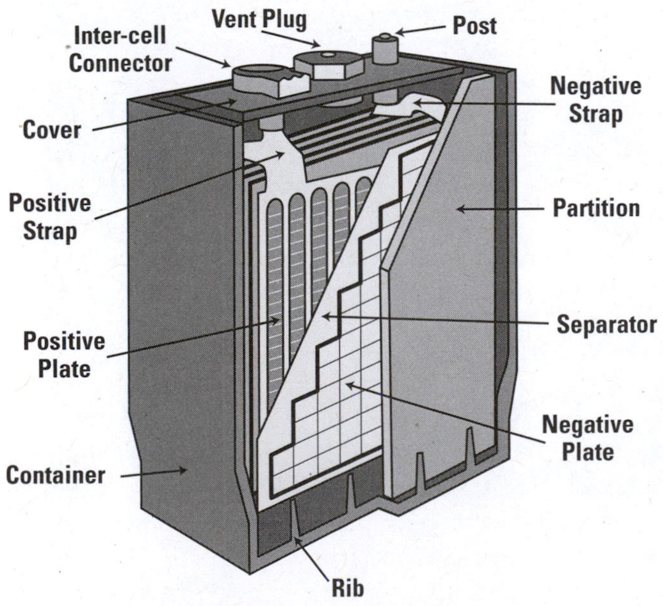

The following types of batteries are commonly used in PV systems:

- Lead-acid batteries
  - Liquid vented
  - Sealed (VRLA- Valve Regulated Lead Acid)
- Alkaline batteries
  - Nickel-cadmium
  - Nickel-iron

## Lead-acid Batteries

In the United States, the battery most commonly used for residential scale photovoltaic applications is the lead-acid battery, which closely resembles an automotive battery. Automotive batteries, however, are not recommended for PV applications because they are not designed to be "deep-cycled." They are designed to discharge large amounts of current for a short duration to start an engine and then be immediately recharged by the vehicle's alternator or generator. Photovoltaic

**Figure 6-1**

**CUT-AWAY OF A STANDARD LEAD ACID BATTERY CELL**

systems often require a battery to discharge small amounts of current over long durations and to be recharged under irregular conditions. Deep cycle batteries can be discharged down as much as 80 percent state of charge. An automotive battery may last for only a few photovoltaic cycles under these conditions. In contrast, deep cycle lead-acid batteries suitable for photovoltaic applications can tolerate these conditions, and, if properly sized and maintained, they will last from three to ten years, or even longer.

This chapter primarily discusses the lead-acid battery system, since these batteries are rechargeable, widely available, relatively inexpensive, and available in a variety of sizes and options. They are also commonly used, easily maintained, and reasonably long lived. Lead-acid batteries may be categorized into liquid electrolyte (liquid vented) and captive electrolyte (sealed or VRLA) subcategories.

**Liquid vented**: Liquid lead-acid batteries are like automobile batteries. The battery is built from positive and negative plates, made of lead and lead alloy placed in an electrolyte solution of sulfuric acid and water. Figure 6-1 shows the cross-section of a common 12-volt liquid lead-acid battery containing 6 individual 2-volt cells. As with the automobile battery, a voltage control is used to regulate the battery voltage. As the battery nears full charge, hydrogen gas is produced and vented out of the battery.

> **Caution! Hydrogen gas is very explosive if contained. No open flames or sparking can be allowed near a battery. Motor generators, gas space heaters, and gas water heaters must be isolated from a battery.**

Water is lost when the battery vents waste gasses, so it must be refilled periodically. Some batteries are equipped with recombinator cell caps that capture the gasses and return them to the battery as water. Deep cycle batteries will last longer if protected from complete discharge. Controls with a low voltage disconnect (LVD) protect batteries from complete discharge. Like an automobile battery, less capacity is available when the battery is cold, and higher temperatures shorten battery life.

**Sealed lead-acid batteries (VRLA)**: Unlike liquid vented batteries, sealed batteries have no caps, and thus no access to the electrolyte. They are not totally sealed — a valve allows excess pressure to escape in case of over charging. This is referred to as a valve regulated lead acid battery (VRLA). Sealed batteries are considered maintenance free because the electrolyte can not be accessed.

The two types of sealed batteries commonly used in photovoltaic systems are **gel cell** and **absorbed glass mat (AGM)**. In traditional gel cell batteries, the electrolyte is gelled by the addition of silica gel that turns the liquid into a gelled mass. AGM batteries use a fibrous silica glass mat to suspend the electrolyte. This mat provides pockets that assist in the recombination of gasses generated during charging, and limit the amount of hydrogen gas produced.

The main advantage of sealed batteries is that they are spill-proof. The gelled electrolyte cannot be spilled, even when broken. This allows them to be safely transported and handled. For this reason, they are a reasonable choice for RV or marine applications. They can be air-shipped in contrast to a commercial liquid lead-acid battery that needs to be shipped dry, then activated on site by the addition of electrolyte. They also do not require periodic maintenance, such as watering or equalization. This makes them a good choice for very remote applications where regular maintenance is unlikely or not cost effective.

Gel cell batteries cost more per unit of capacity compared to liquid lead-acid batteries. They are susceptible to damage from overcharging especially in hot climates and have a shorter life expectancy than other battery types. It is important to remember that most sealed batteries must be charged to lower voltages and at a lower amperage rate to prevent excess gas from damaging cells.

**Using charge controllers with lead acid batteries**: Lead-acid batteries need a controller to prevent overcharging and discharging. These controllers, known as charge controllers, work by monitoring battery voltage, which rises as the battery is charged and falls as the battery discharges. A charge controller is necessary because overcharging causes excessive loss of liquid electrolyte, which increases maintenance requirements and shortens battery life. Also, the deeper a battery is regularly discharged, the shorter its life. Thus, a charge controller with low voltage disconnect (LVD) is often desirable to prevent deep discharge.

In home PV systems, the use of an automatic LVD should not negate the end-user's responsibility to manage battery state-of-charge. LVD only protects the battery from over-discharging from DC loads. AC loads must be managed with an inverter LVD. Many PV system designers consider controllers with LVD to be "the last line of defense" to protect batteries from being overly discharged.

Each battery type has a slightly different charge termination voltage (high voltage disconnect or HVD). With 2-volt nominal cell voltage, the safe charge termination set point for lead acid batteries is 2.35–2.5 volts per cell. The low voltage disconnect (LVD) will also vary depending on the depth of discharge desired. Table 6-1 lists the typical voltage set points for sealed and liquid lead-acid batteries. Always use the manufacturer's specifications if available.

### Alkaline Batteries

Alkaline batteries, such as nickel-cadmium and nickel-iron batteries, also have positive and negative plates in an electrolyte. The plates are made of nickel and cadmium or nickel and iron and the electrolyte is potassium hydroxide. Each cell has a nominal voltage of 1.2 volts and the charge termination point is 1.65-1.8 volts per cell. These batteries are often expensive and may have voltage window compatibility issues with certain inverters and charge controls. An advantage is that they are not as affected by temperature as other types of batteries. For this reason, alkaline batteries are usually only recommended for commercial or industrial applications in locations where extremely cold temperatures (-50°F or less) are anticipated.

In residential PV systems, typically liquid lead-acid batteries are the wisest choice. They usually constitute a significant part of the total system cost. The majority of PV systems and components are designed to use lead-acid batteries. Despite the safety, environmental, and maintenance issues, batteries are necessary to provide the needed flexibility and reliability to a home PV system.

# 6.2 Battery Specifications

A photovoltaic system designer must consider the following variables when specifying and installing battery storage system for a stand-alone photovoltaic system:

- Days of autonomy.
- Battery capacity.
- Rate and depth of discharge.
- Life expectancy.
- Environmental conditions.
- Price and warranty.
- Maintenance schedule.

### Days of Autonomy

Autonomy refers to the number of days a battery system will provide a given load without being recharged by the photovoltaic array or another source. You must consider a system's location, total load, and types of loads to correctly determine the number of days of autonomy.

General weather conditions determine the number of "no sun" days, which is a significant variable in determining autonomy. Local weather patterns and microclimates must also be considered. For example, in the Colorado mountains, storms range from an afternoon summer thundershower to a three-day winter snowstorm. In humid climates, three- to four-week cloudy periods can occur. It may be cost-prohibitive to size a battery system capable of providing power in the most extreme conditions. Consequently, most designers usually opt for a design based on the average number days of cloudy weather or design with a hybrid approach adding a generator or wind turbine.

The most important factors in determining an appropriate autonomy for a system are the size and type of loads that the system services. It's important to answer several questions about each load:

- Is it critical that the load operate at all times?

- Could an important load be "shed" or replaced by alternatives?

- Is the load simply a convenience?

The general range of autonomy is as follows:

- 2 to 3 days for non-essential uses or systems with a generator back-up.

- 5 to 7 days for critical loads with no other power source.

**Note:** The system designer must take into account that the PV array is typically sized to meet the daily load. If autonomy is built into the system and no daily loads are "shed" the PV array might not be capable of fully recharging the battery bank. An alternate means of battery charging should be employed.

## Battery Capacity

Batteries are rated by amp-hour (Ah) capacity. The capacity is based on the amount of power needed to operate the loads and how many days of stored power will be needed due to weather conditions. Using a water analogy, you can think of a battery as a bucket, the stored energy as water and the Ah capacity as the bucket size. The Ah rating will tell you "how large your bucket is".

Most battery manufacturers specify battery capacity in amp-hours. In theory, a 100 Ah battery will deliver one amp for 100 hours or roughly two amps for 50 hours before the battery is considered fully discharged. If more storage capacity is required to meet a specific photovoltaic application requirement then batteries can be connected in parallel. Two 100 amp-hour 12-volt batteries wired in parallel provide 200 amp-hours at 12 volts. Higher voltages are obtained through series wiring. Two 100 amp-hour 12-volt batteries wired in series provide 100 amp-hours at 24 volts.

Many factors can affect battery capacity, including rate of discharge, depth of discharge, temperature, age, and recharging characteristics. Fundamentally, the required capacity is also affected by the size of the load. If the load is reduced, the required battery capacity is also reduced.

Since it is easy to add photovoltaic modules to an existing photovoltaic system, a commonly held misconception is that the entire photovoltaic system is modular as well. However, manufacturers generally advise against adding new batteries to an old battery bank. Older batteries will degrade the performance of new batteries (since internal cell resistance is greater in older batteries) and could result in reduced system voltage when wired in series. In addition, if you were to add batteries to an existing system, you would probably add them in parallel to increase amp-hour capacity and maintain system voltage. It's advisable to minimize excessive "paralleling" because this increases the total number of cells, thereby increasing the potential for failure from a bad cell.

You should initially specify a slightly larger battery capacity than is needed because batteries lose their capacity as they age. However, if you greatly oversize the battery bank, it may remain at a state of partial charge during periods of reduced insolation. This partial charge state can cause shortened battery life, reduced capacity, and increased sulfation. Battery capacity should be determined by the overall load profile.

---

Table 6-1
**Voltage Set Points for Lead-acid Batteries in a 12V System**
for a nominal 12V battery

| Type | Charge Termination | Low Voltage Cutoff |
|---|---|---|
| Sealed (VRLA) | 14.1 volts | 11.6 volts |
| Liquid | 14.6 - 15.0 volts | 11.3 volts |

### Table 6-2
### Effect of Discharge Rate on Battery Capacity

| Model | Volts/Unit | 72 hour AH capacity | 24 hour AH capacity | 16 hour AH capacity | 12 hour AH capacity | 8 hour AH capacity |
|-------|------------|--------------------|--------------------|--------------------|--------------------|--------------------|
| C2 | 2 | 288 | 270 | 259 | 245 | 230 |
| B6 | 6 | 192 | 180 | 173 | 162 | 154 |
| A12 | 12 | 105 | 100 | 95 | 90 | 85 |

## Rate and Depth of Discharge

The rate at which the battery is discharged directly affects its capacity. If a battery is discharged quickly, less capacity is available. Conversely, a battery that is discharged slowly will have a greater capacity. For example, a 6-volt battery may have a 180 Ah capacity if discharged over 24 hours. However, if the same battery is discharged over 72 hours, it will have a 192 Ah capacity.

A common battery specification is the battery's capacity in relation to the number of hours that it is discharged. For example, when a battery is discharged over 20 hours, it is said to have a discharge rate of C/20 or capacity at 20 hours of discharge. If a battery is discharged over 5 hours, the discharge rate is C/5. Note that the C/5 discharge rate is four times faster than the C/20 rate. Most batteries are rated at the C/20 rate. Table 6-2 lists a few batteries and their battery capacity at several discharge rates.

**Note:** Always check manufacturer recommendations.

Similar consideration should be taken when charging batteries. Most flooded lead-acid batteries should not be charged at more than the C/5 rate. If a battery were rated at 220 Ah at the C/20 rate, charging it at a C/10 rate would equal charging at 22 amps (220 ÷ 10). Gel-cells, however, should never be charged at higher than a C/20 rate.

Depth of discharge (DOD) refers to how much capacity will be withdrawn from a battery. Most PV systems are designed for regular discharges of 40 percent to 80 percent. Battery life is directly related to how deep the battery is cycled. For example, if a battery is discharged to 50 percent every day, it will last about twice as long as if it is cycled to 80 percent. Lead-acid batteries should never be completely discharged, even though some deep cycle batteries can survive this condition the voltage will continually decrease. NiCad batteries, on the other hand, can be totally discharged without harming the battery and hold their voltage. When the NiCad is fully discharged it may reverse polarity, potentially harming the load. A manufacturer's specification sheet will list the maximum depth of discharge for any battery.

Shallow cycling systems, discharging the battery only 10 to 20 percent, have two distinct advantages. First, in general, batteries that are shallow cycled will have a longer life span. If a battery is only cycled to 10 percent DOD, it will last about 5 times as long as one cycled to 50 percent. Second, a reserve Ah capacity is designed into the system for extended cloudy weather. This is not to say that a larger battery capacity is always better. As discussed previously, if a battery bank is very large with respect to the capacity of the charging source, the batteries will not be charged quickly enough to return them to a full state of charge. This can result in sulfation and decreased battery life. The most practical number to use when designing a system is 50 percent depth of discharge for the best storage versus cost factor.

---

**Problem:** Using Table 6-2, determine the Ah capacity of the 6V battery (model B6) when discharged over 24 hours versus 8 hours.

**Solution:**

154 Ah at 8 hour rate

180 Ah at 24 hour rate

The battery will have an additional 26 Ah when discharged at the slower 24 hour rate.

---

## Life Expectancy

Most people think of life expectancy in terms of years. Battery manufacturers, however, specify life expectancy in terms of a quantity of cycles. Batteries lose capacity over time and are considered to be at the end of their life when 20 percent of their original capacity is lost, although they can still be used. When sizing a system initially, this should be considered.

Depth of discharge also refers to the percentage of a battery's rated amp-hour capacity that has been used. Battery life (number of daily cycles) versus depth of discharge (a percent of battery capacity) is shown for a lower cost sealed battery in Figure 6-2.

For example, a battery that experienced shallow cycling of only 25 percent DOD would be expected to last 4000 cycles while a battery cycled to an 80 percent DOD would last 1500 cycles. If one cycle equaled one day, the shallowly cycled battery would last for 10.95 years while the deeply cycled battery would last for only 4.12 years.

This is only an estimate. Some batteries are designed to be cycled more than once each day. In addition, batteries degrade over time, affecting life expectancy.

## Environmental Conditions

Batteries are sensitive to their environment and are particularly affected by the temperature of that environment. Higher voltage charge termination points are required to complete charging as a battery's temperature drops (the opposite is true in warmer temperatures). Controllers with a temperature compensation feature can automatically adjust charge voltage based on a battery's temperature.

Manufacturers generally rate batteries at 77°F (25°C). The battery's capacity will decrease at lower temperatures and increase at higher temperatures. A battery at 32°F may be able to achieve only 65 to 85 percent of its fully rated capacity. A battery at -22°F may achieve only 50 percent. Battery capacity is increased at higher temperatures. Figure 6-3 illustrates the effects of temperature on batteries at three discharge rates.

Even though battery capacity decreases at lower temperatures, battery life increases. Likewise, the higher the battery temperature, the shorter the life of the battery. Most manufacturers say there is a 50 percent loss in life for every 15°F increase over the standard 77-degree cell temperature. As far as capacity versus battery life, this tends to even out in most systems, as they will spend part of their lives at higher temperatures, and part at lower temperatures.

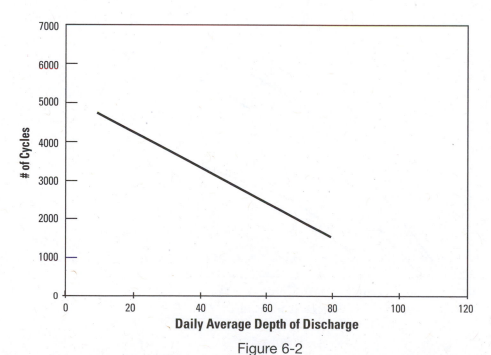

Figure 6-2

**NUMBER OF BATTERY CYCLES IN RELATION TO DAILY DEPTH OF DISCHARGE**

### Table 6-3
### Battery Temperature Multiplier
### at Various Temperatures

| Temperature | Battery Temperature Multiplier |
|---|---|
| 80°F / 26.7°C | 1.00 |
| 70°F / 21.2°C | 1.04 |
| 60°F / 15.6°C | 1.11 |
| 50°F / 10.0°C | 1.19 |
| 40°F / 4.4°C | 1.30 |
| 30°F / -1.1°C | 1.40 |
| 20°F / -6.7°C | 1.59 |

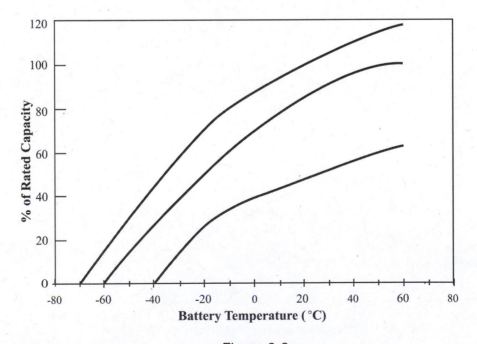

### Figure 6-3
### EFFECTS OF TEMPERATURE ON BATTERY CAPACITY

When sizing a system, you can compensate for the effects of temperature by using a battery temperature multiplier. To find out the adjusted battery capacity needed, multiply the necessary battery capacity by the battery temperature multiplier in Table 6-3.

Colder temperatures affect more than the battery's capacity. In extremely cold environments, the electrolyte can freeze. The temperature at which a battery will freeze is a function of its state of charge. When a liquid electrolyte battery is completely discharged, the electrolyte is principally water. The electrolyte in a fully charged battery has a higher concentration of sulfuric acid, which freezes at a much lower temperature. Typically the electrolyte in a battery consists of approximately 25% sulfuric acid and 75% water. Table 6-4 lists the freezing point at various states of charge. To maintain a constant temperature, lead-acid batteries can be placed in an insulated (R20 extruded polystyrene) battery box. NiCad or sealed batteries are not as susceptible to freeze damage.

Regardless of temperature concerns, batteries should be located in a sturdy enclosure (a battery box). Since liquid electrolyte batteries produce explosive hydrogen gas when charging, the enclosure or area where the batteries are located should be well vented. Even though a battery box helps to contain the gasses, other electric system components should be installed a reasonable distance away from the battery compartment. This reasoning is twofold. One, sparking from the electrical equipment could ignite the gasses. Two, the gasses are corrosive and will attack other system components. Ventilation problems can be addressed by using special re-combinators or catalytic converter cell caps that capture hydrogen vent gas, recombine it with oxygen to create water, and return the liquid water to the battery electrolyte. A battery enclosure should also be used to contain acid in case the batteries leak.

Always try to design systems with batteries as near as is safely possible to the loads and the array to minimize wire runs, thereby saving money on materials and reducing voltage drop.

## Measuring Battery State of Charge

A voltmeter or a hydrometer can be used to measure a battery's state of charge. To properly check voltages, the battery should sit at rest for a few hours (disconnect from charging sources and loads). Table 6-4 can be used to compare a 12V battery's voltage to its state of charge. For a 24-volt system, multiply by 2, and for a 48-volt system, multiply by 4. For gel cell batteries, subtract 0.2 volts from the numbers in the table.

Table 6-4 can also be used to determine a battery's state of charge by measuring the specific gravity of a cell with a hydrometer. (See Chapter 16.)

### Table 6-4
### Liquid Electrolyte Freeze Points, Specific Gravity, and Voltage

| State of Charge | Freeze Point | Specific Gravity | Voltage |
|---|---|---|---|
| 100% | -71 °F | 1.260 | 12.70 |
| 75% | -35 °F | 1.237 | 12.50 |
| 50% | -10 °F | 1.200 | 12.30 |
| 25% | 3 °F | 1.150 | 12.00 |
| 0% | 17°F | 1.100 | 11.70 |

# 6.3 Battery Safety

Batteries are potentially the most dangerous PV system component if improperly handled, installed, or maintained. Dangerous chemicals, heavy weight, and high voltages and currents are potential hazards that can result in electric shock, burns, explosion, or corrosive damage to yourself or your property. Since battery technology is constantly evolving, responsible designers and users should continue their training on the aspects of battery technology beyond what is covered in this chapter to insure proper handling and design. Specific manufacturer's literature can be obtained to help with design, installation, and maintenance decisions.

You should observe the following safety rules to insure proper and safe handling, installing, maintaining, and replacing of photovoltaic system batteries:

## General Battery Safety Rules

- Draw a battery diagram before wiring batteries.
- Remove any jewelry before working around batteries.
- Use proper tools when assembling cells.
- Design the battery area to be properly ventilated.
- Wear protective clothing (especially eye protection) while working on batteries.
- Have baking soda accessible to neutralize acid spills.
- Have fresh water accessible in case electrolyte splashes on skin or eyes; If an accident occurs flush with water for five to 10 minutes, then contact a physician.
- Keep open flames and sparks away from batteries. No smoking near any battery.
- Discharge body static electricity before touching terminal posts.
- Disconnect battery bank from any sources of charging or discharging before working on batteries.
- Do not lift batteries by their terminal posts or by squeezing the sides of the battery. Lift batteries from the bottom or use carrying straps.
- Do not use metal hard hats or non-insulated tools around batteries to avoid possible shock. Use tools with insulated or wrapped handles to avoid accidental short circuits.
- Wash hands immediately after handling batteries.
- Put batteries in a safe enclosure to prevent access from children and inexperienced adults.
- Follow the manufacturer's instructions.
- Use common sense.

## Battery Do's

- Keep batteries out of living space.
- Keep safety gear close to batteries.
- Use safety precautions while working on batteries.
- Keep a regular maintenance and watering schedule.
- Vent the battery box to the outside.
- Keep battery cable lengths the same size.
- Keep number of parallel connections to a minimum.
- Check and record specific gravity on all cells when first receiving batteries.
- Equalize on a regular basis.
- Clean corrosion off posts.
- Cables to Inverter should depart from bottom of battery box (hole should be sealed).
- Connect batteries last.
- Use a spill containment vessel.
- Keep batteries at level temperatures.

## Battery Don'ts

- Don't mix different battery types.
- Don't mix old batteries with new.
- Don't water batteries before equalizing.
- Don't check amps across terminals.

# 6.4 Battery Wiring Configuration

Batteries need to be configured to obtain the desired voltage and amp-hours. Using the design and battery parameters from the example, we can clearly see how a system's batteries should be configured and wired. Two separate six-volt batteries rated at 200 Ah each are wired in series to obtain 12V direct current and 200 Ah. Two of these series strings are wired in parallel to achieve 12V direct current and 400 Ah. Figure 6-4, Figure 6-5, and Figure 6-6 show examples of wiring configurations for 12V, 24V, and 48V battery banks, respectively.

**Note:** To create an equal path length for electron flow through the batteries, you must wire into opposite sides of the battery bank keeping the cables equal length. See Figure 6-4, Figure 6-5, and Figure 6-6.

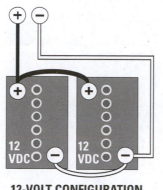

**12-VOLT CONFIGURATION**
with 12-volt batteries in *parallel*

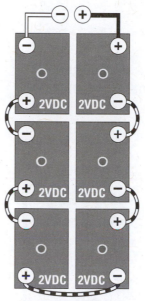

**12-VOLT CONFIGURATION**
with 2-volt batteries in *series*

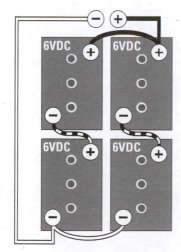

**12-VOLT CONFIGURATION**
with 6-volt batteries in *series/parallel*

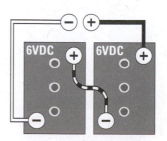

**12-VOLT CONFIGURATION**
with 6-volt batteries in *series*

Figure 6-4

**12 VOLT BATTERY CONFIGURATIONS**

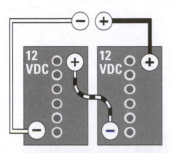

**24-VOLT CONFIGURATION**
with 12-volt batteries in *series*

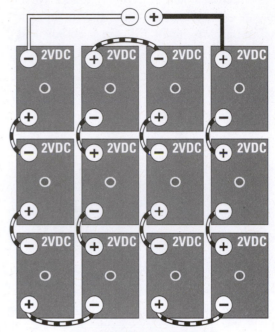

**24-VOLT CONFIGURATION**
with 2-volt batteries in *series*

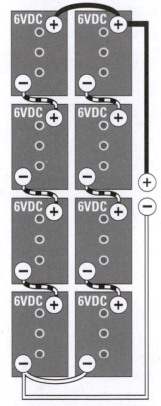

**24-VOLT CONFIGURATION**
with 6-volt batteries in *series/parallel*

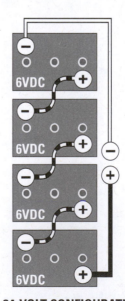

**24-VOLT CONFIGURATION**
with 6-volt batteries in *series*

Figure 6-5

**24-VOLT BATTERY CONFIGURATIONS**

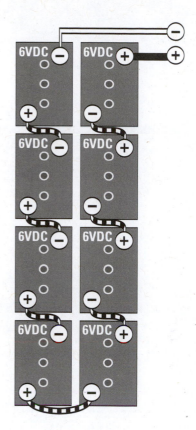

Figure 6-6
**48-VOLT BATTERY CONFIGURATIONS**

# 6.5 Battery Sizing Exercise

## Problem:

Use Table 6-5 to specify a battery bank for the following photovoltaic system. (A complete system sizing worksheet can be found in Appendix D.)

The occupants of a remote home near Ojai, California, are designing a photovoltaic system to meet their 1080 watt hours per day AC electrical load. They have decided on a 12-volt direct current system and feel they need two days of autonomy. The maximum depth of discharge they desire over that two-day period is 50 percent. The occupants have tentatively selected the Model A battery from XYZ Manufacturer, a 6-volt battery rated at 200 amp-hours. The occupants will keep the battery(s) in a conditioned space that will be maintained at 77°F.

Table 6-5
**Battery Sizing Worksheet**

| AC Average Daily Load | ÷ | Inverter Efficiency | + | DC Average Daily Load | ÷ | DC System Voltage | = | Average Amp-hours/ Day |
|---|---|---|---|---|---|---|---|---|
| [( | ÷ | | ) + | | ] ÷ | | = | |
| Average Amp-hours/day | X | Days of Autonomy | ÷ | Discharge Limit | ÷ | Battery AH Capacity | = | Batteries in Parallel |
| | X | | ÷ | | ÷ | | = | |
| DC System Voltage | ÷ | Battery Voltage | = | Batteries in Series | X | Batteries in Parallel | = | Total Batteries |
| | ÷ | | = | | X | | = | |

| Battery Specification | Make: | Model: |
|---|---|---|

**Solution:**

1. To start, de-rate for inverter inefficiency by dividing the AC Average Daily Load (1080 watts) by the standard inverter efficiency figure (90 percent or 0.9).

2. Multiply the resulting Average Amp-Hours/Day (100) by the Days of Autonomy (2) and divide by the Discharge Limit or DOD (50 percent or 0.5) and divide again by the Battery Ah Capacity for the specified battery (200). The resulting figure is the number of batteries in parallel (2).

3. Next determine the number of batteries needed to achieve the system voltage by dividing the DC System Voltage (12) by the Battery Voltage (6). Then multiply this number (2) by the number of batteries in parallel (2) to determine the Total Batteries needed (4).

The complete battery sizing worksheet is listed in Table 6-6.

### Table 6-6
### Answers to the Battery Sizing Exercise

| AC Average Daily Load | ÷ | Inverter Efficiency | + | DC Average Daily Load | ÷ | DC System Voltage | = | Average Amp-hours/ Day |
|---|---|---|---|---|---|---|---|---|
| [( 1080 | ÷ | .9 | ) + | N/A ] | ÷ | 12 | = | 100 |
| Average Amp-hours/day | X | Days of Autonomy | ÷ | Discharge Limit | ÷ | Battery AH Capacity | = | Batteries in parallel |
| 100 | X | 2 | ÷ | .5 | ÷ | 200 | = | 2 |
| DC System Voltage | ÷ | Battery Voltage | = | Batteries in Series | X | Batteries in Parallel | = | Total Batteries |
| 12 | ÷ | 6 | = | 2 | X | 2 | = | 4 |
| **Battery Specification** | | Make: XYZ | | | | Model: A | | |

# Chapter 7
# PV Controllers

**Contents:**

7.1   Controller Types . . . . . . . . . . . . . . . . . . . . . . . . . . . . . . .74
7.2   Controller Features . . . . . . . . . . . . . . . . . . . . . . . . . . . . .75
7.3   Specifying a Controller . . . . . . . . . . . . . . . . . . . . . . . . . .78
7.4   Controller Sizing Exercise . . . . . . . . . . . . . . . . . . . . . . .79

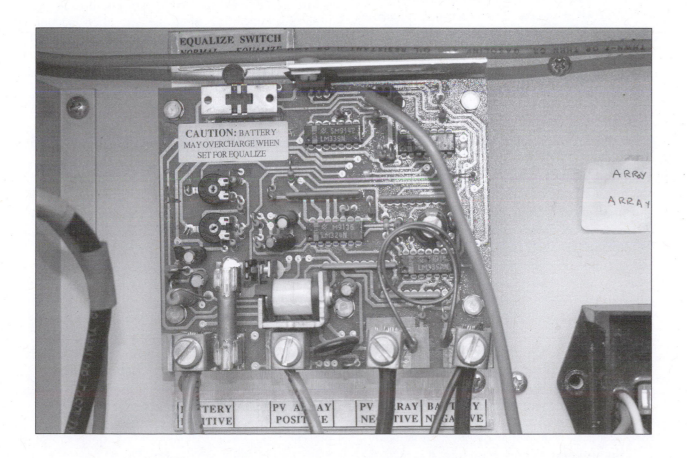

# 7.1 Controller Types

The photovoltaic controller works as a voltage regulator. The primary function of a controller is to prevent the battery from being overcharged by the array. Some PV controllers also protect a battery from being overly discharged by the DC load.

A PV charge controller constantly monitors the battery voltage. When the batteries are fully charged, the controller will stop or decrease the amount of current flowing from the photovoltaic array into the battery. When the batteries are being discharged to a low level, many controllers will shut off the current flowing from the battery to the DC load(s).

Charge controllers come in many sizes, typically from just a few amps to as much as 80 amps. If high currents are required, two or more PV controllers can be used. When using more than one controller, it is necessary to divide the array into sub-arrays. Each sub-array will be wired into its own controller and then they will all be wired into the same battery bank. There are four different types of PV controllers. They are:

- Shunt controllers.

- Single-stage series controllers.

- Diversion controllers.

- Pulse width modulation (PWM) controllers.

## Shunt Controllers

Shunt controllers are designed for very small systems. They prevent overcharging by "shunting" or short-circuiting the PV array when the battery is fully charged. The shunt controller circuitry monitors the battery voltage and switches the PV's current through a power transistor when a pre-set full charge value is reached. The transistor acts like a resistor and converts the PV's power into heat. Shunt controllers have heat sinks with fins that help to dissipate the heat produced.

These controllers also incorporate a blocking diode to prevent current from draining back from the batteries through the solar array at night. Blocking diodes act like one way valves, allowing current to flow into the batteries during charging, and prevent back flow or leakage from batteries to the array at night.

Shunt controllers are simply designed and inexpensive. The circuitry is often completely sealed for environmental protection. They must be exposed to the open air to provide the ventilation required from the cooling fins. Their disadvantages are their limited load handling capability and ventilation requirements.

## Single-stage Series Controllers

Single-stage controllers prevent battery overcharging by switching the PV array off when the battery voltage reaches a pre-set value called the charge termination set point (CTSP). The array and battery are automatically reconnected when the battery reaches a lower preset value called the charge resumption set point (CRSP). Some manufacturers incorporate a built-in timer to cycle the PV array on and off during the end of the charging process to top-off the battery bank.

Single-stage controllers use a relay or transistor to break the circuit and prevent reverse current flow at night, instead of using a blocking diode. These controllers are small and inexpensive, eliminating the need for bulky heat sinks because they do not produce much heat. They have greater load handling capacity than shunt type controllers. Another advantage is they generally do not require significant ventilation.

## Diversion Controllers

These controllers automatically regulate the charging currents depending on the battery's state of charge by diverting excess charging current to a resistive load. The full array current is allowed to flow when the battery is at a low state of charge. As the battery bank approaches full charge, the controller and the load resistors dissipates some of the array power so that less current flows into the batteries. The charging current drops to a trickle charge as the battery bank approaches a fully charged state.

This charging regulation approach is also able to work with other non-PV charging sources like hydro or wind generators — even combinations of these different sources. Like shunt controllers, heat is generated by the dissipation of power, requiring diversion controllers to be properly ventilated. These controllers generally do not have a system that prevents reverse leakage at night, so additional diodes may be required for the PV array.

## Pulse Width Modulation (PWM) Controllers

PWM controllers are the most common residential controller on the market today. These provide a tapering charge by rapidly switching the full charging current on and off when the battery reaches a fully charged state (the pre-set charge termination set point). The length of the charging current pulse gradually decreases as battery voltage rises, reducing the average current into the battery. Most PWM controllers include a built-in method to eliminate night time back-feed losses, so blocking diodes should not be needed with these controllers.

# 7.2 Controller Features

In addition to preventing overcharging, controllers can have many other features that protect the batteries, provide a better user interface, and increase the flexibility of a PV system.

Some controllers provide overdischarge protection to prevent batteries from being overly discharged and damaged. Like a parked car with its lights left on, photovoltaic system loads can easily over-drain batteries, dramatically shortening the life of the battery. Most photovoltaic systems provide protection for the battery against unmanaged discharge.

Controllers prevent over-discharging by:
- Activating lights or buzzers to indicate low battery voltage.
- Temporarily turning off loads at a preset state-of-charge level.
- Turning on a back-up generator.

Turning off loads to prevent further battery discharge (until the photovoltaic modules or other power source recharge the battery to a minimum level) is called load management or load shedding and is accomplished using a low voltage disconnect (LVD). If a controller performs load management, DC loads will automatically be shut down. Therefore, essential DC loads must be wired directly to the battery to avoid unplanned disconnection. In this case, battery over-discharging can still occur since the controller has been bypassed. It is also important to remember that charge controllers only

control DC loads. The inverter LVD needs to be programmed to disconnect the AC loads.

Lights or buzzers may also be used to indicate low battery voltage and prevent cutting off critical loads. If the system is designed for critical loads, such as a vaccine refrigerator in a rural health clinic, warning lights or buzzers might be essential. However, since loads can keep running after the user is warned, there is always the risk of over-discharging and shortening battery life.

Back-up power sources, such as engine generators, can be used to prevent over-discharging. Some controllers automatically start the back-up power source to recharge the battery bank when the batteries reach a low charge state. When the batteries are fully charged, the controller turns off the auxiliary power sources, and the photovoltaic system resumes its charging operation.

Some PV controller manufacturers specify a generic battery voltage at which the controller will begin charging and when it will stop charging. These set points may be fixed or field-adjustable. The system designer must accurately specify the settings if they are different from the manufacturer's settings.

## Maximum Power Point Tracking Controllers

Many controllers available today for residential sized systems are equipped with a function called Maximum Power Point Tracking (MPPT). As the name implies, this feature allows the controller to track the maximum power point of the array throughout the day in order to deliver the maximum available solar energy to the batteries.

Before MPPT was available as an option in controllers, the array voltage would be pulled down to just slightly above the battery voltage while charging a battery. For example, in a 12V battery charging system, an array's peak power point voltage is around 17-18V. Without MPPT, the array would be forced to operate around the voltage of the battery. This results in a loss of the power coming from the array. See Figure 7-1.

The actual benefits of MPPT depend on the operating temperature of the array and on the battery state of charge. When an array operates under colder conditions, it will produce a higher voltage. If an array has a high voltage, there is a larger difference between the array and the battery voltage and thus more potential power gain from MPPT. The same is true with having an extremely discharged battery where the voltage is low. When experiencing both conditions, an extremely discharged battery and a cold day, the system will have the largest voltage difference and thus a significant potential power gain from the MPPT feature. Depending on location, there may be substantial cold weather seasonal gains from MPPT but perhaps minimal gain in the hot summer. For most of the US, the use of MPPT contributes to an approximated annual average energy gain of 10%. This is beneficial as it can often reduce the size of the array and thus the initial cost.

## Voltage Step Down Controllers

The voltage step down feature available in some models of MPPT controllers is relatively new to the industry and allows a higher voltage array to be connected to a lower voltage battery bank. For example, a nominal 48V array connected to a nominal 12V battery (See Figure 7-2). Before this option was available, a 12V battery would require a 12V array and a 12V standard charge controller.

There can be significant economic benefits to using a controller with step down capability. Having a higher voltage array allows for the use of much smaller diameter wire from the array to the controller, thus saving significant wire costs if traveling a long distance. In addition, often times wiring the array at higher voltages results in having more modules in series and fewer panels in parallel and therefore eliminates some of the series fusing PV output breakers.

Another benefit of voltage step down is that it can allow for array expansion without having to increase the size of the wire and conduit. Before step down controllers existed, when adding more PV to the system, the wire and conduit size would need to increase in order to pass the additional current coming from the new PV. Instead, if we use a step down controller and rewire the array to a higher voltage, the amperage can be the same or lower. It is often possible to use the existing wire.

## Recommended Controller Features

Although there are numerous optional features for controllers, system designers should consider using the following:

**Lights:** Indicator lights can tell users and service people how the system is operating. Lights can indicate when the batteries are fully charged, when the battery voltage is low, or when the LVD has shut off the loads.

**Meters:** Meters are used to monitor system performance. Types of meters include indicator lights (LED), digital displays, remote metering, and data logging ports (communication port). Not only do meters allow users to learn about and maintain their system, but also, when problems arise, users can accurately access the system's status. The data supplied by these meters can include: array volts, battery volts, load amps, array amps, array amp-hours, array watts, cumulative watt-hours production and charging history.

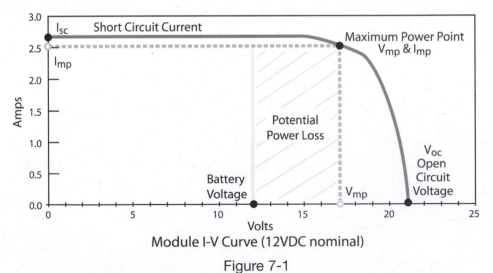

Figure 7-1

**EFFECTS OF A BATTERY ON MODULE POWER PRODUCTION WITHOUT MPPT**

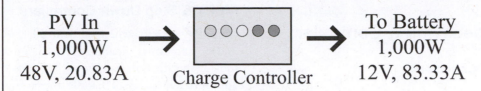

**Note:** This is a simplistic diagram exemplifying how the voltage step down feature works, but does not include efficiency losses. The designer should consult the manufacturer for specific efficiency values.

Figure 7-2

**CONTROLLER STEP DOWN FEATURE**

**Temperature compensation:** When battery temperature is more or less than 25°C, the charging voltage should be adjusted. Some controllers have a temperature compensation sensor to automatically change charging voltage.

Under cold ambient air temperatures, a battery's internal resistance becomes higher. Therefore, charge current causes a greater increase in battery voltage at cold temperatures than at mild temperatures, such as 70°F or 20°C. Under cold conditions, the CTSP will be reached sooner, before the battery has received the amp-hours required to fully charge it.

To resolve this situation, a PV controller with a temperature compensation feature will increase the CTSP approximately ± 5 mV/degrees C per battery cell from 25°C (± 2.77 mV/degrees F from 77°F). Under high ambient temperatures the reverse is true; the PV controller will decrease the CTSP to accommodate for the elevated temperature.

**Power center with disconnect and overcurrent protection:** Many PV controllers come in an integrated package complete with over-current protection, metering, and often an inverter. These integrated units, called **power centers**, are pre-wired and pre-assembled. Most power centers meet *National Electrical Code*® safety standards and use UL approved components. They also may incorporate weatherproof enclosures.

One advantage in using a power center is that it can be installed quickly. If an inverter is included in the package, the user simply connects the batteries, PV array and the AC loads, and the system is operational. Some centers also include input terminals for generator and utility power. Another advantage, besides easy installation, is that power centers containing UL listed components may pass building inspections more easily.

## Other Optional Controller Features

Other optional features that are available for commercially manufactured PV controllers include:

**Low voltage disconnect:** This option automatically cuts off DC loads wired to the control when the battery discharges to a preset level.

**Low voltage warning beeper:** This option sounds an audible alarm when the state-of-charge drops to a preset level.

**Load circuit breaker:** This option can replace a standard load fuse when accessibility to a fuse is difficult or undesirable.

**Generator start control:** This option automatically turns on an auxiliary power source, such as a fuel powered generator.

**Array power diverter:** This option bypasses excess array charging power to non-critical loads.

**Load timers:** This feature consists of mechanical clocks for timing loads requiring pre-set run times, for example security lights.

**Battery charged light:** This option indicates when battery has reached full charge voltage with small LED lights.

**Enclosures:** This option provides weather protection for exterior mounting applications.

**Automatic equalization:** This option equalizes batteries automatically.

# 7.3 Specifying a Controller

## Standard Controllers

In a standard controller, all components of the system have the same voltage. As mentioned earlier, a 12V controller would require the PV array and the battery bank to be wired to 12V. If the 12V controller also controlled the DC loads, it would require the loads to be at 12V. A 24V controller would require all components to be wired to 24 volts. You will recognize this type of controller by the fact that its voltage input and output are the same.

**To specify a standard controller, the following must be considered:**

- *DC System Voltage* – All components must be wired to the same DC voltage.

- *Array Amps (Isc)* – A controller must be able to handle the PV array Isc. Additionally, the designers must multiply the module short circuit current by the number of modules in parallel by a 1.25 safety factor to estimate the minimum charge controller array amps. This is due to the fact that in some circumstances an array may produce more than the STC short circuit current rating.

- *Maximum DC Load Amps* – If the system incorporates DC loads, the controller must be capable of handling the maximum DC load current (amperage) that will pass through the controller.

- *Optional features* – A designer should choose a controller that has all the features to optimize a particular system's performance.

Designers should specify a standard controller using the worksheet in Table 7-1.

## MPPT & Step Down Controllers

In a PV system where a controller with MPPT and/or step down capability is used, a designer must look at specific criteria to verify that the controller is appropriately sized to handle the array. This will ensure the maximum benefit from the MPPT feature. These controllers allow the array to work at the peak power point of the IV curve and convert the excess voltage into amperage. Because of these features, the controller input volts and amps will be different than the output volts and amps.

**To specify a MPPT or step down controller, the following must be considered:**

- *Battery Voltage* – A controller must have an output voltage rating equal to the nominal battery voltage.

- *Maximum Array Voltage* – A controller will have a voltage window stating allowable input voltages from the array. The maximum voltage rating given by the controller must not be exceeded. This value is calculated by multiplying the Voc of the modules by the number of modules in series by a temperature multiplier listed in *NEC® 2005,* Table 690.7. *NEC® 2005,* Article 690.7 states that the lowest expected ambient temperature shall be used to find the temperature multiplier. Remember array voltage will increase in cold temperatures.

- *Maximum Output Amps* – A controller must be capable of handling the maximum output amps going to the battery. To get the maximum benefit from the MPPT feature, the system must be designed so that the input amperage to the controller is lower than the rated output amperage to the battery.

  **Note:** if the designed input amperage to the controller is equal to the controller's rated output amperage, then there is no room to convert the excess voltage into amperage.

- *Maximum Array Watts* – Some manufacturers rate their controllers by specifying the maximum array STC watts allowed at a given nominal battery voltage. For example, 800W at 12 volts.

- *Optional Features* – A designer should choose a controller that has all the features that optimize a particular system's performance.

# 7.4 Controller Sizing Exercise

## Problem:

A client wishes to simultaneously power three 12 volt DC lights (30 watts) and a 12 volt DC television (14 watts) using a 12 volt PV array. Three modules wired in parallel are used in the system. Each module has a peak current of 2.95 amps and a short circuit current of 3.28 amps. Calculate the maximum array output amps used to size a controller.

## Solution:

1. To calculate the maximum short circuit amps, multiply the Isc (3.28 amps) by 3 modules:
   3.28 amps **X** 3 modules = 9.84 amps

2. Next, increase this figure by the safety factor. A controller capable of handling at least 12.3 amps from the PV array must be specified.
   9.84 amps **X** 1.25 safety factor = 12.3 amps

3. To calculate the Maximum DC Load Amps, divide the DC Total Watts (104) by the System Voltage (12 volts).
   [(3 **X** 30 watts) + 14 watts]÷(12 volts) = 8.67 amps

The answers are listed in Table 7-2.

### Table 7-1
### Controller Sizing Worksheet

| Module Short Circuit Current | X | Modules in Parallel | X | 1.25 | = | Array Short Circuit Amps | Controller Array Amps | Listed Desired Features |
|---|---|---|---|---|---|---|---|---|
| | X | | X | 1.25 | = | | | |
| DC Total Connected Watts | ÷ | DC System Voltage | | = | | Maximum DC Load Amps | Controller Load Amps | |
| | ÷ | | | = | | | | |
| Controller Specification | | Make: | | | | Model: | | |

### Table 7-2
### Answers to the Controller Sizing Exercise
A controller capable of handling at least 12.3 amps must be specified.

| Module Short Circuit Current | X | Modules in Parallel | X | 1.25 | = | Array Short Circuit Amps | Controller Array Amps | Listed Desired Features |
|---|---|---|---|---|---|---|---|---|
| 3.28 | X | 3 | X | 1.25 | = | 12.3 | 12.3 | |
| DC Total Connected Watts | ÷ | DC System Voltage | | = | | Maximum DC Load Amps | Controller Load Amps | |
| 104 | ÷ | 12 | | = | | 8.67 | 9 | |
| Controller Specification | | Make: | | | | Model: | | |

## Stand-alone Array & MPPT Controller Sizing Worksheet

**Electric Load Estimation**

1) List the approximate daily average electrical consumption:

AC Average Daily Load: _____ Watt-hrs/day

DC Average Daily Load: _____ Watt-hrs/day

2) Figure out the total daily load (factor in inverter losses – see Notes*):

(_____ AC Avg Daily Load) ÷ (0.9 inverter efficiency – see Notes*) + _____ DC Avg Daily Load

= _____ Total Daily Load (Watt-hrs/day)

**Array Sizing**

3) Figure out the PV array kilowatts needed
(including derate factors for battery losses, temperature losses, and miscellaneous system losses):

Peak Sun Hours per day: _____

(_____ Total Daily Load Wh/day) ÷ (_____ Peak Sun Hours per day) ÷ (*0.80* Battery Efficiency)

÷ (*0.88* PV Temp Losses – see Notes**) ÷ (*0.85* Derate Factor – see Notes***) = _____ PV Array Watts

4) Choose a PV module:

Make: _____ Model: _____

STC watt rating: _____ Voc: _____ Vmp: _____ Isc: _____ Imp: _____

_____ PV Array Watts ÷ _____ STC Watt Rating = _____ # of Modules Needed

5) Factor in Array Nominal Voltage adjust # of Modules Needed if required:

Array Nominal Voltage: _____ PV Module Nominal Voltage: _____

_____ Array Nom. Volts ÷ _____ PV Module Nom. Volts = _____ # of Modules Required in Series

Does the # of Modules Required in Series divide evenly into # of Modules Needed? _____

If YES, then this configuration is compatible with array nominal voltage.

If NO, then the # of Modules Needed (step 7) must be rounded up (or possibly rounded down if backup generator is present) until the # of Modules Required in Series divides evenly into the # of Modules Needed.

Final Number of Modules Needed _____

Manufacturer: _____ Model: _____

## Stand-alone Array & MPPT Controller Sizing Worksheet - Continued

**Charge Controller Sizing**

6) List specific MPPT Charge Controller (CC )to be used:

Manufacturer: _____    Model: _____

Max Array Size Allowed (CC Watt rating) at Nominal Battery Voltage: _____

Max PV Array Open Circuit Voltage Allowed: _____

_____ Final # of Modules Needed **X** _____ STC Watt Rating

= _____ Max Watts Charge Controller(s) Must Pass

7) Calculate how many of these charge controllers the system will require:

_____ Max Watts CC Must Pass ÷ _____ CC Watt rating at Nom Batt Voltage

= _____ # of Charge Controllers Required

**Note:** check to ensure that the maximum array Voc will not exceed the maximum Voc rating (a.k.a "Max PV Array Open Circuit Voltage Allowed") of the charge controller during low temperature conditions at your site. This is especially important if you are utilizing the array voltage step down capability of this charge controller (i.e. wiring array for 60 or 72 Volts nominal). By using the historic lowest temperature for your area, find the correction factor in *NEC® 2005,* Table 690.7 to complete the following equation:

_____ Module Voc **X** _____ # of modules in series **X** _____ correction factor from *NEC®*, Table 690.7

= _____ Maximum PV voltage

The Maximum PV voltage should be less than the Maximum Controller Voltage rating.

## Notes for Stand-alone Array & MPPT Controller Sizing Worksheet

**\*Note:** Instead of using an inverter manufacturer's reported peak efficiency, an inverter efficiency = 0.9 is a more conservative estimate. We choose to use a more conservative value here due to the fact the inverter will be operating under constantly changing AC load requirements.

**\*\*Note:** By using an MPPT controller, the batteries can no longer mask the effects of temperature on the PV array by pulling down PV array voltage. Standard Test Condition ratings where cell temperature = 25° C is not very realistic when solar panels are out in the sun. To account for temperature losses in more realistic situations an average temperature derate value of 0.88 can be used. This assumes an average daytime ambient temperature of 20° C. Also note each module has a slightly different temperature coefficient which is not taken into consideration here.

**\*\*\*Note:** The Derate Factor accounts for other system losses (including module production tolerance, module mismatch, wiring losses, dust/soiling losses, etc.). To account for these losses we use a derate factor = 0.85. Note this value assumes no shading. See textbox below for a summary of how this derate factor is calculated:

### Stand-alone Array & MPPT Controller Sizing

| Derate Values | Range of Acceptable Values | Chosen Value |
|---|---|---|
| PV Module nameplate DC rating | 0.80 - 1.05 | 0.95 |
| Mismatch Modules | 0.97 - 0.995 | 0.98 |
| Diodes and connections | 0.99 - 0.997 | 0.995 |
| DC wiring | 0.97 - 0.99 | 0.98 |
| AC wiring | 0.98 - 0.993 | 0.99 |
| Soiling | 0.30 - 0.995 | 0.95 |
| Shading | 0.00 - 1.00 | 1.00 |
| Age | 0.70 - 1.00 | 1.00 |
| **Total Derate Factor** | | **0.85** |

This value can be adjusted to reflect conditions for a specific site at PV Watts website: ( http://rredc.nrel.gov/solar/codes_algs/PVWATTS/ )

The standard stand-alone PV array sizing (non-MPPT) method automatically accounts for these various efficiency factors (excluding "shading" and "age" factors) via the very conservative assumption that the PV array will operate at the nominal system voltage (rather than actual operating voltage). When using a MPPT charge controller we must incorporate this general Derate Factor.

**Disclaimer:** While this MPPT array sizing method gives a general way to incorporate the potential increase in array power production, it must be noted that this is simply an estimate. The benefits of MPPT are dependent on many factors including battery state of charge and actual array cell temperature which can vary dependent on load usage, actual climate and seasonal temperature variations.

Additionally it must be realized that annual power production of any PV system is largely dependent on how much available sunlight there is and WEATHER PATTERNS VARY YEAR TO YEAR. This means that even though this array sizing method is utilizing long term solar data, each year could have more or less sunshine available which means the actual annual power production of a PV system could exceed or fall short of our expectations. It is not uncommon to experience a variation of (+/-) 10% in annual irradiance from year to year.

# Chapter 8
# Inverters

**Contents:**

8.1  Inverter Operating Principles . . . . . . . . . . . . . . . . . . . . . . . . . . . . .84

8.2  Inverter Types . . . . . . . . . . . . . . . . . . . . . . . . . . . . . . . . . . . . . . . . . .84

8.3  Inverter Features . . . . . . . . . . . . . . . . . . . . . . . . . . . . . . . . . . . . . . .85

8.4  Batteryless Grid-tied Inverters . . . . . . . . . . . . . . . . . . . . . . . . . . . .86

8.5  Grid-tied with Battery Back-up Inverters . . . . . . . . . . . . . . . . . . .87

8.6  Stand-alone Inverters . . . . . . . . . . . . . . . . . . . . . . . . . . . . . . . . . . .88

8.7  AC Coupled Systems . . . . . . . . . . . . . . . . . . . . . . . . . . . . . . . . . . .89

8.8  Stand-alone Inverter Sizing Exercise . . . . . . . . . . . . . . . . . . . . . . .90

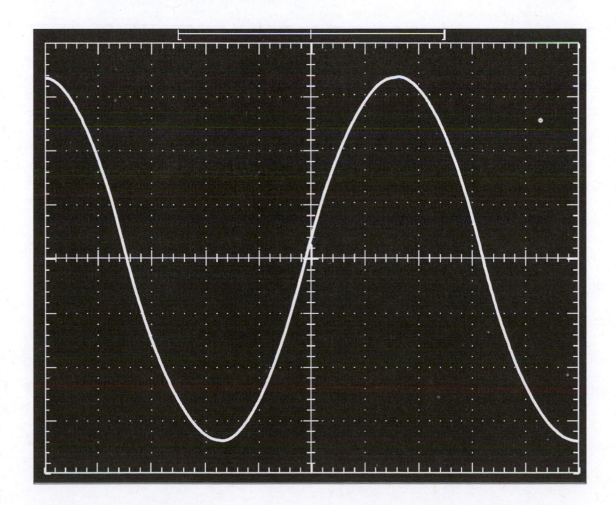

# 8.1 Inverter Operating Principles

Alternating current is easier to transport over a long distance and has become the conventional modern electrical standard. Consequently, most common appliances or loads are designed to operate on alternating current. Photovoltaic modules generate only direct current power. In addition, batteries can only store energy from direct current power sources. Alternating current and direct current are, by nature, fundamentally incompatible. Therefore, a "bridge" — an inverter — is needed between the two.

Historically, inverters have been a weak link in photovoltaic systems. Early inverters were unreliable and inefficient, imposing large penalties on overall system performance. System inefficiencies were compounded by the fact that most alternating current appliances used large amounts of power. Recent improvements in inverters and appliances have reduced this penalty and made inverters a viable "bridge" between direct current power sources and alternating current load requirements.

The fundamental purpose of a photovoltaic system inverter is to change direct current electricity from PV modules (when connected with the utility grid) and batteries (in stand-alone or grid-tied with battery back-up) to alternating current electricity, and finally to power alternating current loads. Inverters can also feed electricity back into the grid. Inverters designed to feed into the utility grid are referred to as grid-tied, line-tied, or utility-connected inverters. These inverters are used in large-scale PV power plants owned by utility companies that generate electricity for the grid, as well as in residential systems that feed electricity to the grid.

## Conversion Methods

Over the years, inverter manufacturers have used different technologies to convert low voltage direct current electricity to higher voltage alternating current. The first inverters used a basic transistor to abruptly switch the polarity of the direct current electricity from positive to negative at close to 60 times per second, creating a square wave form. This relatively crude form of "alternating current" then passes through a **transformer** to increase the voltage.

A transformer increases (or decreases) voltage by passing electricity through a primary transformer coil, inducing flow in the secondary transformer coil. If the number of windings in the secondary coil is greater than the number in the primary coil, then the voltage in the secondary coil will increase directly proportionate to the number of windings in each coil. Stand-alone inverter transformers are designed to increase voltage to 120 or 240 volts alternating current (VAC) depending upon the country in which the inverter will be used.

The advent of sophisticated integrated circuits, field effect transistors, and high-frequency transformers has allowed the creation of lighter, more efficient inverters that produce a waveform closer to a true sine wave. Instead of converting the low voltage DC electricity directly to 120 or 240 VAC, they use a computerized multi-step process with variable timing cycles. For example, 12 volts DC is changed to 160 volts at very high frequency AC (20 kilohertz). Next, high frequency AC is converted to 160 volts DC and finally inverted to the required system voltage and frequency.

# 8.2 Inverter Types

There are three categories of inverters: grid-tied, grid-tied with battery back-up, and stand-alone. The first two inverter types are synchronous or line-tied inverters, which are used with utility-connected photovoltaic systems. The third type is stand-alone or off-grid inverters, which are designed for independent, utility-free power systems and are appropriate for remote photovoltaic installations. Some inverters may be able to function in several of these categories.

Another classification for inverters is the type of waveform they produce. The three most common waveforms include the following:

- square wave
- modified square wave
- sine wave

These three waveforms are illustrated in Figure 8-1.

## Square Wave Inverters

These units switch the direct current input into a step-function or "square" alternating current output. They provide little output voltage control, limited surge capability, and considerable harmonic distortion. Consequently, square wave inverters are only appropriate for small resistive heating loads, some small appliances, and incandescent lights. These inexpensive inverters can actually burn up motors in certain equipment and are not used for residential systems.

## Modified Square Wave Inverters

This type of inverter uses field effect transistors (FET) or silicon-controlled rectifiers (SCR) to switch direct current input to alternating current output. These complex circuits can handle large surges and produce output with much less harmonic distortion. This style of inverter is more appropriate for operating a wide variety of loads, including motors, lights, and standard electronic equipment like televisions and stereos. However, certain electronic devices may pick up inverter noise running on a modified square wave inverter. Also, clocks and microwave ovens that run on digital timekeepers will run either fast or slow on modified square wave inverters. It is also not advised to charge battery packs for cordless tools on modified square wave inverters.

## Sine Wave Inverters

Sine wave inverters are used to operate sensitive electronic hardware that requires a high quality waveform. They are the most common inverters today in residential applications and have many advantages over modified square wave inverters. These inverters are specifically designed to produce output with little harmonic distortion, enabling them to operate even the most sensitive electronic equipment. They have high surge capabilities and can start many types of motors easily. For grid-tied applications you must use a sine wave inverter.

# 8.3 Inverter Features

A system designer should become familiar with the available features of an inverter before choosing one. In later sections we will break out special features for the specific inverter types: Stand Alone, Grid-tied, and Grid-tied with battery back-up.

## Standard Inverter Features

Below are available features that can apply to all inverters:

**High efficiency**: Most inverters today convert 90 percent or more of the direct current input into alternating current output. Many inverter manufacturers claim high efficiency, however, inverters may only be efficient when operated at or near certain outputs. An inverter is often used to power loads at less than its rated capacity. Therefore, it is usually wise to choose a unit rated at a high efficiency over a broad range of loads. Figure 8-2 shows a sample efficiency curve of a 4000-watt inverter.

**Low standby losses**: The inverter should be highly efficient when no loads are operating.

**Frequency regulation**: The inverter should maintain 60 Hz output (in the United States) over a variety of input conditions.

**Harmonic distortion**: The inverter should "smooth out" unwanted output peaks to minimize harmful heating effects on appliances.

**Ease of servicing**: The inverter should contain modular circuitry that is easily replaced in the field.

**Reliability**: The inverter should provide dependable long-term low maintenance.

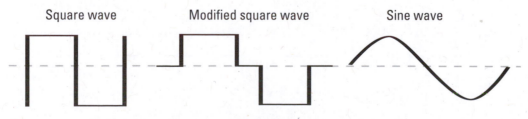

Square wave   Modified square wave   Sine wave

Figure 8-1

**COMMON WAVEFORMS PRODUCED BY INVERTERS**

**Power correction factor:** The inverter should maintain optimum balance between the power source and load requirements.

**Light weight:** The inverter should facilitate convenient installation and service.

### Optional Inverter Features

In addition to the primary functions listed above, the following are desirable features for an inverter:

**Remote control operation and or data-monitoring:** The inverter can be controlled, programmed, and/or monitored from a remote location.

**Load transfer switch:** Manual load switching allows one inverter to meet critical loads in case of failure. This is designed to increase system reliability in systems that have multiple inverters.

**Capability for parallel operation:** In some systems it is advantageous to use multiple inverters. These inverters can be connected in parallel to service more loads at the same time.

**Capability for series operation:** In systems with multiple inverters, this feature enables the inverter to operate higher voltage loads.

# 8.4 Batteryless Grid-tied Inverters

The number of grid-tied PV systems continues to dramatically increase every year. Because of this, the number of available grid-tied inverter options/models continues to increase as well. In order to comply with US standards, all US systems with grid-tied inverters must meet the requirements of IEEE Standard 1547, UL 1741, and FCC part 15.

Most utility-connected inverters don't use a battery bank but instead connect directly to the public utility, using the utility grid as a storage battery. When the sun is shining, electricity comes from the PV array via the inverter. If the PV array is making more power than is being used, the excess is sold to the utility power company through the electric meter. If you use more power than the PV array can supply, the utility makes up the difference. Also, at night and during cloudy weather, all power comes from the grid.

For more information on utility-interactive systems, refer to Chapter 11.

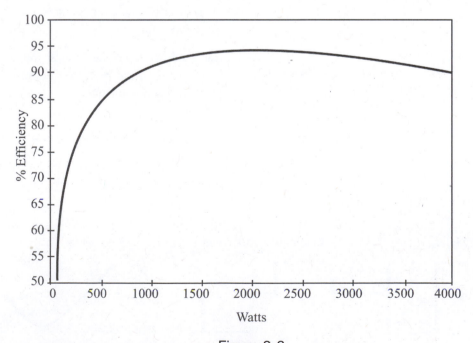

Figure 8-2

**EFFICIENCY OF A 4000 WATT INVERTER**

## Ideal Features for Batteryless Grid-tied Inverters

**Maximum power point tracking (MPPT):** All grid-tied inverters available today will track the peak power point of the array.

**Ground-fault protection (GFP):** Since ground-fault protection is required by *NEC*® for certain systems, most grid-tied inverters today have a GFP built into the inverter. See Chapter 9 for more information on GFP.

**AC/DC disconnects:** Some inverters have built in AC & DC disconnects and/or over-current protection. Additional external disconnects may be needed to safely remove the inverter for repairs.

**Weatherproof enclosure:** Most grid-tied inverters are designed for outdoor installations and have weatherproof enclosures.

### Specifying a Batteryless Grid-tied Inverter

To specify a batteryless grid-tied inverter, the following must be verified:

- *Watts output AC* – When sizing a grid-tied system, the array size will be determined to either match the loads, the client's budget, or space limitations. Once the array size is determined, the inverter will need to pass the total connected watts of the PV array. This value is calculated using the STC rated watts on the DC side of the inverter and multiplying it by the inverter efficiency. This results in the AC watts output from the inverter. Most manufacturers will also give this value as "Watts input DC" to the inverter.

  **Note:** Due to additional derating factors, such as temperature, the connected array watts in hot climates may be designed to exceed the inverters AC output watt rating as per manufacturer's specifications.

- *Input voltage* – Most grid-tied inverters today require a high voltage DC input window of between 75-600V DC. To fall within these higher voltage input windows, the array configuration must utilize multiple modules in series, as needed, to attain these higher voltages. A designer must assure that the voltage of the

array never falls out of or exceeds the inverter's input voltage window throughout the different seasons and throughout the day. The designer will need to be certain that in hot weather when the voltage decreases, the PV array will never drop out of the operating voltage window. If the voltage drops out of the window, the inverter will shut down and there will be no energy production. The designer must also verify that when the weather gets colder and the array's voltage increases, the voltage never exceeds the inverter's window. If the array voltage exceeds the inverter's voltage window, this will void the warranty and potentially damage the unit.

- *Output voltage* – Inverters that are made for residential systems have an output voltage of either 120V or 240V in the US. For grid-tied systems, both of these output voltages are adequate for a home. If the output of an inverter is 120V, the designer needs to verify that there is one spare single pole breaker in the service panel to back-feed. If the output of an inverter is 240V, the designer needs to verify that there is a double pole breaker available for the inverter to back-feed. For a commercial PV system, the designer will need to determine the voltage at which the building is wired. Typical inverter voltages that may apply for commercial systems are 120V, 240V, 208V or 480V. Once the designer verifies the voltage of the building, an inverter needs to be chosen with the correct output voltage.

- *Frequency* – The inverter should maintain 60 Hz output in the United States.

## 8.5 Grid-tied with Battery Back-up Inverters

Battery back-up grid-tied inverters are more complex than batteryless grid-tied inverters because they need to sell power to the grid, supply power to backed-up loads during outages (including surge), and charge batteries from the grid after an outage. These inverters need to have features similar to both a batteryless grid-tied inverter when selling power to the utility, and to a stand-alone inverter when it is feeding the backed-up loads during an outage.

### Ideal Features for Grid-tied with Battery Back-up Inverters

**Battery charge capability:** This type of inverter can act as a battery charger to charge the batteries from an AC source of power after an outage. The battery charging capability of an inverter allows the AC source to charge the batteries through the inverter by converting AC to DC with appropriate voltage. The AC source could be a generator or the utility.

**Automatic warning or shutoff when battery level is low:** The inverter should contain protective circuits that guard the batteries from over-discharging by AC loads. This is often referred to as low-voltage disconnect (LVD).

**High current capability for backed-up loads:** The inverter should be able to provide the high current required to start motors or to run simultaneous loads in the backed-up load panel when the grid is down.

**Generator auto start and stop:** If a generator is used as an AC source during power outages, the inverter can be programmed to automatically start the generator when the batteries get to a low level. The inverter will then shut off the generator once the batteries are sufficiently charged. This is typically done through a built in relay or auxiliary output.

**Power center with disconnects and over-current protection:** Some inverters come with integrated power centers that contain the appropriate disconnects and overcurrent protection devices.

**Sealed or vented:** Inverters can come as sealed or vented units. Sealed units provide protection from harsh climates and conditions, such as dust, bugs, and moisture.

### Specifying Grid-tied with Battery Back-up Inverters

To specify a grid-tied with battery back-up inverter, the following must be verified:

- *Watts output AC* - When sizing a grid-tied system with battery back-up, two wattage ratings must be calculated. The first is similar to how a grid-tied system is sized. The array size will be determined to either match the loads, the client's budget, or space limitations. Once the array size is determined, the inverter will need to pass the total connected watts of the array. This value is calculated using the STC rated watts on the DC side of the inverter and multiplying it by the inverter efficiency. This results in the AC watts output from the inverter. After that, the designer should verify that the inverter will satisfy the backed-up peak load requirements. The inverter must have the capacity to handle all the AC backed-up loads that could be on at one time.

- *DC input voltage from batteries* - An inverter will specify what nominal DC voltage it is rated to accept from the batteries. Typical voltages are 24V and 48V.

- *Output voltage* - An inverter will specify what AC output voltage it will deliver. This is typically 120V in the United States.

- *Frequency* - An inverter should maintain 60 Hz output in the United States.

- *Surge capacity* - Most inverters are able to exceed their rated wattage for limited periods of time. This is necessary to power motors that can draw up to seven times their rated wattage during startup. Consult the manufacturer or measure with an ammeter to determine surge requirements of specific loads in the backed-up load panel.

# 8.6 Stand-alone Inverters

When choosing an inverter for a stand-alone system, you should read and understand the specifications and choose the desired features. A stand-alone inverter needs to be versatile to account for the continuous load rating, generator input, charging batteries, and surging loads.

### Ideal Features for Stand-alone Inverters

**High surge capacity:** The inverter should provide high current required to start motors or run simultaneous loads.

**Automatic warning or shutoff when battery level is low:** The inverter should contain protective circuits that guard the batteries from over-discharging by AC loads, commonly referred to as LVD.

**Sealed or vented**: Inverters can come as sealed or vented units. Sealed units provide protection from harsh climates and conditions, such as dust, bugs, and moisture.

**Battery charge capability**: Many PV systems have a back-up alternating current power source, such as a generator, to charge the batteries, during periods of minimal solar insolation, periods of high energy consumption, and battery equalization. The battery charging capability of an inverter allows the generator to charge the batteries through the inverter, by converting the AC to DC with appropriate voltage, instead of through a separate battery-charging component.

**Generator auto start and stop**: If a generator is used as an AC back-up source of power, the inverter can be programmed to automatically start the generator when the batteries get to a low level. The inverter will then shut off the generator once the batteries are sufficiently charged. This is typically done through a built in relay or auxiliary output.

**Power center with disconnects and overcurrent protection**: Some inverters come with integrated power centers that contain the appropriate disconnects and over-current protection devices.

In stand-alone systems certain loads will cause problems for various inverters. Very small loads may be smaller than the "turn on point" of the inverter. Inverters with a stand-by or sleep mode wait for a certain wattage before they will turn on. Also, certain computers and electronic devices do not present a load until line voltage is available. In other words, the inverter is waiting for a load, and the load is waiting for the inverter.

### Specifying Stand-alone Inverters

**To specify a stand-alone inverter, the following must be verified:**

- *AC output (watts)* - This indicates how many watts of power the inverter can supply during standard operation. It is important to choose an inverter that will satisfy a system's peak load requirements. The inverter must have the capacity to handle all the alternating current loads that could be on at one time. For example, a system user may wish to power a 1000-watt saw and a 500-watt vacuum cleaner at the same time. A minimum of 1500 watts output would be required. System designers should remember that while over-sizing the inverter will allow for system expansion, it could result in reduced system efficiency and increased system cost. See Figure 8-2.

- *DC input voltage from battery* - An inverter will specify what nominal DC voltage it is rated to accept from the batteries. Typical voltages are 12V, 24V and 48V.

- *Output voltage* - An inverter will specify what AC output voltage it will deliver. For a residential stand-alone system this will typically be 120V or 240V. Some inverters only come as 120V units. If a home has 240V appliances, two inverters can be used to attain 240V.

- *Frequency* - An inverter should maintain 60 Hz output in the United States.

- *Surge capacity* - Most inverters are able to exceed their rated wattage for limited periods of time. This is necessary to power motors that can draw up to seven times their rated wattage during startup. Consult the manufacturer or measure with an ammeter to determine surge requirements of specific loads. As a rough "rule of thumb" minimum, surge requirements of a load can be calculated by multiplying the required watts by three.

- *Waveform type* – An inverter should be used with a waveform that can adequately power the required loads.

# 8.7 AC Coupled Systems

Another type of system incorporates inverters that allow renewable energy sources (PV, wind, micro hydro) to combine their outputs on the AC side rather than connecting each individual source to a battery bank. The benefits of this system configuration include the ability to wire the input sources to higher voltages and thus transmit the generated energy over longer distances and on smaller wires to a centralized battery bank. This type of system also opens the door for the potential to create 'mini-grids' in remote locations lacking prior power distribution infrastructure.

| Table 8-1 Inverter Sizing Worksheet | | | |
|---|---|---|---|
| AC Total Connected Watts | DC System Voltage | Estimated Surge Watts | Listed Desired Features |
|  |  |  |  |
| Inverter Specification | Make: | Model: | |

# 8.8 Stand-alone Inverter Sizing Exercise

## Problem:

A client living in the Chuska Mountains near Window Rock, Arizona, wants to power the following 120 volt alternating current loads with his 12-volt direct current photovoltaic system:

- 300-watt blender
- 1000-watt saw
- 640-watt vacuum
- 30-watt VCR

The client also wants to run the saw and the VCR simultaneously. All other loads will be run individually. Use Table 8-1: Inverter Sizing Worksheet to size the inverter.

## Solution:

1. **AC output Watts:** Because the user wants to simultaneously power the saw and VCR, the Alternating Current Total Connected Watts is 1000 watts plus 30 watts or 1030 watts. Thus, an inverter with at least 1030 watts output is required.

2. **DC input voltage from battery:** A 12V battery is a given value in this problem, thus a 12V inverter rated for at least 1030 watts is appropriate, and this size is readily available. Alternating Current Total Connected Watts divided by Direct Current System Voltage equals the Maximum Direct Current Amps Continuous. This value will be used later to size wire.

3. **Output voltage:** Choose an inverter with a 120V output to match the alternating current load voltage.

4. **Frequency:** A unit capable of producing 60 cycles per second alternating current should be specified to match the requirements of the loads.

5. **Surge capacity:** Accounting for load surge requirements, the peak wattage of 1030 is multiplied by 3 to arrive at an Estimated Surge Watts requirement of 3090 watts. Remember that this is only a rule of thumb.

6. **Waveform:** In this case, a modified sine wave inverter will satisfy the requirements of the VCR and large motor loads.

See Table 8-2.

| Table 8-2 Answers to the Inverter Sizing Exercise | | | |
|---|---|---|---|
| AC Total Connected Watts | DC System Voltage | Estimated Surge Watts | Listed Desired Features |
| 1030 | 12 | 3090 | Modified Sine |
| Inverter Specification | Make: | Model: | |

# Chapter 9
# Photovoltaic System Wiring

**Contents:**

9.1  Introduction . . . . . . . . . . . . . . . . . . . . . . . . . . . . . . . . . .92

9.2  Wire Size . . . . . . . . . . . . . . . . . . . . . . . . . . . . . . . . . . . .94

9.3  System Wire Sizing Exercise . . . . . . . . . . . . . . . . . . . . . .100

9.4  Overcurrent Protection . . . . . . . . . . . . . . . . . . . . . . . . .107

9.5  Overcurrent Protection Sizing Exercise . . . . . . . . . . . . . . .107

9.6  Disconnects . . . . . . . . . . . . . . . . . . . . . . . . . . . . . . . . .108

9.7  Grounding . . . . . . . . . . . . . . . . . . . . . . . . . . . . . . . . . .108

9.8  Surge Suppression . . . . . . . . . . . . . . . . . . . . . . . . . . . .113

# 9.1 Introduction

Historically, photovoltaic systems have been installed without referencing and applying the *National Electrical Code®* (*NEC®*). Untrained persons had commonly installed these systems. Today, PV system installations must be in compliance with the *NEC®* to ensure that they will be safe and functional. In this PV textbook the intention is to provide a basic understanding of the *NEC®* requirements and good system design practices for PV systems. However, this textbook does not cover every aspect of electrical wiring, nor does it cover every aspect of *NEC® 2005*, Article 690 (Solar Photovoltaic Systems). This chapter is designed to be used in conjunction with the 2005 edition of the *National Electrical Code®*.

This chapter references the 2005 *National Electrical Code®* book. Reprinted with permission from NFPA 70, *National Electrical Code®*, Copywrite© 2004, National Fire Protection Association, Quincy, MA 02169. This reprinted material is not the complete and official position of the National Fire Protection Association on the referenced subject which is represented only by the standard in its entirety.

Direct current wiring systems are substantially different from conventional household alternating current wiring systems. DC systems generally use lower voltage and the current flows only in one direction. Because DC systems use lower voltage, they often have larger wire sizes compared to AC systems. PV systems often consist of both DC and AC circuits. Alternating current and direct current wiring systems are not compatible and must be separated.

## Wire Types

Wire types differ in conductor material and insulation. The two common conductor materials used in residential and commercial wiring are copper and aluminum. Copper has a greater conductivity than aluminum and therefore can carry more current than aluminum wire of comparable size. Aluminum is less durable than copper in smaller gauges and may break or be weakened during installation; it is not permitted by the *NEC®* for interior home wiring. Aluminum wire is less expensive than copper and is often used in larger gauges for underground or overhead service entrances and for commercial applications.

### Table 9-1
### Wire Types

| Type | Covering Temp | Max. Provisions | Location Covering | Insulation | Outer |
|------|---------------|-----------------|-------------------|------------|-------|
| THHN | Heat Resistant Thermoplastic | 90°C 194°F | Dry or Damp | Flame Retardant & Heat Resistant Thermoplastic | Nylon Jacket |
| THW | Moisture & Heat Resistant Thermoplastic | 75–90° C 167–194°F | Dry or Wet | Flame Retardant & Moisture & Heat Resistant Thermoplastic | None |
| THWN | Moisture & Heat Resistant Thermoplastic | 75°C 167°F | Dry or Wet | Flame Retardant & Moisture & Heat Resistant Thermoplastic | Nylon Jacket |
| TW | Moisture Resistant Thermoplastic | 60°C 140°F | Dry or Wet | Flame Retardant & Moisture Resistant Thermoplastic | None |
| UF | Underground Feeder & Branch Circuit Cable-Single Conductor | 60–75°C 140–167°F | Service Entrance | Moisture and Heat Resistant | Integral with insulation |
| USE | Underground Service Entrance Cable-Single Conductor | 75°C 167°F | Service Entrance | Moisture and Heat Resistant Non-metallic Covering | Moisture Resistant |

A more complete table can be found in the *NEC® 2005*, Table 310.13.

Table 9-2
## Color Coding of Wires

| Alternating Current (AC) Wiring: | | Direct Current (DC) Wiring: | |
|---|---|---|---|
| COLOR | APPLICATION | COLOR | APPLICATION |
| Any color other than green or white/gray | Ungrounded conductor (Hot, L1, L2) | Any color other than green or white/gray allowed (convention is red) | Ungrounded conductor (typically positive) |
| White/gray | Grounded conductor (neutral) | White/gray | Grounded conductor (typically negative) |
| Green or bare | Equipment ground | Green or bare | Equipment ground |
| A more complete table can be found in the *NEC® 2005*, Article 200.6(A) & Article 250.119. | | | |

The conductor itself may be solid or stranded. Stranded conductors consist of many small wires that allow the wire to be highly flexible. This flexibility makes stranded wire the recommended choice when a larger wire size is required.

Insulation covering wire can provide protection from heat, abrasion, moisture, ultraviolet light and/or chemicals. The *NEC®* designates what types of wire may be used for various applications. The following list and Table 9-1 indicate the application of various types of wire:

- THHN is commonly used in dry, indoor locations.

- THW, THWN, and TW can be used indoors or for wet outdoor applications in conduit.

- UF and USE are good for moist or underground applications.

Wires that will be exposed to sunlight must be labeled "Sunlight Resistant."

Electrical wire insulation is color coded to designate its function and use. Technicians should understand the color-coding of conventional electrical wire to ensure safe and efficient installation, troubleshooting, and repair. Disregarding the color coding of wire or using it incorrectly is a safety hazard and violates the *NEC®*. Table 9-2 lists the color codes used in alternating current and direct current systems. When using this table, designers should note whether the wiring is being used for alternating current or direct current. It is also important to know that larger conductors (#4 AWG and larger) usually have only black colored

insulation. Therefore, it is allowable to use colored electrical tape or paint on these larger conductors to color code the wire ends where electrical connections are made.

## Cables and Conduit

Two or more insulated conductors having an overall covering are called a cable (*NEC® 2005*, Article 800.2). As with wire, the protective covering on cables is rated for specific uses, such as resistance to moisture, ultraviolet light, heat, chemicals, or abrasion. The following list and Table 9-3 indicate the application for various types of cable:

- NM is most commonly used for dry, indoor locations (*NEC® 2005*, Article 334.10(A)).

- NMC can be used in dry, moist, or damp locations such as laundry rooms or basements (*NEC® 2005*, Article 334.10(B)).

- UF is permitted for interior wiring in wet, dry, or corrosive locations (*NEC® 2005*, Article 340.10).

- UF and USE cable are permitted for use underground, including direct burial in the earth (*NEC® 2005*, Article 338.2 & Article 340.2).

When exposed to sunlight, Type UF cable identified as Sunlight Resistant must be used (*NEC® 2005*, Article 690.31(B)). For PV module interconnections, you can use types SE, UF, USE, and USE-2 single-conductor cables (*NEC® 2005*, Article 690.31 (B)). Type TC Power and Control Tray Cable are available when flexible two-conductor cable is needed (see *NEC® 2005*, Article 336 for more information). While cable can be used in PV system wiring, installers often choose to use single conductors in conduit instead of cable.

**Conduit** is a metal or plastic pipe that contains wires and offers a protective enclosure for the wires. Conduit is often used in PV systems where wires need to be concealed or protected. Liquidtight, a flexible nonmetallic conduit, is often used to connect module junction boxes. Single conductors, such as THWN-2, are passed through the conduit to make the electrical connections. Another common conduit used in PV systems is polyvinyl chloride (PVC) pipe. This is a rigid and nonmetallic conduit. A typical place to see PVC is between a pole mounted array junction box and a disconnect box in the battery room. PVC can be buried or attached to walls to protect the wires from damage. As with wire, when using PVC, it is important to select a type that is appropriate for the installation.

Tables in the *National Electrical Code®* list the maximum number and size wire which can be run within the given conduit sizes. Using too large a wire or too many wires within conduit will result in overheating and damage to the wire's protective insulation. For *NEC®* references see Table 9-3.

# 9.2 Wire Size

Wire size selection requires you to consider two important criteria:

- Ampacity

- Voltage Drop

### Ampacity

Ampacity refers to the current carrying ability of a wire. The larger a wire is, the greater its capacity to carry current. Using wire with a lower ampacity than wire that carries a larger current will overheat the wire. Overheating means wasted energy and inefficiency, and can result in melted insulation, a short circuit, or fire. The National Electrical Code® has rated various wire sizes and types for the maximum amperage they can safely carry (*NEC®* 2005, Article 310.15). Table 9-4 shows an example of the ampacity of copper wires taken from Table 310.16 and Table 310.17 in the *NEC®* 2005.

Wire size is given in terms of American Wire Gauge (AWG). A larger wire has a greater ampacity and will be designated by a smaller AWG number, up to #1 AWG wire. For example, a #14 AWG wire is smaller than a #10 AWG wire. After #1 AWG wire, wire size increases with higher AWG numbers followed by the /0 symbol (pronounced as "aught"). For example, #2/0 AWG wire is smaller than # 4/0 AWG wire. Beyond #4/0 AWG wire, wire size is measured in kcmils.

|  | Table 9-3 **Cable Types** | |
|---|---|---|
| **Cable Type** | **Name** | ***NEC® 2005*, Article** |
| AC | Armored Cable | 320 |
| MC | Metal-Clad Cable | 330 |
| NM, NM-C | Nonmetallic-Sheathed Cable | 334 |
| UF | Underground Feeder and Branch-Circuit Cable | 340 |
| USE | Underground Service Entrance Cable | 338 |
| TC | Power and Control Tray Cable | 336 |

Section 9.2

| | In Conduit or Cable | | Single Conductors in Free Air | |
| AWG | UF, THW | USE, THWN | UF, THW | USE, THWN |
|---|---|---|---|---|
| 14 | 15 | 15 | 20 | 20 |
| 12 | 20 | 20 | 25 | 25 |
| 10 | 30 | 30 | 40 | 40 |
| 8 | 40 | 50 | 60 | 70 |
| 6 | 55 | 65 | 80 | 95 |
| 4 | 70 | 85 | 105 | 125 |
| 2 | 95 | 115 | 140 | 170 |
| 1/0 | 125 | 150 | 195 | 230 |
| 2/0 | 145 | 175 | 225 | 265 |
| 3/0 | 165 | 200 | 260 | 310 |
| 4/0 | 195 | 230 | 300 | 360 |

**Table 9-4**
**Ampacity of Copper Wire**

For a more complete table see *NEC® 2005*, Table 310.16 and Table 310.17.

**Note:** Table 9-4 is a simplified chart. It does not reflect information for temperature derating. See *NEC® 2005*, Table 310.16 and Table 310.17 for a complete listing of conductor sizes and types, their related ampacity, and temperature deration.

When sizing wire in a PV system, you start by finding out the maximum current load. This is the greatest current going through the circuits of the PV system at one time. For the wire run from the PV panels to the controller or battery, use the module short circuit current multiplied by the number of modules in parallel. For the wire from the battery to the DC service panel, use the total load amps. Once you know the current, multiply it by 125%, so that the conductor never carries more than 80% of its rated capacity. This safety precaution must be done on all wire runs in a PV system. *NEC® 2005*, Article 690.8 requires an additional safety factor to be included in the wire that connects the PV array to the batteries, or the PV array to the inverter in a batteryless system. This is to handle any extra current produced by the panels caused by reflection or exceptionally sunny days. To account for

this additional power, the array short circuit current must be multiplied by an additional 125% to ensure the proper size conductor and to meet code. (Refer to the following for a sample problem.)

## Wire Sizing Exercise

**Problem:** A photovoltaic array has a short circuit current of 40 amps and the system has a DC load of 20 amps. Determine the type and gauge of wire that is required.

**Solution:** Find the total amps for each section of wire.

Wire run from array to battery:
   40 X 1.25 X 1.25 = 62.5.
The wire must have an ampacity that can handle 62.5 amps. Based on Table 9-4, using THWN in conduit, the wire needed is #6 AWG.

Wire run from battery to loads:
   20 X 1.25 = 25.
The wire must have an ampacity that can handle 25 amps. Based on Table 9-4, using THWN in conduit, the wire needed is #10 AWG.

## Voltage Drop

So far we have only discussed wire sizing based on ampacity. The second consideration in choosing the correct wire size is **voltage drop**. Voltage drop is the loss of voltage due to a wire's resistance and length. It is important to consider efficient design practices to minimize energy loss. A PV system's efficiency can be improved when using properly sized wire. This reduces the line loss of the wire.

Voltage drop in wire is a function of the following three parameters:

- Wire gauge
- Length of wire
- Current flow in the wire

The greater a wire's length, the greater resistance to current flow. Excessively long wire runs will result in loss of power to the load and lower system efficiency. It will also reduce the life expectancy of most appliances and equipment. Inductive loads, such as motors, are particularly sensitive to voltage drop. Using a larger wire size, decreasing the current flow, or decreasing the length of wire are all solutions to reduce voltage drop.

As a PV system designer, you will need to choose the appropriate wire size to design efficient systems with low voltage drop in addition to passing *NEC*® safety requirements. A good design practice is to keep the system wiring voltage loss between 2% and 5%. Although 5% loss is acceptable, PV designers often try for 2% losses or less because of the expense of PV panels. Remember that energy lost in wire runs is money lost!

Tables 9-5 through 9-7 list the recommended maximum one-way length for a 2% voltage drop for various wire sizes depending upon the required current and the system voltage. The table values include an allowance for the fact that the wire must travel the distance twice, once to the load and then back.

Note: When using the voltage drop tables to size wire you do not need to include *NEC*® safety factors when calculating the current for each wire run.

For an explanation on using the tables, refer to the Wire Sizing Exercise below. Values given below the stepped line in the lower left corner of these tables must be verified using the *NEC*® *2005,* Table 310.16 and Table 310.17. The conductor may not be large enough to pass *NEC*® requirements. The blank values at the lower left indicate where the wire's ampacity is definitely exceeded.

Note: When designing a system, always consider that the system owner may want to add more loads to the system without running new wire, particularly if the wire is buried or inaccessible.

## Wire Sizing Exercise

**Problem:** A 12-volt battery bank provides power to a 12-volt outdoor security light. The light draws 2 amps and is 40 feet from the direct current load center. What wire size will ensure line losses are not more than 2% voltage drop in the branch circuit?

**Solution:** Table 9-5 is the appropriate table to use for 12 volts at a 2% voltage drop. Find the load current, 2 amps, in the left hand column. Read across to the right until you find the first number that is greater than or equal to the one way distance to the light or 40 feet. This is 48.4 feet. Read straight up to find the minimum wire size. The answer is #10 AWG wire.

## Wire Sizing Charts

**Maximum one-way wire lengths for less than 2% voltage drop:** Tables 9-5, 9-6, and 9-7 indicate the maximum length of wire (in feet) that can be used and have less than 2% voltage drop in one direction. Distances are provided for 12 V, 24 V, and 48 V. These voltage drop tables were created using resistance values at 75°C from *NEC*® *2005,* Chapter 9, Table 8, Conductor Properties.

Table 9-5
## Length (feet) of 12V Copper Wire for 2% Voltage Drop

| Amps | #12 | #10 | #8 | #6 | #4 | #2 | #1/0 | #2/0 | #3/0 | #4/0 |
|---|---|---|---|---|---|---|---|---|---|---|
| 1 | 60.6 | 96.8 | 154.2 | 244.4 | 389.6 | 618.6 | 983.6 | 1241.0 | 1566.6 | 1973.7 |
| 2 | 30.3 | 48.4 | 77.1 | 122.2 | 194.8 | 309.3 | 491.8 | 620.5 | 783.3 | 986.8 |
| 3 | 20.2 | 32.3 | 51.4 | 81.5 | 129.9 | 206.2 | 327.9 | 413.7 | 522.2 | 657.9 |
| 4 | 15.2 | 24.2 | 38.6 | 61.1 | 97.4 | 154.6 | 245.9 | 310.2 | 391.6 | 493.4 |
| 5 | 12.1 | 19.4 | 30.8 | 48.9 | 77.9 | 123.7 | 196.7 | 248.2 | 313.3 | 394.7 |
| 6 | 10.1 | 16.1 | 25.7 | 40.7 | 64.9 | 103.1 | 163.9 | 206.8 | 261.1 | 328.9 |
| 7 | 8.7 | 13.8 | 22.0 | 34.9 | 55.7 | 88.4 | 140.5 | 177.3 | 223.8 | 282.0 |
| 8 | 7.6 | 12.1 | 19.3 | 30.5 | 48.7 | 77.3 | 123.0 | 155.1 | 195.8 | 246.7 |
| 9 | 6.7 | 10.8 | 17.1 | 27.2 | 48.3 | 68.7 | 109.3 | 137.9 | 174.1 | 219.3 |
| 10 | 6.1 | 9.7 | 15.4 | 24.4 | 39.0 | 61.9 | 98.4 | 124.1 | 156.7 | 197.4 |
| 15 | 4.0 | 6.5 | 10.3 | 16.3 | 26.0 | 41.2 | 65.6 | 82.7 | 104.4 | 131.6 |
| 20 | 3.0 | 4.8 | 7.7 | 12.2 | 19.5 | 30.9 | 49.2 | 62.0 | 78.3 | 98.7 |
| 25 | 2.4 | 3.9 | 6.2 | 9.8 | 15.6 | 24.7 | 39.3 | 49.6 | 62.7 | 78.9 |
| 30 | 2.0 | 3.2 | 5.1 | 8.1 | 13.0 | 20.6 | 32.8 | 41.4 | 52.2 | 65.8 |
| 35 | 1.7 | 2.8 | 4.4 | 7.0 | 11.1 | 17.7 | 28.1 | 35.5 | 44.8 | 56.4 |
| 40 | 1.5 | 2.4 | 3.9 | 6.1 | 9.7 | 15.5 | 24.6 | 31.0 | 39.2 | 49.3 |
| 45 | 1.3 | 2.2 | 3.4 | 5.4 | 8.7 | 13.7 | 21.9 | 27.6 | 34.8 | 43.9 |
| 50 | 1.2 | 1.9 | 3.1 | 4.9 | 7.8 | 12.4 | 19.7 | 24.8 | 31.3 | 39.5 |
| 55 | 1.1 | 1.8 | 2.8 | 4.4 | 7.1 | 11.2 | 17.9 | 22.6 | 28.5 | 35.9 |
| 60 | 1.0 | 1.6 | 2.6 | 4.1 | 6.5 | 10.3 | 16.4 | 20.7 | 26.1 | 32.9 |
| 65 | 0.9 | 1.5 | 2.4 | 3.8 | 6.0 | 9.5 | 15.1 | 19.1 | 24.1 | 30.4 |
| 70 | 0.9 | 1.4 | 2.2 | 3.5 | 5.6 | 8.8 | 14.1 | 17.7 | 22.4 | 28.2 |
| 75 | 0.8 | 1.3 | 2.1 | 3.3 | 5.2 | 8.2 | 13.1 | 16.5 | 20.9 | 26.3 |
| 80 | 0.8 | 1.2 | 1.9 | 3.1 | 4.9 | 7.7 | 12.3 | 15.5 | 19.6 | 24.7 |
| 85 | 0.7 | 1.1 | 1.8 | 2.9 | 4.6 | 7.3 | 11.6 | 14.6 | 18.4 | 23.2 |
| 90 | 0.7 | 1.1 | 1.7 | 2.7 | 4.3 | 6.9 | 10.9 | 13.8 | 17.4 | 21.9 |
| 95 | 0.6 | 1.0 | 1.6 | 2.6 | 4.1 | 6.5 | 10.4 | 13.1 | 16.5 | 20.8 |
| 100 | 0.6 | 1.0 | 1.5 | 2.4 | 3.9 | 6.2 | 9.8 | 12.4 | 15.7 | 19.7 |
| 110 | 0.6 | 0.9 | 1.4 | 2.2 | 3.5 | 5.6 | 8.9 | 11.3 | 14.2 | 17.9 |
| 120 | 0.5 | 0.8 | 1.3 | 2.0 | 3.2 | 5.2 | 8.2 | 10.3 | 13.1 | 16.4 |
| 130 | 0.5 | 0.7 | 1.2 | 1.9 | 3.0 | 4.8 | 7.6 | 9.5 | 12.1 | 15.2 |
| 140 | 0.4 | 0.7 | 1.1 | 1.7 | 2.8 | 4.4 | 7.0 | 8.9 | 11.2 | 14.1 |
| 150 | 0.4 | 0.6 | 1.0 | 1.6 | 2.6 | 4.1 | 6.6 | 8.3 | 10.4 | 13.2 |
| 160 | 0.4 | 0.6 | 1.0 | 1.5 | 2.4 | 3.9 | 6.1 | 7.8 | 9.8 | 12.3 |
| 170 | 0.4 | 0.6 | 0.9 | 1.4 | 2.3 | 3.6 | 5.8 | 7.3 | 9.2 | 11.6 |
| 180 | 0.3 | 0.5 | 0.9 | 1.4 | 2.2 | 3.4 | 5.5 | 6.9 | 8.7 | 11.0 |
| 190 | 0.3 | 0.5 | 0.8 | 1.3 | 2.1 | 3.3 | 5.2 | 6.5 | 8.2 | 10.4 |
| 200 | 0.3 | 0.5 | 0.8 | 1.2 | 1.9 | 3.1 | 4.9 | 6.2 | 7.8 | 9.9 |

Values in shaded area may *not* meet *NEC*® requirements.

Table 9-6
## Length (feet) of 24V Copper Wire for 2% Voltage Drop

| Amps | #12 | #10 | #8 | #6 | #4 | #2 | #1/0 | #2/0 | #3/0 | #4/0 |
|------|-----|-----|-----|-----|-----|-----|------|------|------|------|
| 1 | 121.2 | 193.5 | 308.5 | 488.8 | 779.2 | 1237.1 | 1967.2 | 2481.9 | 3133.2 | 3947.4 |
| 2 | 60.6 | 96.8 | 154.2 | 244.4 | 389.6 | 618.6 | 983.6 | 1241.0 | 1566.6 | 1973.7 |
| 3 | 40.4 | 64.5 | 102.8 | 162.9 | 259.7 | 412.4 | 655.7 | 827.3 | 1044.4 | 1315.8 |
| 4 | 30.3 | 48.4 | 77.1 | 122.2 | 194.8 | 309.3 | 491.8 | 620.5 | 783.3 | 986.8 |
| 5 | 24.2 | 38.7 | 61.7 | 97.8 | 155.8 | 247.4 | 393.4 | 496.4 | 626.6 | 789.5 |
| 6 | 20.2 | 32.3 | 51.4 | 81.5 | 129.9 | 206.2 | 327.9 | 413.7 | 522.2 | 657.9 |
| 7 | 17.3 | 27.6 | 44.1 | 69.8 | 111.3 | 176.7 | 281.0 | 354.6 | 447.6 | 563.9 |
| 8 | 15.2 | 24.2 | 38.6 | 61.1 | 97.4 | 154.6 | 245.9 | 310.2 | 391.6 | 493.4 |
| 9 | 13.5 | 21.5 | 34.3 | 54.3 | 86.6 | 137.5 | 218.6 | 275.8 | 348.1 | 438.6 |
| 10 | 12.1 | 19.4 | 30.8 | 48.9 | 77.9 | 123.7 | 196.7 | 248.2 | 313.3 | 394.7 |
| 15 | 8.1 | 12.9 | 20.6 | 32.6 | 51.9 | 82.5 | 131.1 | 165.5 | 208.9 | 263.2 |
| 20 | 6.1 | 9.7 | 15.4 | 24.4 | 39.0 | 61.9 | 98.4 | 124.1 | 156.7 | 197.4 |
| 25 | 4.8 | 7.7 | 12.3 | 19.6 | 31.2 | 49.5 | 78.7 | 99.3 | 125.3 | 157.9 |
| 30 | 4.0 | 6.5 | 10.3 | 16.3 | 26.0 | 41.2 | 65.6 | 82.7 | 104.4 | 131.6 |
| 35 | 3.5 | 5.5 | 8.8 | 14.0 | 22.3 | 35.3 | 56.2 | 70.9 | 89.5 | 112.8 |
| 40 | 3.0 | 4.8 | 7.7 | 12.2 | 19.5 | 30.9 | 49.2 | 62.0 | 78.3 | 98.7 |
| 45 | 2.7 | 4.3 | 6.9 | 10.9 | 17.3 | 27.5 | 43.7 | 55.2 | 69.6 | 87.7 |
| 50 | 2.4 | 3.9 | 6.2 | 9.8 | 15.6 | 24.7 | 39.3 | 49.6 | 62.7 | 78.9 |
| 55 | 2.2 | 3.5 | 5.6 | 8.9 | 14.2 | 22.5 | 35.8 | 45.1 | 57.0 | 71.8 |
| 60 | 2.0 | 3.2 | 5.1 | 8.1 | 13.0 | 20.6 | 32.8 | 41.4 | 52.2 | 65.8 |
| 65 | 1.9 | 3.0 | 4.7 | 7.5 | 12.0 | 19.0 | 30.3 | 38.2 | 48.2 | 60.7 |
| 70 | 1.7 | 2.8 | 4.4 | 7.0 | 11.1 | 17.7 | 28.1 | 35.5 | 44.8 | 56.4 |
| 75 | 1.6 | 2.6 | 4.1 | 6.5 | 10.4 | 16.5 | 26.2 | 33.1 | 41.8 | 52.6 |
| 80 | 1.5 | 2.4 | 3.9 | 6.1 | 9.7 | 15.5 | 24.6 | 31.0 | 39.2 | 49.3 |
| 85 | 1.4 | 2.3 | 3.6 | 5.8 | 9.2 | 14.6 | 23.1 | 29.2 | 36.9 | 46.4 |
| 90 | 1.3 | 2.2 | 3.4 | 5.4 | 8.7 | 13.7 | 21.9 | 27.6 | 34.8 | 43.9 |
| 95 | 1.3 | 2.0 | 3.2 | 5.1 | 8.2 | 13.0 | 20.7 | 26.1 | 33.0 | 41.6 |
| 100 | 1.2 | 1.9 | 3.1 | 4.9 | 7.8 | 12.4 | 19.7 | 24.8 | 31.3 | 39.5 |
| 110 | 1.1 | 1.8 | 2.8 | 4.4 | 7.1 | 11.2 | 17.9 | 22.6 | 28.5 | 35.9 |
| 120 | 1.0 | 1.6 | 2.6 | 4.1 | 6.5 | 10.3 | 16.4 | 20.7 | 26.1 | 32.9 |
| 130 | 0.9 | 1.5 | 2.4 | 3.8 | 6.0 | 9.5 | 15.1 | 19.1 | 24.1 | 30.4 |
| 140 | 0.9 | 1.4 | 2.2 | 3.5 | 5.6 | 8.8 | 14.1 | 17.7 | 22.4 | 28.2 |
| 150 | 0.8 | 1.3 | 2.1 | 3.3 | 5.2 | 8.2 | 13.1 | 16.5 | 20.9 | 26.3 |
| 160 | 0.8 | 1.2 | 1.9 | 3.1 | 4.9 | 7.7 | 12.3 | 15.5 | 19.6 | 24.7 |
| 170 | 0.7 | 1.1 | 1.8 | 2.9 | 4.6 | 7.3 | 11.6 | 14.6 | 18.4 | 23.2 |
| 180 | 0.7 | 1.1 | 1.7 | 2.7 | 4.3 | 6.9 | 10.9 | 13.8 | 17.4 | 21.9 |
| 190 | 0.6 | 1.0 | 1.6 | 2.6 | 4.1 | 6.5 | 10.4 | 13.1 | 16.5 | 20.8 |
| 200 | 0.6 | 1.0 | 1.5 | 2.4 | 3.9 | 6.2 | 9.8 | 12.4 | 15.7 | 19.7 |

Values in shaded area may *not* meet NEC® requirements.

Table 9-7

**Length (feet) of 48V Copper Wire for 2% Voltage Drop**

| Amps | #12 | #10 | #8 | #6 | #4 | #2 | #1/0 | #2/0 | #3/0 | #4/0 |
|---|---|---|---|---|---|---|---|---|---|---|
| 1 | 242.4 | 387.1 | 617.0 | 977.6 | 1558.4 | 2474.2 | 3934.4 | 4963.8 | 6266.3 | 7894.7 |
| 2 | 121.2 | 193.5 | 308.5 | 488.8 | 779.2 | 1237.1 | 1967.2 | 2481.9 | 3133.2 | 3947.4 |
| 3 | 80.8 | 129.0 | 205.7 | 325.9 | 519.5 | 824.7 | 1311.5 | 1654.6 | 2088.8 | 2631.6 |
| 4 | 60.6 | 96.8 | 154.2 | 244.4 | 389.6 | 618.6 | 983.6 | 1241.0 | 1566.6 | 1973.7 |
| 5 | 48.5 | 77.4 | 123.4 | 195.5 | 311.7 | 494.8 | 786.9 | 992.8 | 1253.3 | 1578.9 |
| 6 | 40.4 | 64.5 | 102.8 | 162.9 | 259.7 | 412.4 | 655.7 | 827.3 | 1044.4 | 1315.8 |
| 7 | 34.6 | 55.3 | 88.1 | 139.7 | 222.6 | 353.5 | 562.1 | 709.1 | 895.2 | 1127.8 |
| 8 | 30.3 | 48.4 | 77.1 | 122.2 | 194.8 | 309.3 | 491.8 | 620.5 | 783.3 | 986.8 |
| 9 | 26.9 | 43.0 | 68.6 | 108.6 | 173.2 | 274.9 | 437.2 | 551.5 | 696.3 | 877.2 |
| 10 | 24.2 | 38.7 | 61.7 | 97.8 | 155.8 | 247.4 | 393.4 | 496.4 | 626.6 | 789.5 |
| 15 | 16.2 | 25.8 | 41.1 | 65.2 | 103.9 | 164.9 | 262.3 | 330.9 | 417.8 | 526.3 |
| 20 | 12.1 | 19.4 | 30.8 | 48.9 | 77.9 | 123.7 | 196.7 | 248.2 | 313.3 | 394.7 |
| 25 | 9.7 | 15.5 | 24.7 | 39.1 | 62.3 | 99.0 | 157.4 | 198.6 | 250.7 | 315.8 |
| 30 | 8.1 | 12.9 | 20.6 | 32.6 | 51.9 | 82.5 | 131.1 | 165.5 | 208.9 | 263.2 |
| 35 | 6.9 | 11.1 | 17.6 | 27.9 | 44.5 | 70.7 | 112.4 | 141.8 | 179.0 | 225.6 |
| 40 | 6.1 | 9.7 | 15.4 | 24.4 | 39.0 | 61.9 | 98.4 | 124.1 | 156.7 | 197.4 |
| 45 | 5.4 | 8.6 | 13.7 | 21.7 | 34.6 | 55.0 | 87.4 | 110.3 | 139.3 | 175.4 |
| 50 | 4.8 | 7.7 | 12.3 | 19.6 | 31.2 | 49.5 | 78.7 | 99.3 | 125.3 | 157.9 |
| 55 | 4.4 | 7.0 | 11.2 | 17.8 | 28.3 | 45.0 | 71.5 | 90.3 | 113.9 | 143 |
| 60 | 4.0 | 6.5 | 10.3 | 16.3 | 26.0 | 41.2 | 65.6 | 82.7 | 104.4 | 131.6 |
| 65 | 3.7 | 6.0 | 9.5 | 15.0 | 24.0 | 38.1 | 60.5 | 76.4 | 96.4 | 121.5 |
| 70 | 3.5 | 5.5 | 8.8 | 14.0 | 22.3 | 35.3 | 56.2 | 70.9 | 89.5 | 112.8 |
| 75 | 3.2 | 5.2 | 8.2 | 13.0 | 20.8 | 33.0 | 52.5 | 66.2 | 83.6 | 105.3 |
| 80 | 3.0 | 4.8 | 7.7 | 12.2 | 19.5 | 30.9 | 49.2 | 62.0 | 78.3 | 98.7 |
| 85 | 2.9 | 4.6 | 7.3 | 11.5 | 18.3 | 29.1 | 46.3 | 58.4 | 73.7 | 92.9 |
| 90 | 2.7 | 4.3 | 6.9 | 10.9 | 17.3 | 27.5 | 43.7 | 55.2 | 69.6 | 87.7 |
| 95 | 2.6 | 4.1 | 6.5 | 10.3 | 16.4 | 26.0 | 41.4 | 52.3 | 66.0 | 83.1 |
| 100 | 2.4 | 3.9 | 6.2 | 9.8 | 15.6 | 24.7 | 39.3 | 49.6 | 62.7 | 78.9 |
| 110 | 2.2 | 3.5 | 5.6 | 8.9 | 14.2 | 22.5 | 35.8 | 45.1 | 57.0 | 71.8 |
| 120 | 2.0 | 3.2 | 5.1 | 8.1 | 13.0 | 20.6 | 32.8 | 41.4 | 52.2 | 65.8 |
| 130 | 1.9 | 3.0 | 4.7 | 7.5 | 12.0 | 19.0 | 30.3 | 38.2 | 48.2 | 60.7 |
| 140 | 1.7 | 2.8 | 4.4 | 7.0 | 11.1 | 17.7 | 28.1 | 35.5 | 44.8 | 56.4 |
| 150 | 1.6 | 2.6 | 4.1 | 6.5 | 10.4 | 16.5 | 26.2 | 33.1 | 41.8 | 52.6 |
| 160 | 1.5 | 2.4 | 3.9 | 6.1 | 9.7 | 15.5 | 24.6 | 31.0 | 39.2 | 49.3 |
| 170 | 1.4 | 2.3 | 3.6 | 5.8 | 9.2 | 14.6 | 23.1 | 29.2 | 36.9 | 46.4 |
| 180 | 1.3 | 2.2 | 3.4 | 5.4 | 8.7 | 13.7 | 21.9 | 27.6 | 34.8 | 43.9 |
| 190 | 1.3 | 2.0 | 3.2 | 5.1 | 8.2 | 13.0 | 20.7 | 26.1 | 33.0 | 41.6 |
| 200 | 1.2 | 1.9 | 3.1 | 4.9 | 7.8 | 12.4 | 19.7 | 24.8 | 31.3 | 39.5 |

Values in shaded area may *not* meet *NEC*® requirements.

# 9.3 System Wire Sizing Exercise

**Problem**: Using the sample PV system below, calculate the wire sizes needed for the various portions of the system by answering each of the questions. This system consists of the following specifications and equipment:

- DC system voltage = 24 volts.

- Ten 100-watt modules, each with nominal module voltage of 12 volts. The short circuit current (Isc) of each is 7.2 amps and the maximum power current (Imp) of each is 6.2 amps.

- Eight batteries, each is 6 volts and rated at 350 amp-hours.

- One charge controller that is 24 volts and rated for 60 amps.

- One 2500-watt inverter with an input DC voltage of 24 volts and an output AC voltage of 120 volts.

- Total connected DC load is 500 watts at 24 volts.

- Voltage drop requirement between the PV and battery bank is 2% and the distance is 48 feet.

- Voltage drop requirement between the battery and DC load is 2% and the distance is 12 feet.

- Voltage drop requirement between the battery bank and inverter is 2% and the distance is 7 feet.

Determine the wire size in various circuits on the DC side of a PV system that powers both DC and AC loads.

> **Note**: In this example you will consider the PV to controller and the controller to battery as a single wire run.

## Step 1:

**Question**: How must the 12-volt modules be wired to provide the correct DC system voltage? What is the resulting array nominal voltage and maximum power current?

**Answer**: 2 modules wired in series, then 5 series strings wired in parallel, resulting in a nominal voltage of 24V and an array Imp of 31 amps.

If this is confusing to you, consider breaking the problem into parts. First, you need to determine the voltage. With 12-volt modules, wiring two modules in series will result in one panel with a nominal voltage of 24 volts that delivers 6.2 amps under standard test conditions (STC). Now, if you do this with the remaining eight modules, the result will be 5 sets of 24-volt panels each providing 6.2 amps. To increase the amperage, take each set of 2 modules and wire them in parallel. The final result will be one large array with 24 volts (nominal) that produces 31 amps (STC), wired into a combiner box.

## Step 2:

**Question**: Now, calculate the minimum wire gauge that must be used between the PV array combiner box and battery bank. The wire must be able to safely pass the current provided by the array. Assume you are using type THWN wire in conduit. Disregard voltage drop considerations for the moment. Refer to Table 9-4: Ampacity of Copper Wire.

**Answer**: #6 AWG wire

How did you get this? From Step 1, you simplified the problem to 5 sets of modules in parallel. The short circuit current of the entire array will be 5 X 7.2 amps = 36 amps. Now, using information from Section 9.2: Wire Size, the *NEC*® requires you to multiply the short circuit current by 125% to increase the wire's capacity to handle full current for long durations. Then, you must multiply by another 125% to account for potential excessive current produced by the PV array in conditions of high insolation caused by edge of cloud reflection or snow reflection.

The *NEC*® required ampacity calculation is: 36 amps X 1.25 X 1.25 = 56.25 amps

If you look at Table 9-4, you see that for type THWN wire in conduit, you need to use #6 AWG wire to safely pass up to 65 amps.

## Step 3:

**Question:** Consider voltage drop specifications. If you sized wire solely based on the calculations from Step 2, you may pass the *NEC®* requirement for wire sizing, but you could potentially lose power due to voltage drop. Voltage drop considerations are used to promote efficiency in each circuit. Since PV modules are costly, you want to reduce any power loss due to voltage drop. These calculations must be done in addition to the minimum ampacity (wire gauge) calculations needed above. So, how do you "pass code" and meet the required voltage drop requirements? Using the voltage drop tables (Table 9-5 through Table 9-7), calculate the "normal" operating ampacity of the array. Realistically, you could use the maximum power current (Imp), however using the short circuit current (Isc) is more conservative. So, for this assignment use the Isc of each module in your calculations. Note that you do not need to multiply by 1.56 for voltage drop calculations. Calculate the appropriate array ampacity.

**Answer:** 36 amps

Multiply the short circuit current of each module by the number of sets in parallel.
7.2 amps X 5 (sets) = 36 amps

## Step 4:

**Question:** With the appropriate array ampacity, what size wire do you need between the PV and battery bank wire run for the specified voltage drop?

**Answer:** #1/0 AWG

Using Table 9-6: Length of 24-V wire for 2% voltage drop, for an ampacity of 36 amps, a distance of 48 feet, and a voltage drop of 2%, you need a #1/0 AWG wire.

## Step 5:

**Question:** Given that your answers in Step 2 and Step 4 differ, what wire size will satisfy both the *NEC®* requirements and the 2% voltage drop specification?

**Answer:** #1/0 AWG

#1/0 AWG type THWN in conduit can safely pass up to 150 amps, which satisfies the *NEC®* required ampacity of 56.25 amps. You also know that #1/0 AWG will cause only a 2% voltage drop at the given distance and the "normal" ampacity (36 amps).

## Step 6:

**Question:** Now that you have sized a wire from the PV to battery bank, figure out the wire size for the battery bank to DC load run. Again, you need to first calculate wire based on *NEC®* requirements. What is the minimum ampacity the wire will need to be rated at in order to operate the load safely?

**Answer:** 27 amps

Calculate this number by first figuring out the load ampacity. You know that Watts = Volts X Amps. Use algebra to rearrange this equation to calculate amps.
Watts / Volts = Amps, so:
500 watts / 24 volts = 20.83 amps

Now, multiply this number by 125% to give the wire some excess capacity to handle full current for long durations. This is 20.83 amps X 1.25 = 26.04 amps.

To be safe, always round up. Thus, use 27 amps as your number. You may ask, "Why don't I multiply again by 125% like I did for the PV to battery calculation?" Once the power has reached the battery, excess amperage from the PV modules gets absorbed into the battery bank and is not directly reflected into any other section of the wiring.

## Step 7:

**Question:** What is the minimum wire gauge you must use to satisfy *NEC®* requirements?

**Answer:** #10 AWG

Refer to Table 9-4: Ampacity of Copper Wire. Using THWN in conduit, #10 AWG wire can safely pass up to 30 amps, therefore the 27-amp *NEC®* ampacity requirement is satisfied.

## Step 8:

**Question:** Consider voltage drop specifications. Which wire will satisfy the desired voltage drop? Remember to calculate your "normal" load ampacity.

**Answer:** #8 AWG

Use a "normal" ampacity of 20.83 amps and round up to 21 amps. This calculation is 500 watts/24 volts = 20.83 amps. (You do not need to multiply by 125%.) Table 9-6: Length of 24-V wire for 2% voltage drop indicates that for an ampacity of 21, a distance of 12 feet, and a voltage drop of 2%, you need #8 AWG wire.

## Step 9:

**Question:** Which wire will satisfy both the voltage drop specifications and minimum *NEC®* wire size requirements?

**Answer:** #8 AWG

From Table 9-4: Ampacity of Copper Wire, #8 AWG type THWN wire in conduit can handle 50 amps. The *NEC®* requirement of 27 amps will be more than satisfied. An #8 AWG wire will meet the specified voltage drop of 2% for a distance of 12 feet and the "normal" ampacity of 21 amps.

## Step 10:

**Question:** Specify the wire between the battery bank and the inverter that will satisfy both *NEC®* requirements and voltage drop specifications. Remember, you still are on the DC (24 volt) side of the system.

**Answer:** See inverter manufacturers recommended specification for battery conductor size.

## Voltage Drop Index

Another way to size wires for a PV system uses an equation to calculate the voltage drop index (VDI). With this equation and a VDI chart, you can calculate the wire size for any voltage drop and any nominal system voltage.

$$VDI = \frac{Amps \times feet}{\%Voltage\ drop \times voltage}$$

where:

- amps = maximum number of amps through circuit

- feet = one way wire distance

- % voltage drop = Percentage of voltage drop desired (use 2 for 2%)

- voltage = nominal system voltage

**Calculating Voltage Drop with the VDI Chart:** Using the VDI equation, calculate the voltage drop index for the PV array to the battery for the previous wire sizing exercise, also shown in Figure 9-1. Use a 2% voltage drop and a one way distance of 50 feet. Then use the VDI chart in Table 9-8 to determine the wire size needed. Compare this answer to the answer you got when you used the voltage drop tables in Step 4 of the wire sizing exercise.

### Table 9-8
### Voltage Drop Index Chart

| Wire Size AWG | Copper Wire VDI | Copper Wire Ampacity | Aluminum Wire VDI | Aluminum Wire Ampacity |
|---|---|---|---|---|
| 4/0 | 99 | 230 | 62 | 180 |
| 3/0 | 78 | 200 | 49 | 155 |
| 2/0 | 62 | 175 | 39 | 135 |
| 1/0 | 49 | 150 | 31 | 120 |
| 2 | 31 | 115 | 20 | 94 |
| 4 | 20 | 85 | 12 | 65 |
| 6 | 12 | 65 | | |
| 8 | 8 | 50 | | |
| 10 | 5 | 30 | | |
| 12 | 3 | 20 | | |
| 14 | 2 | 15 | | |

## System Wire Sizing Worksheet

Use the following worksheet to determine system wire sizes.

**PV Combiner Box to Battery**

You can size this section as one wire run from PV to Battery, due to the fact that the controller is basically a pass through device. You can also break this wire run into two sections, PV to Controller and Controller to Battery (see wire sizing worksheet below).

### A. *NEC*® Requirement

$$\frac{\text{Isc of}}{\text{modules}} \; \mathbf{X} \; \frac{\text{\# of modules}}{\text{in parallel}} = \text{Total Amps} \; \mathbf{X} \; 1.25 \; \mathbf{X} \; 1.25 = NEC® \text{ required amps}$$

_____ **X** _____ = _____ **X** 1.25 **X** 1.25 = _____

Amperage satisfying *NEC*® = _____          Wire Size from Table 9-4 = _____

### B. Voltage Drop Requirements

System Voltage: _____          Total Amps: _____

One Way Distance: \_\_\_\_\_          Voltage Drop(%): _____

Wire Size from voltage drop tables (Tables 9-5 through 9-7): \_\_\_\_\_

Is this equal to or greater than the size wire needed for safety? _____

- If yes, this is your answer.          • If no, use the wire size from A.

**PV Combiner Box to Controller**

At times, it can be advantageous to break up the PV to Battery wire run into two separate wire runs, PV to Controller and Controller to Battery. Since the Controller is usually very close to the battery, you can usually size this section with wire smaller than the PV to Controller section as long as it passes the *NEC*® required ampacity from the PV array.

### A. *NEC*® Requirement

$$\frac{\text{Isc of}}{\text{modules}} \; \mathbf{X} \; \frac{\text{\# of modules}}{\text{in parallel}} = \text{Total Amps} \; \mathbf{X} \; 1.25 \; \mathbf{X} \; 1.25 = NEC® \text{ required amps}$$

_____ **X** _____ = _____ **X** 1.25 **X** 1.25 = _____

Amperage satisfying *NEC*® = _____          Wire Size from Table 9-4 = _____

### B. Voltage Drop Requirements

System Voltage: _____          Total Amps: _____

One Way Distance: \_\_\_\_\_          Voltage Drop(%): _____

Wire Size from voltage drop tables (Tables 9-5 through 9-7): \_\_\_\_\_

Is this equal to or greater than the size wire needed for safety? _____

- If yes, this is your answer.          • If no, use the wire size from A.

page 1 of 2

## Controller to Battery

### A. *NEC®* Requirement

$$\frac{\text{Isc of}}{\text{modules}} \ \textbf{X} \ \frac{\text{\# of modules}}{\text{in parallel}} = \text{Total Amps} \ \textbf{X} \ 1.25 \ \textbf{X} \ 1.25 = NEC® \text{ required amps}$$

_____ **X** _____ = _____ **X** 1.25 **X** 1.25 = _____

Amperage satisfying *NEC®* = _____    Wire Size from Table 9-4 = _____

### B. Voltage Drop Requirements

System Voltage: _____        Total Amps: _____
One Way Distance: _____      Voltage Drop(%): _____

Wire Size from voltage drop tables (Tables 9-5 through 9-7): _____
Is this equal to or greater than the size wire needed for safety? _____

- If yes, this is your answer.    • If no, use the wire size from A.

## Battery to DC Load Center

### A. *NEC®* Requirement

$$\text{DC load watts} \div \text{DC voltage} = \text{DC total amps} \ \textbf{X} \ 1.25 = NEC® \text{ required amps}$$

_____ ÷ _____ = _____ **X** 1.25 = _____

Amperage satisfying *NEC®* = _____    Wire Size from Table 9-4 = _____

### B. Voltage Drop Requirements:

System Voltage: _____        Total Amps: _____

One Way Distance: _____      Voltage Drop(%): _____

Wire Size from voltage drop tables (Tables 9-5 through 9-7): _____

Is this equal to or greater than the size wire needed for safety? _____

- If yes, this is your answer.    • If no, use the wire size from A.

## Battery to Inverter

### A. *NEC®* Requirement

$$\frac{\text{Inverter}}{\text{Rated Watts}} \div \frac{\text{Inverter}}{\text{Efficiency}} \div \frac{\text{DC System}}{\substack{\text{(lowest operating)}\\ \text{Voltage}}} = \frac{\text{Inverter}}{\text{Total Amps}} \ \textbf{X} \ 1.25 = \frac{NEC®}{\text{required amps}}$$

_____ ÷ _____ ÷ _____ = _____ **X** 1.25 = _____

Amperage satisfying *NEC®* = _____    Wire Size from Table 9-4 = _____
Verify battery conductor size with inverter manufacturer.

### B. Voltage Drop Requirements

System Voltage: _____        Total Amps: _____

One Way Distance: _____      Voltage Drop(%): _____

Wire Size from voltage drop tables (Tables 9-5 through 9-7): _____

Is this equal to or greater than the size wire needed for safety? _____

- If yes, this is your answer.    • If no, use the wire size from A.

Temperature deration, conduit fill deration, and device terminal compatibility, are not included in the wire sizing worksheets.

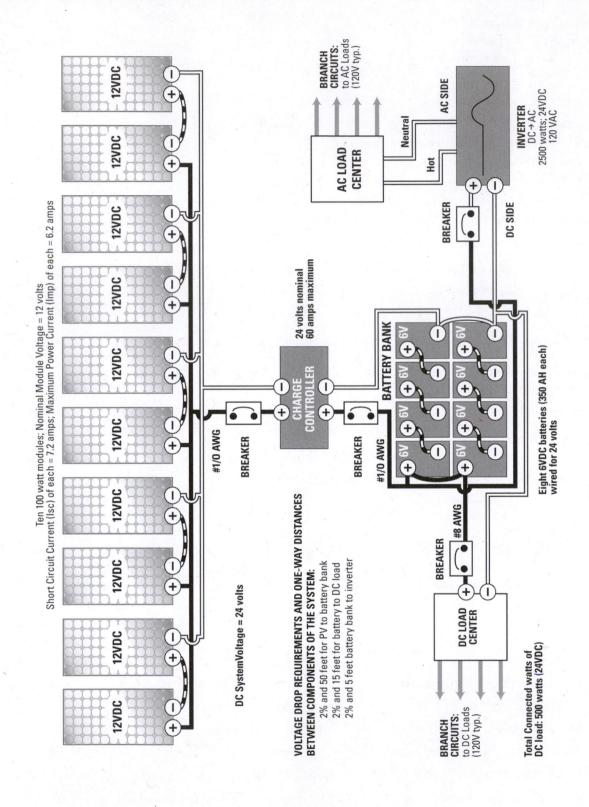

Figure 9-1

**AC AND DC LOAD SCHEMATIC**

# 9.4 Overcurrent Protection

Every circuit must be protected from electrical current that exceeds the wire's ampacity. The *National Electrical Code* specifies the maximum overcurrent protection for each conductor size. Two types of overcurrent protection are:

- Circuit breakers
- Fuses

When the current exceeds a fuse or circuit breaker's rated amperage, the circuit will open and stop all current flow. A fuse that has "blown" must be replaced, while a circuit breaker may simply be reset.

**Circuit breakers:** Circuit breakers must be Underwriters Laboratory (UL) listed and be DC rated if used in direct current circuits. Many circuit breakers commonly used in AC circuits are not suitable for DC systems unless rated specifically for that purpose. Direct current tends to "arc" across the contacts of a breaker as the switch opens the circuit. Consequently, a breaker without an adequate DC rating will soon burn out its contact points. This effect applies to general use switches as well.

**Fuses:** Fuses consist of a wire or metal strip that will burn through when a predetermined maximum current passes through the fuse. This opens the circuit and protects the wire. Fuses, like circuit breakers, must be UL listed and be DC rated if placed in DC circuits.

When a fuse blows or circuit breaker trips, determine the cause before replacing the fuse or resetting the circuit breaker to avoid damaging the PV system wiring or starting a fire. Common causes of fuse failure from excess current are:

- Overload– Operation of too many loads on the same circuit.
- Short circuit or ground fault– Caused by faulty wiring or equipment.

## Overcurrent Protection Placement

The *NEC* requires that every ungrounded conductor (refer to Grounding in this chapter for a definition) be protected by an overcurrent device (*NEC* 2005, Article 240.20). In a DC system, the ungrounded conductor is the positive conductor. In a PV system with multiple sources of power, such as PV panels, batteries, and generators, the overcurrent device must protect the conductor from overcurrent from any power source connected to that conductor. (See *NEC* 2005, Article 690.9(A)) Figure 9-1 shows a PV system with proper overcurrent protection placement.

## Sizing Overcurrent Protection

A common misconception is that breakers and fuses are designed to protect equipment from damage. Remember, their primary task is to protect the wire from overheating and potentially causing a fire. To achieve this, the rating of an overcurrent device must be less than or equal to the ampacity of the wire used, while still passing the full amperage of the power source or power draw, including safety factors.

Overcurrent protection devices (i.e. fuses or circuit breakers) have standard ratings as follows: 1 amp increments from 1-15 amps; 15, 20, 25, 30, 35, 40, 45, 50, 60, 70, 80, 90, 100, 110, 125, 150, 175, 200, 225, 250, 300 amps and higher (*NEC* 2005, Article 240.6(A)). If the rated ampacity of the chosen wire gauge falls between one of the aforementioned standard overcurrent protection values, the next larger overcurrent device shall be used.

# 9.5 Overcurrent Protection Sizing Exercise

This exercise will use the 24V PV system and wire choices made for the Wire Sizing Exercise in Section 9.3. Figure 9-1 shows the placement of the overcurrent protection devices that are sized in this exercise.

**Note:** not all circuit breakers in Figure 9-1 are sized in this example.

**PV Array combiner box to the battery bank:**

**Question:** What is the minimum size breaker or fuse we could use on the 1/0 wire running from the PV panels to the batteries?

**Answer:** 60 Amps

**Explanation:** The minimum overcurrent device is sized to the amperage of the power source or power draw, including safety factors. In the Wire Sizing Example we found that the *NEC* required ampacity for this run is 36 amps x 1.25 x 1.25 = 56.25 amps. (Refer back to the Wire Sizing Exercise, Step 2 for details.)

Viewing the list of standard sizes listed above, we choose the next larger size breaker, rated at 60 amps. If you put a 50-amp breaker on this wire, you may encounter nuisance tripping when you actually see 56.25 amps coming from the array on a clear winter day. Therefore, you need to use a 60-amp breaker to avoid this issue.

In this example, we have already increased the size of our current carrying conductor from #6 AWG wire to #1/0 AWG wire (due to voltage drop considerations). A 60 amp fuse or breaker will easily protect the 1/0 AWG wire that is rated at 150 amps.

**Batteries to the DC loads:**

**Question:** What is the minimum size breaker or fuse we could use on the #8 conductor from the batteries to the DC loads?

**Answer:** 30 amps

**Explanation:** The *NEC* required ampacity for this wire run in 20.83 amps x 1.25 = 26.04 amps. (Refer to Wire Sizing Exercise, Step 6, for details.) Since breakers and fuses come in standard sizes, the next size breaker that would not cause nuisance tripping is 30 amps. In this example, we have already increased the size of our current carrying conductor from #10 AWG wire to #8 AWG wire (due to voltage drop considerations). A 30 amp fuse or breaker will easily protect the #8 AWG wire that is rated at 50 amps.

> **Caution:** Ultimately, a designer must verify that the overcurrent device will protect the conductor under conditions of use. You must also perform an "overcurrent protection check" that takes into account deration factors for the operating temperature of the wire and the number of wires in a single conduit.

**Note:** Also, because you often choose a larger gauge wire (due to voltage drop considerations) you may be able to choose a larger sized fuse or breaker. However, you will still need to perform an overcurrent protection check.

# 9.6 Disconnects

Each piece of equipment in a PV system, such as inverters, batteries, and charge controllers, must be able to be disconnected from all sources of power (*NEC* 2005, Article 690.15). To comply with *NEC* code, disconnects must satisfy the following items:

- They can be switches or circuit breakers.
- They need to be accessible.
- They must not have any exposed live parts.
- They must plainly indicate whether they are in the opened or closed position.
- They must be rated for the nominal system voltage and available current (*NEC* 2005, Article 690.17).

Circuit breakers designed in the system for overcurrent protection can be used as disconnects. Fuses are not considered disconnects unless they are switched fuses.

The total number of disconnecting devices a PV system can have must be six or fewer switches or circuit breakers to shut off all sources of power (*NEC* 2005, Article 690.14). These six disconnects must be grouped together and grouped with other disconnecting means for the system (*NEC* 2005, Article 690.14(C)(5)). Refer to *NEC* for proper labeling.

# 9.7 Grounding

The following list contains the *NEC* definitions (*NEC* 2005, Article 100) for the grounding terms you should be familiar with.

**Grounded:** Connected to the earth or to some conducting body that serves as earth.

**Grounded conductor:** Current carrying conductor that is grounded at one point. Conventionally the white wire.

**Grounding conductor:** A conductor not normally carrying current used to connect the exposed metal portions of equipment or the grounded circuit to the grounding electrode system. Normally bare copper or green wire.

**Grounding electrode conductor:** Bare copper wire connecting grounded conductor and/or equipment grounding conductor to the grounding electrode.

**Grounding electrode:** Usually a ground rod or bare metal well casing.

**Ungrounded conductor:** Current carrying conductor not bonded with ground. Conventionally the red, positive wire on DC; conventionally black, any color besides white, gray, green, or bare copper on the AC side. Refer to Table 9-2 for color-coding.

## Why Ground?

The following is a list of the reasons to ground:

- To limit voltages due to lightning, line surges or unintentional contact with higher voltage lines.

- To stabilize voltages and provide a common reference point being the earth.

- To provide a path in order to facilitate the operation of overcurrent devices.

There are two specific ways to ground a system: equipment grounding and system grounding. It is important to know the difference between the two.

## Equipment Grounding

Equipment grounding provides protection from shock caused by a ground fault and is required in all PV systems by the *NEC*. A ground fault occurs when a current-carrying conductor comes into contact with the frame or chassis of an appliance or an electrical box. A person who touches the frame or chassis of the faulty appliance will complete the circuit and receive a shock. See Figure 9-2. The frame or chassis of an appliance is deliberately wired to a grounding electrode by an equipment grounding wire through the grounding electrode conductor. The wire does not normally carry a current except in the event of a ground fault. The grounding wire must be continuous, connecting every non-current carrying metal part of the installation to ground. It must bond or connect to every metal electrical box,

receptacle, equipment chassis, appliance frame, and photovoltaic panel mounting. The grounding wire is never fused, switched, or interrupted in any way. See Figures 9-3 and 9-4.

When metal conduit or armored cable is used, a separate equipment ground is not usually necessary since the conduit itself acts as the continuous conductor in lieu of the grounding wire. Grounding wires are still needed to connect appliance frames to the conduit.

## System Grounding

System grounding is taking one conductor from a two-wire system and connecting it to ground. The *NEC* requires this for all systems over 50 volts (*NEC 2005*, Article 690.41). In a DC system, this means bonding the negative conductor to ground at one single point in the system (*NEC 2005*, Article 690.42). Locating this grounding connection point as close as practicable to the photovoltaic source better protects the system from voltage surges due to lightning (*NEC 2005*, Article 690.42, FPN). See Figure 9-5 for a schematic with equipment and system grounding.

In grounded systems, the negative becomes our grounded conductor and our positive becomes the ungrounded conductor. If you choose not to system-ground a PV system under 50 volts, both conductors need to have overcurrent protection (*NEC 2005*, Article 240.21), which is often more cumbersome and costly. Most PV installers simply choose to system-ground even if the system operates under 50 volts.

## Ground-fault Protection

Roof-mounted, DC PV arrays located on dwellings must be provided with DC ground-fault protection (*NEC 2005*, Article 690.5). Many grid-tied inverters offer built-in ground fault protection. If a system is to be roof-mounted on a dwelling and the system is not using an inverter package with built-in ground-fault protection, ground fault protection must be wired in separately. Ground-fault protection isolates the grounded conductor (in DC, this is the negative wire) from ground under ground-fault conditions, as well as disconnecting the ungrounded conductor (the positive wire).

## Size of Equipment Grounding Conductor

The size of the equipment grounding wire for the PV source circuits, such as the PV to battery wire run; or for grid-tied systems with no battery back up, the PV to inverter wire run, depends on whether or not the system has ground-fault protection.

If the system has ground-fault protection, the equipment grounding conductors can be as large as the current carrying conductors, the positive and negative wires, but not smaller than specified in *NEC® 2005,* Table 250.122. This table is based on the amperage rating of the overcurrent device protecting that circuit. For example, if the circuit breaker protecting the circuit is rated at or between 30 amps and 60 amps, you can use a #10 AWG copper equipment grounding wire. If the positive and negative conductors have been oversized for voltage drop, the equipment grounding wire also must be oversized proportionally (*NEC® 2005,* Article

250.122(b)). From the example in the Wire Sizing Exercise, you increase the necessary wire size from #6 AWG to #1/0 AWG to satisfy a 2% voltage drop requirement. Here you would have to increase your equipment grounding wire from #10 AWG to #4 AWG.

If the system does not have ground-fault protection, the equipment grounding wire must be sized to carry no less than125% of the PV array short circuit current. For example, if your PV array has a short circuit current of 30 amps, the equipment grounding wire would have to be sized to handle at least 37.5 amps (30 amps X 1.25). Similar to the PV systems with ground-fault protection, if the positive and negative conductors have been oversized for voltage drop, the equipment grounding wire also

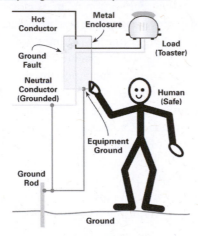

**Proper ground-fault protection**

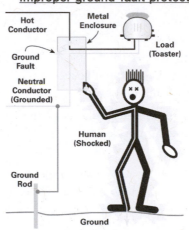

**Improper ground-fault protection**

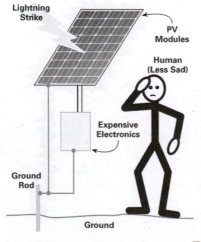

**Proper lightning protection**

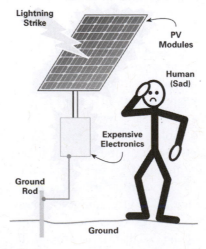

**Improper lightning protection**

Figure 9-2

**GROUNDING** (Courtesy of *Home Power* Magazine, www.homepower.com)

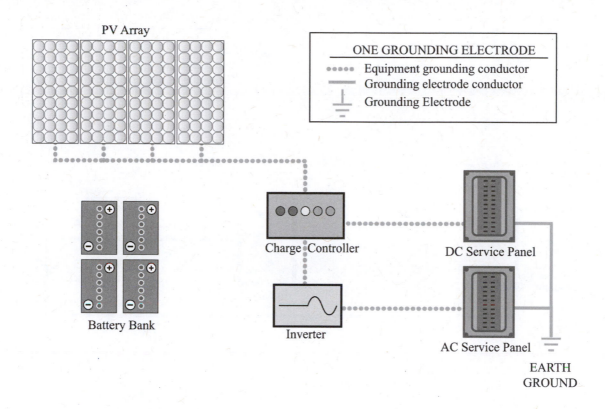

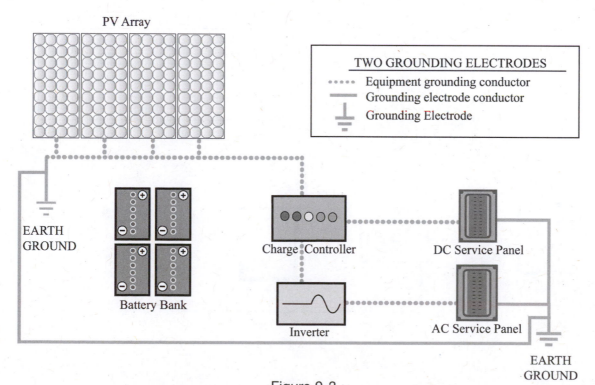

Figure 9-3

EQUIPMENT GROUNDING SCHEMATIC FOR A STAND-ALONE SYSTEM

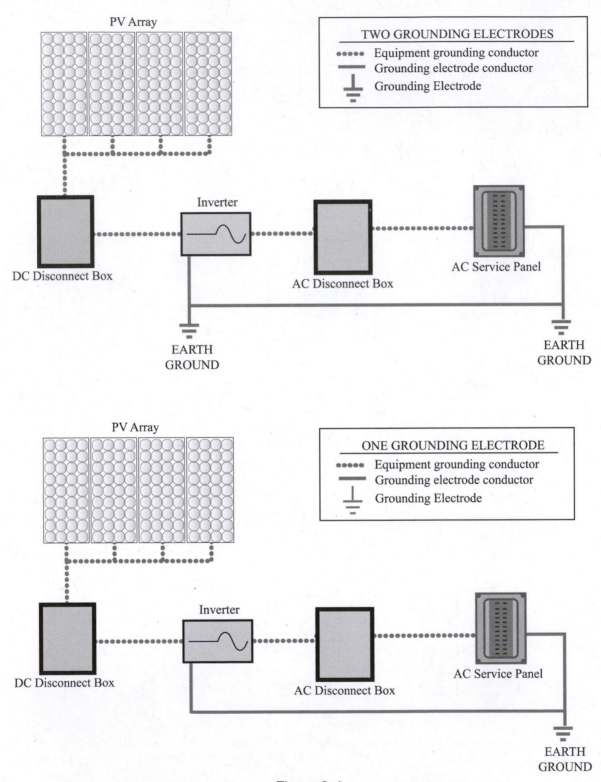

Figure 9-4

**EQUIPMENT GROUNDING SCHEMATIC FOR A GRID-TIED SYSTEM**

must be oversized proportionally (*NEC® 2005,* Article 250.122(b)). From example in the Wire Sizing Exercise, you increase the necessary wire size from #6 AWG to #1/0 AWG to satisfy a 2% voltage drop requirement. Here you would have to also increase the equipment grounding wire from #10 AWG to #4 AWG.

### Size of Grounding Electrode Conductor

The DC system grounding electrode conductor, which is the bare copper wire connecting grounded conductor (the negative wire) and/or equipment grounding conductor to the grounding electrode (the ground rod), cannot be smaller than #6 AWG aluminum or #8 AWG copper or the largest conductor supplied by the system (*NEC® 2005,* Article 250.166). Even though many PV systems have larger conductors in the system (for example, #4/0 inverter cables), they can use #6 AWG copper wire for the grounding electrode conductor if that is the only connection to the grounding electrode (*NEC® 2005,* Article 250.166(C)).

### Grounding Electrodes

Because all PV systems must have equipment grounding, regardless of operating voltage, PV systems must be connected to a grounding electrode. This is usually done by attaching the equipment grounding wire to a ground rod, via a grounding electrode conductor. PV systems often have AC and DC circuits where both sides of the system can use the same grounding electrode. Some PV systems may have two grounding electrodes, which is often the case for pole mounted PV arrays. One electrode for the AC system and one electrode for the DC system at the array. If this is the case, these two grounding electrodes must be bonded together (*NEC® 2005,* Article 690.47).

### Miscellaneous Code Issues

Stand-alone systems must have a plaque or directory permanently installed in a visible area on the exterior of the building or structure used. This sign must indicate that the structure contains a stand-alone electrical power system, and the location of the system's means of disconnection (*NEC® 2005,* Article 690.56).

Alternating current and direct current wiring may be used within the same system, although they may never be installed within the same conduit, or electrical enclosures without some type of physical barrier separating the AC conductors from the DC conductors.

# 9.8 Surge Suppression

Surge suppression is covered only lightly in the *NEC®* because it affects performance more than safety, and it is mainly a utility problem at the transmission line level in AC systems. PV arrays mounted in the open, on the tops of buildings, can act like lightning rods. The PV designer and installer should provide appropriate means to deal with lightning-induced surges coming into the system.

Array frame grounding conductors should be routed directly to supplemental ground rods located as near as possible to the arrays. Metal conduit will add inductance to the array-to-building conductors and slow down any induced surges as well as provide some electromagnetic shielding.

Metal oxide varistors (MOVs) commonly used as surge suppression devices on electronic equipment have several deficiencies. They draw a small amount of current continually. The clamping voltage lowers as they age and may reach the open-circuit voltage of the system. When they fail, they fail in the shorted mode, heat up, and frequently catch fire. In many installations, the MOVs are protected with fast acting fuses to prevent further damage when they fail, but this may limit their effectiveness as surge suppression devices. Other electronic devices are available that do not have these problems.

Silicon oxide surge arrestors do not draw current when they are off. They fail open circuited when overloaded. And while they may split open on overloads; they rarely catch fire. They are not normally protected by fuses and are rated for surge currents up to 100,000 amps. They are rated at voltages of 100 volts and higher and are available from electrical supply houses or Delta Lightning Arrestors, Inc.

Several companies specialize in lightning protection equipment, but much of it is for AC systems. Electronic product directories, such as the Electronic Engineers Master Catalog should be consulted.

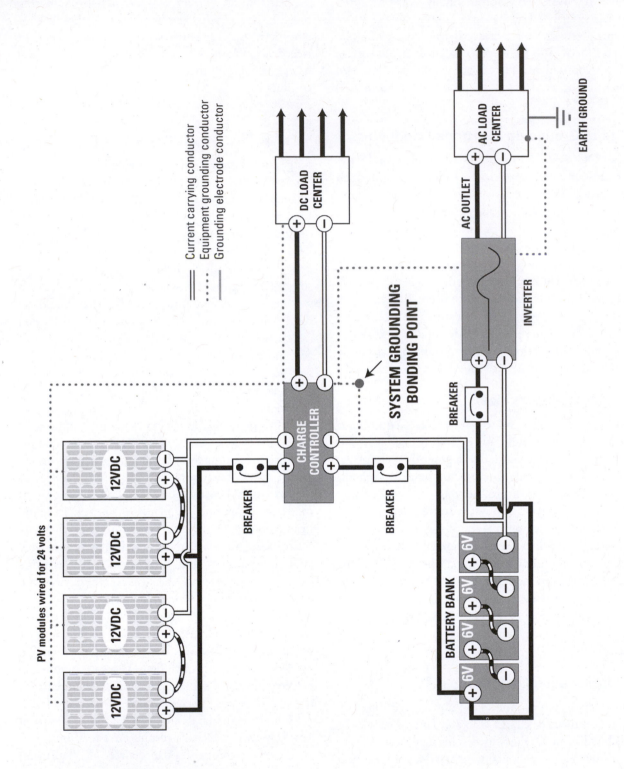

Figure 9-5

SYSTEM AND EQUIPMENT GROUNDING SCHEMATIC

# Chapter 10
# Sizing Stand-alone PV Systems

**Contents:**

10.1 Introduction to Sizing PV Systems . . . . . . . . . . . . . . . . . . . . . . .116

10.2 Design Penalties . . . . . . . . . . . . . . . . . . . . . . . . . . . . . .116

10.3 Sizing Worksheet Explanation . . . . . . . . . . . . . . . . . . . . . .117

10.4 Sample System Exercise . . . . . . . . . . . . . . . . . . . . . . . . . .119

10.5 Hybrid Systems with Generators . . . . . . . . . . . . . . . . . . . . .124

# 10.1 Introduction to Sizing PV Systems

Stand-alone photovoltaic power systems are low-maintenance, versatile solutions to the electric power needs of any off-grid application. They provide electric power for telecommunication stations and water pumping systems throughout the world. Twenty first-century comforts and conveniences can now be provided to remote homes and vacation cabins via photovoltaic systems. These self-contained power stations have proven to be a reliable, cost-effective alternative to conventional power, and frequently replace the noisy, unreliable generators that most remote homes currently use.

Sizing a residential photovoltaic power system is not particularly complex. This chapter illustrates a six-step process to accurately size a system based on the user's projected needs, goals, and budget. Sizing a system includes the following steps:

1. Estimating electric loads.
2. Sizing and specifying batteries.
3. Sizing and specifying an array.
4. Specifying a controller.
5. Sizing and specifying an inverter.
6. Sizing system wiring.

This method is not biased toward any product, but rather will result in generic product specifications for the system. The method uses climate data specific to a location and energy data specific to the user's needs.

Each phase may be broken into smaller, simpler steps. A calculator, sharp pencil, and a lot of common sense are all you're going to need to size a system!

# 10.2 Design Penalties

More people would utilize photovoltaic power systems if it were not for the high initial cost. Since the photovoltaic system industry is competitive, system designers must try to minimize the initial system costs by maximizing the system's energy efficiency. Efficient energy use lowers initial system expenses. For example, reducing the electric lighting load by 75 percent, perhaps by shifting from incandescent to fluorescent lights, will reduce the modules and batteries needed for the system. Eliminating module shading by locating the mounting system in an area with a clear solar window doesn't cost any money and can increase the system's efficiency. Inefficiency caused by excessive voltage drop in the system's wiring can be reduced with proper wire sizing. Intelligent advance planning doesn't cost anything and can drastically reduce a system's initial cost.

Some penalties, such as module efficiency, are out of the realm of system designers and should be left to research scientists. Other penalties are the responsibility of the designer to consider — for example, the fact that some modules perform better in certain climates.

In general, designers should consider the following penalty areas when trying to optimize a system:

**Siting:** The site should be clear of shade to increase the system's efficiency.

**Orientation:** The array orientation with respect to true south and proper inclination is critical for maximizing annual photovoltaic output based on local climate conditions.

**Mounting options:** The optimal mounting system can maximize insolation gain.

**Modules:** PV modules should be selected according to the system's parameters.

**Wiring:** System wiring should be designed to minimize voltage drop, meet safety codes, and provide protection from the environment.

**Controllers:** The controller must operate a system efficiently while meeting the needs of the user.

**Battery storage:** The battery bank must be sized to the specific installation.

**Loads:** The system loads determine the size of the system and should be minimized by intelligent planning.

---

Remember the six P's of photovoltaic system design:

Proper Planning Prevents Poor Photovoltaic Performance

---

Advanced system planning gives the designer the opportunity to quantitatively address these potential areas and minimize their cumulative impact. It is certainly more cost-effective to consider these issues up front rather than trying to provide solutions for a poorly planned system that is already installed.

Designs range from simple analysis to full-scale computer simulations. If continuous power is critical, for example with a life support system, an engineering analysis of the proposed photovoltaic system should be performed. If needs are not critical, a more general sizing method may be adequate.

# 10.3 Sizing Worksheet Explanation

The Photovoltaic Stand-alone Sizing Worksheet is divided into six steps that should be completed sequentially. **For copies of the worksheet, refer to Appendix D.** Each of the following steps correspond to a section of the worksheet and contains detailed procedures for completing the calculations. Complete the worksheet by working through the following steps:

## Step 1: Electric Load Estimation

Complete the Load Estimation Worksheet by inputting the watts, volts, amps, and usage information for each of the loads in a system. Upon completion, you will know the Total Connected Watts (alternating current and/or direct current) and the Average Daily Load (alternating current and/or direct current). If the loads vary significantly on a seasonal or monthly basis or are of a critical nature, use the highest values in designing the system. The method of load analysis in this portion of the worksheet is the same method as described in Chapter 4. Now that you have determined the Average Daily Load the next step is designing an adequate battery bank.

## Step 2: Battery Sizing

Begin by establishing the inverter losses by dividing the AC Average Daily Load by the typical Inverter Efficiency. The inverter efficiency varies with use patterns; generally, 0.9 can be used. Add this figure to the DC Average Daily Load and divide the sum by the DC System Voltage to arrive at the Average Amp-hours Per Day.

To factor in autonomy, multiply the Average Amp-hours per Day by the desired Days of Autonomy to determine the required battery capacity. Divide this total by the Discharge Limit, or the battery's maximum depth of discharge, a number less than 1.0, to determine the total required battery bank capacity.

At this point, you must select a particular battery to be used in the system and use the specifications for that battery. Specify the battery Make and Model at the bottom of the Battery Sizing Worksheet. If you have trouble completing this section, refer to Chapter 6 which lists specific battery information.

Divide the total required battery bank capacity by the Battery Amp-Hour Capacity supplied by the manufacturer to determine the number of Batteries in Parallel needed. If the battery bank includes batteries connected in a series configuration, the required number of Batteries in Series is determined by dividing the Direct Current System Voltage by the Battery Voltage of the battery you have chosen. Multiply Batteries in Series by Batteries in Parallel to obtain Total Batteries Required.

## Step 3: Array Sizing

To begin sizing the array, you must modify the average daily load for the inefficiency of the batteries that have been selected. Divide the Average Amp-hours Per Day from Step 2: Battery Sizing, by the estimated Battery Efficiency, commonly 0.8. Then divide this number by the Peak Sun Hours Per Day available. The resulting figure is the Array Peak Amps. At this stage, you should consider the system's mounting scheme.

> **Note:** Monthly peak sun hours for locations around the world are available in Appendix B. This contains seasonal peak sun hours for tilt angles, azimuth, and tracking options. Also included are worldwide maps that show solar insolation for three tilt angles and four seasons. Peak sun hours may be adjusted to account for other variables, such as added reflectance or shading. You may also consult other sources of solar radiation data for your particular location.

At this point, you must select a particular PV module for the system and use the specifications for that module to complete further calculations. Specify the Make and Model of the selected modules at the bottom of the Array Sizing Worksheet. For more information on modules, refer to Chapter 5. From the module manufacturer's specifications, find the Peak Amps Per Module, which is a tested value at STC (Imp).

**Note:** From the manufacturer's specifications, also write down the Module Short Circuit Current, which will be used in Step 4.

Divide Array Peak Amps by Peak Amps Per Module. The resulting number is the required Modules in Parallel.

To determine the required Modules in Series, divide DC System Voltage by the Nominal Module Voltage. Next, multiply Modules in Series by Modules in Parallel to determine the Total Modules required.

## Step 4: Controller Specification

To begin, multiply Module Short Circuit Current by Modules in Parallel from Step 3. Then multiply this by a safety factor of 1.25. The resulting figure is the Array Short Circuit Amps that the controller must handle under a short circuit condition.

At this point, you must select a controller for the system. Using the Array Short Circuit Amps from the worksheet and the manufacturer's specifications for the desired type of controller, find a controller with Controller Array Amps or Charging Current that meets the required Array Short Circuit Amps. Also consider the other controller features. After you have chosen a controller, specify the Make and Model at the bottom of the Controller Sizing Worksheet. If you choose a controller with LVD make sure it is able

to handle the ampacity of the DC loads connected to it. For more information on controllers, refer to Chapter 7.

Divide the DC Total Connected Watts (from Step 1) by the DC system voltage to calculate the Maximum DC Load Amps the controller will be required to handle. Compare this figure to the manufacturer's specifications for load amperage and enter the load amperage in Controller Load Amps.

## Step 5: Inverter Specification

Divide the Total Connected Watts by the DC System Voltage to calculate the Maximum Direct Current Amps Continuous.

Determine the Maximum Surge Watts required. Remember that electric motors can require from three to seven times their rated wattage during startup. Surge requirements for an appliance are available from the motor manufacturer or can be measured with an ammeter.

Using these figures and the manufacturer's specifications for the desired type of inverter, find an inverter that meets the system's wattage specifications, budget, and other requirements, such as a sine-wave inverter for solid-state equipment. Specify the inverter Make and Model at the bottom of the Inverter Sizing Worksheet. For more information on inverters, refer to Chapter 8.

## Step 6: System Wire Sizing

Refer to Chapter 9 for instructions on completing the system wire sizing section.

# 10.4 Sample System Exercise

The following exercise will design a stand-alone system. The system parameters are listed below followed by a Stand-alone Sizing Worksheet. Review the worksheets to calculate how many batteries and PV modules will be needed. Then select a charge controller and inverter. After sizing this system, a full schematic is displayed.

## Stand-alone PV System Parameters

**Size a system for the following home with consistent year-round use:**
Location: San Antonio, TX

**Electric Load Information:**
(All loads are 120Volts AC)

Fifteen compact fluorescent lights
(15 watts each),
each used an average of 4 hrs/day

One Energy Star 18.9 cu.ft. refrigerator
(127 watts) 9hrs/day

One clothes washer (1450 watts)
4 loads of laundry per week (1 load = 0.5 hr)

One gas clothes dryer (300 watts)
4 loads per week (1 load = 1 hr)

One TV (130 watts), 4 hrs/day

One VCR (40 watts) 2 hours, 3 days/week

One laptop computer (40 watts),
8 hours, 5 days per week

One microwave (1400 watts),
5 minutes/day (0.083hrs/day)

**System specifications:**

DC system voltage: 48 Volts

Days of autonomy: 4 days

Battery Depth of Discharge: 50%

Battery choice: Battery XYZ (350 Ah, 6V)

PV Module choice: Module XYZ
Power = 85 Watts
12 Volt nominal
Peak Amps= 5.02 A
Short Circuit Amps = 5.34 A

PV array mounting is unadjustable and will be set at tilt angle = latitude, year round

There is no back up generator

Controller choice: Controller XYZ
48 Volt nominal
Maximum Pass-through Amperage = 40A

Inverter choice: Inverter XYZ
Efficiency = 90%
Continuous Power Output = 4,000 W
48 Volt nominal
Surge Capacity = 95 Amps AC

## System Sizing Calculations

**Step 1.**

How many watt-hrs/day does this example home consume?

To find the average daily load, complete the load estimation worksheet.

From the worksheet it is calculated that our Average Daily Load is 3527 watt-hours.

**Step 2.**

How many batteries will this system need?

To figure out how many batteries are needed, complete the Battery Sizing Worksheet. In this worksheet the Average Daily Load value (calculated above) is 3527 watt-hrs/day. Also use the information given above: Inverter Efficiency of 90%, 4 Days of Autonomy, 50% Discharge Limit, and Battery XYZ (which is a 6 Volt battery and has a C/20 Battery Capacity of 350Ah). See the Battery Sizing Worksheet for complete calculations.

For this system, there are 16 Batteries required.

**Step 3.**

How many PV modules are required?

To figure out how many panels are needed, complete the Array Sizing Worksheet. To use this worksheet you need to know the Average Amp-hours/day. This was already calculated in the Battery Sizing Worksheet to be 81.6 Amp-hours/day. Assume a Battery Efficiency of 80%, and look up the Peak Sun Hours/day for San Antonio TX, in Appendix B. The array will be set at latitude and there will not be any back-up power source. This dictates using the lowest monthly average value, which is 4.1 Peak Sun Hours/day in December. The system specifications also dictate using XYZ modules, (an 85 Watt, 12 volt module). This module has a Peak Amperage of 5.02 amps and a Short Circuit Current of 5.34 amps. See the Array Sizing Worksheet for complete calculations.

Using this information, 20 modules are required.

**Step 4.**

Choose a charge controller for this array:

Use the module information in the Controller Sizing Worksheet to find out the Array Short Circuit Amps to be 33.38 amps.

We choose Controller XYZ that can pass 40 Amps and can be configured for 48V DC.

**Step 5.**

Choose an inverter for this stand-alone PV system:

From the Example Load Estimation Worksheet, the AC Total Connected Watts was calculated to be 3,712 Watts and the Estimated Surge Watts can conservatively be estimated as three times the AC Total Connected Watts or 11,136 Watts (93 Amps AC). See the Inverter Sizing Worksheet.

The XYZ inverter works as it can pass up to 4,000 Watts, it can accommodate our 48V DC system voltage and it also has a surge capacity of 95 Amps AC.

**Step 6.**

In summary, this system is comprised of:

- 20 modules
- 16 batteries
- 1 40 Amp charge controller
- 1 4,000 Watt, 48V inverter

See the completed worksheet and schematic, Figure 10-1, at the end of this chapter.

## Answers to the Stand-alone Sizing Exercise

### Load Estimation Worksheet (abbreviated)

| Individual Loads | Qty X | Volts X | Amps = | Watts | | X Use hrs/day | X Use days/wk | ÷ 7 days | = Watt Hours | |
|---|---|---|---|---|---|---|---|---|---|---|
| | | | | AC | DC | | | | AC | DC |
| compact fluorescent | 15 | 120 | 0.125 | 225 | | 4 | 7 | 7 | 900 | |
| Energy Star fridge | 1 | 120 | 1.06 | 127 | | 9 | 7 | 7 | 1143 | |
| clothes washer | 1 | 120 | 12 | 1450 | | 0.5 | 4 | 7 | 414 | |
| gas dryer | 1 | 120 | 2.5 | 300 | | 1 | 4 | 7 | 171 | |
| TV | 1 | 120 | 1.1 | 130 | | 4 | 7 | 7 | 520 | |
| VCR | 1 | 120 | 0.33 | 40 | | 2 | 3 | 7 | 34 | |
| laptop computer | 1 | 120 | 0.33 | 40 | | 8 | 5 | 7 | 229 | |
| microwave | 1 | 120 | 11.67 | 1400 | | 0.083 | 7 | 7 | 116 | |
| | | | | | | | | 7 | | |
| | | | | | | | | | | |

**AC Total Connected Watts:** 3712          **AC Average Daily Load:** 3527

**DC Total Connected Watts:** 0          **DC Average Daily Load:** 0

### Battery Sizing Worksheet

| AC Average Daily Load (w-hr/day) | ÷ | Inverter Efficiency | + | DC Average Daily Load (w-hr/day) | ÷ | DC System Voltage | = | Average Amp-hours/ Day |
|---|---|---|---|---|---|---|---|---|
| [( 3527 | ÷ | 0.9 | ) + | 0 | ] ÷ | 48 | = | 81.6 |

| Average Amp-hours/day | X | Days of Autonomy | ÷ | Discharge Limit | ÷ | Battery AH Capacity | = | Batteries in Parallel |
|---|---|---|---|---|---|---|---|---|
| 81.6 | X | 4 | ÷ | 0.5 | ÷ | 350 | = | 2 |

| DC System Voltage | ÷ | Battery Voltage | = | Batteries in Series | X | Batteries in Parallel | = | Total Batteries |
|---|---|---|---|---|---|---|---|---|
| 48 | ÷ | 6 | = | 8 | X | 2 | = | 16 |

**Battery Specification**          Make:          Model:

## Answers to the Stand-alone Sizing Exercise (continued)

### Array Sizing Worksheet

| Average Amp-hrs/day | ÷ | Battery Efficiency | ÷ | Peak Sun Hrs/day | = | Array Peak Amps |
|---|---|---|---|---|---|---|
| 81.6 | ÷ | 0.80 | ÷ | 4.1 | = | 24.9 |

| Array Peak Amps | ÷ | Peak Amps/module | = | Modules in Parallel | | Module Short Circuit Current |
|---|---|---|---|---|---|---|
| 24.9 | ÷ | 5.02 | = | 5 | | 5.34 |

| DC System Voltage | ÷ | Nominal Module Voltage | = | Modules in Series | X | Modules in Parallel | = | Total Modules |
|---|---|---|---|---|---|---|---|---|
| 48 | ÷ | 12 | = | 4 | X | 5 | = | 20 |

| Module Specification | Make: ABC | Model: 2A |
|---|---|---|

### Controller Sizing Worksheet

| Module Short Circuit Current | X | Modules in Parallel | X | 1.25 | = | Array Short Circuit Amps | Controller Array Amps | Listed Desired Features |
|---|---|---|---|---|---|---|---|---|
| 5.34 | X | 5 | X | 1.25 | = | 33.38 | 40 A | Temp. compensation, digital volt meter |

| DC Total Connected Watts | ÷ | DC System Voltage | = | Maximum DC Load Amps | Controller Load Amps |
|---|---|---|---|---|---|
| 0 | ÷ | 48 | = | 0 | |

| Controller Specification | Make: | Model: |
|---|---|---|

### Inverter Sizing Worksheet

| AC Total Connected Watts | DC System Voltage | Estimated Surge Watts | Listed Desired Features |
|---|---|---|---|
| 3712 | 48V | 11,136 | metering, battery charging capability |

| Inverter Specification | Make: | Model: |
|---|---|---|

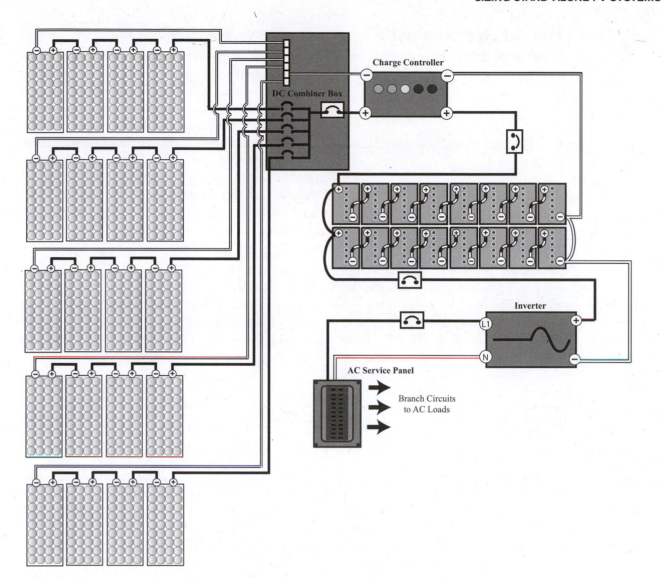

| System Specifications | |
|---|---|
| MODULES | XYZ |
| STC Rating | 85 W |
| Open Circuit Voltage | 21.7 V |
| Max. Power Voltage | 17.4 V |
| Short Circuit Current | 5.34 A |
| Max. Power Current | 5.02 A |
| BATTERIES | XYZ |
| Type | Flooded Lead Acid |
| Nominal Voltage | 6 V |
| Capacity (20 hr) | 350 Ah |
| INVERTER | XYZ |
| Nominal DC Input | 48 V |
| AC Output | 120 VAC |
| Continuous Power | 4000 W |
| Continuous AC Output | 33 A |

Figure 10-1

**SAMPLE SYSTEM EXERCISE SCHEMATIC**

# 10.5 Hybrid Systems with Generators

Most stand-alone PV systems cannot solely satisfy a home's entire electrical needs. A large part of the cost for PV stand-alone systems is due to sizing the array and batteries to support the entire load under worst-case weather conditions. In some instances, this fractional power requirement can be more economically provided by another power source.

A stand-alone PV system with another integrated power source is called a PV hybrid system. The most common auxiliary power source is a gas or diesel-powered engine generator, called a PV-generator system. Although there are many types of hybrid systems, this is the only hybrid system covered in this manual.

The most common configuration for a PV-generator system is one in which both the PV array and the generator charge the same batteries. See Chapter 1, Figure 1-4. The PV array is a slow rate battery charger, and the generator is used primarily as a high-rate battery charger. Generators run more efficiently when operating close to their maximum load, typically at 80 to 90 percent of their rated power. When generators are operating in this range, they can quickly charge batteries to nearly 70 percent state of charge. This allows the generator to operate for a short time at or near its most efficient operating point. As a result, generator maintenance and fuel costs are reduced and its lifetime is prolonged.

The photovoltaic array is designed to complement the generator by supplying the power to the load and completing the battery charging.

## Advantages to a Hybrid System

**Improved economics:** Using the PV array to produce the last 5 percent of system availability is expensive. In regions with a variable climate, where average daily insolation in the winter is two or three times less than in the summer, the use of a hybrid system can be quite economical. For applications with large loads, a generator may also be more economical to provide some power. However, maintenance, logistics, and fuel costs can be quite expensive for generators in remote areas. These factors must be considered in any cost comparisons.

**Lower initial cost:** Meeting the full requirements of the load with photovoltaics may be too expensive for the homeowner. By combining a generator with the PV array, the designer can trade off the high initial cost and low operating cost of PV modules against the generator's low initial cost and high operating cost of a generator.

**Increased reliability:** Because there are two independent charging systems for the battery, there is inherent system redundancy. If the hybrid system is properly maintained and controlled, the overall system reliability is also greater.

**Design flexibility:** The generator backs up the photovoltaic system during periods of cloudy weather and/or heavier than normal electrical use. This is best illustrated in residential systems where the generator is not only used to charge the batteries but also to power large loads, such as washing machines, dryers, and power tools. In a typical home, the owner may use the generator a few hours a week to wash, dry, and iron clothes, vacuum the house, and pump water. While the generator is running, it can also be charging the batteries.

# Chapter 11
# Grid-tied PV Systems

## Contents:

11.1   Introduction . . . . . . . . . . . . . . . . . . . . . . . . . . . . . . . . . .126
11.2   Grid-tied System Types and Advantages . . . . . . . . . . . . . . . . . . .126
11.3   System Sizing and Economics . . . . . . . . . . . . . . . . . . . . . . . . .130
11.4   Obtaining an Interconnection Agreement . . . . . . . . . . . . . . . .131
11.5   Net Metering . . . . . . . . . . . . . . . . . . . . . . . . . . . . . . . . . . .131
11.6   Sample System Exercise . . . . . . . . . . . . . . . . . . . . . . . . . . . .133

# 11.1 Introduction

Grid-tied systems, also called utility-connected or line-tied systems, are solar-based energy systems installed on homes or commercial buildings connected to an electric utility. They are designed to displace all or a portion of the building's total electricity needs. Advances in solar power electronics make it relatively easy to connect a solar electric system to the utility. Energy generated by such a system is first used within the home, and surplus power is "pushed" onto the utility's wires. In most US states, local utilities are required by law to allow "spinning the meter backward" when the electricity being produced by the PV system is greater than what is being used in the home.

The enactment of the Public Utility Regulatory Policies Act of 1978 (PURPA) eliminated the electric utilities' traditional monopoly over electricity generation for the utility grid. Among other things, PURPA required utilities to interconnect non-utility generators to their transmission and distribution networks, which allow small PV systems to be connected to the utility grid.

Most of the non-utility generators developed under PURPA were utility-scale bulk power facilities designed and built to sell power at the wholesale or "avoided cost" price to their utilities who would resell the power to their customers. The most common PURPA facilities were industrial cogeneration facilities sized to produce hundreds of megawatts of power. Most other PURPA facilities, including biomass, geothermal, solar and wind-powered generators, were also megawatt-scale facilities. The interconnection of small-scale facilities sized to serve an individual home, small business, farm, or ranch was relatively unusual.

One of the principal reasons for the scarcity of small-scale generating facilities was the burden of negotiating interconnection requirements with the local utility. Although PURPA established a federal mandate for interconnection of non-utility generation, much of the detailed implementation of PURPA was left to utilities and regulators at the state level. Because utilities historically had exercised primary responsibility for maintaining the safety and integrity of the transmission and distribution network, regulators were inclined to grant the utilities substantial deference and discretion with respect to interconnection requirements. Generally speaking, non-utility generators found utility requirements to be unreasonably and unduly burdensome, and they argued for more streamlined and simplified approaches to resolving interconnection issues.

This chapter discusses the evolution of policymakers' response to interconnection issues, including both technical requirements and non-technical requirements, as well as a system sizing exercise.

# 11.2 Grid-tied System Types and Advantages

There are two types of utility-connected systems, systems without battery back-up and systems with battery back-up. Grid-tied systems without battery back-up consist of just two main components, a PV array and a grid-tied inverter, and have no means of providing power when the utility grid fails. Grid-tied systems with battery back-up also have an array and grid-tied inverter, but include the addition of a battery bank and charge controller. With these components, systems with battery back-up can provide power during utility power outages.

There are many advantages to a utility-connected system, including:

**Improved economics:** It is expensive to get the last 5% of system availability with PV. In regions with variable climate, where average daily insolation in winter is two or three times less than in summer, relying solely on PV requires a large system and can get very expensive. Thus, the use of a utility-connected system may be quite economical. Applications with large loads may also be more economically powered directly by the grid. A batteryless utility-connected system is also more efficient than a battery based system because there are no losses due to the batteries.

**Lower initial cost:** Meeting the full requirements of the load with PV may be too expensive for the homeowner. The start into a utility-connected system with just an inverter and a small PV array is possible for even small budgets. There is no battery, charge controller, control panel, or back-up generator required. More PV modules and/or a battery can be added later on to decrease the grid dependence.

**Increased reliability:** Because there are two independent power systems, there is inherent system redundancy and possibly greater overall reliability.

Adding a battery to the system makes an uninterruptible power system (UPS).

**Design flexibility:** Since the utility provides a permanent power source, the PV system can be designed to the budget and desires of the homeowner.

**Utility:** Interconnected systems have been made possible by advances in inverter technology. These inverters are capable of both converting the DC power from the PV array to standard AC power and synchronizing that power with the utility's electricity. A user-friendly grid-tied inverter includes all components necessary to make a simple, and code compliant, utility-interconnected installation.

## Grid-tied System without Battery Back-up

The advantages and disadvantages of a grid-tied system without battery back-up include the following:

- Cost-effective for net metering.
- Does not provide back-up power in case of grid failure.
- Simple to install.
- No power management opportunities.
- Highest efficiency.

Figure 11-1 displays a grid-tied system without battery back-up.

As long as the PV array produces more power than the house demands, solar power is fed back onto the utility grid. In times where the demands are higher or during the night, the grid helps powering the loads.

## Grid-tied Systems with Battery Back-up

These systems, which are equipped with a battery and charge controller, can provide back-up power during utility power failures. Some battery-based inverters also offer energy management opportunities. They reduce the electrical fees during the time of day when electricity is the most expensive. Peak load shaving is overcoming the utility time of use (TOU) metering by using a battery-based inverter to store energy during the low cost power hours and consume the battery energy during high cost power hours. One disadvantage is that adding batteries can decrease the performance of the system by 10 to 15 percent due to additional efficiency losses in charging the batteries.

The advantages and disadvantages of a battery-based grid-tied system include the following:

- Provides uninterruptible back-up power.
- Batteries are an additional cost.
- Reduces energy costs for utility time of use (TOU) metering.
- Efficiency loss in charging batteries.
- Offers power management opportunities.
- More components to install.

Figure 11-2 displays a grid-tied system with battery back-up.

Under normal operating conditions, the batteries are not cycled and the array maintains a float charge on them. During a power outage, the inverter immediately disconnects from the grid and utilizes battery power to energize the backed-up loads subpanel, which is isolated from the utility.

## Uninterruptible Power Supply

Rolling blackouts have become a regular occurrence for many electric utilities. In many places, the lack of both transmission and generation capacity may plague the residents with blackouts in the near future. An inverter/battery-based uninterruptible power supply (UPS), with or without PV, can provide blackout-proof power to the home. See Figure 11-3. A UPS will keep the electric appliances up and running during utility blackouts.

An inverter-based grid back-up system uses utility power, when it's available, to charge a battery bank. When a blackout occurs, selected loads are automatically backed-up by the batteries. The inverter's main job is to convert a battery's linear DC waveform to a digital representation of an AC sine wave, which the household appliances are designed to run on. The inverter uses the energy stored in the batteries to power household loads.

When the grid comes back online, the loads are automatically transferred back to the grid power. Then, the inverter's battery charger goes to work recharging the batteries to prepare the system for the next blackout. These systems are both modular and expandable; a larger capacity battery bank or renewable energy inputs can easily be added to the system.

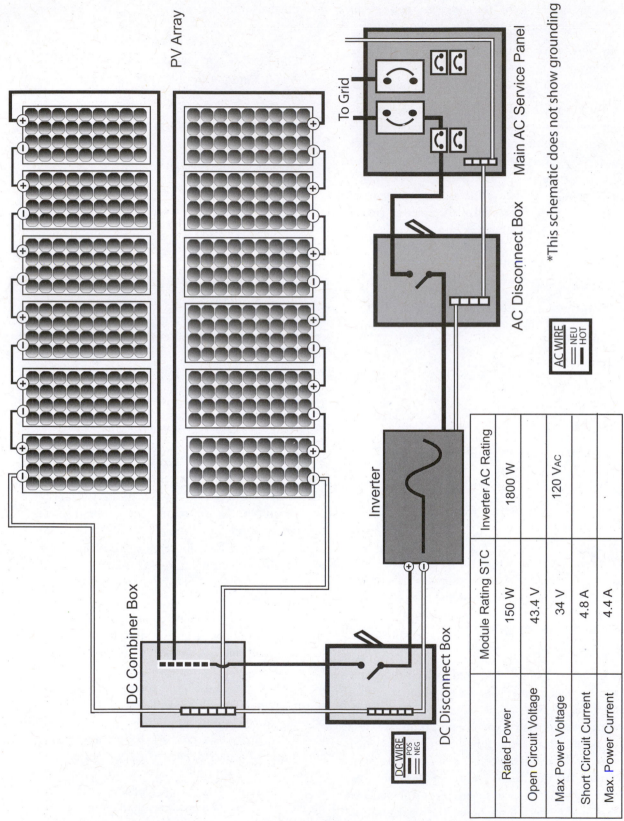

| | Module Rating STC | | Inverter AC Rating |
|---|---|---|---|
| Rated Power | 150 W | | 1800 W |
| Open Circuit Voltage | 43.4 V | | |
| Max Power Voltage | 34 V | | 120 V$_{AC}$ |
| Short Circuit Current | 4.8 A | | |
| Max. Power Current | 4.4 A | | |

Figure 11-1

**GRID-TIED SYSTEM WITHOUT BATTERY BACK-UP**

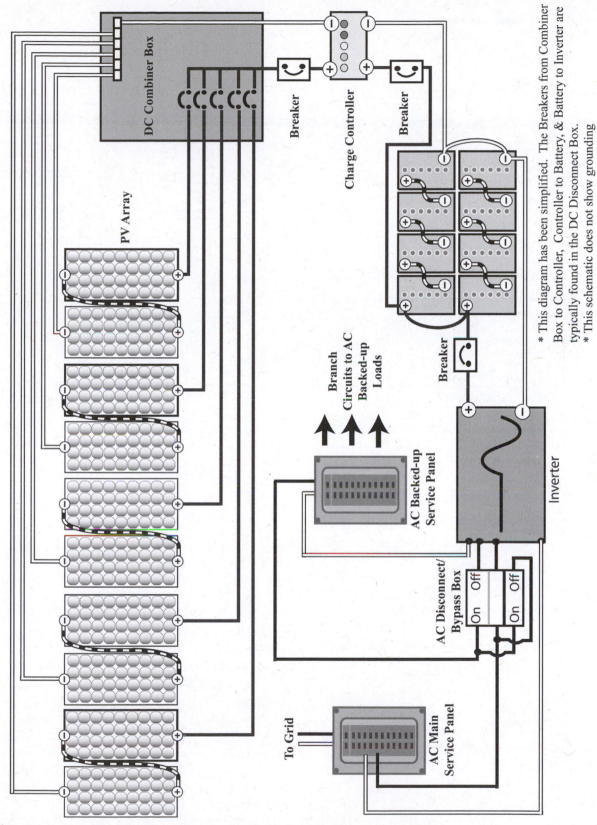

* This diagram has been simplified. The Breakers from Combiner Box to Controller, Controller to Battery, & Battery to Inverter are typically found in the DC Disconnect Box.
* This schematic does not show grounding

Figure 11-2

**GRID-TIED WITH BATTERY BACK-UP SYSTEM**

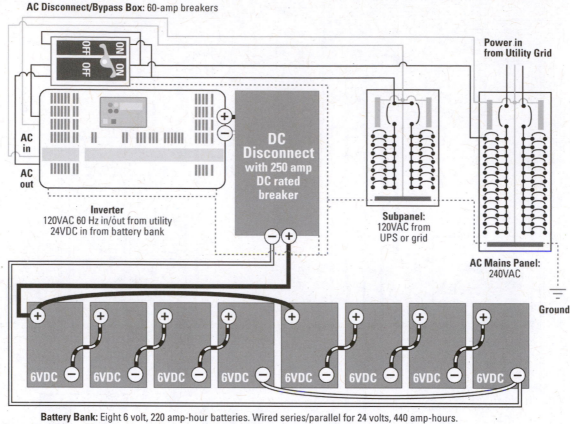

**AC Disconnect/Bypass Box:** 60-amp breakers

**Power in from Utility Grid**

**DC Disconnect** with 250 amp DC rated breaker

AC in

AC out

**Inverter**
120VAC 60 Hz in/out from utility
24VDC in from battery bank

**Subpanel:**
120VAC from
UPS or grid

**AC Mains Panel:**
240VAC

**Ground**

6VDC 6VDC 6VDC 6VDC 6VDC 6VDC 6VDC 6VDC

**Battery Bank:** Eight 6 volt, 220 amp-hour batteries. Wired series/parallel for 24 volts, 440 amp-hours.

Figure 11-3

UNINTERRUPTIBLE POWER SUPPLY SYSTEM

(Courtesy of *Home Power* Magazine)

# 11.3 System Sizing and Economics

Unlike a stand-alone solar electric system, a utility-connected system does not have to provide 100 percent of the daily energy needs. The system can be sized according to the owner's desires and budget. When designing a system, the following considerations are important:

- Budget.
- Available space.
- Percentage of energy to be generated from PV array.
- Availability of tax credits.
- Financing.
- Net metering interconnection rules.
- Utility regulations.

As an example, look at a PV system designed to provide one-third of the electricity needs. If the home uses 8000 kWh per year (obtained from the utility bill), the PV system would have to produce 2667 kWh per year (8000÷3=2667). Assuming the annual delivered energy at that location for a fixed tilt angle is 1400 kWh per 1000 W solar power, a 1900-watt system would be needed (2667÷1400x1000=1900). Using the rate of $10 per watt of installed PV, such a system would cost roughly $19,000.

Depending on the owner's finances, a 1000-watt system might be bought this year and another 900 W could be added in a few years. This is possible because of the scalable nature of PV arrays. However, because inverters come in discrete sizes, a 2000-watt inverter could be bought initially in preparation for future expansion, rather than buying two 1000-watt units separately. The same goes for installation costs; it's cheaper to do it all at once. Initially buying a 2000-watt system will likely cost less time and money than buying in two increments. The larger the system or increment, the less it costs per unit of energy delivered. In addition to the owner's finances, tax credits and utility incentives might also play a role in system sizing.

# 11.4 Obtaining an Interconnection Agreement

Interconnecting a PV system with the utility grid will require entering into an interconnection agreement with the local utility. The interconnection agreement specifies the terms and conditions under which the PV system will be connected to the utility grid. It includes the technical requirements necessary to ensure safety and power quality and other issues, such as the obligation to obtain all necessary permits for the system and having the PV system insured. The key to obtaining an agreement is to involve the utility as early as possible in the installation.

Recently, progress has been made in developing nationally recognized standards for the utility interconnection of PV systems. Although these standards are not necessarily binding on utilities, many utilities are adopting the standards rather than developing their own. The most important standards focus on inverters. Two of these standards are particularly relevant.

> **Note:** The homeowner does not necessarily need to know about these standards, but the PV designer and utility should.

- Institute of Electrical and Electronic Engineers, IEEE Standard 1547: *Recommended Practice for Utility Interface of Photovoltaic Systems*. Institute of Electrical and Electric Engineers, Inc., New York, NY.

- Underwriters Laboratories, UL Subject 1741: *Standard for Static Inverters and Charge Controllers for Use in Photovoltaic Power Systems* (First Edition). Underwriters Laboratories, Inc., Northbrook, IL (December 1997). An inverter listed to UL 1741 with the words "Utility-

Interactive" printed on the listing mark indicates that the unit is fully compliant with IEEE 1547.

The Interstate Renewable Energy Council (IREC) recommends practices and guidelines regarding grid interconnection issues. IREC is a non-profit organization committed to accelerating the sustainable utilization of renewable energy resources and technologies. For more information, refer to their web site at http://www.irecusa.org/connect.htm

## NEC® Requirements

Utility-connected systems present some unique issues for the PV designer and installer in meeting the *NEC*®. All electronic devices and system components need to be UL listed for the proper application and rated to the correct voltage. DC system components must carry a specific DC rating. These items include circuit breakers, fuses, disconnects, mechanical connectors, etc. See *NEC® 2005,* Article 690 for specific labeling details for utility interactive systems.

# 11.5 Net Metering

Net metering allows the exchange of any surplus energy produced by the PV system for utility energy credit to be used during periods when the PV system is not producing enough energy to meet the needs. This means that the electric meter spins "backward" when power is flowing from the building to the utility, and spins "forward" when electricity is flowing from the utility into the building. At the end of the month, only the net consumption is billed. It is the amount of electricity consumed, less the amount of electricity produced. The utility acts much the same as a battery, crediting the energy "account" for later use if production exceeds consumption. See Figure 11-4.

For example, during the middle of the day, the system produces three kilowatt-hours but the building uses only one kilowatt-hour. Thus, the "account" will be credited for two kilowatt-hours. Later that evening, two additional kilowatt-hours might be used and the "account" ends up with a net zero balance, owing the utility nothing for that day.

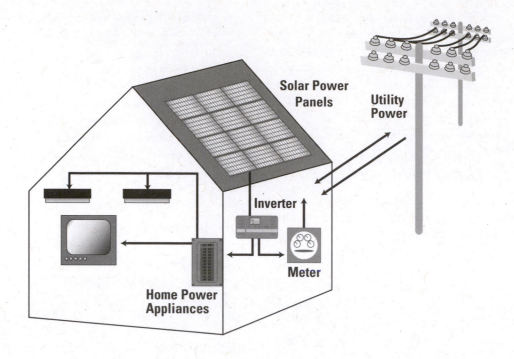

Figure 11-4

**NET METERING: GRID-TIED SYSTEM**

zero balance, owing the utility nothing for that day.

The net metering protocol is a benefit to small renewable energy systems. There are three main reasons net metering is important. First, as increasing numbers of primarily residential customers install renewable energy systems in their homes, net metering provides a simple, standardized protocol for connecting their systems into the electricity grid that ensures safety and power quality. Second, as many residential customers are not at home using electricity during the day when their systems are producing power, net metering allows them to receive full value for the electricity they produce without installing expensive battery storage systems. Third, net metering provides a simple, inexpensive, and easily administered mechanism for encouraging the use of renewable energy systems, enabling important local, national, and global benefits.

## Benefits and Costs

Net metering provides a variety of benefits for both utilities and consumers. Utilities benefit by avoiding the administrative and accounting costs of metering and purchasing the small amounts of excess electricity produced by these small-scale renewable generating

facilities. Net metered PV systems can potentially reduce the utility's peak load demand. Because peak load generation is often expensive ($0.15-0.20 per kwh) this results in significant savings for utilities. Consumers benefit by getting greater value for some of the electricity they generate, being able to interconnect with the utility using their existing utility meter, and being able to interconnect using widely-accepted technical standards.

The only cost associated with net metering is indirect; the customer buys less electricity from the utility, which means the utility collects less revenue from the customer. The reason is that any excess electricity that would have been sold to the utility at the wholesale or 'avoided cost' price is instead being used to offset electricity the customer would have purchased at the retail price.

In most cases, the revenue loss is comparable to the customer reducing their electricity use by investing in energy efficiency measures, such as compact fluorescent lights and efficient appliances. The bill savings for the customer and corresponding revenue loss to the utility depends on a variety of factors, particularly the difference between the 'avoided cost' and retail prices. In general, the difference will be from $10 - $15 a month for a residential-scale PV system (2 kW), and from $25 - $50 a month for a farm scale wind turbine (10 kW). Any revenue losses associated with net metering are, at least partially, offset by the administrative and accounting savings.

## Using the Existing Meter

The standard kilowatt-hour meter used by the vast majority of residential and small commercial customers accurately registers the flow of electricity in either direction. This means the 'netting' process associated with net metering happens automatically. The meter spins forward in the normal direction when the consumer needs more electricity than is being produced and spins backward when the consumer is producing more electricity than they need in the house or building. Some utilities use a meter that records the number of times the meter spins, not registering if it is moving forward or backward. This type of meter will bill the homeowner for PV power produced!

## Current Worldwide Status

Currently, many US states have some form of net metering. Germany, Japan, and Switzerland also have net metering. Many US state net metering rules were enacted by state utility regulators pursuant to state implementation of the federal PURPA statute. In recent years many states have enacted net metering laws legislatively.

For more information about states with net metering legislation and incentives for renewable energy systems, refer to the Database of State Incentives for Renewable Energy (DSIRE) located at: www.dsireusa.org

# 11.6 Sample System Exercise

The following exercise will design a grid-tied system. The system parameters are listed below and are followed by the Grid-tied Sizing Sheets. Review the worksheets to calculate how many PV modules will be needed as well as how to choose an inverter. After sizing this system, a full schematic is displayed. For copies of the Grid-tied sizing worksheet, refer to Appendix D.

### Grid-tied PV System Parameters

**Size a system for the following home:**

Yearly Ave KWH Consumption: 3300 KWH/YR

Location: Albuquerque, NM

**System Specifications:**

Percentage of power to be generated from PV: 100%

Array tilt: Latitude - 15°

Array orientation: True South

Record low temp: -17°F

Average high temp: 92°F

PV module choice: Brand X, Model XYZ
STC rated watts: 170W
Voc: 30.6 V
Vmax: 24.6 V
Isc: 7.38 A
Imp: 6.93 A

Inverter choice: Brand X, Model XYZ
CEC efficiency: 93.5%
Continuous watt rating: 2,500W
DC input voltage: 150-450VDC

See the completed worksheet and schematic, Figure 11-5, at the end of this chapter.

## Answers to the Grid-tied Sizing Exercise

**Electric Load Estimation**

1) Figure out the approximate daily average energy usage and PV System kWh/day:

Yearly average energy consumption: __3300__ kilowatt-hrs/year

__3300__ kilowatt-hrs/yr ÷ 365 days/yr = __9.04__ average kilowatt-hrs/day
(This is our average daily load.)

__100__ % of power to be generated from PV system

__9.04__ Avg. kWh/day X __100__ % of power to be from PV = __9.04__ PV system kilowatt-hrs/day

**Array Sizing**

2) Figure out the PV system kilowatts needed (including derate factors for temperature losses, miscellaneous system losses, and inverter losses):

Average peak sun hours per day: __6.3__

__9.04__ PV System kWh/day ÷ __6.3__ avg. sun hours per day ÷ 0.88 PV Temp Losses (see Notes*)

÷ 0.84 Derate Factor (see Notes**) ÷ __0.94__ inverter efficiency (see Notes***)

= PV array kW needed __2.07__

__2.07__ PV array kW X (1000 watts/kilowatt) = __2070__ PV array watts

3) Choose a PV module:

Make: ____Brand X____ Model: ____XYZ____

STC watt rating: __170W__ Voc: __30.6V__ Vmax: __24.6V__

Isc: __7.38 A__ Imp: __6.93 A__

__2070__ PV array watts ÷ __170__ STC watt rating __12__ # of modules needed

**Answers to the Grid-tied Sizing Exercise (continued)**

Inverter Sizing

4) Choose a specific inverter (or a combination of inverters) that has an appropriate continuous wattage rating:

With grid-tied PV systems an inverter model is chosen based on the maximum amount of watts passing through it from the array (unlike stand-alone PV systems where the inverter size is based on the AC total connected load).

____12____ # of modules needed x __170__ STC watt rating = __2040__ max watts inverter(s) must pass

Manufacturer: ____Brand X____  Model: ____XYZ____

DC (STC) Continuous Power rating: ____2500 watts____

DC input Voltage Range: ____150–450VDC____

5) Calculate how many of these inverters the system will require, and how many modules will be wired into each inverter:

__2040__ max watts inverter must pass ÷ __2500__ inverter watt rating = __1__ #of inverters

__12__ # of PV modules needed ÷ __1__ # of inverters = __12__ # of modules per inverter

6) Find out how many of our modules the chosen inverter requires in series.
Check with the inverter manufacturer to see how many modules this inverter requires in series for its DC input voltage window. Most grid-tied inverters have a string sizing program on their website to calculate how many modules in series are needed.

Array Location (City, State): __Albuquerque, NM__

Record Low Temp: ____–17°F____

Average High Temp: ____92°F____

| **Possible Configurations** | | |
|---|---|---|
| | 1 series string of modules | |
| # M O D U L E S | 8 | Array May be undersized |
| | 9 | Optimal Configuration |
| | 10 | Optimal Configuration |
| | 11 | Optimal Configuration |
| | 12 | Optimal Configuration |
| | 13 | Optimal Configuration |

**INVERTER XYZ STRING SIZING PROGRAM REPORT**

Using the PV modules and the site temperature info, how many modules does the inverter need in series? _8 to 13_

Does this work with the number of modules needed per inverter? _YES! 12 Modules is in this range_

Remember if using more than one inverter, break up the PV array into subarrays that will feed each inverter. Each inverter must have the appropriate number of modules in series to match the inverter's DC input voltage range. If not, either round the number of modules in the array up or down. This will affect the percentage of power to be generated by the PV system. Another option is to choose a different module or inverter to be used in the system.

## Notes for Grid-tied Sizing Exercise

*Note: Standard Test Condition ratings where cell temperature = 25°Celsius is not very realistic when solar modules are in the sun. To account for temperature losses in more realistic situations the sizing sheets use a temperature derate value of 0.88. This assumes an average daytime ambient temperature of 20 degree C. Each module has a slightly different temperature coefficient which is not taken into consideration here.

**Note: The Derate Factor of 0.84 accounts for other system losses (including module production tolerance, module mismatch, wiring losses, dust/soiling losses, etc.). This value assumes no shading. See the table below for a summary of how this derate factor is calculated:

| Derate Values | Range of Acceptable Values | Chosen values |
|---|---|---|
| PV module nameplate DC rating | 0.80 - 1.05 | .95 |
| Mismatch Modules | 0.97 - 0.995 | .98 |
| Diodes and connections | 0.99 - 0.997 | .995 |
| DC wiring | 0.97 - 0.99 | .98 |
| AC wiring | 0.98 - 0.993 | .99 |
| Soiling | 0.30 - 0.995 | .95 |
| System availability | 0.00 - 0.995 | .98 |
| Shading | 0.00 - 1.00 | 1.00 |
| Age | 0.70 - 1.00 | 1.00 |
| **DERATE Factor** | | = **0.84** |

Note: You can adjust this value to reflect conditions for your specific site

Source: http://rredc.nrel.gov/solar/codes_algs/PVWATTS/

***Note: Typically an inverter efficiency of 0.9 is used as a conservative estimate. A specific inverter's average efficiency number can be implemented from the California Consumer Energy Center's List of Eligible Inverters webpage: http://www.consumerenergycenter.org/cgi-bin/eligible_inverters.cgi. This website provides specific inverter's "average" efficiency. (For more information on how the inverter efficiency values are evaluated see the Emerging Renewable Program Guidebook, Appendix 3.)

**Additional Notes:**
The National Renewable Energy Laboratory (NREL) has an online Grid-Connected PV system performance calculator called "PVWatts" which has many features to further customize a PV array.

For example, perhaps the only solar access at a particular site is on an east facing roof. This grid-tied sizing method assumes the array orientation is true south. The NREL PVWatts program allows flexibility to calculate the system production for an east facing array and to adjust the tilt angle of the roof pitch. See PVWatts website: http://rredc.nrel.gov/solar/codes_algs/PVWATTS/version1/

Also realize that the results from this grid-tied sizing sheet will differ slightly from the PVWatts performance calculator primarily due to the fact that it uses a simple temperature derate value, while PVWatts utilizes monthly weather data for each location.

Finally it must be realized that annual power production of a PV system is largely dependent on how much available sunlight there is and **weather patterns vary year to year**. This means that even though this grid-tied sizing method (and PVWatts) is utilizing long term solar data, any year could have more or less sunshine available, which means the actual annual power production of a PV system could exceed or fall short of the expectations.

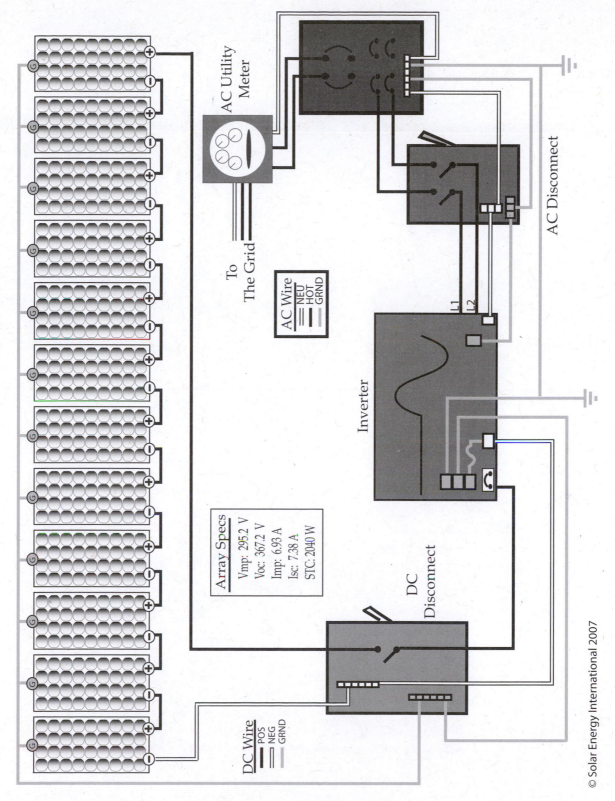

Array Specs

Vmp: 295.2 V
Voc: 367.2 V
Imp: 6.93 A
Isc: 7.38 A
STC: 2040 W

AC Wire
NEU
HOT
GRND

DC Wire
POS
NEG
GRND

AC Utility Meter

To The Grid

AC Disconnect

Inverter

DC Disconnect

© Solar Energy International 2007

Figure 11-5

**SAMPLE SYSTEM EXERCISE SCHEMATIC**

# Chapter 12
# Mounting Photovoltaic Modules

**Contents:**

12.1    Mounting System Types  . . . . . . . . . . . . . . . . . . . . . . . . . . . . . 140
12.2    Building Integrated Photovoltaics  . . . . . . . . . . . . . . . . . . . . . . . 143

# 12.1 Mounting System Types

The photovoltaic system designer must consider many factors when selecting an appropriate site for mounting modules. The location must be oriented toward the sun and be free of shading obstacles throughout the sun's daily and seasonal paths. The site must be in proximity to the power-conditioning center to minimize line losses. The owners or operators of the system should be pleased with the aesthetics of the array and where it is located. Depending on the locale, the site may also need to provide protection from theft and vandalism. Finally, operators and designers should have easy access to perform routine maintenance.

Once you have chosen the site, you can determine the type of mounting system best suited for the site and the system application. There are various systems available for mounting a PV array, from simple bracket systems to complex dual axis trackers. The type of mounting system you choose will depend on the following factors:

- Orientation of the house.

- Shading at the site.

- Weather considerations.

- Roof material.

- Soil and/or roof load bearing capacity .

- System applications.

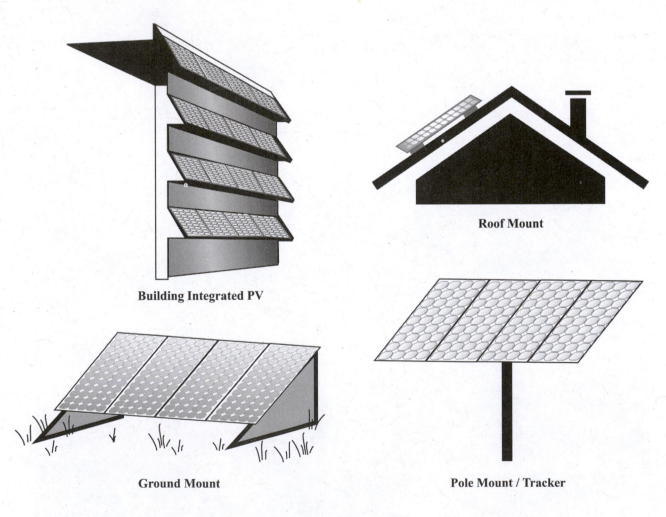

**Building Integrated PV**

**Roof Mount**

**Ground Mount**

**Pole Mount / Tracker**

Figure 12-1
**BASIC MOUNTING STRATEGIES**

We will review the following types of mounting systems (see Figure 12-1):

- Bracket mounts
- Pole mounts
- Ground mounts
- Roof mounts
- Trackers

## Bracket Mounting Systems

In this type of system, two galvanized steel angle brackets are bolted to a building's exterior walls or roof structure. A second pair of compatible brackets is attached to the end frames of the solar module. When the two sets of brackets are mated, they form a simple, durable, cost effective mounting system for a one module photovoltaic system. A simple bracket system can be used to mount a single solar module.

## Pole Mounting Systems

Arrays can also be mounted on a hardware system that bolts directly to a vertical pole placed permanently and securely in the ground. A pole mount is desirable when attaching the array to the building is not an option. This mounting technique can be seasonally adjusted to optimize the system's performance. The diameter and thickness of the pole, as well as how much concrete is needed, will be dictated by the size of the array, the soil type, the terrain and the wind speeds.

## Ground Mounting Systems

A ground mounted array support structure uses a frame that is bolted directly to prepared footings. Standard support structures for four, eight, and twelve modules are commercially available or may be site fabricated.

A mounting frame often consists of two parallel channel bars that form a simple rack. Cross supports are bolted to the frame to increase lateral structural support and to prevent wind damage. Non-adjustable, extruded aluminum legs bolt to the frame to hold the array at a predetermined tilt angle. Adjustable tilting support legs can also be fabricated or purchased to allow manual seasonal tilt adjustments.

You should carefully evaluate local weather characteristics as well as soil load bearing capacity before selecting a final site for a photovoltaic array. Ground mounting systems require level foundations with a sufficient structural integrity to avoid load-bearing failure. The foundation must resist wind uplifting and wind shear (lateral movement). Consult local building codes to help determine specific foundation requirements and to ensure that you meet these standards before installing the mounting system.

## Roof Mounting Systems

Five types of systems are commonly used when roof mounting a photovoltaic array:

- Direct mounts
- Adjustable or fixed rack mount
- Stand-off mounts
- Ballasted mount
- Building integrated mount

**Direct mount:** In direct mounting systems, the photovoltaic modules mount directly to the conventional roof material, eliminating the need for a supporting framework. The modules must not disturb the roof covering's weather tight integrity and be adequately sealed using appropriate sealants.

The direct mounting system does not allow for air circulation around the array's modules, which results in higher operating temperatures and decreased power production compared to other mounting systems. Access to the array's electrical connections is limited, making troubleshooting, repair, and maintenance difficult.

**Rack mount:** With a rack mount, the photovoltaic modules are supported by a metal framework and are set at a predetermined angle. The rack-mounted array is placed on the roof with the rack bolted on the roof's structural members. It should be noted however, that a rack mount does increase the bearing weight of the system on the home's roof and can sometimes pose wind-loading issues. However, since air circulates completely around the modules, they are kept at a cooler, more efficient operating temperature. The array's electrical connections are also easier to access since the rear surfaces of the modules are exposed.

Some rack mounting systems are adjustable, which can increase the photovoltaic system's output throughout the seasons. Many manufacturers have precut, predrilled rack-mounting systems for specific modules. Generic adjustable and fixed rack mounting systems are available from third-party suppliers.

**Stand-off mount**: Stand-off mounting systems place the modules parallel to the roof with an air gap between the two surfaces to gain adequate airflow under the modules. The modules are placed on channeled rails and then clamped down to the rails with 'top down' clips that grasp the aluminum frame. These rails are attached to the roof with mounting feet. Proper lag bolt size is necessary to attach the feet to the rafter or support structure within the roof. Modules should be mounted at least four inches above the roof's surface to allow for adequate airflow. This is the most common roof mounting system.

**Ballasted mount**: Ballasted mounting systems are designed for flat roofs and do not require penetrations. They rely on weight to prevent uplift from wind. This is often accomplished with concrete blocks.

**Building integrated mount**: Building integrated photovoltaic (BIPV) systems will be discussed later in this chapter.

## Tracking Mounting Systems

Solar photovoltaic array mounts that track the sun in its daily path across the sky are a cost-effective, alternative mounting system for some installations. Passive tracking units have no motors, controls, or gears and use the changing weight of a gaseous refrigerant within a sealed frame to track the sun. Sunlight activates the refrigerant, and the frame assembly moves by gravity or is driven by a piston. They can also be seasonally adjusted to optimize altitude angle. Active trackers use motors powered by small, integrated photovoltaic panels to move the array.

Tracking mounts that follow the sun's azimuth but not its altitude are called single-axis trackers. Trackers that follow both the sun's azimuth and altitude are called dual-axis trackers.

Tracking mounts require firm foundations because of their weight and wind loading characteristics. Four to six inch outside diameter pipe is commonly set in reinforced concrete footings to insure long term, safe operation. The tracker stem is then placed over the pipe. The tracking unit needs to be mounted at an adequate height above ground level to allow unobstructed movement above snow or debris.

Tracking units generally enhance a system's annual performance by approximately 25% to 30% but can add a significant cost to a system. The economics of tracking mounts should be carefully evaluated in the initial design phase. Trackers offer different performance gains throughout the year, increasing system performance by 15 percent in the winter and by 40 percent in the summer. Systems requiring larger loads during the summer months are ideal candidates for tracking systems. Longer hours of effective insolation are available in the summer and increase the photovoltaic system's collection potential. In contrast, winter dominated loads are less likely to benefit significantly from trackers. Each system must be carefully evaluated to determine the economic viability of trackers versus fixed mounts.

Remember, trackers add a moving part(s) to a component of the system that is otherwise typically stationary. Since mechanical parts will wear down over time, this potential cost needs to be incorporated into the overall life system cost.

# 12.2 Building Integrated Photovoltaics

Building integrated photovoltaic (BIPV) systems are electric generating systems that are integrated into the building shell. BIPV products can take the place of traditional building materials such as shingles, overhangs, skylights and windows, and provide many auxiliary advantages compared to standard array options. Advantages include:

- Many BIPV products are designed to interface with the building structure and do not require additional racking mechanisms.

- They can provide daylighting, shading, insulation, acoustic control, thermal collection and wind protection while generating electricity.

- As integrated construction components, BIPV system costs can be offset by the avoided costs of the displaced construction materials, and can be included in the initial financing scheme of a building.

- In commercial applications, the electricity savings over time may be applied to offset the initial construction costs of the system.

- The versatility of BIPV products are aesthetically appealing, and can be utilized to increase visibility and educational impact.

BIPV can be integrated in a building via a number of creative methods and products. Awning structures provide shade areas for people to gather or park cars and bicycles. Architectural glass provides natural day-lighting while reducing glare and heat gain. Some BIPV roofing products use plugs designed to provide quick and easy module interconnection, and can be installed by conventional roofers using standard installation practices.

## Roofing

Roofing systems include BIPV shingles, tiles, metal roofing, and atrium or laminate roof systems. These products replace conventional shingles, slates and tiles. Each BIPV product has a method to minimize the potential for leakage. For example, PV shingles are designed to form a double weatherproof layer upon installation, similar to conventional shingles. At the overlap between each shingle an adhesive seal forms a water and weather-tight bond between the consecutive layers.

Some smaller tile and shingle products involve numerous interconnections between modules, increasing the number of "series strings" required in a given area. On larger installations this can create additional design issues regarding wire sizing and wire runs. Be sure to include plans for access during installation and future maintenance. Roof integrated BIPV systems also have less air flow and decreased energy output when compared to traditional module mounting methods.

Some PV roofing modules use an amorphous solar cell that consists of thin-film PV material and laminates on to a metal roof. Amorphous silicon modules are less efficient and thus require more roof space than single and poly-crystalline modules for a similar power output. This product may require trained personnel to adhere the laminate to the roof.

## PV Facade Walls and Awnings

PV awnings provide a large area for generating electricity and reduce solar heating in the summer, which cuts cooling loads and glare. Figure 12-2 shows an example of a sawtooth PV facade consisting of an overhanging PV shade that screens a window. The overhang is positioned to reduce direct sunlight in the summer when the sun appears high in the sky, while allowing passive solar space heating in the winter when the sun appears low in the sky. PV awnings can have greater output than other BIPV systems due to increased airflow around the modules. They can be retrofitted onto existing buildings or integrated into a new building's design.

## PV Skylights and Windows

Because skylight and greenhouse glass is often heavily tinted to minimize glare, semi-transparent PV glazing can make a good substitute. The glazing panels consist of PV material embedded in the glass. Semi-transparent PV units generate electricity while typically allowing about 20% daylighting through the modules.

## Financial Considerations

BIPV system design and installation specifics vary greatly depending on the building design and materials used in construction. While Balance of System components (wiring, inverter, disconnects, etc) will likely be the same as with any other PV system, the array design and installation may require an experienced professional familiar with structural engineering as well as photovoltaic installation. Some BIPV products require specially trained personnel for proper installation and may incur increased installation costs.

The ability to include the costs of PV into a building's mortgage can mitigate one of the most common barriers to the installation of residential and commercial PV systems — out of pocket, up-front costs. Financing a $20,000 grid-tied system with a $200,000 mortgage over 30 years results in a manageable payment over time rather than one large payment after the installation is complete.

## Additional Considerations for BIPV

Criteria to consider when selecting BIPV products:

- Certification (UL, ETL, IEE).
- Fire rating.
- Warranty of power output.
- Installation interface with traditional building products.
- Overall aesthetics.
- Wiring requirements.

Despite the benefits of BIPV technology, these systems are currently under-utilized due to increased cost and lack of education among contractors, architects, builders, and planners. When planned during the early stages of a project, BIPV can be utilized to increase building efficiency while decreasing overall building utility demand. With clever design, the multi-functionality of BIPV systems can indeed make them cost-effective.

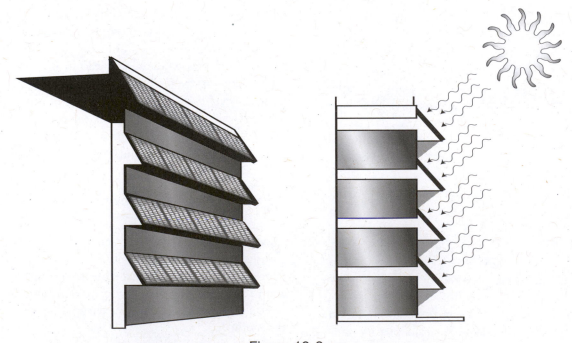

Figure 12-2

**PV FACADE - SAWTOOTH DESIGN**

# Chapter 13
# PV Applications for the Developing World

**Contents:**

13.1   The Need for Reliable Electricity . . . . . . . . . . . . . . . . . . . . . . . .146
13.2   Lighting . . . . . . . . . . . . . . . . . . . . . . . . . . . . . . . . . . . . . . . . . .146
13.3   Television and Radio . . . . . . . . . . . . . . . . . . . . . . . . . . . . . . . .147
13.4   Health Care and Refrigeration . . . . . . . . . . . . . . . . . . . . . . . . .147
13.5   Micro-Enterprises . . . . . . . . . . . . . . . . . . . . . . . . . . . . . . . . . .147
13.6   Water Pumping . . . . . . . . . . . . . . . . . . . . . . . . . . . . . . . . . . . .148
13.7   Determining Solar Access with a Sun Chart . . . . . . . . . . . . . . .152
13.8   Sample Installation Materials . . . . . . . . . . . . . . . . . . . . . . . . . .154

Many people in the industrialized world take electricity for granted. Even people in the developed world who rely on renewable energy technologies to generate electricity often use enough electricity to power an entire community in the developing world. The two billion people throughout the world without access to electricity are not asking for luxuries like dishwashers and air conditioners. In reality, just a small amount of power can have incredible impacts on people's lives.

# 13.1 The Need for Reliable Electricity

Most of the people in the developing world without access to electricity live in remote areas that may never see the electric grid. People in these unelectrified regions can't read or study after the sun sets, have to walk for hours to get drinking or washing water, or have to strain their eyes to work under the dim light of a candle. Others rely on kerosene lanterns, which are polluting, unhealthy, and a fire hazard. Kerosene fumes can cause eye irritation, respiratory problems, and nasal problems.

Other people living in rural areas use batteries, either automotive batteries or small "throw-away" batteries, to provide electricity. In fact, 10% of unelectrified homes throughout the world use car batteries. However, these batteries are used until they're completely discharged, then taken by horseback, canoe, camelback or the sweat of people's own backs, to the nearest electrified town to recharge. This deep discharging and recharging of car batteries leads to very short battery life, and the batteries become toxic waste added to the environment. People are paying exorbitant amounts of money to run their radios and flashlights with small "throw-away" batteries. Even where there is grid power in the developing world, it is often unreliable and frequently unsafe.

Many rural electrification projects throughout the world are installing small solar home systems (SHS) on individual homes. These systems consist of one or two solar electric panels that run some lights and occasionally a black and white TV. Reaching only 10% of the SHS market worldwide would electrify 35 million homes.

# 13.2 Lighting

The largest application for electricity in the developing world is for household light. Although many people in rural areas of the developing world already have access to some form of light, it is usually not a high quality light. Throughout the world, children breathe the fumes of kerosene lanterns as they study, women strain their eyes working by candlelight and doctors perform operations by flashlight. In other areas, people have car batteries that run lights.

Solar home systems with just one or two photovoltaic (PV) modules can provide enough electricity to power lights for studying, cooking, working and socializing. Many of these small SHS also power radios and televisions. The benefits received by a small amount of electricity are numerous. Not only will it displace dangerous and unhealthy kerosene, but also greatly improve education, health, and economies. It particularly improves the lives and health of women and children.

Education can be greatly improved with access to lighting, allowing children to study at night and adults to take evening education classes. A 2002 study by the Energy Sector Management Assistance Program in the Philippines discovered that children from electrified households gained about two years in educational achievement over children from nonelectrified households, due to the improvement in study conditions during the evening.

Today, a wide variety of economical lighting options are available in 12V direct current for small solar home systems and 120V alternating current for larger applications such as schools and health clinics. Excellent 12V direct current and 24V direct current ballasts are now widely available. It often makes sense to run lighting loads on direct current to avoid using an inverter. This creates a more efficient and less expensive system, and also preserves the lighting circuit should an inverter fail. In addition, lights operated from direct current are slightly more efficient and usually last longer than their alternating current counterparts.

See Chapter 4 for more information on lighting.

# 13.3 Television and Radio

After light, the next thing that people usually want is a television or radio. There are many advantages and disadvantages of bringing television into rural homes. TV allows people more access to information and has an important entertainment value. It can be used for adult education and training programs. It has also been shown to decrease the migration rate to the cities, and to lower the birth rate. However, TV can change people's lives for the worse as well. Television can lead to a disenchantment with rural life, less family socializing, a decrease in traditional values, and decreased sleep. In certain cases it has actually increased the migration rate to the cities due to the idealized city life often shown in the media.

# 13.4 Health Care and Refrigeration

Another enormous application for renewable energy throughout the developing world is health care. Photovoltaics can run medical equipment, dental equipment, water purification and desalination units, and vaccine refrigerators. Electricity in a health clinic also means health and medical information can be presented to local families at night, when men are back from the fields and can be present. This is crucial in the war against HIV/AIDS, malaria, tuberculosis, and other diseases. With electric lighting, doctors can perform operations at night, deaths at birth can be reduced by improved delivery conditions, and telecommunications can be made possible. Telecommunications are essential in contacting physicians in the city and in locating and obtaining emergency sources of medicines. Without electricity it is also difficult to attract, much less hold, trained health workers in rural areas.

Refrigeration is also key in providing decent health care to people. Three million children die each year from diseases that are preventable with currently available vaccines. Vaccines need to be kept in a certain temperature range in order to remain potent. This is impossible without refrigeration. A small vaccine refrigerator can be run with one or two solar panels. Photovoltaic refrigerators have proven to be an effective and reliable method for refrigerating medicines and vaccines in remote and rural areas throughout the world.

However, electric refrigeration is not always cost effective for rural homeowners. Homeowners must recognize that refrigeration will constitute a major energy load that significantly adds to the cost of a photovoltaic system. It is important to ask the following questions:

- Is mechanical refrigeration really necessary?
- Would cool storage work sufficiently?
- Would another fuel source be more cost-effective? (i.e. propane)

If a refrigerator is an essential load that cannot fail, for example a vaccine refrigerator, the design must incorporate elements to prevent system failure. Two handbooks available from the World Health Organization, *User's Handbook* and *Fault Finding and Repair of Solar Refrigerators*, are useful reference manuals for users of critical load refrigerators powered by photovoltaic systems.

See Chapter 4 for more details on refrigerators.

# 13.5 Micro-Enterprises

Improving people's economic resources in rural areas is an important factor in improving people's lives. Photovoltaic technology can aid in this by providing power for microenterprises. A microenterprise is a small business that produces goods or services for cash income. In general, microenterprises have limited access to capital, have few employees, and are often home-based. Not all microenterprises are family operated, but when family members do work for the business, they frequently do so without pay. Small cooperatives can also be microenterprises.

Microenterprises can include anything from a PV-powered smoothie stand to a renewable energy electronic repair shop. Most tools can easily be run with renewable energy technologies.

Larger renewable energy systems can run large commercial enterprises such as a village cinema. A common commercial enterprise in the developing world is a battery charging station. As mentioned previously, 10% of unelectrified homes use car batteries to provide electricity. If people had a local place where they could charge their batteries, the batteries could be charged regularly, and would have a longer life. A solar or wind powered battery charging station makes that possible. People pay a small fee to have their batteries recharged and they save the time and effort of hauling batteries to a distant electrified town. In Thailand, a government sponsored program installed 1,000 battery charging stations around the country. Customers only have to pay the charging fees to recharge their car batteries. They can upgrade and buy a solar panel in a later stage. The stations are locally operated, creating employment opportunities and potential business for local entrepreneurs.

# 13.6 Water Pumping

Solar water pumping is done all over the world, for irrigation, domestic use, and livestock, and can greatly alleviate the work load for many rural people. In some areas, women spend hours each day walking to the river and hauling back heavy buckets of water. Compared to many of the alternatives, PV systems are reliable and cost-effective. The simplest method of water delivery, diversion of rain or surface water by gravity, is not possible in many locations. Manual pumps are a common method of water transfer worldwide, but cannot move large volumes of water or pump from deep wells. Mechanical pumps powered by engines or electric motors are expensive, maintenance intensive, and use expensive fuel. Generally, these systems are only used when a community or corporate infrastructure exists to support the costs associated with a larger, more complex distribution mechanism.

Depending on the application, a water pumping system does not need to include batteries. The water tank can act as storage, by pumping water all day while the sun is shining and filling the tank. Then the water is available at night from the tank. This makes water pumping systems even simpler and more cost effective.

## Pump Terminology

The pressure a pump creates, called **head**, is measured in feet. A column of water 2.31 feet in height (or 2.31 feet of head) exerts a pressure of 1 pound per square inch (psi) at its base. If a pump must deliver water to a point 10 feet above the water source, it must create a minimum of 10 feet of head. 10 feet of head is equal to 4.3 psi (10 feet of head divided by 2.31 feet/psi equals 4.3 psi).

The total head a pump must create is a function of the following parameters:

- Head from the water source to the point of discharge or storage,
- Head from the storage point to the delivery point when using one pump,
- Friction loss or the resistance of water flowing through pipes.

When specifying a water pumping system, be familiar with the following terms.

**Suction head:** The vertical distance from the surface of the free water source to the center of the pump when the pump is located above water level.

**Discharge head:** The vertical distance from the center of the pump to the water surface or point of free discharge.

**Static head:** The vertical distance from the water source to the water surface or point of free discharge. Static head is equal to the sum of the suction head and discharge head.

**Service pressure:** The service pressure is the feet of head needed to supply the final discharge pressure. If a final discharge pressure is desired, the pump must supply the required flow at the specified pressure.

**Friction head (FH):** The pressure the pump must provide to overcome the loss of energy due to friction as water moves through a pipe. The smaller the pipe diameter and the faster the flow, the greater the losses. The exact amount of FH can be obtained from friction loss tables, which are out of the scope of this manual. At low flow rates, friction losses are small compared to static head.

**Service head:** The vertical distance from the storage point to the highest delivery point.

**Pressure:** The measure of force exerted on the walls of piping, tanks, and other components by the liquid in a system. Pressure is measured in pounds per square inch where 2.31 feet of head is equal to 1 pound per square inch.

**Flooded section:** The section of pipe between the water source and pump when the water source is higher than the pump. Water flows through this section of pipe to the pump by gravity.

**Flow:** The rate of liquid volume capacity of the pump. Flow is measured as a unit volume per unit of time, such as gallons per minute (gpm) or liters per minute (lpm).

## Pump Types

Specific water pumping applications have dictated the design of water pumps, resulting in a wide range of pumps that serve specific needs. Pumps can be divided into one or both of the following major categories:

**Self-priming:** The ability for a pump to initially run dry and create sufficient suction pressure to draw water from the source to the pump. Pumps that are not self-priming must either be primed prior to operation or installed below the level of the water source.

**Positive displacement:** Any type of pump that moves a liquid by the action of a chamber, plunger, or rotary gear and when discharged, moves another volume of liquid into place that displaces the volume before it. Many positive displacement pumps are self-priming.

Pumps are further divided into one of the following categories:

**Centrifugal:** Any type of pump that moves a liquid by the action of an impeller. The impeller draws the liquid to an intake at the impeller's center and then discharges centrifugally at an outlet at the impeller's perimeter. Centrifugal pumps generally require priming prior to operation or must be installed below the water source level.

**Self-priming centrifugal:** Same as standard centrifugal pumps but with a chamber above the impeller that keeps the pump "primed" for easy restarting.

**Jet pump:** A centrifugal pump that allows some flow to return into a venturi on the input side. This can increase the suction head to as much as 150 feet but results in a decreased flow rate because water is used to move water.

**Submersible:** A pump with a series of centrifugal impellers or diaphragms and a motor in a water tight housing. The entire assembly is submerged near the bottom of the well. These pumps can deliver water from great depths.

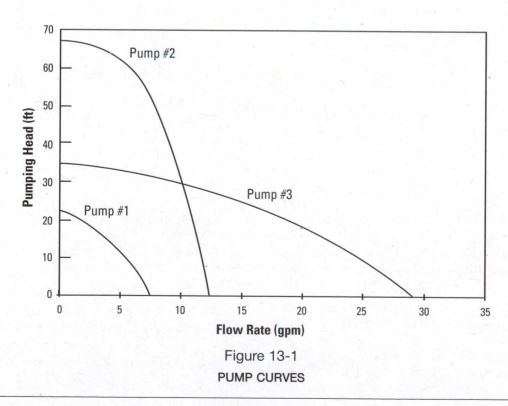

Figure 13-1

**PUMP CURVES**

**Jack pump:** A positive displacement type pump in which the motor operates a reciprocating jack above the ground. The jack pulls a long drive shaft with a plunger at the end to move water in steps. These pumps can deliver water from great depths at low flow rates.

**Rotary vane:** A pump containing two gear-like vanes that rotates within a tight-fitting housing to create positive displacement flow. These pumps are capable of several hundred feet of head and used for shallow well, low volume applications.

## Pump Curves

A pump's performance is measured in terms of flow rate and head. The greater the head a pump must overcome, the lower the flow rate. The relationship of flow rate and head for a particular pump is graphically illustrated by a pump curve. Figure 13-1 shows three sample pump curves similar to the curves supplied by pump manufacturers.

The horizontal axis of the graph represents the pump's flow rate in gallons per minute (gpm). The vertical axis represents feet of head or psi, which is the pressure the pump must overcome. Any point on the curve is a flow rate corresponding to the head that the pump must overcome.

Water pumping feasibility graphs, commonly referred to as $M^4$ charts, have been developed for quickly determining whether solar powered water pumping is applicable for a given set of parameters. These $M^4$ charts plot feasibility ranges by multiplying the required head times the required flow (in metrics units).

Head (m) $\times$ Flow (m³) = Required Power $M^4$

General guidelines are as follows:

- $M^4 > 2000$ - solar pumping not practical
- $50 < M^4 < 1500$ - solar pumping feasible
- $M^4 < 50$ - hand pump is more practical

**Problem:** A pump manufacturer supplies you with pump #2 from Figure 13-1, stating that it is the most appropriate for your specific application. The application requires that the pump overcome 9 feet of head. At what flow rate will the pump perform?

**Solution:** To determine the flow rate of this pump's output, locate 9 feet of head on the vertical scale and

move horizontally until you intersect the pump curve. Move straight down from this point to read the flow rate listed on the horizontal scale. The flow rate is 11.5 gallons per minute (gpm).

**Problem:** You are using pump #3 from Figure 13-1 in an application where a minimum flow of 3 gallons per minute is needed. What is the maximum head this pump can provide and still meet the minimum flow rate?

**Solution:** To determine the head, locate 3 gpm on the horizontal scale and read straight up until you intersect the pump curve. Then from this point move horizontally to find the head listed on the vertical scale. The maximum head is 34 feet.

## Pump Selection Criteria

The following parameters should be considered when specifying a pump:

- *Total head* – This is equal to the sum of suction head, discharge head, friction head, and service or pressure head.

- *Suction head* – For shallow wells, suction head may be a maximum of 20 feet.

- *Required quantity of water* – This is usually specified in gallons per day (gpd) or liters per day (lpd).

To specify a pump, the following conditions of the well site must also be considered:

- Well depth
- Recovery rate
- Static water level

**Note:** For shallow wells, maximum suction head is reduced by 1 foot of suction lift per 1,000 feet elevation above sea level. A shallow well pump system may use a centrifugal pump when the suction lift does not exceed 25 feet (at sea level).

## Storage and Delivery

In some photovoltaic powered water pumping systems, energy can be stored in batteries so that the water that has been pumped from the water source into a storage tank can be distributed during times of no sunlight. In some cases, batteries can also be desirable to provide sufficient surge power for starting the pump.

Some photovoltaic system pump controllers are available that allow the pump to be powered directly from a photovoltaic panel or array. These controllers monitor the photovoltaic system's output and turn the pump on only when there is sufficient sunlight to power the pump. As the sunlight's intensity increases and the photovoltaic system's output increases, the pump's performance increases proportionally.

Elevated or pressurized tanks can provide water storage and delivery. Elevated tanks should be fitted with a float switch that turns the pump on or off according to a preset water level. Water from the tank is delivered to use points below the tank level by gravity pressure. For every 2.31 feet of elevation, 1 psi will be delivered at the discharge point. For example, for the tank to provide 40 psi, it would need to be 92 feet high (40 X 2.31 = 92).

Pressure tanks, such as those commonly used in many well systems, can supply water to points of use above or below the tank. A flexible neoprene bladder within the tank is charged on one side with air pressure. Water is pumped into the tank under pressure against the bladder until a pressure switch turns the pump off at a preset pressure. Water is delivered to the use points by air pressure within the tank. The tank is refilled when the pressure switch activates the pump below a preset pressure setting.

## Sample Water Pumping Systems

You must evaluate every water pumping application individually. The examples in this section contain the calculated electrical load for various pumping applications based on flow and head requirements and the pump manufacturer's specifications. To determine the PV panel and battery requirements for your location, refer to Chapter 10.

### Application 1: Shallow Well
Water Requirements: 200 gallons/day
Required Lift:
15 feet suction head
23 feet discharge head
38 feet total head
Pump Specifications:
Type: 12-volt centrifugal
Maximum suction head (at sea level): 25 feet
Maximum discharge head: 32 feet
Flow rate at 38 feet of head: 8.3 gpm
Power requirements: 300 watts
Hours of operation (at 200 gpd): 0.4 hours
Total daily load (at 200 gpd):
120 watt-hrs/day at 12 volts

### Application 2: Shallow Well
Water Requirements: 400 gallons/day
Required Lift:
5 feet suction head
35 feet discharge head
40 feet total head
Pump Specification:
Type: 12-volt centrifugal
Maximum suction head (at sea level): 25 feet
Maximum discharge head: 44 feet
Flow rate at 40 feet head: 6.7 gpm
Power requirements: 300 watts
Hours of operation (at 400 gpd): 1.0 hours
Total daily load (at 400 gpd):
300 watt-hrs/day at 12 volts

### Application 3: Deep Well
Water Requirements: 200 gallons/day
Required Lift:
30 feet suction head
35 feet discharge head
65 feet total head
Pump Specification:
Type: 12-volt jet pump
Maximum suction head: 60 feet
Maximum discharge head: 67 feet
Flow rate at 65 feet of head: 2.2 gpm
Power requirements: 270 watts
Hours of operation (at 200 gpd): 1.5 hours
Total daily load (at 200 gpd):
405 watt-hrs/day at 12 volts

### Application 4: Deep Well
Water requirements: 600 gallons/day
Required Lift:
0 feet suction head
100 feet discharge head
100 feet total head
Pump Specifications:
Type: 12-volt submersible
Maximum discharge head: 125 feet
Flow rate at 100 feet of head: 2 gpm
Power requirements: 325 watts
Hours of operation (at 600 gpd): 5 hours
Total daily load (at 600 gpd):
1625 watt-hrs/day at 12 volts

**Application 5: Deep Well**

Water requirements: 200 gallons/day

Required Lift:

0 feet suction head

150 feet discharge head

150 feet total head

Pump Specifications:

Type: 24-volt submersible

Maximum discharge head: 175 feet

Flow rate at 150 feet head: 1.3 gpm

Power requirements: 580 watts

Number hours of operations (at 200 gpd)

2.6 hours

Total daily load (at 200 gpd)

1508 watt-hrs/day at 24 volts

# 13.7 Determining Solar Access with a Sun Chart

Chapter 3 discussed "Determining Solar Access" to identify a shade free site for solar modules. Since many remote areas of the world do not have access to, or the financial means to acquire, the solar siting devices commonly used in the US there is another simple way to evaluate the solar access of a site. A compass and a way to measure altitude are needed. A protractor with a small weight hanging from a string attached to the protractor's center is sufficient for simple altitude measurements.

Use the following steps to evaluate the solar access of a site.

1. Place yourself at the proposed center position of the collectors.

2. Use the compass to locate a bearing of 90 degrees east (be sure to account for magnetic declination).

3. Sight along this bearing with the protractor to determine the altitude of any obstructions, including vegetation, buildings, geographic features, and the horizon.

4. Mark the altitude of each obstruction on the chart.

5. Rotate 15 degrees towards the west and repeat steps 3 and 4.

6. Repeat steps 3, 4 and 5 until you have reached 90 degrees west.

7. Connect the obstruction marks on the chart

and shade everything below the line. A good solar site will have no shading between 9 a.m. and 3 p.m. on any day during the year.

Determining solar access for low latitudes: If the site is located between 25 degrees south latitude and 25 degrees north latitude, you can use the following solar site evaluation method. All you need is one or two people, a compass, and the information from Table 13-1.

The following steps guide you through evaluating a site:

**Site information:**

1. Record the site location  _____

2. Record the latitude  _____

3. Record the magnetic declination of the site

   _____

**Establishing the major compass directions:**

4. Stand at the site of the PV array and move into a position where your eyes are level with the bottom of the lowest panel.

5. Locate magnetic south using a compass, and apply the appropriate magnetic declination to determine true south.

6. Establish north-south line. Visually establish the north-south line with extended arms. Set landmarks if necessary.

7. Establish east-west line. Visually establish the east-west line with extended arms. Set landmarks if necessary.

**Visualizing the north and south sides of the solar window:**

8. Create a 90° angle for the north side-Extend one arm horizontally toward the north and the other straight above toward the zenith to create a 90° angle as shown in Figure 13-2. Have a friend stand several steps away to help you adjust your arms.

9. Divide the north arc. Visualize the 90° arc created by your arms and divide the arc into

halves, quarters, and then thirds as shown in Figure 13-2. Practice this with the aid of an observer until you achieve a reasonable accuracy.

10. Create a 90° angle for the south side and divide the arc. Repeat steps 8 and 9 for the south side.

11. Find the solar window. Use Table 13-1 to find the north and south angles for the site's latitude. Mark the values for the north and south sides of the solar window in Figure 13-2.

**Evaluating the solar window:**

12. Evaluate the north-south window. Visually sight the north and south sides of the solar window. This solar window must be clear of obstructions or shading. If either one of these orientations is unacceptable, you must move the site to another location or remove the obstruction. If the north-south window is suitable, then continue with the east and west orientations.

13. Evaluate the 8 a.m. to 4 p.m. east-west window. Visually sight the east and west sides using the 8 a.m. to 4 p.m. window. Use Table 13-1 to find the angles for the site's latitude and mark the values in Figure 13-2. Note any shading in this window and if it could be removed easily. If this window is not suitable, repeat the evaluation for the 9 a.m. to 3 p.m. window.

14. Evaluate the 9 a.m. to 3 p.m. east-west window. Visually sight the east and west sides using the 9 a.m. to 3 p.m. window. Use Table 13-1 to find the angles for the site's latitude and mark the values in Figure 13-2. Note any shading in this window.

15. Assess the options of raising the array to avoid the obstructions or increasing the array capacity by 15 percent. If these options are not feasible, you should consider another site. If absolutely necessary, repeat the evaluation for the 10 a.m. to 2 p.m. window.

16. Evaluate the 10 a.m. to 2 p.m. east-west window. Visually sight the east and west sides using the 10 a.m. to 2 p.m. window. Use Table 13-1 to find the angles for the site's latitude and mark the values in Figure 13-2. If there is any shading in this window, you should locate another site. If this window is clear and you

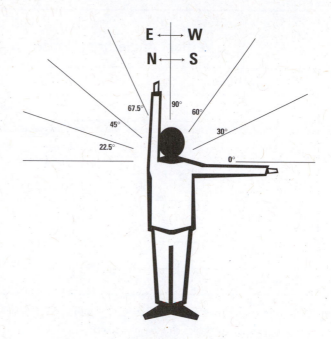

**Figure 13-2**

**VISUALIZING THE FOUR SIDES OF THE SOLAR WINDOW**

choose to use the site, you will need to increase the array capacity by 33 percent. Note that this poses a significant increase in the system cost.

**Evaluating the site:**

17. Assess the site preparations. Using the chosen solar window, determine if the work and/or cost needed to prepare the site, raise the array, or increase the array size is acceptable.

18. If the additional work is not practical or the cost is not feasible, you should locate another site. If the preparations are acceptable, you should continue to step 19.

19. Site recommendations. Make a complete list of the necessary site preparations and/or modifications to the array.

- Remove the obstructions causing shade. How will this be done?

- Increase the height of the PV array. How high above the ground?

- Increase the capacity of the array. What percentage increase and at what cost?

Table 13-1
### Vertical Angle from Horizon to Four Sides of Solar Window

| Latitudes: | 7.5°N - 7.5°S | 7.5° - 17.5° | | 17.5° - 25° | |
|---|---|---|---|---|---|
| | | Hemisphere | | Hemisphere | |
| | | North | South | North | South |
| North | 60° | 67.5° | 45° | 75° | 30° |
| South | 60° | 45° | 67.5° | 30° | 75° |
| East / West 8 a.m. to 4 p.m. | 22.5° | 22.5° | 22.5° | 15° | 15° |
| 9 a.m. to 3 p.m.* | 30° | 30° | 30° | 22.5° | 22.5° |
| 10 a.m. to 2 p.m.** | 45° | 40° | 40° | 30° | 30° |

\* For sites that do not satisfy the 8 a.m. to 4 p.m. solar window, but do satisfy the 9 a.m. to 3 p.m. solar window, increase the array capacity by 15 percent or mount the array with bottom of the lowest panel at least 1/3 the height of the obstruction.

\** For sites that satisfy only the 10 a.m. to 2 p.m. solar window, increase the capacity of the PV array by 33 percent.

# 13.8 Sample Installation Materials

During 30 remote system installations, the Solar Energy International staff compiled the lists included in this section. These contain a combination of the tools and materials required for photovoltaic installations.

All of the installations that were used to generate the following lists required at least one mile of additional travel from a motor vehicle or boat. The sites had no available electrical power before the photovoltaic installation. Table 13-2 has been developed assuming the installation will be a 12-volt, direct current, stand-alone photovoltaic power system requiring a pole mount, battery storage, and controller. It also assumes the system will power incandescent and fluorescent lighting in a previously constructed shelter. Even though the following guidelines have been developed through experience with actual photovoltaic power system installations, you must use them with the understanding that each job is unique and may require

modification of the checklists.

You should check certain installation tools to ensure they are in operational condition and have no missing parts prior to departure. Some tools are also more fragile than others. You should have a back-up for these tools.

Many power tools that would be helpful during an installation are inappropriate for remote system installations. System installers often prefabricate items, especially metal components, prior to going into the field. All alternating current tools are eliminated unless the installation power system has an inverter. The added space and weight requirements of an inverter that would be used only during the installation is usually unwarranted for sites that are inaccessible to motor vehicles. Most power tools can be replaced with hand tools. A variety of direct current power tools are available that can be powered on-site by rechargeable batteries or photovoltaic panels. You should charge all batteries before going into the field.

## Installation – Tools and Materials Checklists

### Basic Tools:

- ☐ pencil
- ☐ volt-ohm meter with spare battery
- ☐ sockets and wrench
- ☐ drill bits
- ☐ hand operated drill
- ☐ screwdriver(s) (1 slotted and 1 #2 Phillips head)
- ☐ tape measure
- ☐ hacksaw blade
- ☐ knife
- ☐ wire cutter
- ☐ wire strippers
- ☐ slip joint pliers
- ☐ torpedo level

### Initial Site Visit Tools:

- ☐ pencil
- ☐ maps
- ☐ inclinometer
- ☐ compass
- ☐ first aid kit
- ☐ paper
- ☐ 50-foot tape measure
- ☐ solar siting device
- ☐ camera
- ☐ personal gear

### Non-motor Transported Installation Tools:

- ☐ pencils
- ☐ rope
- ☐ c-clamps
- ☐ wood chisel
- ☐ drill bits
- ☐ Phillips driver bit
- ☐ drill bit extender
- ☐ expansion bit
- ☐ level
- ☐ prick punch
- ☐ slip joint
- ☐ pliers
- ☐ slotted screwdrivers
- ☐ tape measure
- ☐ string line
- ☐ hole saw
- ☐ utility knife
- ☐ torque wrench
- ☐ collapsible shovel
- ☐ volt-ohm meter
- ☐ wire cutter (8" handle)
- ☐ soldering iron
- ☐ flashlight
- ☐ system operations literature
- ☐ carabiners
- ☐ tool belt
- ☐ drill
- ☐ nut driver bits
- ☐ paddle bits
- ☐ brace and bit
- ☐ uni bit
- ☐ socket set with extender
- ☐ Phillips screwdrivers
- ☐ file
- ☐ adjustable wrench
- ☐ hand saw
- ☐ hack saw
- ☐ caulk gun
- ☐ hammer
- ☐ combination square
- ☐ wire stripper/crimper
- ☐ needle-nose pliers
- ☐ black polyethylene
- ☐ component product literature
- ☐ resealable plastic bags

### Motor Transported Installation Tools:

- ☐ pencils
- ☐ 1000-watt inverter
- ☐ 4-gang outlet with extension cord
- ☐ reciprocating saw with blades
- ☐ full drill bit index
- ☐ jig saw blades
- ☐ chain saw
- ☐ extension ladder
- ☐ shovel
- ☐ pry bar
- ☐ pipe wrenches
- ☐ aviation snips
- ☐ carpenters 6'6" level
- ☐ open end wrenches
- ☐ extension cord
- ☐ circular saw with blades
- ☐ ½" electric drill
- ☐ hole saw bits
- ☐ hole punches
- ☐ step ladder
- ☐ sledge hammer
- ☐ pick
- ☐ nail puller
- ☐ vise grip
- ☐ rasp
- ☐ framing square

Table 13-2

## Sample Installation Materials

This following checklist contains the materials and components needed for installing a 12-volt, direct current, stand-alone PV system with a pole mount, battery storage, and controller.

qty  component/item/material

**Array/Mounting**
2  photovoltaic modules
2  split bolt connectors
8  #8 X ½" bolts, nuts, washers, lock washers
2  photovoltaic panel interconnect cables
1  pole mounting structure
2  ¼" X 3 ½" bolts, nuts, washers, lock washers
4  cable ties
1  2 ½" weatherhead
11'  2 ½" galvanized steel pipe
1  2 ½" x 2 ½" x 2 ½" galvanized tee
1  2 ½" x 90 degree galvanized elbow
2  2 ½" close nipples
2  2 ½" floor flanges
4  ½" X 6" lag bolts and washers
4  wood shims
25'  #10-2 Romex with gd.
3'  #10-uf
2  tubes silicon caulk
1  fused disconnect switch and enclosure
3  30-amp fuses

**Grounding**
15'  #6 bare copper ground wire
1  copper ground rod
1  ground clamp
1  ¼" X 1" bolt, nut, washer, lock washer
1  terminal
1  ½" strain relief connector
1  box Romex staples

**Battery**
1  "No Smoking" sign
1  "Danger" sign
2  batteries
1  prefabricated insulated battery box
2  battery interconnect cables (#2 AWG)
2'  #8 THHN (black)
2'  #8 THHN (red)
2  terminal lugs
4  10-amp fuses
4  ¼" X 1" bolts, nuts, washers, lock washers
6'  1" polyethylene pipe
1  1" tankwall flange
1  1" MPT slip fitting adapter
2  1" slip fitting elbows
1  2" X 2" screen
5  1" hose clamps
2  ¾" bushings

1  ¾" X 2" nipple
4  1" conduit clamps
8  #10 X ¾" screws

**Controller/Load Center**
1  DPST 30-amp safety switch
1  controller with low voltage disconnect, voltmeter, and ammeter
20  #10 spade crimp connectors
1  12" X 12" NEMA Type I electrical enclosure
1  10 ¼" x 10 ¼" enclosure panel
4  #10 X 1½" sheet metal screws with washers
1  ¾" chase nipple with nut and bushing
1  ½" strain relief connector
1  fuse block (6 circuit)
2  boxes 20A glass fuses
2  terminal bus bars (6 circuit)
2  Type T disconnect switches
2  2-gang junction boxes
2  2-gang switch plates
12  #10 X ¾" bolts, nuts, washers, lock washers
6  ½" Romex connectors

**Lighting Load**
7  ½" Romex connectors
50'  #10-2 Romex with gd.
24  #10 X 1½" sheet metal screws with washers
1  round junction box
4  pull chain light fixtures
1  socket adapter #593
3  #1141-21 12V direct current incandescent lights
12  25-watt 12V direct current incandescent bulbs
3  slip-on bulb lampshades
1  swag lamp fixture
1  direct current ballast
2  20-watt fluorescent tubes
1  fluorescent fixture
1  2-junction box
1  Type T switch
1  switchplate cover

**Miscellaneous Wiring Materials**
  wire ties
  solder
  anti-oxidizing compound
  small bottle dish soap or pulling grease
  duct tape
  red and black electrical tape
  assorted wire nuts
  assorted nails and screws
  assorted electrical fittings and screws

# Chapter 14
# System Installation

**Contents:**

14.1   Site Evaluation . . . . . . . . . . . . . . . . . . . . . . . . . . . . . . . . . . . .158

14.2   Photovoltaic Array Installation . . . . . . . . . . . . . . . . . . . . . . . . .158

14.3   Battery Installation . . . . . . . . . . . . . . . . . . . . . . . . . . . . . . . . . .158

14.4   Controller & Inverter Installation . . . . . . . . . . . . . . . . . . . . . .159

14.5   Photovoltaic System Wire Installation . . . . . . . . . . . . . . . . . .160

14.6   PV System Installations Final Checklist . . . . . . . . . . . . . . . . . .162

# 14.1 Site Evaluation

Designers or installers of photovoltaic systems encounter unique challenges in planning the logistics of an installation. Since each site has specific characteristics, you should visit the site prior to the installation. During this initial site visit, you should list all tools and materials that will be required to install the system.

You should bring paper and pencil to record all observations and measurements. A 50-foot tape measure is adequate for most measuring tasks. An inclinometer is useful for leveling and quickly measuring sloped surfaces. A solar siting device is very helpful for assessing site shading. A lightweight camera and compass are also invaluable.

Initial site visits must be thorough enough to provide accurate information needed to completely design and plan the installation. Anything less will result in extra trips, added time, improper equipment choices, installation delays, and less profit.

# 14.2 Photovoltaic Array Installation

The major aspects of installing the photovoltaic array are choosing the most applicable mounting systems and making a proper installation. Once you have installed the array, you should verify that the array is functioning as expected by measuring the output and comparing this figure to the manufacturer's specifications.

## Mounting System Considerations

The first step in completing a safe system installation is carefully selecting the photovoltaic array's location. Electrical equipment should be protected from unnecessary environmental exposure and mounted to facilitate convenient, regular system maintenance. The photovoltaic array should be located as near as possible to the power conditioning equipment to minimize power loss from long wire runs.

Photovoltaic modules are expensive, lightweight and compact, making them vulnerable to theft. Protection systems can be installed to improve the security of photovoltaic arrays. Using specialized screws with unique heads to mount the panels can prevent speedy removal. Padlocking the interconnecting mounting channel to the permanently mounted support frames increases security while allowing access with a key. Commercial padlocking hardware for support structures is also available.

Photovoltaic module support structures should provide a simple, strong, and durable mounting system. Most commercially available photovoltaic modules are manufactured with extruded aluminum frames. These frames are strong, durable, corrosion resistant, and provide adequate support for the module to be incorporated into an array.

Weather-resistant, corrosion-free materials should be used when fabricating a photovoltaic array mounting system. Anodized extruded aluminum, galvanized steel, and stainless steel are the optimal choices. The support structures need to be lightweight so they can be easily transported and installed.

## Measuring Array Output

When the entire installation is complete, you should measure the array output before connecting to the load (battery). You can measure the intensity of the sunlight and cell temperature, then decide if array performance is operating as expected from manufacturer's specification.

> **Note:** Measuring open circuit voltage and short circuit current can be dangerous on arrays. For information about safety procedures and equipment refer to Chapters 5 and 16.

# 14.3 Battery Installation

Remember that batteries are heavy and prone to leakage. Consequently, some carriers will not ship liquid batteries. Even certain sealed batteries contain some liquid called reserve electrolyte that can spill. Upon arrival check the batteries for any damage that may have occurred during shipping.

> **Caution! Remember to think "safety first" when working with batteries. See Chapter 6 for battery safety rules.**

> **Note:** Batteries must be protected from theft, children, temperature extremes, corrosion, and accidental short circuit from falling objects. They also must be protected from open flames and sparks that can cause explosion.

You should protect batteries at all times, including during transport and while boost charging before departure for the site. A boost charge is needed if the batteries are not fully charged; you should charge the batteries before leaving the shop. When boost charging with an AC powered battery charger, make sure you do not exceed the charge termination voltage or charge at a faster rate than the manufacturer specifies. When charging batteries, keep them away from open flames and sparks and open the vent caps to allow explosive hydrogen gas to escape. Keep batteries away from children, unauthorized personnel, dust, and oil. When transporting batteries, pack them to avoid spillage and short circuits. You should cover both battery terminals with an electrical insulator such as wood or insulated connectors.

## Building a Battery Enclosure

The battery bank must be safely located to prevent accidents, yet provide for periodic maintenance. Batteries are often housed in a ventilated battery box that is corrosion resistant and sometimes insulated. Smaller systems can use an insulated, plastic food cooler for a battery box. Larger battery banks can be set on racks that are corrosion resistant, extremely stable, and provide access for maintenance. When building a battery enclosure, you should use the following guidelines:

- Protect batteries from items that can accidentally fall from above and cause a dangerous short circuit.

- Protect batteries from freezing temperatures by insulating the enclosure or locating the enclosure in a heated area.

- Vent to the battery enclosure to provide free airflow and prevent the buildup of explosive hydrogen gas. If vents are used, they must be located high enough to properly vent hydrogen gas, which is lighter than air. Vents should be directed outdoors and screened to prevent insects and animals from blocking them. Note that with free airflow the battery temperature will be about the same as the average ambient air temperature.

- Build the enclosure, box, or compartment so that it can be locked, yet is easily accessible for maintenance. Size the access to the enclosure to allow for easy removal and replacement of

batteries. When placed in the enclosure, the batteries or cells should have air space between them.

- Store the maintenance equipment and manufacturer's information in re-sealable plastic bag or container inside the enclosure.

- Build a strong level floor.

- Build the enclosure with material that will not be damaged by the corrosion from the electrolyte. Use wood, plastic, or painted metal.

When installing batteries, keep in mind the following installation guidelines:

- Use the connector bolt torque specified by the battery manufacturer.

- Adequately size all battery wiring and fuses.

- For wire entries, use connectors that protect the wire from tension and damage.

- Place fuses on the positive wires leaving the battery box.

- Use stainless steel nuts, bolts, and washers on battery terminals.

- Protect all battery terminals and terminal connections from corrosion by coating them with battery terminal coating, petroleum jelly, oxidation protection material, or high temperature grease.

- Provide a one-quarter inch space between batteries.

# 14.4 Controller and Inverter Installation

Photovoltaic charge controllers are intended for specific solar applications, and they should not be used for other system regulation unless specified by the manufacturer. You should read and follow the exact installation procedure specified in the manufacturer's instructions. When installing controllers and inverters, you should use the following guidelines:

- Protect controller hardware from excess dust, dirt, overheating, and rough handling. Remember that electrical equipment is sensitive and requires careful handling. Some controllers require ventilation to prevent components from overheating. Use common sense in locating and mounting controller units; install all system components away from potential hazards, such as overhanging tree limbs, heat sources, snowdrift, and debris build-up areas.

- Use correct wire size and proper terminal fasteners.

- Use overcurrent protection and disconnects between the array to the controller and the controller to the battery. Fuses or circuit breakers protect the controller conductor from too much current in the event of a short circuit. Too much current will overheat the wiring and may cause a fire. Disconnect switching is used to open the circuit between the array, controller, and battery. Disconnect switching also turns off the load. This switching is necessary for servicing the system and in case of emergency. See Chapter 9.

# 14.5 Photovoltaic System Wire Installation

When wiring a PV system, it is very important to use the correct electrical boxes, wiring connections, and switches for each specific application. This section describes the various electrical components and their applications.

## Electrical Boxes

All wiring connections must be made within an accessible electrical box. The box must be securely fastened in place and have a removable cover. Electrical boxes can be either surface mounted or recessed into a wall, ceiling, or floor. Surface mounted boxes, called handy boxes or utility boxes, have rounded corners. Recessed boxes come with a variety of means for fastening to interior framing or surface coverings, such as wallboard and paneling.

Electrical boxes that are exposed to weather must be weather resistant and have weather resistant connectors. Electrical boxes must be an adequate size for the number and size of wires contained within the box.

Electrical boxes used only for making wire connections are called a junction box. They are usually rectangular or octagonal in shape and covered with a blank cover plate, making them accessible when construction is complete.

## Wiring Connections

Electrical connectors perform one of the following connecting functions:

**Wire to wire:** Connections between wire and wire are usually made with either a wire nut, split bolt, terminal bar, reducer or crimped connector. Wire nuts are sized and color-coded. Crimp-type connectors must be crimped using a special tool recommended by the manufacturer and are best used with stranded wire.

**Wire to terminal:** Connections between a wire and terminal are usually made with a ring or spade type connector. These connectors are commonly used when connections are frequently removed and reconnected or where large wire would be difficult to connect.

**Wire, cable, cord, or conduit to electrical box:** Connections between electrical boxes and wires, cables, cords, or conduit must be secure enough to prevent wires from pulling loose. When exposed to weather, these connections must thread into the box and make a watertight seal. A "drip loop" should be used for outdoor wire connections to electrical boxes to prevent water from running down a wire and into the box. Outdoor wire connections are made only from the underside of an electrical box.

## Common Electrical Connectors

**Wire nuts:** Connectors used to join two or more wires. Twist the wires together and screw wire nut on until tight.

**Ring terminals:** Crimp-on terminals that maintain a connection even if the screw loosens.

**Spade terminals:** Crimp-on connectors for use in non-vibrating applications. They allow for quick disconnection.

**Screw lugs:** Bolt-on connectors used to join large or multiple wires to one terminal.

**Flag terminals:** Connectors used for securing wires where no electrical connection is made.

**Cable connectors or Romex connectors:** Connectors used for joining non-metallic sheathed cable to a box.

**Conduit connectors:** Connectors used for attaching conduit to a box for interior or dry applications.

**Conduit couplings:** Connectors used for attaching

lengths of conduit for interior or dry applications.

**Compression couplings:** Connectors used for making watertight connections between lengths of conduit.

**Armored cable connectors:** Connectors used for fastening flexible armored cable to a box.

**Strain relief connectors:** Connectors used with round cord, such as SO, SJ, or TC, where resistance to pulling or weather is necessary.

## Disconnects

Disconnect switches are required for safe direct current system operation. The *National Electrical Code®* requires a means for disconnecting each voltage source. Photovoltaic panels and batteries are voltage sources that need a disconnect switch. Safety switches or circuit breakers are the most appropriate switches to use for disconnecting photovoltaic panels, batteries, and generators. A safety switch is fused, thus providing overcurrent protection between major photovoltaic system components. Circuit breaker function as both a switch and over-current protection.

Disconnects must be rated for a given voltage and for the amount and type of current that will flow through them. Type "T" switches are rated for direct current use. They are snap-action compatible with direct current characteristics. A switch rated for alternating current does not have an adequate "interrupt rating" for direct current. Using alternating current rated switches or "quiet switches" for direct current systems will result in a shortened switch life because the contacts burn out from repeated electrical arcing. Under higher current conditions, the switch contacts may become permanently fused, rendering the switch inoperable.

## Receptacles

Receptacles for direct current wiring are not the same as those commonly used for alternating current wiring. The correct receptacle should be used to prevent damage to appliances, eliminate fire danger and reduce shock hazards. The *National Electrical Code®* requires direct current receptacles to be a twist lock type. More common "bayonet" or "cigarette lighter" type receptacles are unsafe because children can easily insert their fingers or other objects into the outlets. These receptacles should be avoided.

## Wiring Installation Checklist

You should ask the following questions when installing a system to ensure a safe, correct wiring installation. You should be able to answer 'yes' to each question.

- Is the ampacity of the wire adequate for the total of all loads in each circuit plus a 125% safety margin?
- Does the voltage drop not exceed 2% in any branch circuit or 5% from the source to the load?
- Does the overcurrent protection (including the NEC® safety margin) not exceed the wire's ampacity?
- Are the wires properly coded?
- Are the types of wires, cables, cords, and conduits correct for each application?
- Are all conduits the correct size for the number and type of wires that they contain?
- Are all electrical boxes adequately rated, sized, covered, and accessible?
- Do all electrical boxes that are subject to moisture include a "drip loop"?
- Are all electrical connections accessible?
- Are all electrical connections protected from moisture if necessary?
- Are all switches rated for the voltage and current they will switch?
- Are the correct direct current receptacles used?
- Are all receptacles clearly labeled with their correct voltage and current?
- Are all equipment grounds made with green or bare wire or metal conduit?
- Are all ungrounded conductors switched, fused, or interrupted in some way?

**Note:** Grounded conductors should *never* be switched, fused, or interrupted in any way.

- Are all equipment grounds and ground conductors grounded at only one point in the system?
- Is the conduit supported appropriately?
- Is the cable fastened appropriately?

# 14.6 PV System Installations Final Checklist

This section contains a system installation checklist that can be used as a final check for a newly installed system or as a maintenance assessment for an existing system. For additional reference material on system installation checklists, refer to NABCEP's "PV Installers Task Analysis" (available on their website: www.nabcep.org).

---

### Before Testing the System

☐ Use proper safety procedures when working with electricity. See Chapter 16 for safety guidelines and appropriate equipment.

☐ Verify that all disconnects are locked in the open position with a warning label. (This insures that power can not travel further down the line until properly tested, and warns others that there may be live conductors in the box.)

---

### PV Array

☐ Make sure all modules are attached securely to their mounting brackets.

☐ Visually inspect the array for cracked modules, damaged junction boxes, and loose wires.

☐ Visually inspect that all module (quick-connects) are tight.

☐ Open each combiner box and test open circuit voltage on each series string to verify correct voltage and polarity. Recheck torque on all DC terminals.

☐ Before powering up the system, at final array breakers, repeat open circuit voltage tests to verify correct voltage and polarity.

☐ Verify modules are wired so that they can be removed without interrupting the grounded conductor.

☐ Check for labels on the modules. *NEC® 2005,* Article 690.51: "Modules shall be marked with identification of terminals or leads as to polarity, maximum overcurrent device rating for protection, and with rated 1) open-circuit voltage, 2) operating voltage, 3) maximum permissible system voltage, 4) operating current, 5) short-circuit current, and 6) maximum power." See *NEC® 2005,* Article 690.52 for AC module requirements.

---

### Wiring

☐ Check exposed array wiring for correct rating and sunlight resistant insulation.

☐ Check that all wiring and conduit is appropriately rated, neat, and well supported.

☐ Check that strain reliefs/cable clamps are properly installed on all cables and cords by pulling on cables to verify (*NEC® 2005,* Article 300.4, and Article 400.10).

☐ Make sure that all grounded conductors are white and equipment grounding conductors are green or bare (*NEC® 2005,* Article 200.6(A)).

☐ Verify that the conductor rating of the PV circuit is at least 156% of the rated short circuit current (125% X 125% = 156%).

☐ Verify that all junction boxes are accessible.

---

### Overcurrent Protection

☐ Verify that the overcurrent device rating of the PV circuit is at least 156% of the rated short circuit current (125% X 125% = 156%).

☐ Make sure DC voltage and current ratings are clearly marked on overcurrent protection.

## Charge Controllers

☐ Torque all terminations again.

☐ Check that all voltage settings are properly set for the appropriate battery type and proper voltage.

☐ If the system is connected to a utility-interactive inverter, make sure that the settings of the charge controller(s) do not interfere with the proper operation and dispatch of the inverter system.

☐ Verify that charge controller operation matches the programmed settings by forcing the system to the set points and making sure that the unit performs the proper control function. You should test the following points:

– Low voltage disconnect (LVD)

– Low voltage reconnect (LVR)

– High voltage disconnect (HVD)

– High voltage reconnect (HVR)

## Disconnects

☐ Verify that the disconnects are still locked open and the warning label is still intact.

☐ Verify that there are means to disconnect and isolate all pieces of equipment in the system.

☐ If fuses are used, verify means to disconnect the power from both ends.

☐ Ensure switches are accessible and clearly labeled.

☐ Check the continuity of fuses and circuit breakers with power off.

☐ Check voltage drop across switches while operating.

## Batteries

☐ See Chapter 6 for battery safety.

☐ Store safety gear near-by (eye protection, rubber gloves, baking soda and distilled water).

☐ Retorque all battery connections.

☐ Coat each terminal with anticorrosive gel.

☐ Make sure that access to terminals is limited (*NEC® 2005,* Article 690.71(B)).

☐ Make sure that location provides adequate natural ventilation. Well-vented areas include garages, basements, and outbuildings, but not living areas.

☐ If battery contains flooded cells, top off cells with distilled water according to the manufacturers instructions.

☐ If battery contains flooded cells, be sure an eyewash station is accessible.

☐ Once inverter is operational, "equalize charge" the battery to ensure that the battery is properly connected and functioning correctly.

☐ Ideally, run the battery through a few heavy charge-discharge cycles to exercise the battery.

☐ Check individual cell or battery voltages after equalization.

☐ Check the specific gravity of all questionable cells with a hydrometer.

## Inverters in Grid-tied Systems

☐ While disconnects are open, retorque all electrical terminal connections on the inverter to tighten any connections that may have loosened since the initial installation.

☐ Verify in the inverter manual that the array open circuit voltage, under the record lowest temperature, is acceptable to the inverter.

☐ Check utility line voltage to verify that it is within the proper tolerances for inverter. If line voltage is above 124 volts AC before starting inverter, verify that the maximum voltage drop for the inverter output circuit is less than two volts.

☐ If the inverter measures and reports utility or inverter AC voltage on a display, verify that this voltage agrees with a measurement from a high quality, true-RMS AC voltmeter.

☐ For non-battery-based inverters, once inverter has started and is operational, check that the maximum power point tracking (MPPT) circuit is operating. This should be done during clear sky conditions if possible by monitoring array voltage from the open circuit condition until it reaches a point where system power peaks and then starts to drop again. Keep monitoring voltage until you note that the system voltage has been adjusted up and down several times.

☐ Verify that the operating voltage is near the expected peak power voltage for the conditions of the test, this can be found in most manufacturers' literature.

☐ Properly connect the temperature compensation probe to control battery voltage.

☐ Follow inverter-starting procedure from the manufacturer's manual.

☐ Instruct the homeowner on what to do in the event of an inverter failure and provide them with an initial start-up test report.

## Inverters in Battery-based Systems

☐ While disconnects are open, retorque all electrical terminal connections on the inverter to tighten any connections that may have loosened since the initial installation.

☐ For battery-based inverters, use the programming features of the inverter to charge the battery and then connect the battery to the DC source to ensure that these functions are operating properly.

☐ Follow inverter-starting procedure from the manufacturer's manual.

☐ Instruct the homeowner on what to do in the event of an inverter failure and provide them with an initial start-up test report.

## Grounding

☐ Verify that only one connection in the DC circuits and one connection in the AC circuits (grounded conductor to grounding conductor) is being used for system grounding referenced to the same point (*NEC® 2005,* Article 250.21).

☐ Check to see that equipment grounding conductors and system grounding conductors have as short a distance as possible to ground.

☐ Check that non-current carrying metal parts are grounded properly (array frames, racks, metal boxes, etc).

☐ Incorporate ground fault protection on systems required by the *NEC®*.

**Note:** Terminal lugs bolted on an enclosure's finished surface may be insulated because paint/finish at point of contact has not been properly removed.

☐ Check resistance of grounding system to earth ground. *NEC®* allows 25 ohms or less.

☐ Verify that the equipment grounding conductor is a green or bare wire and is properly sized.

## Safety Signs

☐ Label any fuse or circuit breaker that can be energized in either direction (*NEC® 2005,* Article 690.17).

☐ Post an "Interactive Point of Connection" sign for interactive PV systems (*NEC® 2005,* Article 690.54).

☐ Place a sign at the equipment service-entrance that states the type and location of on-site optional standby power sources (*NEC® 2005,* Article 702.8).

☐ Post a "No Smoking" sign near the batteries.

☐ Place a sign at the point of PV system disconnect listing: operating current, operating voltage, maximum system voltage and short-circuit current (*NEC® 2005,* Article 690.53).

☐ Provide any additional documentation that would be helpful to the homeowner, inspector, or fire officials.

# Chapter 15
# Maintenance and Troubleshooting

## Contents:

15.1 Materials and Tools List . . . . . . . . . . . . . . . . . . . . . . . . . .168

15.2 Maintaining PV Components . . . . . . . . . . . . . . . . . . . . . . .168

15.3 Troubleshooting Common System Faults . . . . . . . . . . . . . . . .170

15.4 Troubleshooting Wiring Problems Using a Multimeter . . . . . . . . . .170

15.5 Troubleshooting Specific Problems . . . . . . . . . . . . . . . . . . . .173

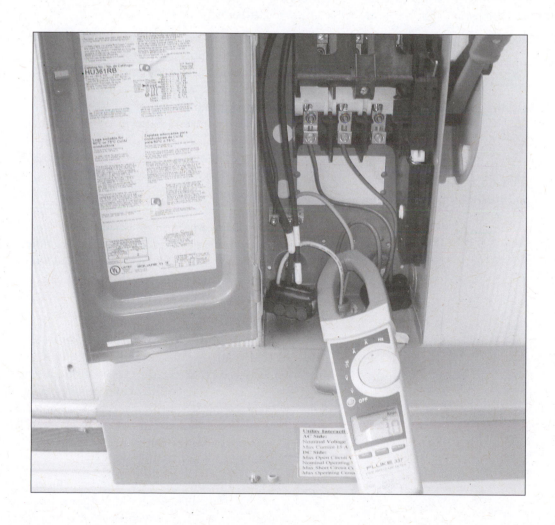

# 15.1 Materials and Tools List

You should bring the following materials and tools on any maintenance trip:

- first aid kit
- soldering iron and solder
- torque wrench
- paper
- pencil
- rags
- screwdrivers (1 Phillips and 2 slotted)
- hydrometers (2)
- safety goggles, rubber gloves and rubber apron
- baking soda (vinegar if you are working with alkaline batteries)
- distilled water
- volt-ohm meter (2)
- adjustable power supply
- anti-oxidizing compound
- spare fuses, batteries, wire nuts, wire
- wire strippers
- lineman's pliers
- manufacturer's literature and troubleshooting guides
- personal gear with pack

# 15.2 Maintaining PV Components

Although photovoltaic power systems require little maintenance compared to other power systems, you should periodically perform a few simple maintenance tasks.

## Photovoltaic Array

The PV array needs very little maintenance. If the system is in a dusty climate with little rain, the array may need to be cleaned off periodically. Clean the modules with water and mild detergent. Avoid solvents or strong detergents.

The junction boxes should be checked periodically for weather protection. And if there are any questions about solar access, check for shading problems caused by new plant growth.

## Batteries

If you are responsible for producing power, then you're responsible for maintenance. Battery maintenance depends largely on battery type, though all batteries require periodic inspections to verify system operation.

To estimate a battery's state of charge, you can measure battery voltage measured with a multimeter. (Refer to Section 15.5 "Troubleshooting Wiring Problems Using a Multimeter") Operating a load for several minutes will stabilize the battery voltage and remove any inaccurate surface charges. Do not measure voltage when the battery is charging or discharging. Disconnect the battery from both the array and load before taking a voltage reading. Table 6-4 in Chapter 6 compares general state of charge to voltage readings. More precise values may be obtained from the battery manufacturer's data.

**Nicad and sealed liquid electrolyte (VRLA) batteries:** Nicads and sealed liquid electrolyte batteries require the least amount of annual maintenance. Terminal connections, casing, venting and wiring should be checked semi-annually. Even so-called "maintenance-free" batteries require inspections of the case, terminal connections, wiring, voltage and any venting strategies.

**Vented liquid lead-acid batteries:** The deep-cycling, liquid electrolyte lead-acid batteries, such as those used in electric vehicles, require the most maintenance. These batteries have a higher amount of gassing than other batteries and require the addition of distilled water. Furthermore, they are prone to acid stratification at the bottom of a cell when continually under-charged. A slight amount of overcharge will assist in de-stratification. This excess charging can also "equalize" all cells in series and/or parallel strings. Equalization is the process of restoring all cells in a battery to an equal state of charge, 100 percent for lead-acid batteries. You can reduce maintenance requirements by using recombinators or catalytic converter battery vent caps that capture hydrogen gas and recombine it with oxygen to form water. This water is automatically returned to the electrolyte.

**Determining state-of charge by measuring specific gravity:** Liquid electrolyte, non-sealed lead-acid batteries can be precisely tested for state of charge by measuring the electrolyte's specific gravity with a **hydrometer**. A hydrometer is a bulb-type syringe that will draw electrolyte from the cell. A glass float in the hydrometer barrel is calibrated to read in terms of specific gravity. The lower the float sinks in the electrolyte, the lower its specific gravity and state of charge. Figure 6-4 in Chapter 6 lists the correlation between specific gravity and state of charge.

> **Note:** Never take a hydrometer reading immediately after water is added to the cell. The water must be thoroughly mixed with the underlying electrolyte by charging the battery before hydrometer readings are reliable.

1. Wear safety glasses, rubber gloves, and a rubber apron. Have some baking soda handy to neutralize any acid spillage and have fresh water available for flushing purposes. If you get acid into your eyes, flush with water immediately for at least 10 minutes and obtain professional medical assistance.

2. Draw the electrolyte in and out of the hydrometer three times to bring the temperature of the hydrometer float and barrel to that of the electrolyte in the cell. Hold the barrel vertically so the float does not rub against the side of it.

3. Draw an amount of electrolyte into the barrel. With the bulb fully expanded, the float should be lifted free, not touching the side, top, or bottom stopper of the barrel.

4. Read the hydrometer with your eye level with the surface of the liquid in the hydrometer barrel. Disregard the curvature of the liquid where the surface rises against the float stem and the barrel due to surface tension.
   **Note:** Keep the float clean. Make certain the hydrometer is not cracked.

5. Adjust the reading for the temperature of the electrolyte. Use the thermometer and directions for temperature compensation that come with the hydrometer.

Hydrometer floats are calibrated to give a true reading at one fixed temperature only, commonly 80°F. A correction factor must be used if the temperature does not match the specific temperature for the thermometer. At increased temperatures, the electrolyte expands and becomes less dense; thus, the float will sink lower in the less dense solutions, resulting in a lower specific gravity reading. Conversely, at lower temperatures the electrolyte shrinks and becomes denser; thus, the float will not sink as deep, resulting in a higher specific gravity.

> **Note:** Remember that some new batteries do not give a full specific gravity reading until they have undergone numerous cycles.

**Adding water:** The only water to use when preparing electrolyte is distilled water. This is also true for routine water additions to the battery. Be sure not to use mineral water or spring water. **USE DISTILLED WATER ONLY!** Avoid using metallic containers. Metal impurities in the water will lower the performance of the battery.

# 15.3 Troubleshooting Common System Faults

The best method for avoiding system failures is to initially install a high quality, properly designed system. Regular maintenance is the second line of defense against failures. The first step in troubleshooting photovoltaic power system problems is to save all of the manufacturer's product literature that comes with each component. This literature should be kept in a handy location that is protected from weather, chemicals, and rodents.

The most common system failures are usually the simplest to fix. You should check the system for fundamental problems first to save a great deal of time. The most common system failures are blown fuses, tripped breakers, or bad connections. The other common problem is a low or empty battery bank. The following general troubleshooting checklist for system operation can be completed visually, with the possible exception of the last two items.

- Has the weather been cloudy for several days?
- Check the array for partial shading or dirt.
- Check all fuses and circuit breakers.
- Check system wiring for loose connections and/or corrosion.
- Check modules and batteries for proper series-parallel configuration.
- Check system wiring for proper polarity.
- Check system for proper system voltage and current.

# 15.4 Troubleshooting Wiring Problems Using a Multimeter

The volt-ohm-milliamp (VOM) meter is essential for troubleshooting wiring problems. You should be familiar with this meter's proper operation to insure your personal safety and protect the meter and system equipment. To acquaint yourself with a meter, refer to the operator's manual for proper use of the meter.

The most useful tasks performed with a VOM meter include:

- Checking for continuity.
- Measuring AC and DC voltage.
- Measuring AC and DC current.
- Checking the polarity of DC voltage.

## General Safety Precautions

These precautions are reminders of specific hazards that should be avoided when using a VOM meter. Always refer to the equipment manual and heed the manufacturer's specific warnings and instructions.

- Always wear safety glasses when working with electrical circuitry.
- Do not handle the instrument, its test leads, or the circuitry while high voltage is being applied.
- Operate the VOM meter only if you are qualified to recognize shock hazards and trained in the safety precautions required to avoid possible injury.
- Do not work alone on electrical circuits. Make certain that someone capable of rendering medical aid is nearby.
- Locate all voltage sources and accessible current paths before making connections to circuitry.
- Turn off all sources of power and discharge any capacitors in the circuit to be measured before connecting to or disconnecting from the circuit.
- Dry your hands, shoes, floor, and workbench before working with electricity.

- Do not change switch settings or test lead connections while the circuit is energized. This could result in damage to the instrument and possible personal injury.

- Check and double check that disconnect switches are open, and verify that the meter leads are in the correct position before applying power to the instrument.

- Make certain that the equipment you are working with is properly grounded and fuses are of the proper type and rating.

- Whenever measuring a current or voltage of unknown magnitude, begin measurements at the highest scale available. Proceed to a lower scale when you are satisfied that the value is within the limits of a lower scale.

## Checking for Continuity

Checking for continuity indicates whether a circuit is open or closed, which is useful when checking for broken wires, short circuits, fuse, or switch operation. Checking for continuity involves circuit resistance. Short circuits have very low resistance. Closed circuits have some resistance depending upon circuit wire and loads. Open circuits exhibit infinite resistance. Use the following procedure for checking continuity:

> **Caution! Resistance and/or continuity measurements should be made with the power off. Review the safety precautions in "Safety and PV installation," Chapter 16.**

**To check the continuity of a circuit:**

1. Turn the power off and discharge all capacitors.

2. Disconnect at least one conductor in the circuit.

3. Choose the Rx100 resistance scale or another resistance scale if more appropriate for your application.

4. Plug the black test lead into the common (-) jack. Plug the red test lead into the positive (+) jack $\Omega$.

5. Connect the ends of the test leads together to short the VOM resistance circuit.

6. Turn the zero ohms control until the needle indicates zero ohms.

7. Disconnect the test leads. You are now ready to check for continuity.

8. Connect one test lead to the disconnected point on the circuit you want to test and the other lead at the opposite end of the circuit.

An open circuit has no continuity and will read infinite resistance. The pointer will not move. A closed circuit has continuity and will read little or no resistance. The needle will move to the right hand side of the scale.

## Measuring Voltage

Measuring voltage is similar to measuring for continuity, with the exception that you are measuring the energy potential in the circuit; the voltage. The following steps are a procedure to measure the voltage in a circuit.

**To measure the voltage of a circuit:**

1. Review Chapter 16 for appropriate personal protective equipment and safety precautions to use while testing voltage.

2. Select the proper type of voltage being measured, alternating current or direct current. To help in measuring direct current voltage some meters have a direct current positive (+) and direct current negative (-) position. When in the direct current positive (+) position, the meter will read a positive value if the wiring is correct and the meter leads are correctly connected. If the meter leads are inadvertently placed on the wrong wires, the polarity will read a negative value and will alert the installer of incorrect wiring or improper meter lead placement.

3. Set the range indicator to the appropriate scale. If the voltage scale is unknown, start at the highest scale and work your way down to prevent meter damage or personal injury

4. Plug the black test lead into common (-) jack. Plug the red test lead into the positive (+) jack (voltage).

5. Turn the power off and discharge any capacitors in the circuit.

6. For direct current circuits, connect the black test lead to the negative side of the circuit. Connect the red lead to the positive side.

7. For alternating current circuits, connect the black test lead to the common or neutral side of the circuit. Connect the red lead to the hot side of the circuit.

8. Turn the power on.

9. Read the voltage on the proper scale. If the needle deflects or moves backwards, the polarity of the wiring or the meter may be reversed.

## Measuring Current

Measuring current is similar to measuring for voltage, with the exception that you are measuring the energy passing through the circuit; the current. The following steps are a procedure to measure the current in a circuit.

**To measure the current of a circuit:**

1. Review Chapter 16 for appropriate personal protective equipment and safety precautions to use while testing current.

2. On the VOM meter, select the type of current being measured, alternating current or direct current. Direct current positive (+) position will indicate proper polarity when red is connected to positive (+) side of the circuit and black is connected to negative (-) side of the circuit. Polarity may be reversed by switching to the direct current negative (-) position.

3. Set the range indicator to the appropriate scale. If the current scale is unknown, start at the highest scale.

4. Plug the black test lead into the common (-) jack. Plug the red test lead into the positive (+) jack (amperage).

5. Turn the power off and discharge all the capacitors in the circuit.

6. Open the ground side of the circuit where the current is being measured.

7. Connect the meter in series (the circuit must be broken then the meter inserted in line with the circuit).

> **Caution! Never connect the meter across a voltage source. Doing so can result in damage to your meter or the device being tested. This does not apply when testing a module, because it is current limited.**

8. Turn power on and read the current on the proper scale.

**Note:** Some technicians like to use clamp on ammeters. They easily measure current without needing to open the circuit. The meter simply clamps on a single conductor and reads current *via inductance.*

## Checking Polarity

Polarity of a circuit refers only to direct current circuits. Alternating current circuits don't have polarity per se, because polarity in an alternating current circuit is reversed sixty times per second.

It is important to check the polarity of a circuit before energizing a system (while the circuit is in the open position). When polarity in a direct current circuit is reversed, direct current motors will run backwards and often overheat. Some direct current appliances simply will not work at all. Others will be destroyed by reverse polarity.

Follow the steps for measuring the voltage of a direct current circuit:

1. Verify disconnect is in the open position.

2. On the VOM meter, select the type of voltage being measured as DC.

3. Connect the test leads with red to the positive (+) side of the circuit and black to negative (-) side.

If the pointer on the meter reads a positive (+) value, the polarity is correct. Polarity is reversed if the pointer deflects below zero on the analog meter scale or a negative value appears on a digital meter.

# 15.5 Troubleshooting Specific Problems

This section contains troubleshooting information for common issues in photovoltaic power systems. The remedies listed may not apply to the equipment supplied with your system. You should familiarize yourself with the manufacturer's specifications for your equipment prior to using this guide.

| Symptom: Load is inoperative | |
| --- | --- |
| **Possible Cause:** | **Remedy:** |
| • Fuse is blown or circuit breaker tripped. | • Investigate for possible short circuit, overload or excessive surge. Replace fuse or reset circuit breaker. |
| • Circuit is open due to break in wire or loose connection. | • Check circuit continuity. Check continuity through load. |
| • Appliance has overheat protection. | • Wait for appliance to cool. Reset thermal protection. |
| • Inverter surge capacity is inadequate. | • Install larger inverter or reduce load. |

| Symptom: Battery is undercharged | |
| --- | --- |
| **Possible Cause:** | **Remedy:** |
| • Period of cloudy weather has not charged batteries. | • Increase system autonomy or reduce electrical energy consumption. |
| • Actual energy consumption has exceeded the estimated load. | • Reduce electrical energy consumption or re-evaluate load and increase system output accordingly. |
| • Battery fluid is low. | • Check fluid level of each cell. Fill with distilled water as needed. |
| • Specific gravity of battery cells is not within 1.1 to 1.4. | • Perform load test. Replace battery if old. |
| • Battery capacity and ability to accept charge has been reduced by age or abuse. | • Replace battery bank. |
| • Excessive voltage drop to battery caused by high current, small wire, and/or long wire runs. | • Check and/or calculate possible excessive voltage drop. |
| • Batteries are too cold and require higher voltage to achieve full charge. | • Insulate battery box or replace controller with a unit with temperature compensation. |
| • If the controller has temperature compensation, damage to the sensor or the sensor wire will cause undercharging. | • Inspect for temperature sensor or wire damage and repair. |
| • Controller not allowed full charge current with charging light on. | • Defective control. Return unit for repair or check setting of high voltage disconnect. |

## Symptom: Battery is overcharging / has excessive water loss

| Possible Cause: | Remedy: |
|---|---|
| • Controller is not receiving proper battery voltage sensing. | • For controllers with temperature compensation, check connection. Adjust high voltage disconnects. |
| • If the controller has temperature compensation, damage to the sensor or the wire will cause overcharging. | • Inspect for temperature sensor or wire damage and repair. |
| • Batteries are too hot. Gassing voltage is lower than normal. | • Insulate battery bank or replace controller with a unit with temperature compensation. |
| • Controller always allows full charge causing batteries to reach too high voltage. | • Check voltage at battery terminals to see when regulator switches. Compare with specs and adjust. |

## Symptom: Relays are buzzing

| Possible Cause: | Remedy: |
|---|---|
| • Incorrect battery voltage. | • Check series-parallel wiring of batteries. Check voltage of controllers for proper match. |
| • Improper battery connection. | • Check for proper connections. |
| • Broken wire(s) from battery. | • Check wiring. |
| • Relay contacts are obstructed. | • Clean contacts. |
| • Batteries are dead. | • Measure battery state-of-charge. If voltage is very low, connect array directly to batteries until charged. Then reconnect controller. |

## Symptom: Erratic operation of controller

| Possible Cause: | Remedy: |
|---|---|
| • First day of operation or the array was disconnected that day. | • Check operation next day. Proper cycle is reinitiated for controllers with timer circuitry. Disconnect array, wait 10 seconds, then reconnect to reset cycle. |
| • Loads, such as inverters, can generate electronic noise. | • Wire inverters directly to battery. Add filtering to loads. |
| • Battery is defective and may be deteriorating, which results in unusual voltage swings. | • Replace battery. |

## Symptom: Loads are disconnecting improperly

| Possible Cause: | Remedy: |
| --- | --- |
| • Controller not receiving proper battery voltage. | • Check connection at battery voltage and temperature compensation sensor terminals. |
| • Inverters can cause this problem. | • Wire inverters directly to battery. Add filtering to load. |
| • Load has high surge. | • Check load specification for battery voltage to drop surge rating. Use larger wire or shorter wire run. Consider larger battery or generator. |
| • Lightning strike. | • Check voltage. |
| • Faulty controller. | • Measure when load control switches, compare with spec sheet. Return unit for repair if incorrect. |
| • Adjustable low voltage disconnect is set too high. | • Reset adjustable low voltage disconnect using variable power supply. |

## Symptom: Loads not disconnecting

| Possible Cause: | Remedy: |
| --- | --- |
| • Load control disabled by switch. | • Check position of manual load switch. |
| • Unit not equipped with low voltage disconnect. | • Controller does not have load control. |
| • Lightning strike or other high voltage source damaged controller. | • Check voltage when load control switches, compare with specifications. Return unit for repair if incorrect. |

## Symptom: Array fuse blows

| Possible Cause: | Remedy: |
| --- | --- |
| • Array short circuit test performed with battery connected. | • Disconnect battery from controller to perform test. |
| • Array exceeds rating of controller. | • Add another controller in parallel if appropriate or replace with controller of greater capacity. |

## Symptom: Load fuse / circuit breaker blows

| Possible Cause: | Remedy: |
| --- | --- |
| • Load exceeds rating of controller, fusing, or circuit breaker. | • Check surge rating on controller. Check for shorts in the load circuit. Check for maximum load amps exceeding over-current protection. |
| • Surge current of load exceeds fuse rating. | • Replace fuse with "slow-blow" type fuse. |

## Symptom: Charge light on at night

| Possible Cause: | Remedy: |
|---|---|
| • Normal operation for some controllers with timer circuitry if less than two hours after sunset. | • Check later. |
| • First day of system installation or array disconnected that day. | • For controllers with timer circuitry, check operation next day. |
| • Faulty controller. | • Return unit for repair. |

## Symptom: Pump cycles on and off

| Possible Cause: | Remedy: |
|---|---|
| • Air in plumbing. | • Vent air. |
| • Restricted pump delivery. | • Check that discharge lines, fittings and valves are not clogged or undersized. |

## Symptom: Pump fails to prime / motor operates but no pump discharge

| Possible Cause: | Remedy: |
|---|---|
| • Yield of water source is inadequate. | • Increase water volume. |
| • Pump has inadequate suction lift. | • Check pump specifications. Replace or plumb another pump in series if necessary. |
| • Direct current pump is running backwards. | • Check polarity and reverse wiring. |
| • Restricted intake or discharge line. | • Open all fixtures, clean clogged lines. |
| • Air leak in intake line. | • Seal air leak. |
| • Punctured pump diaphragm. | • Replace diaphragm. |
| • Defective pump check valve. | • Replace check valve. |
| • Cracked pump housing. | • Replace housing. |

## Symptom: Pump motor fails to turn on

| Possible Cause: | Remedy: |
|---|---|
| • Pump switch is off. | • Turn pump on. |
| • Loose or corroded wiring connection. | • Clean connection. Tighten connection. |
| • Pressure switch failure. | • Replace pressure switch. |
|  | • Replace pump motor. |

## Symptom: Pump fails to turn off after all fixtures are closed

| Possible Cause: | Remedy: |
| --- | --- |
| • Tank not full yet. | • Normal operation. |
| • Punctured pump diaphragm. | • Replace diaphragm. |
| • Discharge line leak. | • Repair leak. |
| • Defective pressure switch. | • Replace pressure switch. |
| • Insufficient voltage to pump. | • Check for loose or corroded wiring connections. |
| | • Check for excessive line losses. |

## Symptom: Fluorescent lights operate erratically

| Possible Cause: | Remedy: |
| --- | --- |
| • Cold temperature affects ballasts. | • Warm rooms before use or relocate. |
| • Open circuit. | • Check all wiring as per previous remedies. |
| • Bad ballast. | • Replace ballast. |
| | • Install safety tubes, if available for lamp size. |

## Symptom: Incandescent lamps fail prematurely

| Possible Cause: | Remedy: |
| --- | --- |
| • Lamp subject to vibrations or shock. | • Remount or relocate fixtures. |
| • Lamp receiving improper voltage. | • Verify and repair. |
| • Incorrect lamp being used. | • Verify and replace. |

## Symptom: Photocontrol malfunctions

| Possible Cause: | Remedy: |
| --- | --- |
| • Line voltage exceptionally high or low. | • Check voltage at photocontrol and take steps to correct condition. |
| • Photocontrol not rated at voltage being used. | • Replace photocontrol with unit of correct voltage rating. |
| • Contacts welded due to excessive load. | • Replace photocontrol and connect only the permissible load and voltage. |
| • Not enough light strikes photocontrol in daytime. | • Reposition photocontrol in the direction of the greatest amount of natural light. |
| • Light from the load is directly or indirectly shining on photocontrol. | • Reposition photocontrol to avoid artificial light sources. |
| • Incorrect wiring. | • Refer to wiring diagram and correct. |

# Chapter 16
# Installation Safety

## Contents:

16.1 Introduction . . . . . . . . . . . . . . . . . . . . . . . . . . . . . . . . . . . . . 180
16.2 Basic Safety . . . . . . . . . . . . . . . . . . . . . . . . . . . . . . . . . . . . . . . 180
16.3 Safely Testing High Voltage . . . . . . . . . . . . . . . . . . . . . . . . . . 182
16.4 Hazards . . . . . . . . . . . . . . . . . . . . . . . . . . . . . . . . . . . . . . . . . . 182
16.5 Safety Equipment . . . . . . . . . . . . . . . . . . . . . . . . . . . . . . . . . . 184
16.6 Site Safety . . . . . . . . . . . . . . . . . . . . . . . . . . . . . . . . . . . . . . . . 185
16.7 First Aid . . . . . . . . . . . . . . . . . . . . . . . . . . . . . . . . . . . . . . . . . . 186

# 16.1 Introduction

As with any activity, safety is a full-time job and the responsibility of everyone working with PV equipment, whether in the design, installation, maintenance, or use of the systems. The following items constitute good, safe practice for any type of job and reduce the potential for accidents and injuries. To work safely, you must have the following:

- Good work habits.

- A clean and orderly work area.

- Proper equipment and training in its use.

- An awareness of potential hazards and how to avoid them.

- Periodic reviews of safety procedures.

- Instruction in basic first aid and cardiopulmonary resuscitation (CPR).

Photovoltaic devices generate electricity, and they should always be considered electrically "hot." Because they generate electricity any time light falls on them, even attempting to cover them, for example with a blanket, is not a safe practice, as light could still reach the PV or the covering could come off. Similarly, batteries are always "hot" and cannot be turned off.

When working with PV modules and systems, you need to be familiar with the basics of safety:

- You are your own best safety system — be alert, check everything, and work carefully.

- Never work on a PV system alone.

- Study and understand the system before you start to work on it.

- Review the safety, testing, and installation steps with everyone involved before starting work.

- Make sure that your tools and test equipment are in proper working order.

- Check your test equipment before going to the job site.

- Wear appropriate clothing, including a hard hat face shield, eye protection, and voltage rated gloves. Also, remove all jewelry that might come in contact with electrical components.

- Measure everything electrical with a digital multi-meter. Measure the conductivity from exposed metal frames and junction boxes to ground. Measure voltage from all conductors (on the PV output circuit) to ground. Measure the operating voltage and current.

- Expect the unexpected. Do not assume that switches always work, that the actual configuration agrees with the electrical diagrams, that current is not flowing in the grounding circuit, etc.

- Working with any size PV system involves a number of potential hazards, both non-electrical and electrical. Consequently, safety must be foremost in the mind of anyone working on a PV system. This chapter provides important and necessary safety information for PV practitioners and others working on or near PV systems.

# 16.2 Basic Safety

Regardless of whether or not the location of the PV system is covered by a local or national electrical safety code, it is important to follow guidelines that ensure safe electrical systems. Examples of codes and standards that provide recommendations and guidelines for electrical safety include:

- *National Electrical Code®* (*NEC®*).

- Underwriters Laboratories (UL) equipment safety testing and certification.

You can find hardware standards from the following organizations:

- Global Approval Program for PV (PV GAP).

- Institute for Electrical and Electronics Engineers (IEEE).

- International Electrotechnical Commission (IEC).

- American Society for Testing and Materials (ASTM).

- International Standards Organization (ISO).

Reference the following materials for procedures when working on or near live parts:

- National Fire Protection Association (NFPA) 70E: Standard for Electrical Safety in the Workplace.
- Occupational Safety and Health Administration (OSHA): Standards for the Construction Industry, Article 1926.400.
- Occupational Safety and Health Administration (OSHA): Standards for General Industry, Article 1910.300.

## System Current and Voltage

When designing a PV system, you should consider the following:

- The rated voltage in any PV source circuit should be the open-circuit voltage.
- Voltages should be less than 600 volts.
- Conductors and overcurrent devices should be able to carry at least 125 percent of the short-circuit current of the source circuit.
- PV source circuit, inverter, and battery conductors should have overcurrent protection.
- A sign indicating PV system operating voltage and the short-circuit current should be placed near the system disconnect point.

## Wiring and Disconnect Requirements

You should be consistent with electrical wiring. There are certain conventions for the color of conductors and specific requirements for disconnecting the power source, including the following:

- The grounded conductor must be white. The convention states that the first ungrounded conductor of a PV system must be red and the second ungrounded conductor must be black.
- Single-conductor cable is allowed for module connections only. Sunlight- or ultra violet (UV)-resistant cable should be used if the cable is exposed.
- Modules must be wired so that they can be removed without interrupting the grounded conductor.
- Any wiring junction boxes must be accessible.
- Connectors must be polarized and guarded to prevent shock.

- Means to disconnect and isolate all PV source circuits must be provided.
- Means to disconnect all ungrounded conductors from the inverter must be provided.
- If fuses are used, means to disconnect the power from both ends must be provided.
- Switches must be accessible and clearly labeled.

## Grounding

The purpose of grounding any electrical system is to prevent unwanted currents from flowing through equipment or people and possibly causing equipment damage, personal injury, or death. Lightning, natural, and man-made ground faults and line surges can cause high voltages in otherwise low-voltage systems. Proper grounding, along with overcurrent protection, limits the possible damage that a ground fault can cause.

You should be familiar with the following and recognize the difference between the equipment grounding conductor and the system ground:

- One conductor of a PV system (>50 V) must be grounded, and the neutral wire of a center-tapped, three-wire system must also be grounded. If these provisions are met, this is considered sufficient for the battery ground, if batteries are included in the system. A ground is achieved by making a solid low-resistance connection to a permanent earth ground, which can be created by driving a metallic rod into the earth, preferably in a moist location.
- A single ground point should be made. This provision will prevent the possibility of potentially dangerous ground fault current flowing between separate grounds. In some PV systems where the PV array is located far from the load, a separate ground can be used at each location. This will better protect the PV array from lightning surges. If multiple ground points are used, they must be bonded together with a grounding conductor.
- All exposed metal parts must be grounded (equipment ground).
- The equipment grounding conductor must be bare wire or green wire and be large enough to handle the highest current that could flow in the circuit.

### PV System Output

Before the PV array is connected to the load, battery, or inverter, there are certain requirements you need to address, including:

- If an inverter is used to interconnect the PV system to a utility, it must disconnect automatically if the utility power goes off. If the inverter is operating in a stand-alone system, it can supply power to the load continuously.

- The output of a single-phase inverter should not be connected to multi-wire branch circuit.

- The AC output from a PV system inverter must be grounded in accordance with requirements for AC systems.

- A circuit breaker or fuse/switch mechanism must be included so that the PV system output can be disconnected.

- The interconnection should be made so that all ground fault interrupters remain active.

- If batteries are used in a system, they must be guarded to prevent unauthorized access if the voltage is greater than 50 Vdc. Otherwise, the voltage must remain below 50 Vdc.

- If batteries are used in a system, charge controllers must be installed in the system.

# 16.3 Safely Testing High Voltage

Batteryless grid-tied inverters on the market today require high voltage input windows. In order to meet these higher input specifications, an array configuration made up of many modules in series (series-string) must be used. These 'series strings' have high voltages, but the output amperage is equal to that of one module within the string. Taking measurements on a high voltage array can be very dangerous. Always use the proper personal protective equipment (PPE) and safety gear when testing high voltage. Relevant PPE includes, but is not limited, to the following: High voltage gloves (lining and outer shell), eye protection, and the appropriate type and size electrical meter.

When measuring array voltage, a Digital Multi-Meter (DMM) internally opens the circuit to read the potential difference between the positive and negative terminal. It is important to verify that the DC voltage rating of the DMM can handle the Voc of the array.

When measuring amperage, a DMM internally shorts the circuit to take the reading. When measuring amperage from an array at high voltages, a large electrical arc can result when pulling the test leads off of the terminals. For this reason, it is not advised to measure the short circuit current with a DMM on high voltage arrays. To measure array current, it is safer to use a Clamp-on Ammeter. This type of meter clamps around one conductor in the circuit in order to measure current via inductance. It is important to verify that the DC amperage rating of the clamp-on Ammeter can handle the short-circuit current of the array.

Measuring current on lower voltage systems (12-48 VDC nominal) can also be dangerous due to high current levels. It is recommended to take these measurements with a clamp-on meter.

Reference the following materials for procedures when working on or near live parts:

- National Fire Protection Association (NFPA) 70E: *Standard for Electrical Safety in the Workplace.*

- Occupational Safety and Health Administration (OSHA): *Standards for the Construction Industry*, Article 1926.400.

- Occupational Safety and Health Administration (OSHA): *Standards for General Industry*, Article 1910.300.

# 16.4 Hazards

When installing or working with PV systems, you should be aware of the many potential physical, electrical, and chemical hazards.

### Physical Hazards (non-electrical, non-chemical)

When working on a PV system you will be working outdoors (possibly in remote areas) using hand and power tools on electrical equipment. In many systems, you will also be working with batteries, which pose their own sets of burn, shock, and physical hazards. Take the necessary precautions to use these tools safely and appropriately.

## Exposure

When designed properly, PV systems are installed where the sun is brightest and no shade exists. When working on a PV system, you should wear a hat, keep yourself covered, and use plenty of sunscreen to protect yourself from the sun. In hot weather, drink plenty of fluids, preferably water, and never alcohol. Take regular breaks in the shade for a few minutes each hour. In the wintertime, dress warmly, and wear gloves whenever possible.

## Insects, Snakes, and Other Creatures

Spiders and many insects, including wasps, will often move in and inhabit junction boxes, array framing, and other enclosures of a PV system. Snakes use the shade provided by the array. Also, ants are commonly found under arrays or near battery boxes. Always be prepared for the unexpected when you open junction boxes and other enclosures. Look carefully before you crawl under or move behind the array.

## Cuts and Bumps

Most PV systems consist of components that have sharp edges and can cause injury if you are not careful. These include metal framing, junction boxes, bolts, nuts, guy wires, and anchor bolts. Wear gloves when handling metal, particularly if you are drilling or sawing. Metal slivers from a drill bit often remain around a hole, and these can cause severe cuts to a bare hand. Wear a dielectric hard hat any time you are working under an array or on a system with hardware higher than your head.

## Falls, Sprains, and Strains

Many PV systems are installed in remote areas and in rough terrain. Walking to the site and around it, particularly carrying systems components and test equipment, can result in falls and sprains. Wear comfortable shoes, preferably with soft soles. Steel toe reinforced shoes should not be worn around PV systems because they lower the resistance of a potential current path, increasing your risk of becoming a conductor. Be careful when lifting and carrying heavy equipment, particularly batteries. To avoid back strains, lift with your legs and not with your back. If climbing is required, have a partner hold the ladder firmly anchored and assist with handling equipment. Always be sure to employ proper fall protection as required by OSHA. Also, remember that a PV module can act as a wind sail and knock you off a ladder on windy days.

## Thermal Burns

Metal exposed to the sun can reach temperatures of 80°C (176°F). This is too hot to handle but unlikely to cause burns if you break contact quickly. To be safe, wear gloves at all times when working on PV systems in the summertime. Survey the system to be aware of elements that might get hot.

## Electrical Hazards

Common electrical accidents result in shocks and burns, which can cause muscle contractions and traumatic injuries resulting from falls. These injuries can occur any time electric current flows through the human body. The amount of current that will flow is determined by the difference in potential (voltage) and the resistance in the current path. At low frequencies (60 Hz or less), the human body acts like a resistor, but the value of resistance varies with conditions. It is difficult to estimate when current will flow through the body or the severity of the injury that might occur because the resistivity of human skin varies from just under a thousand ohms to several hundred thousand ohms, depending primarily on skin moisture.

If a current greater than 0.02 amperes (only 20 milliamperes) flows through your body, you are in serious jeopardy because you may not be able to let go of the current-carrying wire. This small amount of current can be forced through sweaty hands with a voltage as low as 20 volts, and the higher the voltage, the higher the probability that current will flow. High voltage shock (greater than 400 volts) may burn away the protective layer of outer skin at the entry and exit points. When this occurs, the body resistance is lowered and lethal currents can cause instant death.

Electrical shock is painful, and potentially minor injuries are often aggravated by the reflex reaction of jumping back away from the source of the shock.

The best way to avoid shock is to always use a clamp-on ammeter to measure the current flowing in the wires and never disconnect a wire before you have checked the voltage and current. Do not presume that everything is connected and working as designed. Do not trust switches to operate perfectly and do not "believe" schematics. A digital multi-meter is a wonderful instrument and using it could save your life.

## AC Power Hazards

If alternating current power is to be supplied, a power conditioning unit is required to convert the direct current power from the PV system to AC power. This equipment may have high voltage at both input and output when it is operating. The output is nominally 120 Vac or 240 Vac, which is enough current to kill a person.

## Chemical Hazards

Batteries are potentially the most dangerous PV system component if improperly handled, installed, or maintained. Dangerous chemicals, heavy weight, and high voltages and currents are potential hazards that can result in electric shock, burns, explosion, or corrosive damage to yourself or your property. Please reference the following materials regarding chemical hazards:

- Section 6.3, Battery Safety.
- NFPA 70E *Standard for Electrical Safety in the Workplace,* Article 240.
- OSHA *Standards for the Construction Industry,* Article 1926.441.

## Acid Burns

Most stand-alone PV systems use batteries, which are typically the most dangerous component in PV systems. The most common type of battery is the lead-acid battery that uses sulfuric acid as the electrolyte. Sulfuric acid is extremely hazardous; it can spill when handling a battery and spray as a fine mist when a battery is charging. If acid makes contact with an unprotected part of your body, you will receive a chemical burn; your eyes are particularly vulnerable. It will also burn holes in your clothing. Any time you are working around lead-acid batteries, you should wear non-absorbent gloves, protective eyewear (chemical goggles), and a neoprene-coated apron.

## Gas Explosion or Fire

Most types of batteries used in PV systems release hydrogen gas as a result of the charging process. This flammable gas is a hazard, and flames, sparks, and any equipment that could create a spark, such as controllers with relays, should be kept away from the batteries. Batteries should also be located in a well-ventilated area to prevent a buildup of hydrogen gas.

# 16.5 Safety Equipment

This section lists the recommended safety equipment that you should have available. You should make sure that this equipment is in working order before beginning a job.

**Personal Safety Resources:**

- Work partner (never work alone!).
- Understanding of safety practices, equipment, and emergency procedures.
- Safety checklist.
- Hard hat.
- Eye protection and face shield.
- Rubber gloves for working with batteries.
- Apron for working with batteries.
- Appropriate harnesses, if working on roofs or other elevated sites.
- Proper measuring equipment, electrical and dimensional.
- Tape and wire nuts (never leave wire ends exposed!).
- High voltage gloves.
- Clamp-on ammeter.

**Job-Site Safety Resources:**

- Safety plan.
- Eyewash solution and/or station.
- First-aid kit.
- Fire extinguisher.
- Distilled water.
- Baking soda.
- Appropriate ladders.
- Appropriate lifting equipment.
- Proper labels on all components, such as boxes and wiring.

**Note:** Safety equipment standards can also refer to what you should not wear. Remove all jewelry that might come in contact with electrical components!

# 16.6 Site Safety

Sometimes, you will need to troubleshoot a PV system that is not working correctly. Safety should be your main concern, both in planning to go to the site and during the actual testing. Before working with any PV system, you should become familiar with the electrical configuration.

Before traveling to the site, you should be able to answer the following questions:

- Who will assist you? (Always work with a trained partner or team.)
- Where is the nearest medical help located, and what phone number should be called in an emergency?
- How many modules make up a source circuit?
- Are batteries involved?
- What are the system voltages?
- What are the system currents?
- How many circuits are there?
- How can the system be disconnected?
- What safety equipment is available?
- What equipment will you need to bring?

At the PV system site, you should take the following safety precautions:

- Remove jewelry.
- Walk around the PV system and record any apparent hazards in the system logbook or a notebook. Take photographs of the system and any hazards.

- Locate the safety equipment, such as a fire extinguisher and check the condition of all equipment before starting work. Locate the nearest telephone.
- Check the actual system configuration against the electrical schematics.
- Locate and inspect all subsystems, such as the batteries, inverter, and the load.
- Determine if, how, and where the system is grounded. Check to see if the AC and DC grounds are common.
- Locate and inspect all disconnect switches and fuses. Determine if the switches are designed to interrupt both positive and negative conductors.
- Disconnect the source circuits and measure the open-circuit voltage to verify the proper operation of the disconnected switch.
- Measure the voltage from each conductor to ground and from line to line.
  **Note:** Only when you are sure that you understand the circuit should you proceed with testing.
- Keep the work area clear of obstacles, particularly the area behind where you are working.
- Never disconnect a wire before measuring voltages.
- Keep your hands dry and/or wear gloves.
- Work with only one hand, if possible.
- Have your partner or team member stationed near the disconnect switches.
- Once a wire is disconnected, don't leave the end exposed — tape it or use a wire nut for temporary covering.
- Reconnect the wires from one source circuit before disconnecting a second source circuit.

# 16.7 First Aid

The following is a review of the first-aid procedures that everyone working on PV systems should be familiar with. Each person working on the installation or maintenance of PV systems should also complete a cardiopulmonary resuscitation (CPR) course or equivalent training. The following information is a summary of the first-aid you should understand and be capable of performing, but it is not intended to replace formal training in first-aid or CPR.

> Note: Both electrical and non-electrical injuries can occur when working around/with PV systems.

If you witness an accident or are the first person to arrive at the scene, perform the following first-aid actions:

- Survey the scene for potential hazards. The first rule is personal safety. The worst thing that could happen is that you, the rescuer, get injured or killed in an attempt to provide assistance to the victim. Try to determine if a hazard still exists. Is a live conductor still lying on or near the victim's body? Is the victim still holding a live conductor? Are there other hazards, such as fire or spilled caustic material that would put you in jeopardy? You will be safer in assisting a victim if you are with someone else, but do not delay to wait for a partner.

> Note: Also, be aware that some otherwise competent people may not react well or as expected in an emergency situation, everyone reacts differently. You are on your own to protect yourself and help the victim.

- Check the victim for an open airway, adequate breathing, and adequate pulse. Determine the victim's condition.

- Call for help and give the victim's condition and vital information. During an emergency, do anything you can to quickly attract attention to the scene. Call an ambulance, get someone else to do it, or even pull a fire alarm, but get qualified emergency personnel to the scene as quickly as possible, then attend to the victim using accepted first aid and CPR techniques. Although in remote areas, you may need to provide the necessary initial care, again, call for emergency help. They can meet you half way to the hospital if necessary.

## Electrical, Chemical, and Thermal Injuries

The number one priority in assisting injured people should always be your (the rescuer's) safety. This is especially important in situations involving electrical hazards. Avoid becoming a second victim. Electrical injuries consist mainly of shocks, burns, muscle contractions, and traumatic injuries associated with falls after electrical shocks. Burns can result from electrical, chemical, and thermal exposure.

## Electrical Injuries

**Electric shock** is a general term indicating any situation where electric current flows through the body. The intensity of a shock can vary from a barely perceptible tingle to a strong shock to near-instant death. A stabbing pain or intense tingling and burning is usually associated with electric shock. The points of entry and exit are often badly burned.

Frequently, a shock involves involuntary muscle contraction. If the strong muscles of the back and legs contract, this can lead to falls, broken bones, or worse. The large muscles of the chest, throat, and diaphragm can contract, causing respiratory arrest.

When electric current passes through the heart, it can cause a spasmodic contraction and relaxation of the ventricles, called ventricular fibrillation. This is one of the major causes of death associated with shocks. Once a person's heart has begun fibrillating, it is difficult to stop. Sometimes, another electric shock, administered by a trained technician using a defibrillator can restore the heart to its normal beating cycle. Victims of fibrillation need qualified paramedic help within minutes to survive.

If you are at the scene of a suspected electrical accident, you must survey the scene for hazards before you rush in to help the victim. If the victim is holding a live conductor, chances are that they may be physically unable to let go. You must find some way to disconnect the power so that you can help them. This is one more reason that familiarity with the system is very important. If there is no way to switch off the power, you have to find a way to remove the conductor from the victim's body or vice versa. A properly equipped PV site should have a grounding stick or non-conducting wooden cane near possible electrical hazards. Use one of these to move the conductor from the victim. You can use a rope or belt to drag the victim away from the live wire or even cut the live wire with a wooden handled ax. Be creative with what you have available. Remember that the victim's life is in danger and time is of the essence.

In the case of spinal injuries, possibly resulting from a fall after being shocked, you may possibly cause more injury to the victim by moving them. Do not move a victim unless it is absolutely necessary. However, if the person is likely to die unless you do move them, possible spinal injury may be a small price to pay for a life. You have to decide.

Once you and the victim are free from the shock hazard, you can begin assessing injuries and treating the victim. Remember the ABCs of CPR: Airway, Breathing, and Circulation. Determine if the victim is conscious or unconscious. If they are unconscious, open the airway and check for breathing. Put your cheek close to their mouth and feel for breath as you watch for the chest to rise and fall. You should take 5-10 seconds to check for a neck pulse at this time as well. Check closely, it may be faint. If they are not breathing, using mouth-to-mouth, with a clean sterile mouth shield if available, give two breaths. If the air doesn't go in, check to be sure the victim's airway is clear. It could be blocked by their tongue.

Once you've cleared the airway, if they are still not breathing, begin artificial respiration. In addition, if there is no pulse, begin CPR.

**Artificial respiration and CPR should be performed in accordance with current first-aid standards.**

Hopefully, the victim will begin to breathe and their heart will beat. Only when this happened should you stop CPR. If you stop sooner, they may die. If they do breathe and their heart beats, watch them closely until the ambulance arrives, as they may need your help again.

If the victim is breathing, has a pulse, and is conscious, the victim should be treated for ordinary shock, which is the body's attempt to correct a failing circulatory system. To treat for shock perform the following first-aid:

- Have the victim lie down.

- Raise the feet to help keep the blood flowing to the vital organs.

- If they are cool, cover the victim to keep them warm.

## Chemical, Electrical, and Thermal Burns

Minor burns or red skin with no blistering should be flushed with cool water and a loose dressing and bandage should be applied. This will protect the burn from possible infection. Deep burns with blistering and charred skin are life threatening and an ambulance must be called immediately. The biggest problem is contamination, which causes infection. Do not put water on a deep burn, unless it is a chemical burn, such as from battery electrolyte, which should be flushed with clean water. Use the following first-aid for deep burns:

- Carefully remove any large pieces of debris.

- Prevent further contamination, if possible, by covering with a dry, loose dressing (gauze pad) and then bandage. Apply as little pressure as possible. If possible, use sterile dressings.

- Treat for shock.

- Call for help and stay with the victim until medical professionals arrive to take charge.

The batteries typically used in PV systems are some variant of a lead-acid design. These batteries are filled with highly concentrated sulfuric acid and give off hydrogen gas, which could explode if concentrated and exposed to a spark or flame. In addition to the potential for explosion, the acid could splash on your skin, clothing, in your eyes, or in your mouth. Consequently, always wear proper clothing and protective gear, and be prepared with the proper first aid materials to treat victims involved in accidents with acids.

For chemical burns, including in the eyes, you should perform the following first-aid actions:

- Flush immediately with large amounts of water for fifteen to thirty minutes.

- Remove any affected clothing or jewelry.

- Call an ambulance.

- If the chemical burn is from acid, such as battery acid, flush with water and apply baking soda to neutralize the acid. Cover with a loose, dry, sterile dressing and bandage as loosely as possible.

- If the burn is in an eye, cover both eyes. Then, treat for shock.

- If acid is somehow taken internally, drink large

quantities of water or milk, followed with milk of magnesia, beaten egg, or vegetable oil, and seek immediate medical attention.

### Non-electrical Injuries

These injuries include cuts, sprains, broken bones, exposure, and insect and snake bites. Most of the time, these situations are not life threatening, but in some cases if care is not provided immediately, the victim may go into shock and potentially die. Respond quickly.

### Cuts

If someone receives a cut, you should stop the bleeding by using the following methods, in this order:

- Direct pressure — Apply direct pressure with a sterile dressing (gauze pad) between the wound and your hand. Use a clean cloth if a sterile dressing is not available.

- Elevation — Elevate the wound if it continues to bleed.

- Pressure points — Apply direct pressure to a nearby pressure point if the wound continues to bleed. For example, if the lower arm is cut, apply pressure with the fingers on the middle inside of the upper arm where the pulse can be felt.

- Pressure bandage — Wrap the wound with a roller bandage using overlapping turns to completely cover the wound. Apply additional sterile dressing before wrapping if necessary.

### Sprains, Strains, Dislocations, and Fractures

These injuries are sometimes hard to differentiate, so treat them all as you would a fracture. Help the victim move into the shade and to a comfortable position with as little movement to the injured area as possible. The injury, usually to an arm or leg, will need to be splinted to lessen the pain and prevent further injury. Splints can be made from rolled-up newspaper, magazines, pieces of wood, blankets, or pillows. The splint can be tied up with bandages or cloth, such as a shirt torn into strips. The following principles should be followed when splinting:

- Splint only if you can do it without causing more pain.

- Splint an injury in the position you find it.

- Immobilize the limb and joints above and below the injury.

- Check the blood circulation by pinching nail beds of the fingers or toes. Red color should return within two seconds, if not, loosen the splint.

- If the injury is a closed fracture, no bone extruding, apply a cold pack to it.
  **Note:** Do not apply a cold pack to an open or compound fracture.

### Heat Exposure

Heat exposure is a common hazard to system installation and maintenance personnel because of the location of the systems. If you or your partner has cramps, heavy sweating, cool and pale skin, dilated pupils, headaches, nausea, or dizziness, you may be nearing heat exhaustion. You should perform the following first-aid:

- Get the victim to the shade.

- Give them one half of a glass of water every 15 minutes (if they can tolerate it).

- If heavy sweating occurs, have the victim lie down and raise their feet, loosen clothing, and put wet towels or sheets over them.

- If the victim has red, dry skin, they may have heat stroke, which is life-threatening. Immerse them in cool water, if possible, or wrap their body with wet sheets, and fan them. Do not give them anything to drink. Call an ambulance.

### Cold Exposure

Persons exposed to extended periods of cold may suffer from hypothermia. Possible symptoms are shivering, feeling dizzy, confusion, or numbness. You should perform the following first-aid:

- Take the victim to a warm place.

- Remove wet clothing.

- Warm the body slowly.

- Call an ambulance.

- If fully conscious, give them a warm drink a little at a time. Check the temperature of the liquid. (Don't add a scalded tongue to their injuries.)

## Insect and Snake Bites

A small number of people may have an allergic reaction to an insect bite or sting. If so, this situation could be life threatening. Signs of an allergic reaction include pain, swelling of the throat, redness or discoloration, itching, hives, decreased consciousness, and difficulty in breathing. If these symptoms occur, perform the following first-aid:

- Call an ambulance immediately.

- If a stinger from an insect is embedded into the flesh, remove it (do not squeeze it) with tweezers or scrape it away with a credit card, rigid strip of plastic, or a playing card.

- Wash the area.

- Put on a cold pack with a cloth between the skin and the ice.

- Arrange the victim so the affected area is below the heart.

- If it is a snake bite, immediately call for medical help. Keep the victim still and the affected area below the heart to slow absorption of the snake venom. A splint can be used if the bite is on an arm or leg. Try to remember what the snake looked like. Do not cut a snake bite and try to suck the venom out. This only increases the chances of infection. Few people die from snake bites.

# Appendix A: Glossary

## -A-

**absorbed glass mat (AGM):** A fibrous silica glass mat to suspend the electrolyte in batteries. This mat provides pockets that assist in the recombination gasses generated during charging back into water.

**alternating current (AC):** Electric current in which the direction of flow is reversed at frequent intervals, usually 100 or 120 times per second (50 or 60 cycles per second or 50//60 Hz).

**altitude:** The angle between the horizon (a horizontal plane) and the sun's position in the sky, measured in degrees.

**amorphous silicon:** A non-crystalline semiconductor material that has no long-range order, often used in thin film photovoltaic modules.

**ampere (A) or amp:** The unit for the electric current; the flow of electrons. One amp is 1 coulomb passing in one second. One amp is produced by an electric force of 1 volt acting across a resistance of 1 ohm. Sometimes this is abbreviated as I for intensity.

**ampere-hour (Ah):** Quantity of electrical energy equal to the flow of one ampere of current for one hour. Typically used to quantify battery bank capacity.

**angle of incidence:** Angle which references the sun's radiation striking a surface. A "normal" angle of incidence refers to the sun striking a surface at a 90-degree angle.

**array:** Any number of photovoltaic modules connected together to provide a single electrical output at a specified voltage. Arrays are often designed to produce significant amounts of electricity.

**autonomous system:** A stand-alone PV system that has no back-up generating source. May or may not include storage batteries.

**avoided cost:** The minimum amount an electric utility is required to pay an independent power producer, under the PURPA regulations of 1978, equal to the costs the utility calculates it avoids in not having to produce that power (usually substantially less than the retail price charged by the utility for power it sells to customers).

**azimuth:** Angle between true south and the point directly below the location of the sun. Measured in degrees east or west of true south in northern latitudes.

## -B-

**balance of system (BOS):** All system components and costs other than the PV modules. It includes design costs, land, site preparation, system installation, support structures, power conditioning, operation and maintenance costs, indirect storage, and related costs.

**barrier energy:** The energy given up by an electron in penetrating the cell barrier, a measure of the electrostatic potential of the barrier.

**base power:** Power generated by a utility unit that operates at a very high capacity factor.

**baseline performance value:** Initial values of Isc, Voc, Pmp, Imp measured by the accredited laboratory and corrected to Standard Test Conditions, used to validate the manufacturer's performance measurements provided with the qualification modules per IEEE 1262.

**battery:** Two or more "cells" electrically connected for storing electrical energy. Common usage permits this designation to be applied also to a single cell used independently, as in a flashlight battery.

**battery capacity:** The total number of ampere-hours that can be withdrawn from a fully charged cell or battery.

**battery cell:** A galvanic cell for storage of electrical energy. This cell, after being discharged, may be restored to a fully charged condition by an electric current.

**battery cycle life:** The number of cycles, to a specified depth of discharge, that a cell or battery can undergo before failing to meet its specified capacity or efficiency performance criteria.

**battery self-discharge:** The rate at which a battery, without a load, will lose its charge.

**battery state of charge:** Percentage of full charge or 100 percent minus the depth of discharge.

**building-integrated photovoltaics (BIPV):** A term for the design and integration of PV into the building envelope, typically replacing conventional building materials. This integration may be in vertical facades, replacing view glass, spandrel glass, or other facade material; into semitransparent skylight systems; into roofing systems, replacing traditional roofing materials; into shading "eyebrows" over windows; or other building envelope systems.

**blocking diode:** A semi-conductor device connected in series with a PV module and a storage battery to prevent a reverse current discharge of the battery through the module when there is no output, or low output from the cells. When connected in series to a PV string; it protects its modules from a reverse power flow preventing against the risk of thermal destruction of solar cells.

**boron (B):** A chemical element, atomic number 5, semi-metallic in nature, used as a dopant to make p-semiconductor layers.

**British thermal unit (Btu):** The amount of heat energy required to raise the temperature of one pound of water from 60 degrees F to 61 degrees F at one atmosphere pressure. Roughly equivalent to the amount of energy released by burning one stick match.

**bypass diode:** A diode connected across one or more solar cells in a photovoltaic module such that the diode will conduct if the cell(s) become reverse biased. Alternatively, a diode connected anti-parallel across a part of the solar cells of a PV module. It protects these solar cells from thermal destruction in case of total or partial shading of individual solar cells while other cells are exposed to full light.

## -C-

**cadmium (Cd):** A chemical element, atomic number 48, used in making certain types of solar cells and batteries.

**cadmium telluride (CdTe):** A polycrystalline, thin-film photovoltaic material.

**capacity factor:** The amount of energy that the system produces at a particular site as a percentage of the total amount that it would produce if it operated at rated capacity during the entire year. For example, the capacity factor for a wind farm ranges from 20% to 35%.

**cathodic protection:** A method of preventing oxidation (rusting) of exposed metal structures, such as bridges and pipelines, by imposing between the structure and the ground a small electrical voltage that opposes the flow of electrons and that is greater than the voltage present during oxidation.

**cell:** The basic unit of a photovoltaic module. This word is also commonly used to describe the basic unit of batteries (ie. a 6-volt battery has 3 2-volt cells).

**cell barrier:** A very thin region of static electric charge along the interface of the positive and negative layers in a photovoltaic cell. The barrier inhibits the movement of electrons from one layer to the other, so that higher-energy electrons from one side diffuse preferentially through it in one direction, creating a current and thus a voltage across the cell. Also called depletion zone, cell junction, or space charge.

**cell junction:** The area of immediate contact between two layers (positive and negative) of a photovoltaic cell. The junction lies at the center of the cell barrier or depletion zone.

**central power:** The generation of electricity in large power plants with distribution through a network of transmission lines (grid) for sale to a number of users. Opposite of distributed power.

**charge controller:** A device that controls the charging rate and/or state of charge for batteries.

**charge rate:** The current applied to a cell or battery to restore its available capacity.

**chemical vapor deposition (CVD):** A method of depositing thin semiconductor films. With this method, a substrate is exposed to one or more vaporized compounds, one or more of which contain desirable constituents. A chemical reaction is initiated, at or near the substrate surface, to produce the desired material that will condense on the substrate.

**cleavage of lateral epitaxial films for transfer (CLEFT):** A process for making inexpensive GaAs photovoltaic cells in which a thin film of GaAs is grown atop a thick, single-crystal GaAs (or other suitable material) substrate and then is cleaved from the substrate and incorporated into a cell, allowing the substrate to be reused to grow more thin-film GaAs.

**coal:** A black, solid fossil fuel, usually found underground. Coal is often burned to make electricity in utility scale production.

**combined collector:** A photovoltaic device or module that provides useful heat energy in addition to electricity.

**compact fluorescent lights:** Lights that use a lot less energy than regular light bulbs. We can use compact fluorescent lights for reading lights and ceiling lights.

**concentrator:** A PV module that uses optical elements to increase the amount of sunlight incident on a PV cell. Concentrating arrays must track the sun and use only the direct sunlight because the diffuse portion cannot be focused onto the PV cells.

**conversion efficiency:** The ratio of the electric energy produced by a photovoltaic device (under full sun conditions) to the energy from sunlight incident upon the cell.

**copper indium diselenide (CuInSe2, or CIS):** A polycrystalline thin-film photovoltaic material (sometimes incorporating gallium (CIGS) and/or sulfur).

**crystalline silicon:** A type of PV cell made from a single crystal or polycrystalline slice of silicon.

**current:** The flow of electric charge in a conductor between two points having a difference in potential (voltage).

**current at maximum power (Imp):** The current at which maximum power is available from a module. [UL 1703]

**cycle life:** Number of discharge-charge cycles that a battery can tolerate under specified conditions before it fails to meet specified criteria as to performance (e.g., capacity decreases to 80-percent of the nominal capacity).

**Czochralski process:** A method of growing large size, high quality semiconductor crystal by slowly lifting a seed crystal from a molten bath of the material under careful cooling conditions.

## -D-

**days of autonomy:** The number of consecutive days a stand-alone system battery bank will meet a defined load without solar energy input.

**DC to DC converter:** Electronic circuit to convert DC voltages (e.g., PV module voltage) into other levels (e.g., load voltage). Can be part of a maximum power point tracker (MPPT).

**deep cycle battery:** Type of battery that can be discharged to a large fraction of capacity many times without damaging the battery.

**deep discharge:** Discharging a battery to 50 percent or less of its full charge.

**depth of discharge (DOD):** The amount of ampere-hours removed from a fully charged cell or battery, expressed as a percentage of rated capacity.

**design month:** The month having the combination of insolation and load that requires the maximum energy from the array.

**diffuse insolation:** Sunlight received indirectly as a result of scattering due to clouds, fog, haze, dust, or other obstructions in the atmosphere. Opposite of direct insolation.

**diode:** Electronic component that allows current flow in one direction only.

**direct current (DC):** Electric current in which electrons flow in one direction only. Opposite of alternating current.

**direct insolation:** Full sunlight falling directly upon a collector. Opposite of diffuse insolation.

**discharge rate:** The rate, usually expressed in amperes over time, at which electrical current is taken from the battery.

**disconnect:** Switch gear used to connect or disconnect components of a PV system for safety or service.

**distributed power:** Generic term for any power supply located near the point where the power is used. Opposite of central power. See 'stand-alone'; 'remote site.'

**dopant:** A chemical element (impurity) added in small amounts to an otherwise pure semiconductor material to modify the electrical properties of the material. An n-dopant introduces more electrons. A p-dopant creates electron vacancies (holes).

**doping:** The addition of dopants to a semi-conductor.

**duty cycle:** The ratio of active time to total time. Used to describe the operating regime of appliances or loads.

## -E-

**edge-defined film-fed growth (EFG):** A method for making sheets of polycrystalline silicon in which molten silicon is drawn upward by capillary action through a mold.

**efficiency:** The ratio of output power to input power. Expressed as a percent.

**electric circuit:** Path followed by electrons from a power source (generator or battery) through an external line (including devices that use the electricity) and returning through another line to the source.

**electric current:** A flow of electrons; electricity.

**electrical grid:** An integrated system of electricity distribution, usually covering a large area.

**electrodeposition:** Electrolytic process in which a metal is deposited at the cathode from a solution of its ions.

**electrolyte:** A liquid conductor of electricity in which flow of current takes place by migration of ions. The electrolyte for a lead-acid storage cell is an aqueous solution of sulfuric acid.

**electron volt:** An energy unit equal to the energy an electron acquires when it passes through a potential difference of one volt; it is equal to $1.602 \times 10^{-19}$ volt.

**energy:** The ability to do work. Stored energy becomes working energy when we use it.

**energy audit:** A survey that shows how much energy you use in your house, apartment, or business. It can indicate your most intensive energy consuming appliances and even identify heating and cooling leaks that will help you find ways to use less energy.

**energy density:** The ratio of energy available from a battery to its volume (Wh/1) or mass (Wh/kg).

**energy pay back time:** The time required for any energy producing system or device to produce as much energy as was required in its manufacture.

**equalization:** The process of mixing the electrolyte in batteries by periodically overcharging the batteries for a short period to "refresh" cell capacity.

## -F-

**fill factor:** The ratio of a photovoltaic cell's actual power to its power if both current and voltage were at their maxima. A key characteristic in evaluating cell performance.

**flat-plate PV:** Refers to a PV array or module that consists of nonconcentrating elements. Flat-plate arrays and modules use direct and diffuse sunlight, but if the array is fixed in position, some portion of the direct sunlight is lost because of oblique sun-angles in relation to the array.

**float charge:** Float charge is the voltage required to counteract the self-discharge of the battery at a certain temperature.

**float life:** Number of years that a battery can keep its stated capacity when it is kept at float charge (see float charge).

**fossil fuels:** Fuels formed in the ground from the remains of dead plants and animals. It takes millions of years to form fossil fuels. Oil, natural gas, and coal are fossil fuels.

**fuel:** Any material that can be burned to make energy.

## -G-

**gassing current:** Portion of charge current that goes into electrolytical production of hydrogen and oxygen from the electrolytic liquid in the battery. This current increases with increasing voltage and temperature.

**gel-type battery:** Lead-acid battery in which the electrolyte is composed of a silica gel matrix.

**gigawatt (GW):** One billion watts. One million kilowatts. One thousand megawatts.

**glazings:** Clear materials (such as glass or plastic) that allow sunlight to pass into solar collectors and solar buildings, trapping heat inside.

**grain boundaries:** The boundaries where crystallites in a multicrystalline material meet.

**grid:** See 'Electrical grid.'

**grid-connected:** A PV system in which the PV array acts like a central generating plant, supplying power to the grid.

**grid-interactive:** See 'grid-connected (PV system).'

## -H-

**hybrid system:** A PV system that includes other sources of electricity generation, such as wind or fossil fuel generators.

## -I-

**incident light:** Light that shines onto the surface of a solar cell or module.

**infrared radiation:** Electromagnetic radiation whose wavelengths lie in the range from 0.75 micrometer to 1000 micrometers.

**insolation:** Sunlight, direct or diffuse; from 'incident solar radiation.' Usually expressed in watts per square meter. Not to be confused with 'insulation.'

**insulation:** Materials that reduce the rate or slow down the movement of heat.

**interconnect:** A conductor within a module or other means of connection which provides an electrical interconnection between the solar cells.

**inverters:** Devices that convert DC electricity into AC electricity (single or multiphase), either for stand-alone systems (not connected to the grid) or for utility-interactive systems.

**I-V curve:** A graphical presentation of the current versus the voltage from a photovoltaic device as the load is increased from the short circuit (no load) condition to the open circuit (maximum voltage) condition. Typically measured at 1000 watts per square meter of solar insolation at a specific cell temperature. The shape of the curve characterizes cell performance.

## -J-

**junction box:** An electrical box designed to be a safe enclosure in which to make proper electrical connections. On PV modules this is where PV strings are electrically connected.

## -K-

**kilowatt (kW):** 1000 watts.

**kilowatt-hour (kWh):** One thousand watt hours. The kWh is a unit of energy. 1 kWh=3600 kJ.

## -L-

**life cycle cost:** An estimate of the cost of owning and operating a system for the period of its useful life; usually expressed in terms of the present value of all lifetime costs.

**line-commutated inverter:** An inverter that is tied to a power grid or line. The commutation of power (conversion from DC to AC) is controlled by the power line, so that, if there is a failure in the power grid, the PV system cannot feed power into the line.

**load:** Anything in an electrical circuit that, when the circuit is turned on, draws power from that circuit.

## -M-

**maximum power point (MPP):** The point on the current-voltage (I-V) curve of a module under illumination, where the product of current and voltage is maximum. For a typical silicon cell, this is at about 0.45 V.

**maximum power point tracker (MPPT):** Means of a power conditioning unit that automatically operates the PV generator at its MPP under all conditions.

**megawatt (MW):** One million watts; 1000 kilowatts.

**module:** See 'photovoltaic module.'

**multicrystalline**: Material that is solidified at such as rate that many small crystals (crystallites) form. The atoms within a single crystallite are symmetrically arranged, whereas crystallites are jumbled together. These numerous grain boundaries reduce the device efficiency. A material composed of variously oriented, small individual crystals. (Sometimes referred to as polycrystalline or semicrystalline.)

## -N-

**NEC**: An abbreviation for the National Electrical Code® which contains safety guidelines and required practices for all types of electrical installations. Article 690 pertains to solar photovoltaic systems.

**nominal operating cell temperature (NOCT)**: The reference cell (module) operating temperature presented on manufacturer's literature. Generally the NOCT is referenced at 25°C, 77°F.

**nominal voltage**: A reference voltage used to describe batteries, modules, or systems (ie. a 12-, 24-, or 48-volt battery, module or system).

**nonrenewable fuels**: Fuels that cannot be easily made or "renewed." We can use up nonrenewable fuels. Oil, natural gas, and coal are nonrenewable fuels.

**n-type semiconductor**: A semiconductor produced by doping an intrinsic semiconductor with an electron-donor impurity, for example phosphorous in silicon.

## -O-

**ohm**: The unit of resistance to the flow of an electric current.

**one-axis tracking**: A system capable of rotating about one axis, also referred to as single axis. These tracking systems usually follow the sun from east to west throughout the day.

**open-circuit voltage (Voc)**: The maximum possible voltage across a photovoltaic cell or module; the voltage across the cell in sunlight when no current is flowing.

**orientation**: Placement according to the compass directions, north, south, east, west.

## -P-

**panel**: See 'photovoltaic panel.'

**parallel connection**: A way of joining two or more electricity-producing devices such as PV cells or modules, or batteries by connecting positive leads together and negative leads together; such a configuration increases the current but the voltage is constant.

**passive solar building**: A building that utilizes non-mechanical, non-electrical methods for heating , cooling and/or lighting.

**peak load; peak demand**: The maximum load, or usage, of electrical power occurring in a given period of time, typically a day.

**peak power**: Power generated by a utility unit that operates at a very low capacity factor; generally used to meet short-lived and variable high demand periods.

**peak sun hours**: The equivalent number of hours per day when solar irradiance averages 1000 w/m$^2$ (full sun).

**phosphorous (P)**: A chemical element, atomic number 15, used as a dopant in making n-semiconductor layers.

**photon**: A particle of light that acts as an individual unit of energy.

**photovoltaic (PV)**: Pertaining to the direct conversion of photons of sunlight into electricity.

**photovoltaic array**: An interconnected system of PV modules that function as a single electricity-producing unit. The modules are assembled as a discrete structure, with common support or mounting. In smaller systems, an array can consist of a single module.

**photovoltaic cell**: The smallest semiconductor element within a PV module to perform the immediate conversion of light into electrical energy (DC voltage and current).

**photovoltaic conversion efficiency**: The ratio of the electric power produced by a photovoltaic device to the power of the sunlight incident on the device.

**photovoltaic module**: The smallest environmentally protected, essentially planar assembly of solar cells and ancillary parts, such as interconnections, terminals, and protective devices such as diodes intended to generate DC power under unconcentrated sunlight. The structural (load carrying) member of a module can either be the top layer (superstrate) or the back layer (substrate).

**photovoltaic panel**: Often used interchangeably with PV module (especially in one-module systems), but more accurately used to refer to a physically connected collection of modules (i.e., a laminate string of modules used to achieve a required voltage and current).

**photovoltaic peak watt**: Maximum "rated" output of a cell, module, or system. Typical rating conditions are 0.645 watts per square inch (1000 watts per square meter) of sunlight, 68 degrees F (20 degrees C) ambient air temperature and $6.2 \times 10^{-3}$ mi/s (1 m/s) wind speed.

**photovoltaic system**: A complete set of components for converting sunlight into electricity by the photovoltaic process, including the array and balance of system components.

**physical vapor deposition**: A method of depositing thin semiconductor films. With this method, physical processes, such as thermal evaporation or bombardment of ions, are used to deposit elemental semiconductor material on a substrate.

**p/n**: A semiconductor device structure in which the junction is formed between a p-type layer and an n-type layer.

**polycrystalline**: See 'multicrystalline.'

**power conditioning equipment**: Electrical equipment, or power electronics, used to convert power from a photovoltaic array into a form suitable for subsequent use. A collective term for inverter, converter, battery charge regulator, and blocking diode.

**power factor**: The ratio of the average power and the apparent volt-amperes.

**pulse-width-modulated wave inverter (PWM)**: PWM inverters are the most expensive, but produce a high quality of output signal at minimum current harmonics. The output voltage is very close to sinusoidal.

**PV**: Abbreviation for photovoltaic.

**P-Type silicon**: Semi-conductor grade silicon doped with the element boron giving it a positive bias.

## -Q-

**quad**: A measure of energy equal to one trillion BTUs; an energy equivalent to approximately 172 million barrels of oil.

**qualification test**: A procedure applied to a selected set of PV modules involving the application of defined electrical, mechanical, or thermal stress in a prescribed manner and amount. Test results are subject to a list of defined requirements.

## -R-

**rectifier**: A device that converts AC to DC. See "inverter."

**remote site**: Site which is not located near the utility grid.

**remote systems**: Systems located away from the utility grid.

**resistance (R)**: The property of a conductor which opposes the flow of an electric current resulting in the generation of heat in the conducting material. The unit of resistance is ohms.

## -S-

**satellite power system (SPS)**: Concept for providing large amounts of electricity for use on the Earth from one or more satellites in geosynchronous Earth orbit. A very large array of solar cells on each satellite would provide electricity, which would be converted to microwave energy and beamed to a receiving antenna on the ground. There, it would be reconverted into electricity and distributed the same as any other centrally generated power, through a grid.

**semiconductor**: Any material that has a limited capacity for conducting an electric current. Certain semiconductors, including silicon, gallium arsenide, copper indium dislenide, and cadmium telluride, are uniquely suited to the photovoltaic conversion process.

**semicrystalline**: See 'multicrystalline.'

**series connection:** A way of joining electrical equipment by connecting positive leads to negative leads; such a configuration increases the voltage while current remains the same.

**series regulator:** Type of battery charge regulator where the charging current is controlled by a switch connected in series with the PV module or array.

**shelf life of batteries:** The length of time, under specified conditions, that a battery can be stored so that it keeps its guaranteed capacity.

**short-circuit current (Isc):** The current flowing freely from a photovoltaic cell through an external circuit that has no load or resistance; the maximum current possible.

**shunt regulator:** Type of a battery charge regulator where the charging current is controlled by a switch connected in parallel with the PV generator. Overcharging of the battery is prevented by shorting the PV generator.

**silicon (Si):** A chemical element, atomic number 14, semimetallic in nature, dark gray, an excellent semiconductor material. A common constituent of sand and quartz (as the oxide). Crystallizes in face-centered cubic lattice-like a diamond. The most common semiconductor material used in making photovoltaic devices.

**sine wave inverter:** An inverter that produces utility-quality, sine wave power forms.

**single-crystal material:** A material that is composed of a single crystal or a few large crystals.

**solar cell:** See 'photovoltaic cell.'

**solar constant:** The strength of sunlight; 1353 watts per square meter in space and about 1000 watts per square meter at sea level at the equator at solar noon.

**solar energy:** Energy from the sun. For example, the heat that builds up in your car when the windows are closed is solar energy.

**solar-grade silicon:** Intermediate-grade silicon used in the manufacture of solar cells. Less expensive than electronic-grade silicon.

**solar noon:** That moment of the day that divides the daylight hours for that day exactly in half. To determine solar noon, calculate the length of the day from the time of sunset and sunrise and divide by two. The moment the sun is highest in the sky.

**solar spectrum:** The total distribution of electromagnetic radiation emanating from the sun.

**solar thermal electric:** Method of producing electricity from solar energy by using focused sunlight to heat a working fluid, which in turn drives a turbogenerator.

**square wave inverter:** The inverter consists of a DC source, four switches, and the load. The switches are power semiconductors that can carry a large current and withstand a high voltage rating. The switches are turned on and off at a correct sequence, at a certain frequency. The square wave inverter is the simplest and the least expensive to purchase, but it produces the lowest quality of power.

**Staebler-Wronski effect:** The tendency of amorphous silicon photovoltaic devices to lose efficiency upon initial exposure to light; named for Dr. David Staebler and Dr. Christopher Wronski; work performed at RCA.

**stand-alone:** An autonomous or hybrid photovoltaic system not connected to a grid. Some stand-alone systems require batteries or some other form of storage. Also called, "stand-alone PV system."

**stand-off mounting:** Technique for mounting a PV array on a sloped roof, which involves mounting the modules a short distance above the pitched roof and tilting them to the optimum angle. This promotes air flow to cool the modules.

**standard reporting conditions (SRC):** A fixed set of conditions (including meteorological) to which the electrical performance data of a photovoltaic module is translated from the set of actual test conditions [ASTM E 1036].

**standard test conditions (STC):** Conditions under which a module is typically tested in a laboratory: (1) Irradiance intensity of 1000 W/square meter (0.645 watts per square inch), AM1.5 solar reference spectrum, and (3) a cell (module) temperature of 25 °C, plus or minus 2 °C (77 °F, plus or minus 3.6 °F).

**state of charge (SOC):** The available capacity remaining in a cell or battery, expressed as a percentage of the rated capacity. For example, if 25 amp-hours have been removed from a fully charged 100 amp-hour cell, the state of charge is 75 percent.

**substrate:** The physical material upon which a photovoltaic cell is made.

**sulfation:** A condition that afflicts unused and discharged batteries; large crystals of lead sulfate grow on the plate, instead of the usual tiny crystals, making the battery extremely difficult to recharge.

**superconductivity:** The pairing of electrons in certain materials that, when cooled below a critical temperature, cause the material to lose all resistance to electricity flow. Superconductors can carry electric current without any energy losses.

**superstrate:** The covering on the sun side of a PV module, providing protection for the PV materials from impact and environmental degradation while allowing maximum transmission of the appropriate wavelengths of the solar spectrum.

**surge:** The momentary start-up condition of a motor requiring a large amount of electrical current.

**surge capacity:** The ability of an inverter or generator to deliver high currents momentarily required when starting a motor.

## -T-

**temperature compensation:** An allowance made in charge controller set points for changing battery temperatures.

**thermal electric:** Electric energy derived from heat energy, usually by heating a working fluid, which drives a turbogenerator See also 'solar thermal electric.'

**thermal mass:** Materials, typically masonry, that store heat in a passive solar home.

**thin film:** A layer of semiconductor material, such as copper indium diselenide, cadmium telluride, gallium arsenide, or amorphous silicon, a few microns or less in thickness, used to make photovoltaic cells.

**tilt angle:** Angle of inclination of collector as measured in degrees from the horizontal. For maximum performance solar collectors/modules should be set at a perpendicular to the sun.

**total harmonic distortion (thd):** The measure of closeness in shape between a waveform and its fundamental component.

**tracking PV array:** PV array that follows the path of the sun to maximize the solar radiation incident on the PV surface. The two most common orientations are (1) one axis where the array tracks the sun east to west and (2) two-axis tracking where the array points directly at the sun at all times. Tracking arrays use both the direct and diffuse sunlight. Two-axis tracking arrays capture the maximum possible daily energy.

**transformer:** An electromagnetic device used to convert AC electricity, either to increase or decrease the voltage.

**transmission lines:** Conductors used to transmit high-voltage electricity from the transformer to the electric distribution system.

**trickle charge:** A charge at a low rate, balancing through self-discharge losses, to maintain a cell or battery in a fully charged condition.

**two-axis tracking:** A system capable of rotating independently about two axes and following the sun's orientation and height in the sky (e.g., vertical and horizontal).

## -U-

**ultraviolet (UV):** Electromagnetic radiation in the wavelength range of 4 to 400 nanometers.

**uninterruptible power supply (UPS):** The designation of a power supply providing continuous uninterruptible service when a main power source is lost.

**utility-interactive inverter:** An inverter that can function only when tied to the utility grid, and uses the prevailing line-voltage frequency on the utility line as a control parameter to ensure that the PV system's output is fully synchronized with the utility power.

## -V-

**Vac:** Volts AC.

**Vdc:** Volts DC.

**Voc:** Open-circuit voltage.

**vacuum deposition**: Method of depositing thin coatings of a substance by heating it in a vacuum system.

**vacuum evaporation**: The deposition of thin films of semiconductor material by the evaporation of elemental sources in a vacuum.

**volt (V)**: A unit of measure of the force, or 'push,' given the electrons in an electric circuit. One volt produces one ampere of current when acting against a resistance of one ohm.

**voltage at maximum power (Vmp)**: The voltage at which maximum power is available from a module.

## -W-

**wafer**: A thin sheet of semiconductor material made by mechanically sawing it from a single-crystal or multicrystal ingot or casting.

**watt (W)**: The unit of electric power, or amount of work. One ampere of current flowing at a potential of one volt produces one watt of power.

**watt-hour (Wh)**: A quantity of electrical energy when one watt is used for one hour.

**waveform**: The shape of the curve graphically representing the change in the AC signal voltage and current amplitude, with respect to time.

# Appendix B: Solar Data

The U.S. solar data contained in this appendix is from the *Solar Radiation Data Manual for Flat-Plate and Concentrating Collectors* and was provided by the National Renewable Energy Laboratory (NREL). The data was compiled from the National Solar Radiation Database, a database of hourly solar radiation data collected by the National Weather Service from 1961 to 1990. There are 239 sites recorded.

The international solar data was provided by Vern Risser of Daystar, Inc. and was compiled for Sandia National Laboratories. There are 46 sites.

The solar radiation data is displayed as monthly and yearly averages, expressed as $kWh/m^2/day$. Each site has data for seven configurations:

- Modules facing south with a tilt equal to latitude.

- Modules facing south with a tilt equal to latitude + 15°.

- Modules facing south with a tilt equal to latitude - 15°.

- Single axis tracker with a tilt equal to latitude.

- Single axis tracker with a tilt equal to latitude + 15°.

- Single axis tracker with a tilt equal to latitude - 15°.

- Dual axis tracker.

Single axis trackers pivot on one axis to track the sun, facing east in the morning and west in the afternoon. The data presented assumes continuous tracking of the sun throughout the day.

Data for dual axis trackers represents the maximum solar radiation at a site available to a PV module. Tracking the sun in both azimuth and elevation; the trackers keep the sun's rays normal to the module surface.

The website portal to NREL's Solar Resource information is: http://rredc.nrel.gov/solar

# United States Daily Insolation Data (KWh/m²)

## ANCHORAGE  AK  Latitude: 61.17 degrees  Elevation: 35 meters

|  | Jan | Feb | Mar | Apr | May | Jun | Jul | Aug | Sep | Oct | Nov | Dec | Avg |
|---|---|---|---|---|---|---|---|---|---|---|---|---|---|
| Fixed array |  |  |  |  |  |  |  |  |  |  |  |  |  |
| Lat - 15 | 0.9 | 2.1 | 3.8 | 4.7 | 4.9 | 5.0 | 4.8 | 4.1 | 3.1 | 2.0 | 1.1 | 0.5 | 3.1 |
| Latitude | 1.0 | 2.2 | 3.9 | 4.6 | 4.6 | 4.5 | 4.4 | 3.8 | 3.1 | 2.1 | 1.2 | 0.6 | 3.0 |
| Lat + 15 | 1.0 | 2.3 | 3.9 | 4.3 | 4.0 | 3.9 | 3.8 | 3.4 | 2.9 | 2.0 | 1.3 | 0.6 | 2.8 |
| Single axis tracker |  |  |  |  |  |  |  |  |  |  |  |  |  |
| Lat - 15 | 0.9 | 2.4 | 4.6 | 6.1 | 6.6 | 6.6 | 6.3 | 5.2 | 3.8 | 2.2 | 1.2 | 0.5 | 3.9 |
| Latitude | 1.0 | 2.5 | 4.8 | 6.1 | 6.4 | 6.3 | 6.1 | 5.1 | 3.8 | 2.3 | 1.3 | 0.6 | 3.9 |
| Lat + 15 | 1.1 | 2.6 | 4.7 | 5.8 | 6 | 5.9 | 3.7 | 4.8 | 3.6 | 2.3 | 1.4 | 0.6 | 3.7 |
| Dual axis tracker |  |  |  |  |  |  |  |  |  |  |  |  |  |
|  | 1.1 | 2.6 | 4.8 | 6.1 | 6.7 | 6.8 | 6.5 | 5.2 | 3.8 | 2.4 | 1.4 | 0.7 | 4.0 |

## ANNETTE  AK  Latitude: 55.03 degrees  Elevation: 34 meters

|  | Jan | Feb | Mar | Apr | May | Jun | Jul | Aug | Sep | Oct | Nov | Dec | Avg |
|---|---|---|---|---|---|---|---|---|---|---|---|---|---|
| Fixed array |  |  |  |  |  |  |  |  |  |  |  |  |  |
| Lat - 15 | 1.2 | 2.1 | 3.1 | 4.1 | 4.9 | 4.9 | 4.9 | 4.4 | 3.5 | 2.1 | 1.3 | 0.9 | 3.1 |
| Latitude | 1.4 | 2.2 | 3.1 | 4.0 | 4.5 | 4.5 | 4.5 | 4.2 | 3.5 | 2.1 | 1.5 | 1.1 | 3.0 |
| Lat + 15 | 1.4 | 2.2 | 3.0 | 3.6 | 4.0 | 3.9 | 3.9 | 3.7 | 3.2 | 2.1 | 1.5 | 1.1 | 2.8 |
| Single axis tracker |  |  |  |  |  |  |  |  |  |  |  |  |  |
| Lat - 15 | 1.4 | 2.4 | 3.7 | 5.3 | 6.4 | 6.5 | 6.4 | 5.7 | 4.3 | 2.3 | 1.5 | 1.0 | 3.9 |
| Latitude | 1.5 | 2.5 | 3.7 | 5.2 | 6.2 | 6.2 | 6.2 | 5.6 | 4.3 | 2.4 | 1.6 | 1.1 | 3.9 |
| Lat + 15 | 1.6 | 2.5 | 3.6 | 4.9 | 5.9 | 5.8 | 5.8 | 5.3 | 4.2 | 2.4 | 1.6 | 1.2 | 3.7 |
| Dual axis tracker |  |  |  |  |  |  |  |  |  |  |  |  |  |
|  | 1.6 | 2.5 | 3.8 | 5.3 | 6.6 | 6.7 | 6.6 | 5.8 | 4.4 | 2.4 | 1.7 | 1.2 | 4.0 |

## BARROW  AK  Latitude: 71.30 degrees  Elevation: 4 meters

|  | Jan | Feb | Mar | Apr | May | Jun | Jul | Aug | Sep | Oct | Nov | Dec | Avg |
|---|---|---|---|---|---|---|---|---|---|---|---|---|---|
| Fixed array |  |  |  |  |  |  |  |  |  |  |  |  |  |
| Lat - 15 | 0.0 | 1.1 | 3.8 | 5.8 | 5.2 | 4.8 | 4.6 | 2.8 | 1.7 | 0.9 | 0.1 | 0.0 | 2.6 |
| Latitude | 0.0 | 1.1 | 4.0 | 5.8 | 4.9 | 4.4 | 4.2 | 2.6 | 1.6 | 1.0 | 0.1 | 0.0 | 2.5 |
| Lat + 15 | 0.0 | 1.2 | 4.0 | 5.6 | 4.6 | 3.8 | 3.6 | 2.3 | 1.5 | 1.0 | 0.1 | 0.0 | 2.3 |
| Single axis tracker |  |  |  |  |  |  |  |  |  |  |  |  |  |
| Lat - 15 | 0.0 | 1.2 | 4.9 | 8.2 | 7.0 | 6.9 | 6.6 | 3.5 | 1.9 | 1.0 | 0.1 | 0.0 | 3.5 |
| Latitude | 0.0 | 1.2 | 5.0 | 8.2 | 6.9 | 6.7 | 6.4 | 3.4 | 1.9 | 1.1 | 0.1 | 0.0 | 3.4 |
| Lat + 15 | 0 | 1.3 | 5 | 8 | 6.7 | 6.4 | 6.1 | 3.2 | 1.8 | 1.1 | 0.1 | 0 | 3.3 |
| Dual axis tracker |  |  |  |  |  |  |  |  |  |  |  |  |  |
|  | 0.0 | 1.3 | 5.0 | 8.3 | 7.1 | 7.3 | 6.9 | 3.6 | 1.9 | 1.1 | 0.1 | 0.0 | 3.6 |

## BETHEL  AK  Latitude: 60.78 degrees  Elevation: 46 meters

|  | Jan | Feb | Mar | Apr | May | Jun | Jul | Aug | Sep | Oct | Nov | Dec | Avg |
|---|---|---|---|---|---|---|---|---|---|---|---|---|---|
| Fixed array |  |  |  |  |  |  |  |  |  |  |  |  |  |
| Lat - 15 | 1.2 | 2.7 | 4.3 | 5.1 | 4.8 | 4.7 | 4.2 | 3.4 | 3.0 | 2.1 | 1.3 | 0.8 | 3.1 |
| Latitude | 1.4 | 3.0 | 4.5 | 5.0 | 4.5 | 4.2 | 3.8 | 3.2 | 2.9 | 2.2 | 1.4 | 1.0 | 3.1 |
| Lat + 15 | 1.5 | 3.1 | 4.4 | 4.7 | 4.0 | 3.6 | 3.3 | 2.9 | 2.8 | 2.1 | 1.5 | 1.0 | 2.9 |
| Single axis tracker |  |  |  |  |  |  |  |  |  |  |  |  |  |
| Lat - 15 | 1.3 | 3.3 | 5.5 | 6.9 | 6.6 | 6.3 | 5.5 | 4.3 | 3.7 | 2.4 | 1.4 | 0.9 | 4.0 |
| Latitude | 1.5 | 3.5 | 5.6 | 6.9 | 6.4 | 6.0 | 5.3 | 4.1 | 3.7 | 2.5 | 1.6 | 1.0 | 4.0 |
| Lat + 15 | 1.6 | 3.6 | 5.6 | 6.7 | 6 | 5.6 | 4.9 | 3.9 | 3.5 | 2.5 | 1.6 | 1.1 | 3.9 |
| Dual axis tracker |  |  |  |  |  |  |  |  |  |  |  |  |  |
|  | 1.6 | 3.6 | 5.6 | 6.9 | 6.7 | 6.5 | 5.7 | 4.3 | 3.7 | 2.5 | 1.6 | 1.1 | 4.2 |

## BETTLES AK    Latitude: 66.92 degrees    Elevation: 205 meters

| | Jan | Feb | Mar | Apr | May | Jun | Jul | Aug | Sep | Oct | Nov | Dec | Avg |
|---|---|---|---|---|---|---|---|---|---|---|---|---|---|
| Fixed array | | | | | | | | | | | | | |
| Lat - 15 | 0.4 | 2.0 | 4.2 | 5.8 | 6.2 | 5.8 | 5.2 | 4.2 | 3.3 | 1.9 | 0.7 | 0.1 | 3.3 |
| Latitude | 0.5 | 2.2 | 4.4 | 5.8 | 5.9 | 5.3 | 4.8 | 4.0 | 3.2 | 2.0 | 0.8 | 0.1 | 3.2 |
| Lat + 15 | 0.5 | 2.3 | 4.4 | 5.5 | 5.2 | 4.5 | 4.1 | 3.5 | 3.0 | 2.0 | 0.8 | 0.1 | 3.0 |
| Single axis tracker | | | | | | | | | | | | | |
| Lat - 15 | 0.5 | 2.3 | 5.4 | 8.1 | 9.4 | 9.0 | 7.7 | 5.8 | 4.2 | 2.1 | 0.8 | 0.1 | 4.6 |
| Latitude | 0.5 | 2.5 | 5.5 | 8.1 | 9.2 | 8.8 | 7.5 | 5.6 | 4.2 | 2.2 | 0.8 | 0.1 | 4.6 |
| Lat + 15 | 0.6 | 2.6 | 5.5 | 7.9 | 8.8 | 8.3 | 7.1 | 5.3 | 4 | 2.2 | 0.9 | 0.1 | 4.5 |
| Dual axis tracker | | | | | | | | | | | | | |
| | 0.6 | 2.6 | 5.5 | 8.1 | 9.5 | 9.4 | 8.0 | 5.8 | 4.2 | 2.3 | 0.9 | 0.1 | 4.8 |

## BIG DELTA AK    Latitude: 64.00 degrees    Elevation: 388 meters

| | Jan | Feb | Mar | Apr | May | Jun | Jul | Aug | Sep | Oct | Nov | Dec | Avg |
|---|---|---|---|---|---|---|---|---|---|---|---|---|---|
| Fixed array | | | | | | | | | | | | | |
| Lat - 15 | 1.0 | 2.4 | 4.4 | 5.5 | 5.7 | 5.6 | 5.4 | 4.7 | 3.6 | 2.1 | 1.2 | 0.5 | 3.5 |
| Latitude | 1.1 | 2.6 | 4.6 | 5.4 | 5.3 | 5.1 | 5.0 | 4.4 | 3.6 | 2.2 | 1.3 | 0.6 | 3.4 |
| Lat + 15 | 1.2 | 2.7 | 4.6 | 5.1 | 4.7 | 4.4 | 4.3 | 4.0 | 3.4 | 2.2 | 1.4 | 0.6 | 3.2 |
| Single axis tracker | | | | | | | | | | | | | |
| Lat - 15 | 1.1 | 2.8 | 5.7 | 7.6 | 8.3 | 8.1 | 7.8 | 6.4 | 4.6 | 2.4 | 1.3 | 0.5 | 4.7 |
| Latitude | 1.2 | 3.0 | 5.9 | 7.6 | 8.1 | 7.8 | 7.5 | 6.3 | 4.6 | 2.5 | 1.5 | 0.6 | 4.7 |
| Lat + 15 | 1.3 | 3.9 | 5.9 | 7.3 | 7.7 | 7.4 | 7.1 | 5.9 | 4.5 | 2.5 | 1.5 | 0.6 | 4.6 |
| Dual axis tracker | | | | | | | | | | | | | |
| | 1.3 | 3.1 | 5.9 | 7.6 | 8.5 | 8.4 | 8.0 | 6.5 | 4.7 | 2.5 | 1.5 | 0.6 | 4.9 |

## COLD BAY AK    Latitude: 55.20 degrees    Elevation: 29 meters

| | Jan | Feb | Mar | Apr | May | Jun | Jul | Aug | Sep | Oct | Nov | Dec | Avg |
|---|---|---|---|---|---|---|---|---|---|---|---|---|---|
| Fixed array | | | | | | | | | | | | | |
| Lat - 15 | 1.2 | 2.0 | 3.0 | 3.3 | 3.6 | 3.7 | 3.5 | 3.0 | 2.5 | 1.9 | 1.2 | 0.9 | 2.5 |
| Latitude | 1.4 | 2.1 | 3.0 | 3.2 | 3.3 | 3.4 | 3.2 | 2.8 | 2.4 | 2.0 | 1.3 | 1.0 | 2.4 |
| Lat + 15 | 1.4 | 2.1 | 2.9 | 2.9 | 2.9 | 2.9 | 2.8 | 2.5 | 2.2 | 1.9 | 1.4 | 1.1 | 2.3 |
| Single axis tracker | | | | | | | | | | | | | |
| Lat - 15 | 1.4 | 2.2 | 3.5 | 3.9 | 4.2 | 4.3 | 4.0 | 3.4 | 2.8 | 2.1 | 1.3 | 1.0 | 2.9 |
| Latitude | 1.5 | 2.3 | 3.6 | 3.8 | 4.1 | 4.1 | 3.8 | 3.2 | 2.8 | 2.2 | 1.4 | 1.1 | 2.8 |
| Lat + 15 | 1.6 | 2.4 | 3.5 | 3.7 | 3.8 | 3.8 | 3.5 | 3 | 2.7 | 2.2 | 1.5 | 1.1 | 2.7 |
| Dual axis tracker | | | | | | | | | | | | | |
| | 1.6 | 2.4 | 3.6 | 4.0 | 4.4 | 4.5 | 4.2 | 3.4 | 2.9 | 2.2 | 1.5 | 1.1 | 3.0 |

## FAIRBANKS AK    Latitude: 64.82 degrees    Elevation: 138 meters

| | Jan | Feb | Mar | Apr | May | Jun | Jul | Aug | Sep | Oct | Nov | Dec | Avg |
|---|---|---|---|---|---|---|---|---|---|---|---|---|---|
| Fixed array | | | | | | | | | | | | | |
| Lat - 15 | 0.7 | 2.2 | 4.5 | 5.6 | 5.7 | 5.7 | 5.4 | 4.5 | 3.4 | 1.9 | 1.0 | 0.2 | 3.4 |
| Latitude | 0.7 | 2.4 | 4.7 | 5.6 | 5.3 | 5.2 | 4.9 | 4.2 | 3.4 | 2.0 | 1.1 | 0.3 | 3.3 |
| Lat + 15 | 0.8 | 2.5 | 4.7 | 5.3 | 4.6 | 4.5 | 4.3 | 3.8 | 3.2 | 2.0 | 1.1 | 0.3 | 3.1 |
| Single axis tracker | | | | | | | | | | | | | |
| Lat - 15 | 0.7 | 2.6 | 5.7 | 7.7 | 8.2 | 8.3 | 7.6 | 6.0 | 4.4 | 2.2 | 1.1 | 0.2 | 4.6 |
| Latitude | 0.8 | 2.7 | 5.9 | 7.7 | 8.0 | 8.0 | 7.4 | 5.8 | 4.4 | 2.3 | 1.2 | 0.3 | 4.5 |
| Lat + 15 | 0.8 | 2.8 | 5.8 | 7.4 | 7.6 | 7.6 | 6.9 | 5.5 | 4.2 | 2.3 | 1.2 | 0.3 | 4.4 |
| Dual axis tracker | | | | | | | | | | | | | |
| | 0.8 | 2.8 | 5.8 | 7.7 | 8.4 | 8.7 | 7.9 | 6.0 | 4.4 | 2.3 | 1.2 | 0.3 | 4.7 |

## GULKANA  AK  Latitude: 62.15 degrees  Elevation: 481 meters

|              | Jan | Feb | Mar | Apr | May | Jun | Jul | Aug | Sep | Oct | Nov | Dec | Avg |
|--------------|-----|-----|-----|-----|-----|-----|-----|-----|-----|-----|-----|-----|-----|
| Fixed array  |     |     |     |     |     |     |     |     |     |     |     |     |     |
| Lat - 15     | 1.1 | 2.5 | 4.5 | 5.6 | 5.6 | 5.5 | 5.5 | 5.0 | 3.9 | 2.4 | 1.3 | 0.6 | 3.6 |
| Latitude     | 1.2 | 2.7 | 4.7 | 5.6 | 5.2 | 5.0 | 5.1 | 4.7 | 3.9 | 2.5 | 1.4 | 0.7 | 3.6 |
| Lat + 15     | 1.3 | 2.8 | 4.7 | 5.3 | 4.6 | 4.3 | 4.4 | 4.3 | 3.6 | 2.5 | 1.5 | 0.8 | 3.3 |
| Single axis tracker |  |  |   |     |     |     |     |     |     |     |     |     |     |
| Lat - 15     | 1.2 | 2.9 | 5.8 | 7.8 | 8.1 | 8.1 | 7.9 | 6.8 | 5.0 | 2.8 | 1.4 | 0.7 | 4.9 |
| Latitude     | 1.3 | 3.1 | 6.0 | 7.8 | 7.9 | 7.8 | 7.6 | 6.7 | 5.0 | 2.9 | 1.5 | 0.8 | 4.9 |
| Lat + 15     | 1.4 | 3.2 | 6   | 7.6 | 7.5 | 7.4 | 7.2 | 6.4 | 4.8 | 2.9 | 1.6 | 0.8 | 4.7 |
| Dual axis tracker |  |   |     |     |     |     |     |     |     |     |     |     |     |
|              | 1.4 | 3.2 | 6.0 | 7.9 | 8.3 | 8.4 | 8.1 | 6.9 | 5.0 | 2.9 | 1.6 | 0.8 | 5.1 |

## KING SALMON  AK  Latitude: 58.68 degrees  Elevation: 15 meters

|              | Jan | Feb | Mar | Apr | May | Jun | Jul | Aug | Sep | Oct | Nov | Dec | Avg |
|--------------|-----|-----|-----|-----|-----|-----|-----|-----|-----|-----|-----|-----|-----|
| Fixed array  |     |     |     |     |     |     |     |     |     |     |     |     |     |
| Lat - 15     | 1.4 | 2.5 | 3.7 | 4.4 | 4.5 | 4.5 | 4.3 | 3.6 | 3.0 | 2.4 | 1.5 | 1.0 | 3.1 |
| Latitude     | 1.5 | 2.7 | 3.9 | 4.3 | 4.2 | 4.1 | 3.9 | 3.4 | 2.9 | 2.5 | 1.6 | 1.2 | 3.0 |
| Lat + 15     | 1.6 | 2.8 | 3.8 | 3.9 | 3.7 | 3.6 | 3.4 | 3.0 | 2.8 | 2.5 | 1.7 | 1.3 | 2.8 |
| Single axis tracker |  |  |   |     |     |     |     |     |     |     |     |     |     |
| Lat - 15     | 1.5 | 3.0 | 4.7 | 5.8 | 6.0 | 5.8 | 5.4 | 4.5 | 3.6 | 2.8 | 1.7 | 1.1 | 3.8 |
| Latitude     | 1.7 | 3.2 | 4.8 | 5.7 | 5.8 | 5.6 | 5.2 | 4.3 | 3.6 | 2.9 | 1.8 | 1.3 | 3.8 |
| Lat + 15     | 1.8 | 3.2 | 4.8 | 5.5 | 5.4 | 5.2 | 4.8 | 4.1 | 3.5 | 2.9 | 1.9 | 1.3 | 3.7 |
| Dual axis tracker |  |   |     |     |     |     |     |     |     |     |     |     |     |
|              | 1.8 | 3.3 | 4.8 | 5.8 | 6.1 | 6.0 | 5.6 | 4.5 | 3.7 | 2.9 | 1.9 | 1.3 | 4.0 |

## KODIAK  AK  Latitude: 57.75 degrees  Elevation: 34 meters

|              | Jan | Feb | Mar | Apr | May | Jun | Jul | Aug | Sep | Oct | Nov | Dec | Avg |
|--------------|-----|-----|-----|-----|-----|-----|-----|-----|-----|-----|-----|-----|-----|
| Fixed array  |     |     |     |     |     |     |     |     |     |     |     |     |     |
| Lat - 15     | 1.2 | 2.2 | 3.5 | 4.3 | 4.4 | 4.5 | 4.5 | 4.2 | 3.3 | 2.6 | 1.6 | 1.0 | 3.1 |
| Latitude     | 1.4 | 2.4 | 3.7 | 4.2 | 4.1 | 4.1 | 4.1 | 4.0 | 3.3 | 2.8 | 1.8 | 1.2 | 3.1 |
| Lat + 15     | 1.5 | 2.5 | 3.6 | 3.9 | 3.6 | 3.5 | 3.6 | 3.6 | 3.1 | 2.8 | 1.8 | 1.2 | 2.9 |
| Single axis tracker |  |  |   |     |     |     |     |     |     |     |     |     |     |
| Lat - 15     | 1.4 | 2.6 | 4.5 | 5.7 | 5.9 | 5.9 | 5.9 | 5.5 | 4.2 | 3.2 | 1.8 | 1.1 | 4.0 |
| Latitude     | 1.5 | 2.8 | 4.6 | 5.7 | 5.7 | 5.7 | 5.7 | 5.4 | 4.2 | 3.3 | 2.0 | 1.2 | 4.0 |
| Lat + 15     | 1.6 | 2.8 | 4.5 | 5.4 | 5.3 | 5.3 | 5.3 | 5.1 | 4   | 3.3 | 2   | 1.3 | 3.8 |
| Dual axis tracker |  |   |     |     |     |     |     |     |     |     |     |     |     |
|              | 1.6 | 2.8 | 4.6 | 5.8 | 6.0 | 6.2 | 6.1 | 5.6 | 4.2 | 3.3 | 2.0 | 1.3 | 4.1 |

## KOTZEBUE  AK  Latitude: 66.87 degrees  Elevation: 5 meters

|              | Jan | Feb | Mar | Apr | May | Jun | Jul | Aug | Sep | Oct | Nov | Dec | Avg |
|--------------|-----|-----|-----|-----|-----|-----|-----|-----|-----|-----|-----|-----|-----|
| Fixed array  |     |     |     |     |     |     |     |     |     |     |     |     |     |
| Lat - 15     | 0.4 | 2.1 | 4.3 | 6.0 | 6.4 | 5.7 | 5.0 | 3.9 | 3.1 | 2.1 | 0.7 | 0.1 | 3.3 |
| Latitude     | 0.5 | 2.3 | 4.5 | 6.0 | 6.1 | 5.2 | 4.6 | 3.7 | 3.1 | 2.2 | 0.7 | 0.1 | 3.2 |
| Lat + 15     | 0.5 | 2.3 | 4.5 | 5.7 | 5.6 | 4.4 | 3.9 | 3.3 | 2.9 | 2.2 | 0.8 | 0.1 | 3.0 |
| Single axis tracker |  |  |   |     |     |     |     |     |     |     |     |     |     |
| Lat - 15     | 0.4 | 2.4 | 5.5 | 8.4 | 9.5 | 8.7 | 7.3 | 5.3 | 3.9 | 2.4 | 0.7 | 0.1 | 4.6 |
| Latitude     | 0.5 | 2.5 | 5.7 | 8.4 | 9.4 | 8.4 | 7.1 | 5.2 | 3.9 | 2.5 | 0.8 | 0.1 | 4.5 |
| Lat + 15     | 0.5 | 2.6 | 5.7 | 8.2 | 9   | 8   | 6.7 | 4.9 | 3.8 | 2.5 | 0.8 | 0.1 | 4.4 |
| Dual axis tracker |  |   |     |     |     |     |     |     |     |     |     |     |     |
|              | 0.5 | 2.6 | 5.7 | 8.4 | 9.6 | 9.0 | 7.6 | 5.4 | 4.0 | 2.5 | 0.8 | 0.1 | 4.7 |

## MCGRATH AK    Latitude: 62.97 degrees    Elevation: 103 meters

| | Jan | Feb | Mar | Apr | May | Jun | Jul | Aug | Sep | Oct | Nov | Dec | Avg |
|---|---|---|---|---|---|---|---|---|---|---|---|---|---|
| Fixed array | | | | | | | | | | | | | |
| Lat - 15 | 1.0 | 2.5 | 4.4 | 5.6 | 5.3 | 5.1 | 4.7 | 4.0 | 3.1 | 2.0 | 1.1 | 0.6 | 3.3 |
| Latitude | 1.2 | 2.8 | 4.7 | 5.6 | 4.9 | 4.6 | 4.3 | 3.7 | 3.1 | 2.1 | 1.3 | 0.6 | 3.2 |
| Lat + 15 | 1.2 | 2.9 | 4.6 | 5.4 | 4.4 | 4.0 | 3.7 | 3.3 | 2.9 | 2.1 | 1.3 | 0.7 | 3.0 |
| Single axis tracker | | | | | | | | | | | | | |
| Lat - 15 | 1.1 | 3.0 | 5.7 | 7.7 | 7.5 | 7.2 | 6.6 | 5.3 | 4.0 | 2.3 | 1.3 | 0.6 | 4.3 |
| Latitude | 1.2 | 3.2 | 5.9 | 7.7 | 7.3 | 7.0 | 6.3 | 5.1 | 4.0 | 2.4 | 1.4 | 0.7 | 4.3 |
| Lat + 15 | 1.3 | 3.3 | 5.9 | 7.5 | 7 | 6.6 | 5.9 | 4.9 | 3.8 | 2.4 | 1.4 | 0.7 | 4.2 |
| Dual axis tracker | | | | | | | | | | | | | |
| | 1.3 | 3.3 | 5.9 | 7.7 | 7.7 | 7.5 | 6.8 | 5.3 | 4.0 | 2.4 | 1.4 | 0.7 | 4.5 |

## NOME AK    Latitude: 64.50 degrees    Elevation: 7 meters

| | Jan | Feb | Mar | Apr | May | Jun | Jul | Aug | Sep | Oct | Nov | Dec | Avg |
|---|---|---|---|---|---|---|---|---|---|---|---|---|---|
| Fixed array | | | | | | | | | | | | | |
| Lat - 15 | 0.8 | 2.5 | 4.5 | 5.9 | 5.9 | 5.5 | 4.7 | 3.7 | 3.2 | 2.2 | 1.0 | 0.4 | 3.4 |
| Latitude | 0.9 | 2.8 | 4.7 | 5.9 | 5.6 | 5.0 | 4.3 | 3.5 | 3.2 | 2.4 | 1.1 | 0.5 | 3.3 |
| Lat + 15 | 0.9 | 2.9 | 4.7 | 5.6 | 5.0 | 4.3 | 3.7 | 3.1 | 3.0 | 2.4 | 1.2 | 0.5 | 3.1 |
| Single axis tracker | | | | | | | | | | | | | |
| Lat - 15 | 0.8 | 3.0 | 5.8 | 8.3 | 8.7 | 8.3 | 6.7 | 5.0 | 4.1 | 2.7 | 1.1 | 0.4 | 4.6 |
| Latitude | 0.9 | 3.2 | 5.9 | 8.3 | 8.5 | 8.0 | 6.4 | 4.8 | 4.1 | 2.8 | 1.2 | 0.5 | 4.6 |
| Lat + 15 | 1 | 3.2 | 5.9 | 8.1 | 8.1 | 7.6 | 6.1 | 4.6 | 4 | 2.8 | 1.3 | 0.5 | 4.4 |
| Dual axis tracker | | | | | | | | | | | | | |
| | 1.0 | 3.2 | 5.9 | 8.4 | 8.8 | 8.6 | 6.9 | 5.0 | 4.1 | 2.8 | 1.3 | 0.5 | 4.7 |

## ST PAUL IS. AK    Latitude: 57.15 degrees    Elevation: 7 meters

| | Jan | Feb | Mar | Apr | May | Jun | Jul | Aug | Sep | Oct | Nov | Dec | Avg |
|---|---|---|---|---|---|---|---|---|---|---|---|---|---|
| Fixed array | | | | | | | | | | | | | |
| Lat - 15 | 1.0 | 2.0 | 3.3 | 4.1 | 3.9 | 3.8 | 3.3 | 2.8 | 2.6 | 1.8 | 1.1 | 0.7 | 2.5 |
| Latitude | 1.1 | 2.2 | 3.4 | 4.0 | 3.6 | 3.4 | 3.0 | 2.6 | 2.5 | 1.9 | 1.1 | 0.8 | 2.5 |
| Lat + 15 | 1.1 | 2.2 | 3.3 | 3.7 | 3.2 | 3.0 | 2.6 | 2.3 | 2.3 | 1.8 | 1.2 | 0.8 | 2.3 |
| Single axis tracker | | | | | | | | | | | | | |
| Lat - 15 | 1.1 | 2.3 | 4.0 | 4.9 | 4.6 | 4.3 | 3.6 | 3.1 | 3.0 | 2.0 | 1.1 | 0.8 | 2.9 |
| Latitude | 1.2 | 2.5 | 4.1 | 4.9 | 4.4 | 4.1 | 3.4 | 2.9 | 2.9 | 2.1 | 1.2 | 0.8 | 2.9 |
| Lat + 15 | 1.2 | 2.5 | 4 | 4.7 | 4.1 | 3.8 | 3.1 | 2.7 | 2.8 | 2.1 | 1.2 | 0.9 | 2.8 |
| Dual axis tracker | | | | | | | | | | | | | |
| | 1.2 | 2.5 | 4.1 | 5.0 | 4.7 | 4.5 | 3.8 | 3.1 | 3.0 | 2.1 | 1.3 | 0.9 | 3.0 |

## TALKEETNA AK    Latitude: 62.30 degrees    Elevation: 105 meters

| | Jan | Feb | Mar | Apr | May | Jun | Jul | Aug | Sep | Oct | Nov | Dec | Avg |
|---|---|---|---|---|---|---|---|---|---|---|---|---|---|
| Fixed array | | | | | | | | | | | | | |
| Lat - 15 | 1.2 | 2.5 | 4.1 | 5.5 | 5.2 | 5.0 | 4.8 | 4.2 | 3.3 | 2.3 | 1.4 | 0.7 | 3.3 |
| Latitude | 1.3 | 2.7 | 4.3 | 5.4 | 4.9 | 4.5 | 4.4 | 3.9 | 3.3 | 2.4 | 1.6 | 0.8 | 3.3 |
| Lat + 15 | 1.4 | 2.8 | 4.2 | 5.2 | 4.4 | 3.9 | 3.8 | 3.5 | 3.1 | 2.4 | 1.7 | 0.9 | 3.1 |
| Single axis tracker | | | | | | | | | | | | | |
| Lat - 15 | 1.3 | 2.9 | 5.2 | 7.5 | 7.3 | 6.8 | 6.6 | 5.5 | 4.2 | 2.7 | 1.6 | 0.8 | 4.4 |
| Latitude | 1.4 | 3.1 | 5.3 | 7.5 | 7.1 | 6.6 | 6.3 | 5.3 | 4.2 | 2.8 | 1.8 | 0.9 | 4.4 |
| Lat + 15 | 1.5 | 3.2 | 5.3 | 7.3 | 6.8 | 6.2 | 5.9 | 5 | 4.1 | 2.8 | 1.8 | 0.9 | 4.2 |
| Dual axis tracker | | | | | | | | | | | | | |
| | 1.5 | 3.2 | 5.3 | 7.5 | 7.5 | 7.1 | 6.8 | 5.5 | 4.2 | 2.8 | 1.8 | 0.9 | 4.5 |

## YAKUTAT   AK   Latitude: 59.52 degrees   Elevation: 9 meters

| | Jan | Feb | Mar | Apr | May | Jun | Jul | Aug | Sep | Oct | Nov | Dec | Avg |
|---|---|---|---|---|---|---|---|---|---|---|---|---|---|
| Fixed array | | | | | | | | | | | | | |
| Lat - 15 | 1.0 | 1.9 | 3.2 | 4.2 | 4.2 | 4.2 | 4.1 | 3.7 | 2.8 | 1.7 | 1.2 | 0.7 | 2.8 |
| Latitude | 1.2 | 2.1 | 3.3 | 4.2 | 3.9 | 3.8 | 3.7 | 3.5 | 2.7 | 1.8 | 1.3 | 0.8 | 2.7 |
| Lat + 15 | 1.2 | 2.1 | 3.3 | 3.9 | 3.5 | 3.3 | 3.2 | 3.1 | 2.5 | 1.7 | 1.3 | 0.8 | 2.5 |
| Single axis tracker | | | | | | | | | | | | | |
| Lat - 15 | 1.1 | 2.2 | 4.0 | 5.5 | 5.5 | 5.4 | 5.2 | 4.7 | 3.4 | 2.0 | 1.3 | 0.8 | 3.4 |
| Latitude | 1.2 | 2.3 | 4.0 | 5.5 | 5.3 | 5.1 | 4.9 | 4.5 | 3.4 | 2.0 | 1.4 | 0.8 | 3.4 |
| Lat + 15 | 1.3 | 2.4 | 4 | 5.3 | 5 | 4.8 | 4.6 | 4.3 | 3.2 | 2 | 1.4 | 0.9 | 3.3 |
| Dual axis tracker | | | | | | | | | | | | | |
| | 1.3 | 2.4 | 4.1 | 5.6 | 5.7 | 5.6 | 5.3 | 4.7 | 3.4 | 2.0 | 1.4 | 0.9 | 3.5 |

## BIRMINGHAM   AL   Latitude: 33.57 degrees   Elevation: 192 meters

| | Jan | Feb | Mar | Apr | May | Jun | Jul | Aug | Sep | Oct | Nov | Dec | Avg |
|---|---|---|---|---|---|---|---|---|---|---|---|---|---|
| Fixed array | | | | | | | | | | | | | |
| Lat - 15 | 3.3 | 4.0 | 4.9 | 5.7 | 6.0 | 6.1 | 5.8 | 5.8 | 5.2 | 4.8 | 3.6 | 3.1 | 4.9 |
| Latitude | 3.7 | 4.4 | 5.1 | 5.6 | 5.6 | 5.6 | 5.4 | 5.6 | 5.3 | 5.2 | 4.1 | 3.5 | 4.9 |
| Lat + 15 | 3.9 | 4.5 | 5.0 | 5.2 | 5.0 | 4.9 | 4.8 | 5.1 | 5.1 | 5.2 | 4.3 | 3.8 | 4.7 |
| Single axis tracker | | | | | | | | | | | | | |
| Lat - 15 | 4.1 | 5.0 | 6.2 | 7.3 | 7.6 | 7.7 | 7.2 | 7.2 | 6.5 | 6.1 | 4.5 | 3.8 | 6.1 |
| Latitude | 4.4 | 5.3 | 6.3 | 7.3 | 7.4 | 7.4 | 6.9 | 7.1 | 6.6 | 6.3 | 4.8 | 4.1 | 6.2 |
| Lat + 15 | 4.6 | 5.4 | 6.2 | 7 | 6.9 | 6.9 | 6.5 | 6.7 | 6.4 | 6.4 | 5 | 4.3 | 6.1 |
| Dual axis tracker | | | | | | | | | | | | | |
| | 4.6 | 5.4 | 6.3 | 7.4 | 7.7 | 7.8 | 7.3 | 7.2 | 6.6 | 6.4 | 5.0 | 4.4 | 6.4 |

## HUNTSVILLE   AL   Latitude: 34.65 degrees   Elevation: 190 meters

| | Jan | Feb | Mar | Apr | May | Jun | Jul | Aug | Sep | Oct | Nov | Dec | Avg |
|---|---|---|---|---|---|---|---|---|---|---|---|---|---|
| Fixed array | | | | | | | | | | | | | |
| Lat - 15 | 3.1 | 3.8 | 4.7 | 5.6 | 5.9 | 6.2 | 6.0 | 5.9 | 5.2 | 4.7 | 3.5 | 2.9 | 4.8 |
| Latitude | 3.5 | 4.2 | 4.8 | 5.5 | 5.6 | 5.7 | 5.6 | 5.7 | 5.3 | 5.0 | 3.9 | 3.3 | 4.8 |
| Lat + 15 | 3.7 | 4.3 | 4.7 | 5.1 | 5.0 | 4.9 | 4.9 | 5.2 | 5.0 | 5.1 | 4.1 | 3.5 | 4.6 |
| Single axis tracker | | | | | | | | | | | | | |
| Lat - 15 | 3.8 | 4.7 | 5.9 | 7.3 | 7.6 | 7.9 | 7.6 | 7.5 | 6.5 | 5.9 | 4.2 | 3.4 | 6.0 |
| Latitude | 4.1 | 5.0 | 6.0 | 7.2 | 7.4 | 7.6 | 7.3 | 7.3 | 6.6 | 6.2 | 4.5 | 3.8 | 6.1 |
| Lat + 15 | 4.3 | 5.1 | 5.9 | 6.9 | 6.9 | 7.1 | 6.8 | 7 | 6.4 | 6.2 | 4.7 | 3.9 | 5.9 |
| Dual axis tracker | | | | | | | | | | | | | |
| | 4.3 | 5.1 | 6.0 | 7.3 | 7.7 | 8.1 | 7.7 | 7.5 | 6.6 | 6.3 | 4.7 | 4.0 | 6.3 |

## MOBILE   AL   Latitude: 30.68 degrees   Elevation: 67 meters

| | Jan | Feb | Mar | Apr | May | Jun | Jul | Aug | Sep | Oct | Nov | Dec | Avg |
|---|---|---|---|---|---|---|---|---|---|---|---|---|---|
| Fixed array | | | | | | | | | | | | | |
| Lat - 15 | 3.3 | 4.1 | 4.9 | 5.6 | 5.8 | 5.8 | 5.5 | 5.3 | 5.1 | 4.9 | 3.8 | 3.2 | 4.8 |
| Latitude | 3.7 | 4.5 | 5.0 | 5.6 | 5.5 | 5.4 | 5.1 | 5.2 | 5.1 | 5.2 | 4.3 | 3.6 | 4.9 |
| Lat + 15 | 4.0 | 4.6 | 4.9 | 5.2 | 4.9 | 4.7 | 4.5 | 4.7 | 4.9 | 5.3 | 4.5 | 3.9 | 4.7 |
| Single axis tracker | | | | | | | | | | | | | |
| Lat - 15 | 4.1 | 5.2 | 6.1 | 7.2 | 7.4 | 7.3 | 6.7 | 6.6 | 6.3 | 6.2 | 4.7 | 3.9 | 6.0 |
| Latitude | 4.4 | 5.5 | 6.2 | 7.1 | 7.2 | 7.0 | 6.5 | 6.5 | 6.4 | 6.5 | 5.1 | 4.2 | 6.1 |
| Lat + 15 | 4.6 | 5.5 | 6.2 | 6.9 | 6.7 | 6.5 | 6.1 | 6.2 | 6.2 | 6.6 | 5.3 | 4.4 | 5.9 |
| Dual axis tracker | | | | | | | | | | | | | |
| | 4.6 | 5.6 | 6.3 | 7.2 | 7.5 | 7.4 | 6.8 | 6.6 | 6.4 | 6.6 | 5.3 | 4.5 | 6.2 |

## MONTGOMERY  AL  Latitude: 32.30 degrees  Elevation: 62 meters

| | Jan | Feb | Mar | Apr | May | Jun | Jul | Aug | Sep | Oct | Nov | Dec | Avg |
|---|---|---|---|---|---|---|---|---|---|---|---|---|---|
| **Fixed array** | | | | | | | | | | | | | |
| Lat - 15 | 3.4 | 4.2 | 5.0 | 5.9 | 6.2 | 6.3 | 6.0 | 5.9 | 5.3 | 4.9 | 3.8 | 3.3 | 5.0 |
| Latitude | 3.8 | 4.6 | 5.2 | 5.8 | 5.8 | 5.8 | 5.6 | 5.7 | 5.4 | 5.3 | 4.3 | 3.7 | 5.1 |
| Lat + 15 | 4.0 | 4.7 | 5.1 | 5.4 | 5.2 | 5.1 | 4.9 | 5.2 | 5.2 | 5.4 | 4.5 | 4.0 | 4.9 |
| **Single axis tracker** | | | | | | | | | | | | | |
| Lat - 15 | 4.2 | 5.3 | 6.3 | 7.6 | 7.8 | 7.8 | 7.3 | 7.2 | 6.6 | 6.2 | 4.7 | 4.0 | 6.3 |
| Latitude | 4.5 | 5.5 | 6.5 | 7.5 | 7.6 | 7.5 | 7.0 | 7.1 | 6.7 | 6.5 | 5.1 | 4.4 | 6.3 |
| Lat + 15 | 4.7 | 5.6 | 6.4 | 7.2 | 7.1 | 7 | 6.5 | 6.7 | 6.5 | 6.6 | 5.2 | 4.6 | 6.2 |
| **Dual axis tracker** | | | | | | | | | | | | | |
| | 4.7 | 5.6 | 6.5 | 7.6 | 7.9 | 8.0 | 7.4 | 7.2 | 6.7 | 6.6 | 5.3 | 4.7 | 6.5 |

## LITTLE ROCK  AR  Latitude: 34.73 degrees  Elevation: 81 meters

| | Jan | Feb | Mar | Apr | May | Jun | Jul | Aug | Sep | Oct | Nov | Dec | Avg |
|---|---|---|---|---|---|---|---|---|---|---|---|---|---|
| **Fixed array** | | | | | | | | | | | | | |
| Lat - 15 | 3.4 | 4.1 | 4.9 | 5.6 | 6.1 | 6.4 | 6.3 | 6.1 | 5.3 | 4.8 | 3.5 | 3.0 | 5.0 |
| Latitude | 3.8 | 4.5 | 5.1 | 5.5 | 5.7 | 5.9 | 5.9 | 5.9 | 5.4 | 5.1 | 4.0 | 3.5 | 5.0 |
| Lat + 15 | 4.1 | 4.6 | 5.0 | 5.1 | 5.1 | 5.1 | 5.2 | 5.4 | 5.2 | 5.2 | 4.2 | 3.7 | 4.8 |
| **Single axis tracker** | | | | | | | | | | | | | |
| Lat - 15 | 4.1 | 5.1 | 6.2 | 7.2 | 7.8 | 8.3 | 8.3 | 8.0 | 6.8 | 6.1 | 4.4 | 3.7 | 6.3 |
| Latitude | 4.5 | 5.4 | 6.4 | 7.2 | 7.6 | 8.0 | 8.0 | 7.8 | 6.8 | 6.4 | 4.7 | 4.1 | 6.4 |
| Lat + 15 | 4.7 | 5.5 | 6.3 | 6.9 | 7.1 | 7.4 | 7.5 | 7.5 | 6.7 | 6.4 | 4.8 | 4.3 | 6.3 |
| **Dual axis tracker** | | | | | | | | | | | | | |
| | 4.7 | 5.5 | 6.4 | 7.3 | 7.9 | 8.5 | 8.4 | 8.0 | 6.9 | 6.4 | 4.9 | 4.3 | 6.6 |

## FORT SMITH  AR  Latitude: 35.33 degrees  Elevation: 141 meters

| | Jan | Feb | Mar | Apr | May | Jun | Jul | Aug | Sep | Oct | Nov | Dec | Avg |
|---|---|---|---|---|---|---|---|---|---|---|---|---|---|
| **Fixed array** | | | | | | | | | | | | | |
| Lat - 15 | 3.6 | 4.2 | 5.0 | 5.7 | 6.0 | 6.3 | 6.5 | 6.2 | 5.3 | 4.8 | 3.6 | 3.2 | 5.1 |
| Latitude | 4.1 | 4.6 | 5.2 | 5.6 | 5.7 | 5.8 | 6.0 | 6.0 | 5.4 | 5.1 | 4.1 | 3.7 | 5.1 |
| Lat + 15 | 4.3 | 4.8 | 5.1 | 5.2 | 5.0 | 5.0 | 5.3 | 5.5 | 5.2 | 5.2 | 4.3 | 4.0 | 4.9 |
| **Single axis tracker** | | | | | | | | | | | | | |
| Lat - 15 | 4.4 | 5.3 | 6.3 | 7.3 | 7.8 | 8.3 | 8.7 | 8.2 | 6.8 | 6.1 | 4.5 | 3.9 | 6.5 |
| Latitude | 4.8 | 5.6 | 6.4 | 7.2 | 7.6 | 8.0 | 8.4 | 8.0 | 6.9 | 6.4 | 4.8 | 4.3 | 6.5 |
| Lat + 15 | 5 | 5.7 | 6.4 | 7 | 7.1 | 7.5 | 7.8 | 7.6 | 6.7 | 6.4 | 5 | 4.6 | 6.4 |
| **Dual axis tracker** | | | | | | | | | | | | | |
| | 5.1 | 5.7 | 6.5 | 7.3 | 7.9 | 8.5 | 8.8 | 8.2 | 6.9 | 6.4 | 5.0 | 4.6 | 6.8 |

## FLAGSTAFF  AZ  Latitude: 35.13 degrees  Elevation: 2135 meters

| | Jan | Feb | Mar | Apr | May | Jun | Jul | Aug | Sep | Oct | Nov | Dec | Avg |
|---|---|---|---|---|---|---|---|---|---|---|---|---|---|
| **Fixed array** | | | | | | | | | | | | | |
| Lat - 15 | 4.4 | 5.2 | 5.9 | 6.8 | 7.2 | 7.4 | 6.2 | 6.1 | 6.1 | 5.6 | 4.7 | 4.2 | 5.8 |
| Latitude | 5.2 | 5.8 | 6.2 | 6.7 | 6.7 | 6.7 | 5.8 | 5.9 | 6.3 | 6.1 | 5.4 | 4.9 | 6.0 |
| Lat + 15 | 5.6 | 6.1 | 6.2 | 6.2 | 5.9 | 5.7 | 5.0 | 5.4 | 6.0 | 6.3 | 5.8 | 5.4 | 5.8 |
| **Single axis tracker** | | | | | | | | | | | | | |
| Lat - 15 | 5.8 | 6.9 | 8.0 | 9.5 | 10.2 | 10.7 | 8.6 | 8.4 | 8.5 | 7.6 | 6.2 | 5.4 | 8.0 |
| Latitude | 6.4 | 7.4 | 8.3 | 9.4 | 9.9 | 10.3 | 8.3 | 8.3 | 8.6 | 8.0 | 6.7 | 6.0 | 8.1 |
| Lat + 15 | 6.5 | 7.6 | 8.2 | 9.1 | 9.4 | 9.6 | 7.8 | 7.9 | 8.4 | 8.1 | 7 | 6.4 | 8 |
| **Dual axis tracker** | | | | | | | | | | | | | |
| | 6.8 | 7.6 | 8.3 | 9.5 | 10.3 | 11.0 | 8.7 | 8.5 | 8.6 | 8.1 | 7.1 | 6.5 | 8.4 |

## PHOENIX  AZ  Latitude: 33.43 degrees  Elevation: 339 meters

|  | Jan | Feb | Mar | Apr | May | Jun | Jul | Aug | Sep | Oct | Nov | Dec | Avg |
|---|---|---|---|---|---|---|---|---|---|---|---|---|---|
| Fixed array | | | | | | | | | | | | | |
| Lat - 15 | 4.4 | 5.4 | 6.4 | 7.5 | 8.0 | 8.1 | 7.5 | 7.3 | 6.8 | 6.0 | 4.9 | 4.2 | 6.4 |
| Latitude | 5.1 | 6.0 | 6.7 | 7.4 | 7.5 | 7.3 | 6.9 | 7.1 | 7.0 | 6.5 | 5.6 | 4.9 | 6.5 |
| Lat + 15 | 5.5 | 6.2 | 6.6 | 6.9 | 6.6 | 6.3 | 6.0 | 6.4 | 6.7 | 6.7 | 5.9 | 5.3 | 6.3 |
| Single axis tracker | | | | | | | | | | | | | |
| Lat - 15 | 5.6 | 7.1 | 8.5 | 10.3 | 11.1 | 11.3 | 10.0 | 9.8 | 9.2 | 8.0 | 6.3 | 5.3 | 8.5 |
| Latitude | 6.2 | 7.5 | 8.7 | 10.3 | 10.7 | 10.8 | 9.6 | 9.6 | 9.3 | 8.4 | 6.8 | 5.8 | 8.6 |
| Lat + 15 | 6.6 | 7.4 | 8.2 | 9.4 | 9.7 | 10.1 | 8.4 | 8.4 | 8.9 | 8.5 | 7.2 | 6.3 | 8.3 |
| Dual axis tracker | | | | | | | | | | | | | |
|  | 6.6 | 7.7 | 8.7 | 10.4 | 11.2 | 11.6 | 10.1 | 9.8 | 9.3 | 8.5 | 7.1 | 6.3 | 8.9 |

## PRESCOTT  AZ  Latitude: 34.65 degrees  Elevation: 1531 meters

|  | Jan | Feb | Mar | Apr | May | Jun | Jul | Aug | Sep | Oct | Nov | Dec | Avg |
|---|---|---|---|---|---|---|---|---|---|---|---|---|---|
| Fixed array | | | | | | | | | | | | | |
| Lat - 15 | 4.4 | 5.1 | 5.9 | 7.0 | 7.5 | 7.7 | 6.7 | 6.5 | 6.5 | 5.8 | 4.8 | 4.1 | 6.0 |
| Latitude | 5.1 | 5.7 | 6.2 | 6.9 | 7.0 | 7.0 | 6.2 | 6.3 | 6.6 | 6.4 | 5.5 | 4.9 | 6.1 |
| Lat + 15 | 5.5 | 5.9 | 6.1 | 6.4 | 6.1 | 6.0 | 5.4 | 5.7 | 6.4 | 6.5 | 5.9 | 5.4 | 5.9 |
| Single axis tracker | | | | | | | | | | | | | |
| Lat - 15 | 5.7 | 6.8 | 8.1 | 9.8 | 10.6 | 11.2 | 9.3 | 9.0 | 9.0 | 8.0 | 6.3 | 5.4 | 8.3 |
| Latitude | 6.3 | 7.2 | 8.3 | 9.8 | 10.3 | 10.7 | 8.9 | 8.8 | 9.1 | 8.4 | 6.9 | 6.0 | 8.4 |
| Lat + 15 | 6.6 | 7.4 | 8.2 | 9.4 | 9.7 | 10.1 | 8.4 | 8.4 | 8.9 | 8.5 | 7.2 | 6.3 | 8.3 |
| Dual axis tracker | | | | | | | | | | | | | |
|  | 6.7 | 7.4 | 8.3 | 9.9 | 10.8 | 11.5 | 9.4 | 9.0 | 9.1 | 8.5 | 7.2 | 6.5 | 8.7 |

## TUCSON  AZ  Latitude: 32.12 degrees  Elevation: 779 meters

|  | Jan | Feb | Mar | Apr | May | Jun | Jul | Aug | Sep | Oct | Nov | Dec | Avg |
|---|---|---|---|---|---|---|---|---|---|---|---|---|---|
| Fixed array | | | | | | | | | | | | | |
| Lat - 15 | 4.6 | 5.5 | 6.4 | 7.5 | 7.8 | 7.8 | 6.9 | 6.9 | 6.6 | 6.1 | 5.0 | 4.3 | 6.3 |
| Latitude | 5.4 | 6.2 | 6.7 | 7.3 | 7.3 | 7.1 | 6.4 | 6.6 | 6.8 | 6.6 | 5.8 | 5.1 | 6.5 |
| Lat + 15 | 5.9 | 6.4 | 6.6 | 6.8 | 6.4 | 6.1 | 5.6 | 6.0 | 6.6 | 6.8 | 6.2 | 5.6 | 6.3 |
| Single axis tracker | | | | | | | | | | | | | |
| Lat - 15 | 6.1 | 7.4 | 8.7 | 10.4 | 11.1 | 11.1 | 9.1 | 9.2 | 9.0 | 8.2 | 6.7 | 5.6 | 8.6 |
| Latitude | 6.7 | 7.8 | 9.0 | 10.4 | 10.7 | 10.6 | 8.8 | 9.1 | 9.1 | 8.6 | 7.3 | 6.2 | 8.7 |
| Lat + 15 | 7 | 8 | 8.9 | 10 | 10.1 | 9.9 | 8.2 | 8.6 | 9 | 8.7 | 7.6 | 6.6 | 8.6 |
| Dual axis tracker | | | | | | | | | | | | | |
|  | 7.1 | 8.1 | 9.0 | 10.5 | 11.2 | 11.3 | 9.3 | 9.2 | 9.2 | 8.7 | 7.6 | 6.7 | 9.0 |

## ARCATA  CA  Latitude: 40.98 degrees  Elevation: 69 meters

|  | Jan | Feb | Mar | Apr | May | Jun | Jul | Aug | Sep | Oct | Nov | Dec | Avg |
|---|---|---|---|---|---|---|---|---|---|---|---|---|---|
| Fixed array | | | | | | | | | | | | | |
| Lat - 15 | 2.7 | 3.3 | 4.3 | 5.4 | 5.9 | 5.9 | 5.8 | 5.3 | 5.1 | 3.9 | 2.9 | 2.5 | 4.4 |
| Latitude | 3.0 | 3.5 | 4.4 | 5.3 | 5.5 | 5.4 | 5.4 | 5.0 | 5.1 | 4.1 | 3.2 | 2.8 | 4.4 |
| Lat + 15 | 3.2 | 3.6 | 4.3 | 4.9 | 4.9 | 4.7 | 4.7 | 4.6 | 4.9 | 4.1 | 3.3 | 3.0 | 4.2 |
| Single axis tracker | | | | | | | | | | | | | |
| Lat - 15 | 3.2 | 3.9 | 5.3 | 6.8 | 7.5 | 7.5 | 7.4 | 6.5 | 6.4 | 4.8 | 3.4 | 2.9 | 5.5 |
| Latitude | 3.5 | 4.1 | 5.3 | 6.7 | 7.2 | 7.2 | 7.1 | 6.4 | 6.4 | 5.0 | 3.7 | 3.2 | 5.5 |
| Lat + 15 | 3.6 | 4.2 | 5.2 | 6.4 | 6.8 | 6.7 | 6.6 | 6 | 6.3 | 5 | 3.8 | 3.4 | 5.3 |
| Dual axis tracker | | | | | | | | | | | | | |
|  | 3.6 | 4.2 | 5.4 | 6.8 | 7.5 | 7.7 | 7.5 | 6.5 | 6.5 | 5.0 | 3.8 | 3.4 | 5.7 |

## BAKERSFIELD    CA      Latitude: 35.42 degrees      Elevation: 150 meters

|  | Jan | Feb | Mar | Apr | May | Jun | Jul | Aug | Sep | Oct | Nov | Dec | Avg |
|---|---|---|---|---|---|---|---|---|---|---|---|---|---|
| Fixed array | | | | | | | | | | | | | |
| Lat - 15 | 3.0 | 4.2 | 5.4 | 6.6 | 7.4 | 7.8 | 7.8 | 7.5 | 6.8 | 5.5 | 3.8 | 2.8 | 5.7 |
| Latitude | 3.3 | 4.5 | 5.6 | 6.5 | 6.9 | 7.1 | 7.2 | 7.3 | 6.9 | 6.0 | 4.3 | 3.2 | 5.7 |
| Lat + 15 | 3.5 | 4.7 | 5.5 | 6.0 | 6.1 | 6.1 | 6.2 | 6.6 | 6.7 | 6.1 | 4.5 | 3.4 | 5.4 |
| Single axis tracker | | | | | | | | | | | | | |
| Lat - 15 | 3.5 | 5.1 | 6.8 | 8.7 | 10.2 | 11.0 | 11.1 | 10.5 | 9.1 | 7.2 | 4.7 | 3.3 | 7.6 |
| Latitude | 3.8 | 5.4 | 7.0 | 8.7 | 9.9 | 10.5 | 10.7 | 10.3 | 9.2 | 7.5 | 5.0 | 3.6 | 7.6 |
| Lat + 15 | 3.9 | 5.5 | 6.9 | 8.3 | 9.3 | 9.8 | 10 | 9.8 | 9 | 7.6 | 5.2 | 3.8 | 7.4 |
| Dual axis tracker | | | | | | | | | | | | | |
| | 4.0 | 5.5 | 7.0 | 8.8 | 10.3 | 11.2 | 11.3 | 10.5 | 9.2 | 7.6 | 5.2 | 3.8 | 7.9 |

## DAGGETT    CA      Latitude: 34.87 degrees      Elevation: 588 meters

|  | Jan | Feb | Mar | Apr | May | Jun | Jul | Aug | Sep | Oct | Nov | Dec | Avg |
|---|---|---|---|---|---|---|---|---|---|---|---|---|---|
| Fixed array | | | | | | | | | | | | | |
| Lat - 15 | 4.6 | 5.4 | 6.5 | 7.5 | 7.9 | 8.1 | 7.8 | 7.6 | 7.1 | 6.2 | 5.0 | 4.4 | 6.5 |
| Latitude | 5.3 | 6.0 | 6.8 | 7.4 | 7.4 | 7.4 | 7.2 | 7.3 | 7.3 | 6.8 | 5.8 | 5.2 | 6.6 |
| Lat + 15 | 5.7 | 6.2 | 6.7 | 6.8 | 6.5 | 6.3 | 6.2 | 6.6 | 7.0 | 6.9 | 6.2 | 5.6 | 6.4 |
| Single axis tracker | | | | | | | | | | | | | |
| Lat - 15 | 5.9 | 7.1 | 8.8 | 10.4 | 11.2 | 11.7 | 11.1 | 10.8 | 10.0 | 8.4 | 6.6 | 5.6 | 9.0 |
| Latitude | 6.5 | 7.5 | 9.0 | 10.3 | 10.9 | 11.2 | 10.7 | 10.6 | 10.1 | 8.8 | 7.2 | 6.3 | 9.1 |
| Lat + 15 | 6.8 | 7.7 | 8.9 | 10 | 10.3 | 10.5 | 10.1 | 10.1 | 9.9 | 8.9 | 7.5 | 6.6 | 8.9 |
| Dual axis tracker | | | | | | | | | | | | | |
| | 6.9 | 7.7 | 9.0 | 10.4 | 11.3 | 12.0 | 11.4 | 10.8 | 10.1 | 9.0 | 7.5 | 6.8 | 9.4 |

## FRESNO    CA      Latitude: 36.77 degrees      Elevation: 100 meters

|  | Jan | Feb | Mar | Apr | May | Jun | Jul | Aug | Sep | Oct | Nov | Dec | Avg |
|---|---|---|---|---|---|---|---|---|---|---|---|---|---|
| Fixed array | | | | | | | | | | | | | |
| Lat - 15 | 2.8 | 4.1 | 5.5 | 6.8 | 7.6 | 7.8 | 7.9 | 7.5 | 6.8 | 5.5 | 3.6 | 2.5 | 5.7 |
| Latitude | 3.1 | 4.4 | 5.7 | 6.7 | 7.1 | 7.2 | 7.3 | 7.3 | 6.9 | 6.0 | 4.1 | 2.8 | 5.7 |
| Lat + 15 | 3.2 | 4.5 | 5.6 | 6.2 | 6.3 | 6.1 | 6.3 | 6.6 | 6.7 | 6.1 | 4.2 | 3.0 | 5.4 |
| Single axis tracker | | | | | | | | | | | | | |
| Lat - 15 | 3.2 | 5.0 | 7.0 | 9.0 | 10.4 | 10.9 | 11.2 | 10.4 | 9.1 | 7.1 | 4.4 | 2.9 | 7.5 |
| Latitude | 3.4 | 5.2 | 7.2 | 8.9 | 10.1 | 10.5 | 10.8 | 10.3 | 9.2 | 7.4 | 4.7 | 3.1 | 7.6 |
| Lat + 15 | 3.5 | 5.3 | 7.1 | 8.6 | 9.5 | 9.8 | 10.1 | 9.8 | 9 | 7.5 | 4.9 | 3.2 | 7.4 |
| Dual axis tracker | | | | | | | | | | | | | |
| | 3.6 | 5.3 | 7.2 | 9.0 | 10.5 | 11.2 | 11.4 | 10.5 | 9.2 | 7.5 | 4.9 | 3.3 | 7.8 |

## LONG BEACH    CA      Latitude: 33.82 degrees      Elevation: 17 meters

|  | Jan | Feb | Mar | Apr | May | Jun | Jul | Aug | Sep | Oct | Nov | Dec | Avg |
|---|---|---|---|---|---|---|---|---|---|---|---|---|---|
| Fixed array | | | | | | | | | | | | | |
| Lat - 15 | 3.8 | 4.5 | 5.4 | 6.4 | 6.4 | 6.5 | 7.2 | 6.9 | 6.0 | 5.0 | 4.1 | 3.6 | 5.5 |
| Latitude | 4.3 | 4.9 | 5.6 | 6.3 | 6.1 | 6.0 | 6.7 | 6.7 | 6.1 | 5.4 | 4.7 | 4.2 | 5.6 |
| Lat + 15 | 4.6 | 5.1 | 5.5 | 5.8 | 5.4 | 5.2 | 5.8 | 6.1 | 5.8 | 5.5 | 5.0 | 4.5 | 5.4 |
| Single axis tracker | | | | | | | | | | | | | |
| Lat - 15 | 4.7 | 5.6 | 6.8 | 8.2 | 8.1 | 8.3 | 9.3 | 8.9 | 7.6 | 6.3 | 5.1 | 4.4 | 6.9 |
| Latitude | 5.1 | 5.9 | 7.0 | 8.1 | 7.9 | 7.9 | 8.9 | 8.8 | 7.6 | 6.6 | 5.6 | 4.9 | 7.0 |
| Lat + 15 | 5.3 | 6 | 6.9 | 7.8 | 7.4 | 7.4 | 8.3 | 8.3 | 7.5 | 6.7 | 5.8 | 5.1 | 6.9 |
| Dual axis tracker | | | | | | | | | | | | | |
| | 5.4 | 6.0 | 7.0 | 8.2 | 8.2 | 8.4 | 9.4 | 8.9 | 7.7 | 6.7 | 5.8 | 5.2 | 7.3 |

## LOS ANGELES CA  Latitude: 33.93 degrees  Elevation: 32 meters

| | Jan | Feb | Mar | Apr | May | Jun | Jul | Aug | Sep | Oct | Nov | Dec | Avg |
|---|---|---|---|---|---|---|---|---|---|---|---|---|---|
| **Fixed array** | | | | | | | | | | | | | |
| Lat - 15 | 3.8 | 4.5 | 5.5 | 6.4 | 6.4 | 6.4 | 7.1 | 6.8 | 5.9 | 5.0 | 4.2 | 3.6 | 5.5 |
| Latitude | 4.4 | 5.0 | 5.7 | 6.3 | 6.1 | 6.0 | 6.6 | 6.6 | 6.0 | 5.4 | 4.7 | 4.2 | 5.6 |
| Lat + 15 | 4.7 | 5.1 | 5.6 | 5.9 | 5.4 | 5.2 | 5.8 | 6.0 | 5.7 | 5.5 | 5.0 | 4.5 | 5.4 |
| **Single axis tracker** | | | | | | | | | | | | | |
| Lat - 15 | 4.7 | 5.6 | 6.9 | 8.2 | 8.0 | 8.0 | 9.0 | 8.6 | 7.3 | 6.3 | 5.2 | 4.4 | 6.9 |
| Latitude | 5.1 | 6.0 | 7.1 | 8.2 | 7.8 | 7.7 | 8.7 | 8.4 | 7.4 | 6.6 | 5.6 | 4.9 | 7.0 |
| Lat + 15 | 5.4 | 6.1 | 7 | 7.8 | 7.3 | 7.1 | 8.1 | 8 | 7.2 | 6.6 | 5.8 | 5.2 | 6.8 |
| **Dual axis tracker** | | | | | | | | | | | | | |
| | 5.4 | 6.1 | 7.1 | 8.3 | 8.1 | 8.2 | 9.1 | 8.6 | 7.4 | 6.7 | 5.8 | 5.3 | 7.2 |

## SACRAMENTO CA  Latitude: 38.52 degrees  Elevation: 8 meters

| | Jan | Feb | Mar | Apr | May | Jun | Jul | Aug | Sep | Oct | Nov | Dec | Avg |
|---|---|---|---|---|---|---|---|---|---|---|---|---|---|
| **Fixed array** | | | | | | | | | | | | | |
| Lat - 15 | 2.6 | 3.9 | 5.2 | 6.5 | 7.3 | 7.6 | 7.8 | 7.5 | 6.7 | 5.3 | 3.3 | 2.4 | 5.5 |
| Latitude | 2.9 | 4.2 | 5.4 | 6.3 | 6.8 | 7.0 | 7.2 | 7.2 | 6.9 | 5.7 | 3.7 | 2.7 | 5.5 |
| Lat + 15 | 3.1 | 4.3 | 5.2 | 5.9 | 6.0 | 6.0 | 6.3 | 6.5 | 6.6 | 5.8 | 3.9 | 2.9 | 5.2 |
| **Single axis tracker** | | | | | | | | | | | | | |
| Lat - 15 | 3.0 | 4.7 | 6.6 | 8.6 | 10.1 | 10.8 | 11.2 | 10.4 | 9.1 | 6.8 | 4.0 | 2.8 | 7.3 |
| Latitude | 3.3 | 4.9 | 6.7 | 8.5 | 9.8 | 10.3 | 10.8 | 10.2 | 9.2 | 7.1 | 4.3 | 3.0 | 7.4 |
| Lat + 15 | 3.4 | 5 | 6.6 | 8.2 | 9.2 | 9.7 | 10.1 | 9.8 | 9 | 7.2 | 4.5 | 3.2 | 7.2 |
| **Dual axis tracker** | | | | | | | | | | | | | |
| | 3.4 | 5.0 | 6.7 | 8.6 | 10.2 | 11.0 | 11.4 | 10.4 | 9.2 | 7.2 | 4.5 | 3.2 | 7.6 |

## SAN DIEGO CA  Latitude: 32.73 degrees  Elevation: 9 meters

| | Jan | Feb | Mar | Apr | May | Jun | Jul | Aug | Sep | Oct | Nov | Dec | Avg |
|---|---|---|---|---|---|---|---|---|---|---|---|---|---|
| **Fixed array** | | | | | | | | | | | | | |
| Lat - 15 | 4.1 | 4.8 | 5.6 | 6.4 | 6.3 | 6.3 | 6.8 | 6.7 | 6.0 | 5.3 | 4.5 | 3.9 | 5.6 |
| Latitude | 4.7 | 5.3 | 5.8 | 6.3 | 5.9 | 5.8 | 6.4 | 6.5 | 6.1 | 5.7 | 5.1 | 4.6 | 5.7 |
| Lat + 15 | 5.1 | 5.5 | 5.7 | 5.9 | 5.2 | 5.1 | 5.6 | 5.9 | 5.8 | 5.8 | 5.4 | 5.0 | 5.5 |
| **Single axis tracker** | | | | | | | | | | | | | |
| Lat - 15 | 5.2 | 6.1 | 7.2 | 8.3 | 7.7 | 7.8 | 8.7 | 8.6 | 7.6 | 6.7 | 5.7 | 5.0 | 7.1 |
| Latitude | 5.7 | 6.5 | 7.4 | 8.2 | 7.5 | 7.4 | 8.4 | 8.4 | 7.7 | 7.1 | 6.2 | 5.5 | 7.2 |
| Lat + 15 | 6 | 6.6 | 7.3 | 7.9 | 7 | 6.9 | 7.8 | 8 | 7.5 | 7.1 | 6.5 | 5.8 | 7 |
| **Dual axis tracker** | | | | | | | | | | | | | |
| | 6.0 | 6.6 | 7.4 | 8.3 | 7.8 | 7.9 | 8.9 | 8.6 | 7.7 | 7.2 | 6.5 | 5.9 | 7.4 |

## SAN FRANCISCO CA  Latitude: 37.62 degrees  Elevation: 5 meters

| | Jan | Feb | Mar | Apr | May | Jun | Jul | Aug | Sep | Oct | Nov | Dec | Avg |
|---|---|---|---|---|---|---|---|---|---|---|---|---|---|
| **Fixed array** | | | | | | | | | | | | | |
| Lat - 15 | 3.1 | 3.9 | 5.0 | 6.2 | 6.8 | 7.0 | 7.3 | 6.9 | 6.2 | 5.0 | 3.5 | 2.9 | 5.3 |
| Latitude | 3.5 | 4.2 | 5.2 | 6.1 | 6.4 | 6.5 | 6.8 | 6.7 | 6.4 | 5.4 | 3.9 | 3.4 | 5.4 |
| Lat + 15 | 3.7 | 4.4 | 5.1 | 5.6 | 5.7 | 5.6 | 5.9 | 6.1 | 6.1 | 5.5 | 4.1 | 3.6 | 5.1 |
| **Single axis tracker** | | | | | | | | | | | | | |
| Lat - 15 | 3.7 | 4.7 | 6.3 | 8.0 | 8.9 | 9.2 | 9.7 | 9.0 | 8.1 | 6.3 | 4.3 | 3.5 | 6.8 |
| Latitude | 4.0 | 5.0 | 6.5 | 8.0 | 8.7 | 8.9 | 9.4 | 8.8 | 8.2 | 6.6 | 4.6 | 3.9 | 6.9 |
| Lat + 15 | 4.2 | 5.1 | 6.4 | 7.7 | 8.1 | 8.2 | 8.7 | 8.4 | 8 | 6.7 | 4.8 | 4.1 | 6.7 |
| **Dual axis tracker** | | | | | | | | | | | | | |
| | 4.2 | 5.1 | 6.5 | 8.1 | 9.0 | 9.4 | 9.9 | 9.0 | 8.2 | 6.7 | 4.8 | 4.1 | 7.1 |

## SANTA MARIA CA — Latitude: 34.90 degrees — Elevation: 72 meters

|  | Jan | Feb | Mar | Apr | May | Jun | Jul | Aug | Sep | Oct | Nov | Dec | Avg |
|---|---|---|---|---|---|---|---|---|---|---|---|---|---|
| Fixed array |  |  |  |  |  |  |  |  |  |  |  |  |  |
| Lat - 15 | 4.0 | 4.7 | 5.7 | 6.6 | 7.0 | 7.2 | 7.4 | 7.1 | 6.3 | 5.4 | 4.4 | 3.9 | 5.8 |
| Latitude | 4.6 | 5.2 | 5.9 | 6.5 | 6.6 | 6.6 | 6.9 | 6.9 | 6.4 | 5.8 | 5.0 | 4.5 | 5.9 |
| Lat + 15 | 4.9 | 5.4 | 5.8 | 6.0 | 5.8 | 5.7 | 6.0 | 6.2 | 6.2 | 6.0 | 5.3 | 4.9 | 5.7 |
| Single axis tracker |  |  |  |  |  |  |  |  |  |  |  |  |  |
| Lat - 15 | 5.0 | 6.0 | 7.3 | 8.7 | 9.1 | 9.4 | 9.6 | 9.1 | 8.1 | 6.9 | 5.6 | 4.9 | 7.5 |
| Latitude | 5.5 | 6.4 | 7.5 | 8.6 | 8.8 | 9.0 | 9.2 | 8.9 | 8.2 | 7.3 | 6.0 | 5.4 | 7.6 |
| Lat + 15 | 5.7 | 6.5 | 7.4 | 8.3 | 8.3 | 8.3 | 8.6 | 8.5 | 8 | 7.3 | 6.3 | 5.7 | 7.4 |
| Dual axis tracker |  |  |  |  |  |  |  |  |  |  |  |  |  |
|  | 5.8 | 6.5 | 7.5 | 8.7 | 9.2 | 9.6 | 9.7 | 9.1 | 8.2 | 7.4 | 6.3 | 5.8 | 7.8 |

## ALAMOSA CO — Latitude: 37.45 degrees — Elevation: 2297 meters

|  | Jan | Feb | Mar | Apr | May | Jun | Jul | Aug | Sep | Oct | Nov | Dec | Avg |
|---|---|---|---|---|---|---|---|---|---|---|---|---|---|
| Fixed array |  |  |  |  |  |  |  |  |  |  |  |  |  |
| Lat - 15 | 4.7 | 5.5 | 6.2 | 6.9 | 7.1 | 7.4 | 7.0 | 6.8 | 6.5 | 5.9 | 4.9 | 4.4 | 6.1 |
| Latitude | 5.5 | 6.2 | 6.5 | 6.8 | 6.6 | 6.8 | 6.5 | 6.5 | 6.7 | 6.4 | 5.6 | 5.2 | 6.3 |
| Lat + 15 | 6.0 | 6.5 | 6.4 | 6.3 | 5.8 | 5.7 | 5.6 | 5.9 | 6.4 | 6.5 | 6.0 | 5.7 | 6.1 |
| Single axis tracker |  |  |  |  |  |  |  |  |  |  |  |  |  |
| Lat - 15 | 6.1 | 7.4 | 8.4 | 9.6 | 10.1 | 10.7 | 9.8 | 9.4 | 9.0 | 8.0 | 6.4 | 5.7 | 8.4 |
| Latitude | 6.8 | 7.9 | 8.7 | 9.6 | 9.8 | 10.3 | 9.5 | 9.2 | 9.1 | 8.4 | 7.0 | 6.3 | 8.5 |
| Lat + 15 | 7.2 | 8.1 | 8.2 | 9.2 | 9.3 | 9.6 | 8.9 | 8.8 | 9 | 8.5 | 7.3 | 6.7 | 8.4 |
| Dual axis tracker |  |  |  |  |  |  |  |  |  |  |  |  |  |
|  | 7.2 | 8.1 | 8.7 | 9.7 | 10.2 | 11.0 | 10.0 | 9.4 | 9.2 | 8.5 | 7.3 | 6.8 | 8.8 |

## BOULDER CO — Latitude: 40.02 degrees — Elevation: 1634 meters

|  | Jan | Feb | Mar | Apr | May | Jun | Jul | Aug | Sep | Oct | Nov | Dec | Avg |
|---|---|---|---|---|---|---|---|---|---|---|---|---|---|
| Fixed array |  |  |  |  |  |  |  |  |  |  |  |  |  |
| Lat - 15 | 3.8 | 4.6 | 5.4 | 6.1 | 6.2 | 6.6 | 6.6 | 6.3 | 5.9 | 5.1 | 4.0 | 3.5 | 5.4 |
| Latitude | 4.4 | 5.1 | 5.6 | 6.0 | 5.9 | 6.1 | 6.1 | 6.1 | 6.0 | 5.6 | 4.6 | 4.2 | 5.5 |
| Lat + 15 | 4.8 | 5.3 | 5.6 | 5.6 | 5.2 | 5.2 | 5.3 | 5.5 | 5.8 | 5.7 | 4.8 | 4.5 | 5.3 |
| Single axis tracker |  |  |  |  |  |  |  |  |  |  |  |  |  |
| Lat - 15 | 4.8 | 5.9 | 7.0 | 8.1 | 8.4 | 9.1 | 9.1 | 8.6 | 7.9 | 6.7 | 5.0 | 4.4 | 7.1 |
| Latitude | 5.2 | 6.2 | 7.2 | 8.0 | 8.1 | 8.8 | 8.7 | 8.4 | 7.9 | 7.1 | 5.5 | 4.9 | 7.2 |
| Lat + 15 | 5.5 | 6.4 | 7.1 | 7.7 | 7.7 | 8.2 | 8.2 | 8 | 7.8 | 7.1 | 5.7 | 5.2 | 7.1 |
| Dual axis tracker |  |  |  |  |  |  |  |  |  |  |  |  |  |
|  | 5.6 | 6.4 | 7.2 | 8.1 | 8.5 | 9.4 | 9.2 | 8.6 | 8.0 | 7.1 | 5.7 | 5.3 | 7.4 |

## COLORADO SPRINGS CO — Latitude: 38.82 degrees — Elevation: 1881 meters

|  | Jan | Feb | Mar | Apr | May | Jun | Jul | Aug | Sep | Oct | Nov | Dec | Avg |
|---|---|---|---|---|---|---|---|---|---|---|---|---|---|
| Fixed array |  |  |  |  |  |  |  |  |  |  |  |  |  |
| Lat - 15 | 4.0 | 4.7 | 5.5 | 6.2 | 6.2 | 6.7 | 6.6 | 6.4 | 6.0 | 5.4 | 4.2 | 3.7 | 5.5 |
| Latitude | 4.6 | 5.2 | 5.7 | 6.1 | 5.9 | 6.2 | 6.1 | 6.1 | 6.1 | 5.8 | 4.8 | 4.4 | 5.6 |
| Lat + 15 | 5.0 | 5.4 | 5.6 | 5.6 | 5.2 | 5.3 | 5.3 | 5.6 | 5.9 | 5.9 | 5.1 | 4.8 | 5.4 |
| Single axis tracker |  |  |  |  |  |  |  |  |  |  |  |  |  |
| Lat - 15 | 5.1 | 6.1 | 7.2 | 8.3 | 8.5 | 9.3 | 9.0 | 8.6 | 8.1 | 7.1 | 5.4 | 4.7 | 7.3 |
| Latitude | 5.6 | 6.5 | 7.4 | 8.3 | 8.2 | 9.0 | 8.7 | 8.5 | 8.1 | 7.4 | 5.8 | 5.3 | 7.4 |
| Lat + 15 | 5.9 | 6.6 | 7.3 | 7.9 | 7.7 | 8.4 | 8.1 | 8.1 | 8 | 7.5 | 6.1 | 5.6 | 7.3 |
| Dual axis tracker |  |  |  |  |  |  |  |  |  |  |  |  |  |
|  | 5.9 | 6.7 | 7.4 | 8.4 | 8.6 | 9.6 | 9.2 | 8.6 | 8.2 | 7.5 | 6.1 | 5.7 | 7.7 |

## EAGLE CO    Latitude: 39.65 degrees    Elevation: 1985 meters

|  | Jan | Feb | Mar | Apr | May | Jun | Jul | Aug | Sep | Oct | Nov | Dec | Avg |
|---|---|---|---|---|---|---|---|---|---|---|---|---|---|
| Fixed array | | | | | | | | | | | | | |
| Lat - 15 | 3.7 | 4.6 | 5.3 | 6.1 | 6.4 | 7.0 | 6.9 | 6.5 | 6.1 | 5.2 | 3.8 | 3.4 | 5.4 |
| Latitude | 4.3 | 5.2 | 5.6 | 6.0 | 6.0 | 6.4 | 6.3 | 6.3 | 6.2 | 5.6 | 4.3 | 3.9 | 5.5 |
| Lat + 15 | 4.6 | 5.4 | 5.5 | 5.6 | 5.3 | 5.5 | 5.5 | 5.7 | 5.9 | 5.7 | 4.5 | 4.3 | 5.3 |
| Single axis tracker | | | | | | | | | | | | | |
| Lat - 15 | 4.6 | 5.9 | 6.9 | 8.2 | 8.9 | 9.9 | 9.6 | 9.0 | 8.3 | 6.8 | 4.7 | 4.2 | 7.3 |
| Latitude | 5.1 | 6.3 | 7.1 | 8.1 | 8.6 | 9.5 | 9.3 | 8.8 | 8.4 | 7.2 | 5.1 | 4.6 | 7.3 |
| Lat + 15 | 5.3 | 6.5 | 7.1 | 7.8 | 8.1 | 8.9 | 8.7 | 8.4 | 8.2 | 7.2 | 5.3 | 4.9 | 7.2 |
| Dual axis tracker | | | | | | | | | | | | | |
|  | 5.4 | 6.5 | 7.1 | 8.2 | 9.0 | 10.2 | 9.8 | 9.0 | 8.4 | 7.3 | 5.4 | 5.0 | 7.6 |

## GRAND JUNCTION CO    Latitude: 39.12 degrees    Elevation: 1475 meters

|  | Jan | Feb | Mar | Apr | May | Jun | Jul | Aug | Sep | Oct | Nov | Dec | Avg |
|---|---|---|---|---|---|---|---|---|---|---|---|---|---|
| Fixed array | | | | | | | | | | | | | |
| Lat - 15 | 3.8 | 4.7 | 5.5 | 6.5 | 7.0 | 7.5 | 7.3 | 7.0 | 6.5 | 5.4 | 4.1 | 3.6 | 5.7 |
| Latitude | 4.4 | 5.2 | 5.7 | 6.4 | 6.6 | 6.8 | 6.7 | 6.7 | 6.6 | 5.9 | 4.6 | 4.1 | 5.8 |
| Lat + 15 | 4.7 | 5.4 | 5.6 | 6.0 | 5.8 | 5.8 | 5.8 | 6.1 | 6.4 | 6.0 | 4.9 | 4.5 | 5.6 |
| Single axis tracker | | | | | | | | | | | | | |
| Lat - 15 | 4.7 | 6.0 | 7.1 | 8.7 | 9.6 | 10.6 | 10.1 | 9.5 | 8.9 | 7.1 | 5.1 | 4.4 | 7.7 |
| Latitude | 5.2 | 6.4 | 7.3 | 8.7 | 9.3 | 10.1 | 9.8 | 9.4 | 9.0 | 7.5 | 5.6 | 4.9 | 7.8 |
| Lat + 15 | 5.4 | 6.6 | 7.2 | 8.3 | 8.8 | 9.5 | 9.2 | 8.9 | 8.8 | 7.5 | 5.8 | 5.2 | 7.6 |
| Dual axis tracker | | | | | | | | | | | | | |
|  | 5.5 | 6.6 | 7.3 | 8.8 | 9.7 | 10.8 | 10.3 | 9.6 | 9.0 | 7.6 | 5.8 | 5.2 | 8.0 |

## PUEBLO CO    Latitude: 38.28 degrees    Elevation: 1439 meters

|  | Jan | Feb | Mar | Apr | May | Jun | Jul | Aug | Sep | Oct | Nov | Dec | Avg |
|---|---|---|---|---|---|---|---|---|---|---|---|---|---|
| Fixed array | | | | | | | | | | | | | |
| Lat - 15 | 4.1 | 4.9 | 5.7 | 6.5 | 6.7 | 7.2 | 7.1 | 6.9 | 6.3 | 5.6 | 4.4 | 3.9 | 5.8 |
| Latitude | 4.8 | 5.4 | 6.0 | 6.4 | 6.3 | 6.6 | 6.6 | 6.6 | 6.4 | 6.0 | 5.0 | 4.6 | 5.9 |
| Lat + 15 | 5.2 | 5.6 | 5.9 | 6.0 | 5.6 | 5.6 | 5.7 | 6.0 | 6.2 | 6.2 | 5.3 | 5.0 | 5.7 |
| Single axis tracker | | | | | | | | | | | | | |
| Lat - 15 | 5.3 | 6.3 | 7.5 | 8.9 | 9.1 | 9.9 | 9.7 | 9.3 | 8.5 | 7.4 | 5.6 | 4.9 | 7.7 |
| Latitude | 5.8 | 6.7 | 7.7 | 8.8 | 8.8 | 9.5 | 9.4 | 9.1 | 8.6 | 7.7 | 6.1 | 5.5 | 7.8 |
| Lat + 15 | 6.1 | 6.9 | 7.7 | 8.5 | 8.3 | 8.9 | 8.8 | 8.7 | 8.4 | 7.8 | 6.3 | 5.8 | 7.7 |
| Dual axis tracker | | | | | | | | | | | | | |
|  | 6.2 | 6.9 | 7.8 | 8.9 | 9.2 | 10.2 | 9.9 | 9.3 | 8.6 | 7.8 | 6.4 | 5.9 | 8.1 |

## BRIDGEPORT CT    Latitude: 41.17 degrees    Elevation: 2 meters

|  | Jan | Feb | Mar | Apr | May | Jun | Jul | Aug | Sep | Oct | Nov | Dec | Avg |
|---|---|---|---|---|---|---|---|---|---|---|---|---|---|
| Fixed array | | | | | | | | | | | | | |
| Lat - 15 | 2.9 | 3.7 | 4.4 | 5.1 | 5.5 | 5.8 | 5.8 | 5.5 | 4.9 | 4.0 | 2.8 | 2.4 | 4.4 |
| Latitude | 3.3 | 4.0 | 4.6 | 5.0 | 5.2 | 5.3 | 5.4 | 5.3 | 4.9 | 4.3 | 3.1 | 2.8 | 4.4 |
| Lat + 15 | 3.5 | 4.1 | 4.5 | 4.6 | 4.6 | 4.6 | 4.7 | 4.8 | 4.7 | 4.3 | 3.3 | 2.9 | 4.2 |
| Single axis tracker | | | | | | | | | | | | | |
| Lat - 15 | 3.5 | 4.5 | 5.5 | 6.4 | 6.9 | 7.3 | 7.4 | 7.0 | 6.1 | 5.0 | 3.4 | 2.8 | 5.5 |
| Latitude | 3.8 | 4.8 | 5.6 | 6.3 | 6.7 | 7.0 | 7.1 | 6.8 | 6.1 | 5.2 | 3.6 | 3.1 | 5.5 |
| Lat + 15 | 4 | 4.9 | 5.5 | 6.1 | 6.3 | 6.5 | 6.6 | 6.5 | 6 | 5.2 | 3.7 | 3.3 | 5.4 |
| Dual axis tracker | | | | | | | | | | | | | |
|  | 4.0 | 4.9 | 5.6 | 6.4 | 7.0 | 7.5 | 7.5 | 7.0 | 6.2 | 5.3 | 3.8 | 3.3 | 5.7 |

## HARTFORD CT — Latitude: 41.93 degrees — Elevation: 55 meters

| | Jan | Feb | Mar | Apr | May | Jun | Jul | Aug | Sep | Oct | Nov | Dec | Avg |
|---|---|---|---|---|---|---|---|---|---|---|---|---|---|
| Fixed array | | | | | | | | | | | | | |
| Lat - 15 | 2.9 | 3.7 | 4.4 | 5.1 | 5.5 | 5.8 | 5.8 | 5.4 | 4.8 | 3.9 | 2.7 | 2.3 | 4.4 |
| Latitude | 3.3 | 4.1 | 4.6 | 4.9 | 5.2 | 5.3 | 5.4 | 5.2 | 4.8 | 4.1 | 2.9 | 2.7 | 4.4 |
| Lat + 15 | 3.5 | 4.2 | 4.5 | 4.6 | 4.6 | 4.6 | 4.8 | 4.7 | 4.6 | 4.2 | 3.0 | 2.8 | 4.2 |
| Single axis tracker | | | | | | | | | | | | | |
| Lat - 15 | 3.4 | 4.5 | 5.5 | 6.3 | 6.9 | 7.3 | 7.4 | 6.9 | 5.9 | 4.7 | 3.1 | 2.7 | 5.4 |
| Latitude | 3.7 | 4.8 | 5.6 | 6.2 | 6.7 | 7.0 | 7.1 | 6.7 | 6.0 | 4.9 | 3.3 | 3.0 | 5.4 |
| Lat + 15 | 3.9 | 4.9 | 5.5 | 6 | 6.2 | 6.5 | 6.6 | 6.4 | 5.8 | 5 | 3.4 | 3.1 | 5.3 |
| Dual axis tracker | | | | | | | | | | | | | |
| | 4.0 | 4.9 | 5.6 | 6.3 | 7.0 | 7.5 | 7.5 | 6.9 | 6.0 | 5.0 | 3.4 | 3.2 | 5.6 |

## WILMINGTON DE — Latitude: 39.67 degrees — Elevation: 24 meters

| | Jan | Feb | Mar | Apr | May | Jun | Jul | Aug | Sep | Oct | Nov | Dec | Avg |
|---|---|---|---|---|---|---|---|---|---|---|---|---|---|
| Fixed array | | | | | | | | | | | | | |
| Lat - 15 | 3.0 | 3.8 | 4.6 | 5.3 | 5.7 | 6.1 | 6.0 | 5.8 | 5.0 | 4.3 | 3.1 | 2.6 | 4.6 |
| Latitude | 3.4 | 4.2 | 4.8 | 5.2 | 5.4 | 5.6 | 5.6 | 5.5 | 5.1 | 4.5 | 3.5 | 3.0 | 4.6 |
| Lat + 15 | 3.6 | 4.3 | 4.7 | 4.8 | 4.7 | 4.9 | 4.9 | 5.0 | 4.9 | 4.6 | 3.6 | 3.2 | 4.4 |
| Single axis tracker | | | | | | | | | | | | | |
| Lat - 15 | 3.6 | 4.7 | 5.9 | 6.7 | 7.2 | 7.8 | 7.7 | 7.3 | 6.3 | 5.3 | 3.7 | 3.1 | 5.8 |
| Latitude | 4.0 | 5.0 | 6.0 | 6.7 | 7.0 | 7.5 | 7.4 | 7.2 | 6.3 | 5.5 | 4.0 | 3.4 | 5.8 |
| Lat + 15 | 4.1 | 5.1 | 5.9 | 6.4 | 6.6 | 7 | 6.9 | 6.8 | 6.2 | 5.5 | 4.1 | 3.6 | 5.7 |
| Dual axis tracker | | | | | | | | | | | | | |
| | 4.2 | 5.1 | 6.0 | 6.8 | 7.3 | 8.0 | 7.8 | 7.3 | 6.4 | 5.6 | 4.2 | 3.6 | 6.0 |

## DAYTONA BEACH FL — Latitude: 29.18 degrees — Elevation: 12 meters

| | Jan | Feb | Mar | Apr | May | Jun | Jul | Aug | Sep | Oct | Nov | Dec | Avg |
|---|---|---|---|---|---|---|---|---|---|---|---|---|---|
| Fixed array | | | | | | | | | | | | | |
| Lat - 15 | 3.8 | 4.5 | 5.5 | 6.4 | 6.4 | 6.0 | 5.9 | 5.8 | 5.2 | 4.7 | 4.1 | 3.6 | 5.2 |
| Latitude | 4.3 | 4.9 | 5.7 | 6.3 | 6.0 | 5.5 | 5.5 | 5.6 | 5.3 | 5.0 | 4.6 | 4.1 | 5.2 |
| Lat + 15 | 4.6 | 5.1 | 5.6 | 5.9 | 5.4 | 4.8 | 4.9 | 5.1 | 5.1 | 5.1 | 4.8 | 4.4 | 5.1 |
| Single axis tracker | | | | | | | | | | | | | |
| Lat - 15 | 4.8 | 5.7 | 7.0 | 8.3 | 8.2 | 7.4 | 7.4 | 7.2 | 6.5 | 5.9 | 5.1 | 4.4 | 6.5 |
| Latitude | 5.2 | 6.0 | 7.2 | 8.3 | 7.9 | 7.1 | 7.1 | 7.0 | 6.5 | 6.1 | 5.5 | 4.9 | 6.6 |
| Lat + 15 | 5.4 | 6.1 | 7.1 | 8 | 7.4 | 6.6 | 6.6 | 6.7 | 6.4 | 6.2 | 5.7 | 5.1 | 6.4 |
| Dual axis tracker | | | | | | | | | | | | | |
| | 5.5 | 6.1 | 7.2 | 8.4 | 8.2 | 7.5 | 7.5 | 7.2 | 6.5 | 6.2 | 5.7 | 5.2 | 6.8 |

## JACKSONVILLE FL — Latitude: 30.50 degrees — Elevation: 9 meters

| | Jan | Feb | Mar | Apr | May | Jun | Jul | Aug | Sep | Oct | Nov | Dec | Avg |
|---|---|---|---|---|---|---|---|---|---|---|---|---|---|
| Fixed array | | | | | | | | | | | | | |
| Lat - 15 | 3.6 | 4.3 | 5.2 | 6.1 | 6.1 | 5.8 | 5.7 | 5.5 | 5.0 | 4.5 | 3.9 | 3.4 | 4.9 |
| Latitude | 4.2 | 4.7 | 5.5 | 6.0 | 5.7 | 5.4 | 5.4 | 5.3 | 5.0 | 4.9 | 4.4 | 3.9 | 5.0 |
| Lat + 15 | 4.4 | 4.9 | 5.4 | 5.6 | 5.1 | 4.7 | 4.7 | 4.9 | 4.8 | 4.9 | 4.7 | 4.2 | 4.9 |
| Single axis tracker | | | | | | | | | | | | | |
| Lat - 15 | 4.6 | 5.5 | 6.7 | 8.0 | 7.8 | 7.3 | 7.2 | 6.9 | 6.2 | 5.7 | 4.9 | 4.2 | 6.3 |
| Latitude | 5.0 | 5.8 | 6.9 | 8.0 | 7.6 | 7.0 | 6.9 | 6.8 | 6.2 | 6.0 | 5.3 | 4.6 | 6.3 |
| Lat + 15 | 5.2 | 5.9 | 6.8 | 7.6 | 7.1 | 6.5 | 6.5 | 6.4 | 6.1 | 6 | 5.5 | 4.9 | 6.2 |
| Dual axis tracker | | | | | | | | | | | | | |
| | 5.2 | 5.9 | 6.9 | 8.0 | 7.9 | 7.5 | 7.3 | 6.9 | 6.3 | 6.0 | 5.5 | 4.9 | 6.5 |

## KEY WEST    FL    Latitude: 24.55 degrees    Elevation: 1 meters

| | Jan | Feb | Mar | Apr | May | Jun | Jul | Aug | Sep | Oct | Nov | Dec | Avg |
|---|---|---|---|---|---|---|---|---|---|---|---|---|---|
| Fixed array | | | | | | | | | | | | | |
| Lat - 15 | 4.2 | 4.9 | 5.8 | 6.5 | 6.3 | 6.0 | 6.0 | 5.9 | 5.4 | 5.0 | 4.4 | 4.0 | 5.4 |
| Latitude | 4.9 | 5.5 | 6.1 | 6.4 | 6.0 | 5.5 | 5.6 | 5.7 | 5.5 | 5.4 | 5.0 | 4.7 | 5.5 |
| Lat + 15 | 5.3 | 5.7 | 6.0 | 6.0 | 5.3 | 4.8 | 5.0 | 5.2 | 5.3 | 5.5 | 5.3 | 5.1 | 5.4 |
| Single axis tracker | | | | | | | | | | | | | |
| Lat - 15 | 5.5 | 6.5 | 7.7 | 8.5 | 8.1 | 7.5 | 7.5 | 7.3 | 6.7 | 6.3 | 5.6 | 5.2 | 6.9 |
| Latitude | 6.0 | 6.9 | 7.9 | 8.5 | 7.8 | 7.1 | 7.2 | 7.2 | 6.8 | 6.6 | 6.1 | 5.7 | 7.0 |
| Lat + 15 | 6.3 | 7.1 | 7.8 | 8.2 | 7.3 | 6.6 | 6.7 | 6.8 | 6.6 | 6.7 | 6.3 | 6 | 6.9 |
| Dual axis tracker | | | | | | | | | | | | | |
| | 6.4 | 7.1 | 7.9 | 8.6 | 8.1 | 7.6 | 7.6 | 7.3 | 6.8 | 6.7 | 6.4 | 6.2 | 7.2 |

## MIAMI    FL    Latitude: 25.80 degrees    Elevation: 2 meters

| | Jan | Feb | Mar | Apr | May | Jun | Jul | Aug | Sep | Oct | Nov | Dec | Avg |
|---|---|---|---|---|---|---|---|---|---|---|---|---|---|
| Fixed array | | | | | | | | | | | | | |
| Lat - 15 | 4.1 | 4.7 | 5.5 | 6.2 | 5.9 | 5.5 | 5.7 | 5.6 | 5.1 | 4.7 | 4.2 | 3.9 | 5.1 |
| Latitude | 4.7 | 5.2 | 5.7 | 6.1 | 5.6 | 5.1 | 5.4 | 5.5 | 5.1 | 5.1 | 4.7 | 4.5 | 5.2 |
| Lat + 15 | 5.0 | 5.4 | 5.6 | 5.7 | 5.0 | 4.5 | 4.8 | 5.0 | 4.9 | 5.1 | 4.9 | 4.9 | 5.1 |
| Single axis tracker | | | | | | | | | | | | | |
| Lat - 15 | 5.2 | 6.1 | 7.0 | 7.9 | 7.4 | 6.6 | 7.0 | 6.9 | 6.2 | 5.9 | 5.2 | 4.9 | 6.4 |
| Latitude | 5.7 | 6.4 | 7.2 | 7.8 | 7.2 | 6.3 | 6.7 | 6.7 | 6.2 | 6.1 | 5.6 | 5.4 | 6.5 |
| Lat + 15 | 5.9 | 6.5 | 7.1 | 7.5 | 6.7 | 5.9 | 6.3 | 6.4 | 6.1 | 6.2 | 5.8 | 5.7 | 6.3 |
| Dual axis tracker | | | | | | | | | | | | | |
| | 6.0 | 6.6 | 7.2 | 7.9 | 7.4 | 6.7 | 7.1 | 6.9 | 6.2 | 6.2 | 5.9 | 5.8 | 6.7 |

## TALLAHASSEE    FL    Latitude: 30.38 degrees    Elevation: 21 meters

| | Jan | Feb | Mar | Apr | May | Jun | Jul | Aug | Sep | Oct | Nov | Dec | Avg |
|---|---|---|---|---|---|---|---|---|---|---|---|---|---|
| Fixed array | | | | | | | | | | | | | |
| Lat - 15 | 3.6 | 4.3 | 5.2 | 6.1 | 6.2 | 6.0 | 5.7 | 5.6 | 5.3 | 5.0 | 4.1 | 3.4 | 5.0 |
| Latitude | 4.0 | 4.7 | 5.4 | 6.0 | 5.9 | 5.6 | 5.4 | 5.4 | 5.3 | 5.4 | 4.6 | 4.0 | 5.1 |
| Lat + 15 | 4.3 | 4.9 | 5.3 | 5.6 | 5.2 | 4.9 | 4.7 | 4.9 | 5.1 | 5.5 | 4.8 | 4.2 | 5.0 |
| Single axis tracker | | | | | | | | | | | | | |
| Lat - 15 | 4.4 | 5.5 | 6.6 | 7.9 | 7.9 | 7.4 | 7.1 | 6.9 | 6.5 | 6.4 | 5.1 | 4.2 | 6.3 |
| Latitude | 4.8 | 5.8 | 6.7 | 7.8 | 7.7 | 7.1 | 6.8 | 6.8 | 6.6 | 6.7 | 5.5 | 4.6 | 6.4 |
| Lat + 15 | 5 | 5.9 | 6.6 | 7.5 | 7.2 | 6.6 | 6.3 | 6.4 | 6.4 | 6.7 | 5.7 | 4.9 | 6.3 |
| Dual axis tracker | | | | | | | | | | | | | |
| | 5.1 | 5.9 | 6.7 | 7.9 | 8.0 | 7.6 | 7.2 | 6.9 | 6.6 | 6.7 | 5.7 | 5.0 | 6.6 |

## TAMPA    FL    Latitude: 27.97 degrees    Elevation: 3 meters

| | Jan | Feb | Mar | Apr | May | Jun | Jul | Aug | Sep | Oct | Nov | Dec | Avg |
|---|---|---|---|---|---|---|---|---|---|---|---|---|---|
| Fixed array | | | | | | | | | | | | | |
| Lat - 15 | 3.9 | 4.6 | 5.5 | 6.4 | 6.4 | 5.9 | 5.7 | 5.5 | 5.2 | 5.0 | 4.2 | 3.8 | 5.2 |
| Latitude | 4.5 | 5.1 | 5.8 | 6.3 | 6.0 | 5.5 | 5.3 | 5.4 | 5.2 | 5.4 | 4.8 | 4.4 | 5.3 |
| Lat + 15 | 4.8 | 5.3 | 5.7 | 5.9 | 5.3 | 4.8 | 4.7 | 4.9 | 5.0 | 5.5 | 5.1 | 4.7 | 5.1 |
| Single axis tracker | | | | | | | | | | | | | |
| Lat - 15 | 5.0 | 5.9 | 7.2 | 8.4 | 8.2 | 7.4 | 7.1 | 6.9 | 6.5 | 6.4 | 5.4 | 4.8 | 6.6 |
| Latitude | 5.4 | 6.3 | 7.3 | 8.4 | 8.0 | 7.1 | 6.8 | 6.8 | 6.5 | 6.7 | 5.9 | 5.2 | 6.7 |
| Lat + 15 | 5.6 | 6.4 | 7.3 | 8.1 | 7.5 | 6.6 | 6.3 | 6.4 | 6.4 | 6.8 | 6.1 | 5.5 | 6.6 |
| Dual axis tracker | | | | | | | | | | | | | |
| | 5.7 | 6.4 | 7.4 | 8.5 | 8.3 | 7.6 | 7.2 | 6.9 | 6.5 | 6.8 | 6.1 | 5.6 | 6.9 |

## WEST PALM BEACH FL — Latitude: 26.68 degrees — Elevation: 6 meters

| | Jan | Feb | Mar | Apr | May | Jun | Jul | Aug | Sep | Oct | Nov | Dec | Avg |
|---|---|---|---|---|---|---|---|---|---|---|---|---|---|
| Fixed array | | | | | | | | | | | | | |
| Lat - 15 | 3.8 | 4.5 | 5.3 | 6.1 | 5.9 | 5.6 | 5.8 | 5.6 | 5.1 | 4.6 | 4.0 | 3.7 | 5.0 |
| Latitude | 4.4 | 5.0 | 5.6 | 6.0 | 5.6 | 5.2 | 5.4 | 5.4 | 5.1 | 4.9 | 4.5 | 4.3 | 5.1 |
| Lat + 15 | 4.7 | 5.1 | 5.5 | 5.6 | 5.0 | 4.5 | 4.8 | 5.0 | 4.9 | 5.0 | 4.7 | 4.7 | 5.0 |
| Single axis tracker | | | | | | | | | | | | | |
| Lat - 15 | 4.9 | 5.8 | 6.9 | 7.9 | 7.5 | 6.8 | 7.2 | 7.0 | 6.2 | 5.8 | 5.0 | 4.7 | 6.3 |
| Latitude | 5.3 | 6.2 | 7.1 | 7.9 | 7.3 | 6.6 | 6.9 | 6.8 | 6.3 | 6.0 | 5.4 | 5.2 | 6.4 |
| Lat + 15 | 5.6 | 6.3 | 7 | 7.6 | 6.9 | 6.1 | 6.4 | 6.5 | 6.1 | 6.1 | 5.6 | 5.4 | 6.3 |
| Dual axis tracker | | | | | | | | | | | | | |
| | 5.6 | 6.3 | 7.1 | 8.0 | 7.6 | 7.0 | 7.3 | 7.0 | 6.3 | 6.1 | 5.6 | 5.5 | 6.6 |

## ATHENS GA — Latitude: 33.95 degrees — Elevation: 244 meters

| | Jan | Feb | Mar | Apr | May | Jun | Jul | Aug | Sep | Oct | Nov | Dec | Avg |
|---|---|---|---|---|---|---|---|---|---|---|---|---|---|
| Fixed array | | | | | | | | | | | | | |
| Lat - 15 | 3.5 | 4.2 | 5.1 | 5.9 | 6.1 | 6.2 | 6.0 | 5.8 | 5.3 | 4.8 | 3.8 | 3.2 | 5.0 |
| Latitude | 3.9 | 4.6 | 5.2 | 5.8 | 5.7 | 5.7 | 5.6 | 5.6 | 5.4 | 5.2 | 4.3 | 3.7 | 5.1 |
| Lat + 15 | 4.2 | 4.8 | 5.2 | 5.4 | 5.1 | 5.0 | 4.9 | 5.1 | 5.1 | 5.3 | 4.5 | 3.9 | 4.9 |
| Single axis tracker | | | | | | | | | | | | | |
| Lat - 15 | 4.3 | 5.4 | 6.5 | 7.7 | 7.8 | 7.9 | 7.5 | 7.3 | 6.6 | 6.1 | 4.7 | 3.9 | 6.3 |
| Latitude | 4.7 | 5.7 | 6.6 | 7.7 | 7.6 | 7.6 | 7.3 | 7.1 | 6.7 | 6.4 | 5.1 | 4.3 | 6.4 |
| Lat + 15 | 4.9 | 5.8 | 6.5 | 7.4 | 7.1 | 7 | 6.8 | 6.8 | 6.5 | 6.5 | 5.3 | 4.5 | 6.3 |
| Dual axis tracker | | | | | | | | | | | | | |
| | 4.9 | 5.8 | 6.6 | 7.8 | 7.9 | 8.1 | 7.6 | 7.3 | 6.7 | 6.5 | 5.3 | 4.6 | 6.6 |

## ATLANTA GA — Latitude: 33.65 degrees — Elevation: 315 meters

| | Jan | Feb | Mar | Apr | May | Jun | Jul | Aug | Sep | Oct | Nov | Dec | Avg |
|---|---|---|---|---|---|---|---|---|---|---|---|---|---|
| Fixed array | | | | | | | | | | | | | |
| Lat - 15 | 3.4 | 4.2 | 5.1 | 6.0 | 6.2 | 6.3 | 6.1 | 5.9 | 5.3 | 4.9 | 3.8 | 3.2 | 5.0 |
| Latitude | 3.8 | 4.6 | 5.3 | 5.8 | 5.8 | 5.8 | 5.7 | 5.7 | 5.4 | 5.2 | 4.2 | 3.7 | 5.1 |
| Lat + 15 | 4.1 | 4.7 | 5.1 | 5.4 | 5.2 | 5.1 | 5.0 | 5.2 | 5.1 | 5.3 | 4.5 | 3.9 | 4.9 |
| Single axis tracker | | | | | | | | | | | | | |
| Lat - 15 | 4.2 | 5.3 | 6.5 | 7.7 | 7.9 | 8.0 | 7.6 | 7.4 | 6.6 | 6.2 | 4.7 | 3.9 | 6.3 |
| Latitude | 4.5 | 5.5 | 6.6 | 7.6 | 7.7 | 7.6 | 7.3 | 7.2 | 6.7 | 6.4 | 5.0 | 4.3 | 6.4 |
| Lat + 15 | 4.7 | 5.6 | 6.5 | 7.3 | 7.2 | 7.1 | 6.8 | 6.9 | 6.5 | 6.5 | 5.2 | 4.5 | 6.2 |
| Dual axis tracker | | | | | | | | | | | | | |
| | 4.8 | 5.7 | 6.6 | 7.7 | 8.0 | 8.1 | 7.7 | 7.4 | 6.7 | 6.5 | 5.3 | 4.5 | 6.6 |

## AUGUSTA GA — Latitude: 33.37 degrees — Elevation: 45 meters

| | Jan | Feb | Mar | Apr | May | Jun | Jul | Aug | Sep | Oct | Nov | Dec | Avg |
|---|---|---|---|---|---|---|---|---|---|---|---|---|---|
| Fixed array | | | | | | | | | | | | | |
| Lat - 15 | 3.4 | 4.3 | 5.1 | 6.0 | 6.1 | 6.2 | 6.0 | 5.7 | 5.2 | 4.9 | 3.8 | 3.3 | 5.0 |
| Latitude | 3.9 | 4.7 | 5.3 | 5.9 | 5.8 | 5.7 | 5.6 | 5.5 | 5.3 | 5.3 | 4.3 | 3.8 | 5.1 |
| Lat + 15 | 4.1 | 4.8 | 5.2 | 5.5 | 5.1 | 5.0 | 4.9 | 5.0 | 5.1 | 5.3 | 4.6 | 4.1 | 4.9 |
| Single axis tracker | | | | | | | | | | | | | |
| Lat - 15 | 4.2 | 5.4 | 6.5 | 7.8 | 7.8 | 7.8 | 7.5 | 7.1 | 6.5 | 6.2 | 4.8 | 4.0 | 6.3 |
| Latitude | 4.6 | 5.7 | 6.7 | 7.8 | 7.6 | 7.5 | 7.2 | 7.0 | 6.6 | 6.5 | 5.2 | 4.4 | 6.4 |
| Lat + 15 | 4.8 | 5.8 | 6.6 | 7.5 | 7.1 | 7 | 6.7 | 6.6 | 6.4 | 6.5 | 5.3 | 4.7 | 6.3 |
| Dual axis tracker | | | | | | | | | | | | | |
| | 4.9 | 5.9 | 6.7 | 7.9 | 7.9 | 8.0 | 7.6 | 7.1 | 6.6 | 6.6 | 5.4 | 4.8 | 6.6 |

## COLUMBUS      GA      Latitude: 32.52 degrees      Elevation: 136 meters

| | Jan | Feb | Mar | Apr | May | Jun | Jul | Aug | Sep | Oct | Nov | Dec | Avg |
|---|---|---|---|---|---|---|---|---|---|---|---|---|---|
| Fixed array | | | | | | | | | | | | | |
| Lat - 15 | 3.5 | 4.3 | 5.2 | 6.0 | 6.2 | 6.2 | 5.9 | 5.8 | 5.3 | 4.9 | 3.9 | 3.3 | 5.0 |
| Latitude | 3.9 | 4.7 | 5.3 | 5.9 | 5.8 | 5.7 | 5.5 | 5.6 | 5.4 | 5.3 | 4.4 | 3.8 | 5.1 |
| Lat + 15 | 4.2 | 4.8 | 5.2 | 5.5 | 5.2 | 5.0 | 4.9 | 5.1 | 5.2 | 5.4 | 4.6 | 4.1 | 4.9 |
| Single axis tracker | | | | | | | | | | | | | |
| Lat - 15 | 4.3 | 5.4 | 6.6 | 7.8 | 8.0 | 7.8 | 7.3 | 7.2 | 6.6 | 6.3 | 4.8 | 4.1 | 6.4 |
| Latitude | 4.7 | 5.7 | 6.7 | 7.8 | 7.7 | 7.5 | 7.1 | 7.1 | 6.7 | 6.6 | 5.2 | 4.5 | 6.4 |
| Lat + 15 | 4.8 | 5.8 | 6.6 | 7.5 | 7.3 | 7 | 6.6 | 6.7 | 6.5 | 6.6 | 5.7 | 4.7 | 6.3 |
| Dual axis tracker | | | | | | | | | | | | | |
| | 4.9 | 5.8 | 6.7 | 7.9 | 8.1 | 8.0 | 7.4 | 7.3 | 6.7 | 6.7 | 5.4 | 4.8 | 6.6 |

## MACON      GA      Latitude: 32.70 degrees      Elevation: 110 meters

| | Jan | Feb | Mar | Apr | May | Jun | Jul | Aug | Sep | Oct | Nov | Dec | Avg |
|---|---|---|---|---|---|---|---|---|---|---|---|---|---|
| Fixed array | | | | | | | | | | | | | |
| Lat - 15 | 3.4 | 4.3 | 5.1 | 6.0 | 6.2 | 6.2 | 5.9 | 5.8 | 5.2 | 4.9 | 3.9 | 3.3 | 5.0 |
| Latitude | 3.9 | 4.7 | 5.3 | 5.9 | 5.8 | 5.7 | 5.5 | 5.6 | 5.3 | 5.3 | 4.4 | 3.8 | 5.1 |
| Lat + 15 | 4.1 | 4.8 | 5.2 | 5.5 | 5.2 | 5.0 | 4.9 | 5.1 | 5.1 | 5.4 | 4.6 | 4.0 | 4.9 |
| Single axis tracker | | | | | | | | | | | | | |
| Lat - 15 | 4.2 | 5.4 | 6.5 | 7.8 | 7.9 | 7.8 | 7.3 | 7.2 | 6.5 | 6.3 | 4.8 | 4.0 | 6.3 |
| Latitude | 4.6 | 5.7 | 6.7 | 7.7 | 7.7 | 7.5 | 7.0 | 7.1 | 6.6 | 6.5 | 5.2 | 4.4 | 6.4 |
| Lat + 15 | 4.8 | 5.8 | 6.6 | 7.4 | 7.2 | 6.9 | 6.6 | 6.7 | 6.4 | 6.6 | 5.4 | 4.6 | 6.3 |
| Dual axis tracker | | | | | | | | | | | | | |
| | 4.8 | 5.8 | 6.7 | 7.8 | 8.0 | 8.0 | 7.4 | 7.2 | 6.6 | 6.6 | 5.5 | 4.7 | 6.6 |

## SAVANNAH      GA      Latitude: 32.13 degrees      Elevation: 16 meters

| | Jan | Feb | Mar | Apr | May | Jun | Jul | Aug | Sep | Oct | Nov | Dec | Avg |
|---|---|---|---|---|---|---|---|---|---|---|---|---|---|
| Fixed array | | | | | | | | | | | | | |
| Lat - 15 | 3.5 | 4.3 | 5.2 | 6.1 | 6.2 | 6.1 | 6.0 | 5.6 | 5.1 | 4.8 | 3.9 | 3.4 | 5.0 |
| Latitude | 4.0 | 4.7 | 5.4 | 6.0 | 5.8 | 5.7 | 5.6 | 5.4 | 5.1 | 5.1 | 4.4 | 3.9 | 5.1 |
| Lat + 15 | 4.3 | 4.8 | 5.3 | 5.6 | 5.2 | 4.9 | 4.9 | 5.0 | 4.9 | 5.2 | 4.6 | 4.2 | 4.9 |
| Single axis tracker | | | | | | | | | | | | | |
| Lat - 15 | 4.4 | 5.4 | 6.7 | 8.0 | 7.9 | 7.7 | 7.4 | 6.9 | 6.3 | 6.0 | 4.9 | 4.1 | 6.3 |
| Latitude | 4.8 | 5.7 | 6.8 | 7.9 | 7.7 | 7.4 | 7.1 | 6.8 | 6.3 | 6.3 | 5.3 | 4.5 | 6.4 |
| Lat + 15 | 5 | 5.8 | 6.7 | 7.6 | 7.2 | 6.8 | 6.6 | 6.5 | 6.2 | 6.3 | 5.4 | 4.8 | 6.2 |
| Dual axis tracker | | | | | | | | | | | | | |
| | 5.0 | 5.8 | 6.8 | 8.0 | 8.0 | 7.8 | 7.5 | 7.0 | 6.4 | 6.4 | 5.5 | 4.9 | 6.6 |

## HILO      HI      Latitude: 19.72 degrees      Elevation: 11 meters

| | Jan | Feb | Mar | Apr | May | Jun | Jul | Aug | Sep | Oct | Nov | Dec | Avg |
|---|---|---|---|---|---|---|---|---|---|---|---|---|---|
| Fixed array | | | | | | | | | | | | | |
| Lat - 15 | 4.0 | 4.4 | 4.7 | 4.8 | 5.1 | 5.3 | 5.1 | 5.3 | 5.1 | 4.5 | 3.9 | 3.7 | 4.7 |
| Latitude | 4.5 | 4.9 | 4.8 | 4.7 | 4.8 | 4.9 | 4.8 | 5.1 | 5.2 | 4.8 | 4.3 | 4.3 | 4.8 |
| Lat + 15 | 4.9 | 5.0 | 4.7 | 4.4 | 4.3 | 4.3 | 4.3 | 4.7 | 5.0 | 4.8 | 4.5 | 4.6 | 4.6 |
| Single axis tracker | | | | | | | | | | | | | |
| Lat - 15 | 5.1 | 5.6 | 5.7 | 5.8 | 6.2 | 6.5 | 6.2 | 6.6 | 6.4 | 5.6 | 4.8 | 4.7 | 5.8 |
| Latitude | 5.5 | 5.9 | 5.8 | 5.7 | 6.0 | 6.2 | 6.0 | 6.5 | 6.4 | 5.8 | 5.1 | 5.1 | 5.8 |
| Lat + 15 | 5.7 | 6 | 5.8 | 5.5 | 5.6 | 5.8 | 5.6 | 6.7 | 6.3 | 5.9 | 5.3 | 5.4 | 5.7 |
| Dual axis tracker | | | | | | | | | | | | | |
| | 5.8 | 6.1 | 5.9 | 5.8 | 6.2 | 6.6 | 6.3 | 6.6 | 6.4 | 5.9 | 5.3 | 5.4 | 6.0 |

## HONOLULU     HI     Latitude: 21.33 degrees     Elevation: 5 meters

| | Jan | Feb | Mar | Apr | May | Jun | Jul | Aug | Sep | Oct | Nov | Dec | Avg |
|---|---|---|---|---|---|---|---|---|---|---|---|---|---|
| **Fixed array** | | | | | | | | | | | | | |
| Lat - 15 | 4.3 | 5.0 | 5.6 | 5.9 | 6.3 | 6.4 | 6.5 | 6.5 | 6.1 | 5.3 | 4.5 | 4.1 | 5.5 |
| Latitude | 4.9 | 5.5 | 5.8 | 5.9 | 5.9 | 5.9 | 6.0 | 6.2 | 6.2 | 5.7 | 5.1 | 4.8 | 5.7 |
| Lat + 15 | 5.3 | 5.8 | 5.8 | 5.5 | 5.3 | 5.1 | 5.3 | 5.7 | 6.0 | 5.8 | 5.4 | 5.2 | 5.5 |
| **Single axis tracker** | | | | | | | | | | | | | |
| Lat - 15 | 5.6 | 6.6 | 7.3 | 7.7 | 8.3 | 8.5 | 8.6 | 8.7 | 8.1 | 7.0 | 5.8 | 5.3 | 7.3 |
| Latitude | 6.1 | 7.0 | 7.5 | 7.7 | 8.0 | 8.1 | 8.3 | 8.5 | 8.2 | 7.3 | 6.2 | 5.9 | 7.4 |
| Lat + 15 | 6.3 | 7.2 | 7.5 | 7.4 | 7.5 | 7.5 | 7.7 | 8.1 | 8 | 7.4 | 6.5 | 6.2 | 7.3 |
| **Dual axis tracker** | | | | | | | | | | | | | |
| | 6.4 | 7.2 | 7.5 | 7.8 | 8.3 | 8.6 | 8.8 | 8.7 | 8.2 | 7.4 | 6.5 | 6.3 | 7.7 |

## KAHULUI     HI     Latitude: 20.90 degrees     Elevation: 15 meters

| | Jan | Feb | Mar | Apr | May | Jun | Jul | Aug | Sep | Oct | Nov | Dec | Avg |
|---|---|---|---|---|---|---|---|---|---|---|---|---|---|
| **Fixed array** | | | | | | | | | | | | | |
| Lat - 15 | 4.4 | 5.0 | 5.6 | 5.9 | 6.4 | 6.6 | 6.6 | 6.5 | 6.2 | 5.4 | 4.6 | 4.2 | 5.6 |
| Latitude | 5.1 | 5.6 | 5.9 | 5.9 | 5.9 | 6.0 | 6.0 | 6.3 | 6.3 | 5.9 | 5.3 | 5.0 | 5.8 |
| Lat + 15 | 5.5 | 5.8 | 5.8 | 5.5 | 5.3 | 5.1 | 5.2 | 5.7 | 6.1 | 6.0 | 5.6 | 5.5 | 5.6 |
| **Single axis tracker** | | | | | | | | | | | | | |
| Lat - 15 | 5.9 | 6.8 | 7.5 | 7.9 | 8.5 | 8.9 | 8.9 | 8.9 | 8.5 | 7.3 | 6.2 | 5.7 | 7.6 |
| Latitude | 6.4 | 7.2 | 7.7 | 7.8 | 8.3 | 8.5 | 8.6 | 8.7 | 8.6 | 7.6 | 6.7 | 6.2 | 7.7 |
| Lat + 15 | 6.7 | 7.3 | 7.6 | 7.5 | 7.8 | 7.9 | 8 | 8.3 | 8.4 | 7.7 | 6.9 | 6.6 | 7.6 |
| **Dual axis tracker** | | | | | | | | | | | | | |
| | 6.8 | 7.4 | 7.7 | 7.9 | 8.7 | 9.1 | 9.1 | 9.0 | 8.6 | 7.7 | 7.0 | 6.7 | 8.0 |

## LIHUE     HI     Latitude: 21.98 degrees     Elevation: 45 meters

| | Jan | Feb | Mar | Apr | May | Jun | Jul | Aug | Sep | Oct | Nov | Dec | Avg |
|---|---|---|---|---|---|---|---|---|---|---|---|---|---|
| **Fixed array** | | | | | | | | | | | | | |
| Lat - 15 | 4.0 | 4.6 | 5.0 | 5.4 | 5.8 | 5.9 | 5.9 | 5.9 | 5.8 | 4.9 | 4.1 | 3.8 | 5.1 |
| Latitude | 4.6 | 5.1 | 5.2 | 5.3 | 5.5 | 5.5 | 5.5 | 5.7 | 5.9 | 5.3 | 4.6 | 4.4 | 5.2 |
| Lat + 15 | 5.0 | 5.3 | 5.2 | 4.9 | 4.9 | 4.8 | 4.8 | 5.2 | 5.7 | 5.4 | 4.8 | 4.8 | 5.1 |
| **Single axis tracker** | | | | | | | | | | | | | |
| Lat - 15 | 5.2 | 6.0 | 6.4 | 6.7 | 7.4 | 7.6 | 7.5 | 7.7 | 7.6 | 6.3 | 5.1 | 4.9 | 6.5 |
| Latitude | 5.6 | 6.4 | 6.5 | 6.7 | 7.2 | 7.3 | 7.2 | 7.5 | 7.7 | 6.6 | 5.5 | 5.4 | 6.6 |
| Lat + 15 | 5.9 | 6.5 | 6.5 | 6.4 | 6.7 | 6.7 | 6.7 | 7.2 | 7.5 | 6.7 | 5.7 | 5.6 | 6.5 |
| **Dual axis tracker** | | | | | | | | | | | | | |
| | 6.0 | 6.5 | 6.5 | 6.7 | 7.5 | 7.8 | 7.6 | 7.7 | 7.7 | 6.7 | 5.7 | 5.7 | 6.8 |

## DES MOINES     IA     Latitude: 41.53 degrees     Elevation: 294 meters

| | Jan | Feb | Mar | Apr | May | Jun | Jul | Aug | Sep | Oct | Nov | Dec | Avg |
|---|---|---|---|---|---|---|---|---|---|---|---|---|---|
| **Fixed array** | | | | | | | | | | | | | |
| Lat - 15 | 3.2 | 3.9 | 4.6 | 5.3 | 5.9 | 6.4 | 6.5 | 6.0 | 5.2 | 4.4 | 3.1 | 2.6 | 4.8 |
| Latitude | 3.6 | 4.3 | 4.7 | 5.2 | 5.5 | 5.8 | 6.0 | 5.8 | 5.3 | 4.7 | 3.4 | 3.0 | 4.8 |
| Lat + 15 | 3.9 | 4.4 | 4.7 | 4.8 | 4.9 | 5.1 | 5.2 | 5.3 | 5.1 | 4.7 | 3.6 | 3.2 | 4.6 |
| **Single axis tracker** | | | | | | | | | | | | | |
| Lat - 15 | 3.8 | 4.9 | 5.8 | 6.9 | 7.7 | 8.6 | 8.7 | 8.0 | 6.8 | 5.5 | 3.7 | 3.1 | 6.1 |
| Latitude | 4.2 | 5.2 | 5.9 | 6.8 | 7.5 | 8.2 | 8.4 | 7.9 | 6.8 | 5.8 | 4.0 | 3.4 | 6.2 |
| Lat + 15 | 4.4 | 5.3 | 5.8 | 6.5 | 7.1 | 7.7 | 7.9 | 7.5 | 6.7 | 5.8 | 4.1 | 3.6 | 6 |
| **Dual axis tracker** | | | | | | | | | | | | | |
| | 4.5 | 5.3 | 5.9 | 6.9 | 7.8 | 8.8 | 8.9 | 8.1 | 6.9 | 5.8 | 4.1 | 3.7 | 6.4 |

## MASON CITY · IA · Latitude: 43.15 degrees · Elevation: 373 meters

| | Jan | Feb | Mar | Apr | May | Jun | Jul | Aug | Sep | Oct | Nov | Dec | Avg |
|---|---|---|---|---|---|---|---|---|---|---|---|---|---|
| Fixed array | | | | | | | | | | | | | |
| Lat - 15 | 3.1 | 3.9 | 4.6 | 5.1 | 5.9 | 6.2 | 6.3 | 5.9 | 5.1 | 4.1 | 2.8 | 2.5 | 4.6 |
| Latitude | 3.5 | 4.3 | 4.7 | 5.0 | 5.5 | 5.7 | 5.8 | 5.6 | 5.1 | 4.4 | 3.1 | 2.8 | 4.6 |
| Lat + 15 | 3.8 | 4.4 | 4.7 | 4.6 | 4.9 | 4.9 | 5.1 | 5.1 | 4.9 | 4.4 | 3.2 | 3.0 | 4.4 |
| Single axis tracker | | | | | | | | | | | | | |
| Lat - 15 | 3.7 | 4.8 | 5.7 | 6.6 | 7.7 | 8.3 | 8.5 | 7.7 | 6.5 | 5.1 | 3.3 | 2.9 | 5.9 |
| Latitude | 4.1 | 5.1 | 5.8 | 6.5 | 7.5 | 8.0 | 8.2 | 7.6 | 6.6 | 5.4 | 3.5 | 3.2 | 6.0 |
| Lat + 15 | 4.3 | 5.2 | 5.3 | 6.2 | 7 | 7.5 | 7.7 | 7.2 | 6.4 | 5.4 | 3.6 | 3.3 | 5.8 |
| Dual axis tracker | | | | | | | | | | | | | |
| | 4.3 | 5.2 | 5.8 | 6.6 | 7.8 | 8.5 | 8.6 | 7.8 | 6.6 | 5.4 | 3.6 | 3.4 | 6.2 |

## SIOUX CITY · IA · Latitude: 42.40 degrees · Elevation: 336 meters

| | Jan | Feb | Mar | Apr | May | Jun | Jul | Aug | Sep | Oct | Nov | Dec | Avg |
|---|---|---|---|---|---|---|---|---|---|---|---|---|---|
| Fixed array | | | | | | | | | | | | | |
| Lat - 15 | 3.1 | 3.9 | 4.7 | 5.4 | 5.9 | 6.4 | 6.5 | 6.1 | 5.2 | 4.4 | 3.0 | 2.6 | 4.8 |
| Latitude | 3.6 | 4.3 | 4.9 | 5.3 | 5.5 | 5.9 | 6.0 | 5.8 | 5.3 | 4.7 | 3.4 | 3.0 | 4.8 |
| Lat + 15 | 3.9 | 4.4 | 4.8 | 4.9 | 4.9 | 5.1 | 5.3 | 5.3 | 5.1 | 4.7 | 3.5 | 3.2 | 4.6 |
| Single axis tracker | | | | | | | | | | | | | |
| Lat - 15 | 3.8 | 4.8 | 6.0 | 7.0 | 7.8 | 8.7 | 8.9 | 8.1 | 6.8 | 5.5 | 3.7 | 3.1 | 6.2 |
| Latitude | 4.2 | 5.1 | 6.1 | 6.9 | 7.6 | 8.3 | 8.5 | 8.0 | 6.9 | 5.8 | 3.9 | 3.5 | 6.2 |
| Lat + 15 | 4.4 | 5.2 | 6 | 6.6 | 7.1 | 7.8 | 8 | 7.6 | 6.7 | 5.8 | 4.1 | 3.6 | 6.1 |
| Dual axis tracker | | | | | | | | | | | | | |
| | 4.4 | 5.2 | 6.1 | 7.0 | 7.9 | 8.9 | 9.0 | 8.1 | 6.9 | 5.8 | 4.1 | 3.7 | 6.4 |

## WATERLOO · IA · Latitude: 42.55 degrees · Elevation: 265 meters

| | Jan | Feb | Mar | Apr | May | Jun | Jul | Aug | Sep | Oct | Nov | Dec | Avg |
|---|---|---|---|---|---|---|---|---|---|---|---|---|---|
| Fixed array | | | | | | | | | | | | | |
| Lat - 15 | 3.0 | 3.8 | 4.4 | 5.1 | 5.8 | 6.2 | 6.3 | 5.9 | 5.0 | 4.1 | 2.7 | 2.5 | 4.6 |
| Latitude | 3.4 | 4.2 | 4.6 | 5.0 | 5.4 | 5.7 | 5.8 | 5.6 | 5.1 | 4.3 | 3.0 | 2.8 | 4.6 |
| Lat + 15 | 3.7 | 4.3 | 4.5 | 4.6 | 4.8 | 5.0 | 5.1 | 5.1 | 4.8 | 4.4 | 3.1 | 3.0 | 4.4 |
| Single axis tracker | | | | | | | | | | | | | |
| Lat - 15 | 3.6 | 4.7 | 5.5 | 6.6 | 7.5 | 8.3 | 8.4 | 7.7 | 6.5 | 5.1 | 3.2 | 2.9 | 5.8 |
| Latitude | 4.0 | 5.0 | 5.6 | 6.5 | 7.3 | 7.9 | 8.1 | 7.6 | 6.5 | 5.3 | 3.4 | 3.2 | 5.9 |
| Lat + 15 | 4.2 | 5.1 | 5.6 | 6.2 | 6.9 | 7.4 | 7.6 | 7.2 | 6.3 | 5.3 | 3.5 | 3.3 | 5.7 |
| Dual axis tracker | | | | | | | | | | | | | |
| | 4.2 | 5.1 | 5.6 | 6.6 | 7.6 | 8.5 | 8.6 | 7.8 | 6.5 | 5.3 | 3.6 | 3.4 | 6.1 |

## BOISE · ID · Latitude: 43.57 degrees · Elevation: 874 meters

| | Jan | Feb | Mar | Apr | May | Jun | Jul | Aug | Sep | Oct | Nov | Dec | Avg |
|---|---|---|---|---|---|---|---|---|---|---|---|---|---|
| Fixed array | | | | | | | | | | | | | |
| Lat - 15 | 2.5 | 3.5 | 4.7 | 5.9 | 6.7 | 7.1 | 7.6 | 7.1 | 6.3 | 4.9 | 2.9 | 2.3 | 5.1 |
| Latitude | 2.8 | 3.8 | 4.9 | 5.8 | 6.2 | 6.5 | 7.0 | 6.8 | 6.5 | 5.2 | 3.2 | 2.6 | 5.1 |
| Lat + 15 | 2.9 | 3.9 | 4.8 | 5.4 | 5.5 | 5.5 | 6.0 | 6.2 | 6.2 | 5.3 | 3.3 | 2.8 | 4.8 |
| Single axis tracker | | | | | | | | | | | | | |
| Lat - 15 | 2.9 | 4.3 | 6.1 | 7.9 | 9.2 | 10.0 | 11.0 | 10.1 | 8.7 | 6.3 | 3.4 | 2.7 | 6.9 |
| Latitude | 3.1 | 4.5 | 6.2 | 7.8 | 9.0 | 9.6 | 10.7 | 9.9 | 6.6 | 3.7 | 2.9 | 6.9 | 8.1 |
| Lat + 15 | 3.2 | 4.6 | 6.1 | 7.5 | 8.5 | 9 | 10 | 9.5 | 8.6 | 6.7 | 3.8 | 3.1 | 6.7 |
| Dual axis tracker | | | | | | | | | | | | | |
| | 3.3 | 4.6 | 6.2 | 7.9 | 9.4 | 10.3 | 11.3 | 10.1 | 8.8 | 6.7 | 3.8 | 3.1 | 7.1 |

## POCATELLO ID    Latitude: 42.92 degrees    Elevation: 1365 meters

| | Jan | Feb | Mar | Apr | May | Jun | Jul | Aug | Sep | Oct | Nov | Dec | Avg |
|---|---|---|---|---|---|---|---|---|---|---|---|---|---|
| Fixed array | | | | | | | | | | | | | |
| Lat - 15 | 2.6 | 3.6 | 4.7 | 5.6 | 6.3 | 6.8 | 7.3 | 6.8 | 6.1 | 4.9 | 2.9 | 2.3 | 5.0 |
| Latitude | 2.9 | 3.9 | 4.9 | 5.5 | 5.9 | 6.2 | 6.7 | 6.6 | 6.2 | 5.3 | 3.2 | 2.6 | 5.0 |
| Lat + 15 | 3.0 | 4.0 | 4.8 | 5.1 | 5.2 | 5.4 | 5.8 | 6.0 | 6.0 | 5.3 | 3.4 | 2.8 | 4.7 |
| Single axis tracker | | | | | | | | | | | | | |
| Lat - 15 | 3.0 | 4.4 | 6.1 | 7.5 | 8.6 | 9.6 | 10.4 | 9.6 | 8.4 | 6.3 | 3.5 | 2.7 | 6.7 |
| Latitude | 3.3 | 4.6 | 6.2 | 7.4 | 8.3 | 9.3 | 10.1 | 9.5 | 8.5 | 6.6 | 3.8 | 3.0 | 6.7 |
| Lat + 15 | 3.4 | 4.7 | 6.1 | 7.1 | 7.8 | 8.7 | 9.5 | 9 | 8.3 | 6.7 | 3.9 | 3.1 | 6.5 |
| Dual axis tracker | | | | | | | | | | | | | |
| | 3.4 | 4.7 | 6.2 | 7.5 | 8.7 | 9.9 | 10.7 | 9.7 | 8.5 | 6.7 | 3.9 | 3.2 | 6.9 |

## CHICAGO IL    Latitude: 41.78 degrees    Elevation: 190 meters

| | Jan | Feb | Mar | Apr | May | Jun | Jul | Aug | Sep | Oct | Nov | Dec | Avg |
|---|---|---|---|---|---|---|---|---|---|---|---|---|---|
| Fixed array | | | | | | | | | | | | | |
| Lat - 15 | 2.7 | 3.5 | 4.1 | 5.0 | 5.8 | 6.1 | 6.1 | 5.7 | 4.9 | 3.9 | 2.5 | 2.2 | 4.4 |
| Latitude | 3.1 | 3.8 | 4.2 | 4.9 | 5.4 | 5.7 | 5.6 | 5.5 | 4.9 | 4.2 | 2.8 | 2.4 | 4.4 |
| Lat + 15 | 3.3 | 3.9 | 4.1 | 4.5 | 4.8 | 4.9 | 4.9 | 5.0 | 4.7 | 4.2 | 2.9 | 2.6 | 4.1 |
| Single axis tracker | | | | | | | | | | | | | |
| Lat - 15 | 3.2 | 4.2 | 5.0 | 6.2 | 7.4 | 8.0 | 7.9 | 7.3 | 6.2 | 4.8 | 2.9 | 2.5 | 5.5 |
| Latitude | 3.5 | 4.5 | 5.1 | 6.1 | 7.2 | 7.7 | 7.7 | 7.2 | 6.2 | 5.0 | 3.1 | 2.7 | 5.5 |
| Lat + 15 | 3.7 | 4.5 | 5 | 5.9 | 6.8 | 7.2 | 7.2 | 6.8 | 6 | 5 | 3.2 | 2.9 | 5.3 |
| Dual axis tracker | | | | | | | | | | | | | |
| | 3.7 | 4.6 | 5.1 | 6.2 | 7.5 | 8.2 | 8.1 | 7.4 | 6.2 | 5.0 | 3.2 | 2.9 | 5.7 |

## ROCKFORD IL    Latitude: 42.20 degrees    Elevation: 221 meters

| | Jan | Feb | Mar | Apr | May | Jun | Jul | Aug | Sep | Oct | Nov | Dec | Avg |
|---|---|---|---|---|---|---|---|---|---|---|---|---|---|
| Fixed array | | | | | | | | | | | | | |
| Lat - 15 | 2.9 | 3.8 | 4.3 | 5.0 | 5.7 | 6.1 | 6.1 | 5.7 | 5.0 | 4.0 | 2.6 | 2.3 | 4.5 |
| Latitude | 3.3 | 4.1 | 4.4 | 4.9 | 5.4 | 5.6 | 5.7 | 5.5 | 5.0 | 4.2 | 2.9 | 2.6 | 4.5 |
| Lat + 15 | 3.6 | 4.2 | 4.3 | 4.5 | 4.8 | 4.9 | 4.9 | 5.0 | 4.8 | 4.3 | 3.0 | 2.8 | 4.3 |
| Single axis tracker | | | | | | | | | | | | | |
| Lat - 15 | 3.5 | 4.6 | 5.3 | 6.3 | 7.4 | 8.1 | 8.1 | 7.4 | 6.3 | 4.9 | 3.1 | 2.7 | 5.7 |
| Latitude | 3.8 | 4.9 | 5.4 | 6.2 | 7.2 | 7.7 | 7.8 | 7.3 | 6.3 | 5.1 | 3.3 | 3.0 | 5.7 |
| Lat + 15 | 4 | 5 | 5.3 | 6 | 6.8 | 7.2 | 7.3 | 6.9 | 6.2 | 5.1 | 3.4 | 3.1 | 5.5 |
| Dual axis tracker | | | | | | | | | | | | | |
| | 4.1 | 5.0 | 5.4 | 6.3 | 7.5 | 8.3 | 8.2 | 7.5 | 6.4 | 5.2 | 3.4 | 3.2 | 5.9 |

## SPRINGFIELD IL    Latitude: 39.83 degrees    Elevation: 187 meters

| | Jan | Feb | Mar | Apr | May | Jun | Jul | Aug | Sep | Oct | Nov | Dec | Avg |
|---|---|---|---|---|---|---|---|---|---|---|---|---|---|
| Fixed array | | | | | | | | | | | | | |
| Lat - 15 | 3.1 | 3.9 | 4.4 | 5.4 | 6.0 | 6.4 | 6.4 | 6.0 | 5.4 | 4.5 | 3.1 | 2.6 | 4.8 |
| Latitude | 3.5 | 4.2 | 4.5 | 5.3 | 5.6 | 5.9 | 5.9 | 5.8 | 5.4 | 4.8 | 3.4 | 2.9 | 4.8 |
| Lat + 15 | 3.8 | 4.4 | 4.4 | 4.9 | 5.0 | 5.1 | 5.1 | 5.3 | 5.2 | 4.8 | 3.5 | 3.1 | 4.5 |
| Single axis tracker | | | | | | | | | | | | | |
| Lat - 15 | 3.8 | 4.8 | 5.5 | 6.9 | 7.9 | 8.5 | 8.5 | 8.0 | 6.9 | 5.6 | 3.7 | 3.0 | 6.1 |
| Latitude | 4.1 | 5.1 | 5.6 | 6.8 | 7.6 | 8.2 | 8.2 | 7.8 | 7.0 | 5.9 | 3.9 | 3.3 | 6.1 |
| Lat + 15 | 4.3 | 5.2 | 5.6 | 6.5 | 7.2 | 7.6 | 7.7 | 7.5 | 6.8 | 5.9 | 4.1 | 3.5 | 6 |
| Dual axis tracker | | | | | | | | | | | | | |
| | 4.4 | 5.2 | 5.6 | 6.9 | 8.0 | 8.7 | 8.7 | 8.0 | 7.0 | 5.9 | 4.1 | 3.5 | 6.3 |

## EVANSVILLE      IN      Latitude: 38.05 degrees      Elevation: 118 meters

| | Jan | Feb | Mar | Apr | May | Jun | Jul | Aug | Sep | Oct | Nov | Dec | Avg |
|---|---|---|---|---|---|---|---|---|---|---|---|---|---|
| Fixed array | | | | | | | | | | | | | |
| Lat − 15 | 2.9 | 3.7 | 4.5 | 5.3 | 5.9 | 6.3 | 6.2 | 6.0 | 5.3 | 4.5 | 3.1 | 2.5 | 4.7 |
| Latitude | 3.3 | 4.0 | 4.6 | 5.2 | 5.6 | 5.8 | 5.8 | 5.8 | 5.3 | 4.8 | 3.4 | 2.9 | 4.7 |
| Lat + 15 | 3.5 | 4.1 | 4.5 | 4.8 | 4.9 | 5.0 | 5.1 | 5.3 | 5.1 | 4.9 | 3.6 | 3.0 | 4.5 |
| Single axis tracker | | | | | | | | | | | | | |
| Lat − 15 | 3.5 | 4.5 | 5.6 | 6.8 | 7.7 | 8.4 | 8.2 | 7.8 | 6.7 | 5.7 | 3.7 | 3.0 | 6.0 |
| Latitude | 3.8 | 4.7 | 5.7 | 6.7 | 7.5 | 8.1 | 7.9 | 7.7 | 6.8 | 5.9 | 4.0 | 3.3 | 6.0 |
| Lat + 15 | 3.9 | 4.8 | 5.6 | 6.4 | 7.1 | 7.5 | 7.4 | 7.3 | 6.6 | 5.9 | 4.1 | 3.4 | 5.9 |
| Dual axis tracker | | | | | | | | | | | | | |
| | 4.0 | 4.8 | 5.7 | 6.8 | 7.8 | 8.6 | 8.4 | 7.9 | 6.8 | 6.0 | 4.1 | 3.5 | 6.2 |

## MOLINE      IL      Latitude: 41.45 degrees      Elevation: 181 meters

| | Jan | Feb | Mar | Apr | May | Jun | Jul | Aug | Sep | Oct | Nov | Dec | Avg |
|---|---|---|---|---|---|---|---|---|---|---|---|---|---|
| Fixed array | | | | | | | | | | | | | |
| Lat − 15 | 2.9 | 3.8 | 4.3 | 5.1 | 5.8 | 6.2 | 6.2 | 5.9 | 5.1 | 4.2 | 2.8 | 2.4 | 4.6 |
| Latitude | 3.3 | 4.1 | 4.4 | 5.0 | 5.4 | 5.7 | 5.8 | 5.6 | 5.1 | 4.5 | 3.1 | 2.7 | 4.6 |
| Lat + 15 | 3.5 | 4.2 | 4.3 | 4.6 | 4.8 | 4.9 | 5.1 | 5.1 | 4.9 | 4.5 | 3.3 | 2.9 | 4.4 |
| Single axis tracker | | | | | | | | | | | | | |
| Lat − 15 | 3.5 | 4.6 | 5.3 | 6.5 | 7.6 | 8.3 | 8.3 | 7.7 | 6.6 | 5.2 | 3.4 | 2.8 | 5.8 |
| Latitude | 3.8 | 4.9 | 5.4 | 6.5 | 7.3 | 7.9 | 8.0 | 7.6 | 6.6 | 5.5 | 3.6 | 3.1 | 5.9 |
| Lat + 15 | 4 | 5 | 5.3 | 6.2 | 5.9 | 7.4 | 7.5 | 7.2 | 6.4 | 5.5 | 3.7 | 3.3 | 5.7 |
| Dual axis tracker | | | | | | | | | | | | | |
| | 4.1 | 5.0 | 5.4 | 6.6 | 7.7 | 8.5 | 8.5 | 7.7 | 6.6 | 5.5 | 3.7 | 3.3 | 6.1 |

## PEORIA      IL      Latitude: 40.67 degrees      Elevation: 199 meters

| | Jan | Feb | Mar | Apr | May | Jun | Jul | Aug | Sep | Oct | Nov | Dec | Avg |
|---|---|---|---|---|---|---|---|---|---|---|---|---|---|
| Fixed array | | | | | | | | | | | | | |
| Lat − 15 | 2.9 | 3.7 | 4.2 | 5.2 | 5.8 | 6.3 | 6.2 | 5.9 | 5.2 | 4.3 | 2.9 | 2.4 | 4.6 |
| Latitude | 3.3 | 4.1 | 4.4 | 5.1 | 5.5 | 5.8 | 5.8 | 5.7 | 5.2 | 4.6 | 3.2 | 2.7 | 4.6 |
| Lat + 15 | 3.5 | 4.2 | 4.2 | 4.7 | 4.8 | 5.0 | 5.1 | 5.1 | 5.0 | 4.6 | 3.3 | 2.9 | 4.4 |
| Single axis tracker | | | | | | | | | | | | | |
| Lat − 15 | 3.5 | 4.6 | 5.2 | 6.6 | 7.6 | 8.3 | 8.3 | 7.7 | 6.6 | 5.3 | 3.4 | 2.8 | 5.8 |
| Latitude | 3.8 | 4.8 | 5.3 | 6.5 | 7.4 | 8.0 | 8.0 | 7.5 | 6.7 | 5.6 | 3.6 | 3.1 | 5.9 |
| Lat + 15 | 4 | 4.9 | 5.2 | 6.2 | 6.9 | 7.4 | 7.5 | 7.2 | 6.5 | 5.6 | 3.7 | 3.2 | 5.7 |
| Dual axis tracker | | | | | | | | | | | | | |
| | 4.0 | 5.0 | 5.3 | 6.6 | 7.7 | 8.5 | 8.4 | 7.7 | 6.7 | 5.6 | 3.8 | 3.3 | 6.1 |

## FORT WAYNE      IN      Latitude: 41.00 degrees      Elevation: 252 meters

| | Jan | Feb | Mar | Apr | May | Jun | Jul | Aug | Sep | Oct | Nov | Dec | Avg |
|---|---|---|---|---|---|---|---|---|---|---|---|---|---|
| Fixed array | | | | | | | | | | | | | |
| Lat − 15 | 2.5 | 3.4 | 4.1 | 5.0 | 5.7 | 6.1 | 6.0 | 5.7 | 5.0 | 3.9 | 2.5 | 2.0 | 4.3 |
| Latitude | 2.8 | 3.7 | 4.2 | 4.9 | 5.4 | 5.6 | 5.6 | 5.4 | 5.0 | 4.1 | 2.7 | 2.2 | 4.3 |
| Lat + 15 | 3.0 | 3.8 | 4.1 | 4.5 | 4.8 | 4.9 | 4.9 | 4.9 | 4.8 | 4.1 | 2.8 | 2.3 | 4.1 |
| Single axis tracker | | | | | | | | | | | | | |
| Lat − 15 | 2.9 | 4.0 | 4.9 | 6.2 | 7.3 | 7.9 | 7.8 | 7.3 | 6.2 | 4.8 | 2.9 | 2.3 | 5.4 |
| Latitude | 3.2 | 4.3 | 5.0 | 6.2 | 7.1 | 7.6 | 7.6 | 7.1 | 6.3 | 4.9 | 3.1 | 2.4 | 5.4 |
| Lat + 15 | 3.3 | 4.3 | 4.9 | 5.9 | 6.7 | 7.1 | 7.1 | 6.8 | 6.1 | 5 | 3.1 | 2.5 | 5.2 |
| Dual axis tracker | | | | | | | | | | | | | |
| | 3.4 | 4.4 | 5.0 | 6.3 | 7.4 | 8.1 | 8.0 | 7.3 | 6.3 | 5.0 | 3.2 | 2.6 | 5.6 |

## INDIANAPOLIS IN    Latitude: 39.73 degrees    Elevation: 246 meters

|  | Jan | Feb | Mar | Apr | May | Jun | Jul | Aug | Sep | Oct | Nov | Dec | Avg |
|---|---|---|---|---|---|---|---|---|---|---|---|---|---|
| Fixed array | | | | | | | | | | | | | |
| Lat - 15 | 2.8 | 3.6 | 4.3 | 5.2 | 5.9 | 6.3 | 6.2 | 5.9 | 5.2 | 4.2 | 2.8 | 2.3 | 4.6 |
| Latitude | 3.1 | 3.9 | 4.4 | 5.1 | 5.6 | 5.8 | 5.8 | 5.7 | 5.3 | 4.5 | 3.1 | 2.6 | 4.6 |
| Lat + 15 | 3.3 | 4.0 | 4.3 | 4.7 | 4.9 | 5.0 | 5.1 | 5.2 | 5.1 | 4.5 | 3.2 | 2.7 | 4.3 |
| Single axis tracker | | | | | | | | | | | | | |
| Lat - 15 | 3.3 | 4.3 | 5.2 | 6.5 | 7.6 | 8.2 | 8.1 | 7.7 | 6.6 | 5.2 | 3.3 | 2.7 | 5.7 |
| Latitude | 3.6 | 4.6 | 5.3 | 6.5 | 7.4 | 7.9 | 7.8 | 7.5 | 6.7 | 5.5 | 3.5 | 2.9 | 5.8 |
| Lat + 15 | 3.7 | 4.7 | 5.2 | 6.2 | 6.9 | 7.3 | 7.3 | 7.3 | 6.5 | 5.5 | 3.6 | 3 | 5.6 |
| Dual axis tracker | | | | | | | | | | | | | |
|  | 3.7 | 4.7 | 5.3 | 6.6 | 7.7 | 8.4 | 8.3 | 7.7 | 6.7 | 5.5 | 3.6 | 3.1 | 5.9 |

## SOUTH BEND IN    Latitude: 41.70 degrees    Elevation: 236 meters

|  | Jan | Feb | Mar | Apr | May | Jun | Jul | Aug | Sep | Oct | Nov | Dec | Avg |
|---|---|---|---|---|---|---|---|---|---|---|---|---|---|
| Fixed array | | | | | | | | | | | | | |
| Lat - 15 | 2.4 | 3.3 | 4.0 | 5.0 | 5.7 | 6.1 | 6.0 | 5.6 | 4.8 | 3.7 | 2.3 | 1.9 | 4.2 |
| Latitude | 2.7 | 3.5 | 4.1 | 4.8 | 5.4 | 5.6 | 5.6 | 5.4 | 4.8 | 3.9 | 2.5 | 2.0 | 4.2 |
| Lat + 15 | 2.8 | 3.6 | 4.0 | 4.5 | 4.7 | 4.8 | 4.9 | 4.9 | 4.6 | 3.9 | 2.6 | 2.1 | 4.0 |
| Single axis tracker | | | | | | | | | | | | | |
| Lat - 15 | 2.7 | 3.9 | 4.8 | 6.2 | 7.3 | 7.9 | 7.8 | 7.2 | 6.0 | 4.5 | 2.6 | 2.1 | 5.3 |
| Latitude | 3.0 | 4.1 | 4.9 | 6.1 | 7.1 | 7.6 | 7.5 | 7.1 | 6.0 | 4.7 | 2.8 | 2.2 | 5.3 |
| Lat + 15 | 3.1 | 4.1 | 4.8 | 5.8 | 6.6 | 7 | 7 | 6.7 | 5.9 | 4.7 | 2.9 | 2.3 | 5.1 |
| Dual axis tracker | | | | | | | | | | | | | |
|  | 3.1 | 4.2 | 4.9 | 6.2 | 7.4 | 8.1 | 7.9 | 7.3 | 6.1 | 4.7 | 2.9 | 2.3 | 5.4 |

## DODGE CITY KS    Latitude: 37.77 degrees    Elevation: 787 meters

|  | Jan | Feb | Mar | Apr | May | Jun | Jul | Aug | Sep | Oct | Nov | Dec | Avg |
|---|---|---|---|---|---|---|---|---|---|---|---|---|---|
| Fixed array | | | | | | | | | | | | | |
| Lat - 15 | 4.0 | 4.8 | 5.5 | 6.3 | 6.5 | 7.0 | 7.1 | 6.6 | 5.9 | 5.2 | 4.0 | 3.6 | 5.5 |
| Latitude | 4.6 | 5.3 | 5.8 | 6.2 | 6.1 | 6.4 | 6.5 | 6.4 | 6.0 | 5.6 | 4.6 | 4.2 | 5.6 |
| Lat + 15 | 5.0 | 5.5 | 5.7 | 5.8 | 5.4 | 5.5 | 5.7 | 5.8 | 5.7 | 5.7 | 4.8 | 4.6 | 5.4 |
| Single axis tracker | | | | | | | | | | | | | |
| Lat - 15 | 5.1 | 6.1 | 7.2 | 8.4 | 8.7 | 9.4 | 9.6 | 8.8 | 7.7 | 6.7 | 5.0 | 4.5 | 7.3 |
| Latitude | 5.6 | 6.5 | 7.4 | 8.3 | 8.4 | 9.0 | 9.2 | 8.7 | 7.8 | 7.0 | 5.5 | 5.0 | 7.4 |
| Lat + 15 | 5.8 | 6.6 | 7.3 | 8 | 7.9 | 8.4 | 8.6 | 8.2 | 7.6 | 7.1 | 5.7 | 5.3 | 7.2 |
| Dual axis tracker | | | | | | | | | | | | | |
|  | 5.9 | 6.6 | 7.4 | 8.4 | 8.8 | 9.7 | 9.7 | 8.9 | 7.8 | 7.1 | 5.7 | 5.4 | 7.6 |

## GOODLAND KS    Latitude: 39.37 degrees    Elevation: 1124 meters

|  | Jan | Feb | Mar | Apr | May | Jun | Jul | Aug | Sep | Oct | Nov | Dec | Avg |
|---|---|---|---|---|---|---|---|---|---|---|---|---|---|
| Fixed array | | | | | | | | | | | | | |
| Lat - 15 | 3.9 | 4.6 | 5.4 | 6.2 | 6.4 | 7.0 | 7.0 | 6.7 | 6.0 | 5.3 | 4.0 | 3.6 | 5.5 |
| Latitude | 4.5 | 5.1 | 5.7 | 6.1 | 6.0 | 6.4 | 6.5 | 6.4 | 6.1 | 5.7 | 4.6 | 4.2 | 5.6 |
| Lat + 15 | 4.9 | 5.3 | 5.6 | 5.7 | 5.3 | 5.5 | 5.6 | 5.8 | 5.8 | 5.8 | 4.8 | 4.6 | 5.4 |
| Single axis tracker | | | | | | | | | | | | | |
| Lat - 15 | 4.9 | 5.9 | 7.1 | 8.3 | 8.6 | 9.6 | 9.7 | 9.1 | 8.0 | 6.9 | 5.0 | 4.5 | 7.3 |
| Latitude | 5.4 | 6.3 | 7.3 | 8.3 | 8.3 | 9.2 | 9.4 | 8.9 | 8.0 | 7.2 | 5.5 | 4.9 | 7.4 |
| Lat + 15 | 5.7 | 6.4 | 7.2 | 8 | 7.8 | 8.6 | 8.8 | 8.5 | 7.9 | 7.3 | 5.7 | 5.3 | 7.3 |
| Dual axis tracker | | | | | | | | | | | | | |
|  | 5.7 | 6.4 | 7.3 | 8.4 | 8.7 | 9.9 | 9.9 | 9.1 | 8.1 | 7.3 | 5.7 | 5.3 | 7.7 |

## TOPEKA · KS · Latitude: 39.07 degrees · Elevation: 270 meters

| | Jan | Feb | Mar | Apr | May | Jun | Jul | Aug | Sep | Oct | Nov | Dec | Avg |
|---|---|---|---|---|---|---|---|---|---|---|---|---|---|
| **Fixed array** | | | | | | | | | | | | | |
| Lat - 15 | 3.4 | 4.0 | 4.7 | 5.5 | 5.9 | 6.3 | 6.5 | 6.1 | 5.3 | 4.6 | 3.4 | 2.9 | 4.9 |
| Latitude | 3.9 | 4.4 | 4.9 | 5.4 | 5.5 | 5.8 | 6.0 | 5.9 | 5.4 | 4.9 | 3.8 | 3.4 | 4.9 |
| Lat + 15 | 4.2 | 4.5 | 4.7 | 5.0 | 4.9 | 5.0 | 5.3 | 5.3 | 5.2 | 4.9 | 4.0 | 3.6 | 4.7 |
| **Single axis tracker** | | | | | | | | | | | | | |
| Lat - 15 | 4.2 | 5.0 | 5.9 | 7.1 | 7.7 | 8.4 | 8.8 | 8.1 | 6.9 | 5.8 | 4.1 | 3.5 | 6.3 |
| Latitude | 4.6 | 5.3 | 6.1 | 7.0 | 7.5 | 8.0 | 8.5 | 8.0 | 7.0 | 6.1 | 4.5 | 3.9 | 6.4 |
| Lat + 15 | 4.8 | 5.4 | 6 | 6.7 | 7 | 7.5 | 7.9 | 7.6 | 6.8 | 6.1 | 4.6 | 4.1 | 6.2 |
| **Dual axis tracker** | | | | | | | | | | | | | |
| | 4.9 | 5.5 | 6.1 | 7.1 | 7.8 | 8.5 | 8.9 | 8.2 | 7.0 | 6.1 | 4.7 | 4.2 | 6.6 |

## WICHITA · KS · Latitude: 37.65 degrees · Elevation: 408 meters

| | Jan | Feb | Mar | Apr | May | Jun | Jul | Aug | Sep | Oct | Nov | Dec | Avg |
|---|---|---|---|---|---|---|---|---|---|---|---|---|---|
| **Fixed array** | | | | | | | | | | | | | |
| Lat - 15 | 3.6 | 4.3 | 5.0 | 5.8 | 6.1 | 6.5 | 6.8 | 6.4 | 5.5 | 4.8 | 3.7 | 3.3 | 5.2 |
| Latitude | 4.2 | 4.7 | 5.2 | 5.7 | 5.7 | 6.0 | 6.3 | 6.1 | 5.6 | 5.2 | 4.2 | 3.8 | 5.2 |
| Lat + 15 | 4.5 | 4.9 | 5.1 | 5.3 | 5.1 | 5.2 | 5.5 | 5.6 | 5.4 | 5.3 | 4.4 | 4.1 | 5.0 |
| **Single axis tracker** | | | | | | | | | | | | | |
| Lat - 15 | 4.5 | 5.4 | 6.4 | 7.6 | 8.0 | 8.7 | 9.1 | 8.5 | 7.2 | 6.2 | 4.6 | 4.0 | 6.7 |
| Latitude | 5.0 | 5.7 | 6.6 | 7.5 | 7.8 | 8.4 | 8.8 | 8.4 | 7.3 | 6.5 | 5.0 | 4.4 | 6.8 |
| Lat + 15 | 5.2 | 5.8 | 6.5 | 7.2 | 7.3 | 7.8 | 8.3 | 8 | 7.3 | 6.5 | 5.1 | 4.7 | 6.6 |
| **Dual axis tracker** | | | | | | | | | | | | | |
| | 5.3 | 5.9 | 6.6 | 7.6 | 8.1 | 8.9 | 9.3 | 8.6 | 7.3 | 6.6 | 5.2 | 4.7 | 7.0 |

## COVINGTON · KY · Latitude: 39.07 degrees · Elevation: 271 meters

| | Jan | Feb | Mar | Apr | May | Jun | Jul | Aug | Sep | Oct | Nov | Dec | Avg |
|---|---|---|---|---|---|---|---|---|---|---|---|---|---|
| **Fixed array** | | | | | | | | | | | | | |
| Lat - 15 | 2.7 | 3.5 | 4.2 | 5.1 | 5.7 | 6.1 | 5.9 | 5.8 | 5.1 | 4.2 | 2.8 | 2.3 | 4.5 |
| Latitude | 3.0 | 3.7 | 4.3 | 5.0 | 5.4 | 5.6 | 5.5 | 5.6 | 5.2 | 4.5 | 3.1 | 2.5 | 4.5 |
| Lat + 15 | 3.2 | 3.8 | 4.2 | 4.6 | 4.8 | 4.9 | 4.8 | 5.1 | 4.9 | 4.5 | 3.2 | 2.7 | 4.2 |
| **Single axis tracker** | | | | | | | | | | | | | |
| Lat - 15 | 3.2 | 4.2 | 5.1 | 6.4 | 7.3 | 7.9 | 7.7 | 7.5 | 6.5 | 5.2 | 3.3 | 2.6 | 5.6 |
| Latitude | 3.5 | 4.4 | 5.2 | 6.4 | 7.1 | 7.6 | 7.4 | 7.3 | 6.5 | 5.4 | 3.5 | 2.8 | 5.6 |
| Lat + 15 | 3.6 | 4.5 | 5.1 | 6.1 | 6.7 | 7.1 | 7 | 7 | 6.4 | 5.5 | 3.6 | 2.9 | 5.4 |
| **Dual axis tracker** | | | | | | | | | | | | | |
| | 3.6 | 4.5 | 5.2 | 6.5 | 7.4 | 8.1 | 7.9 | 7.5 | 6.6 | 5.5 | 3.6 | 3.0 | 5.8 |

## LEXINGTON · KY · Latitude: 38.03 degrees · Elevation: 301 meters

| | Jan | Feb | Mar | Apr | May | Jun | Jul | Aug | Sep | Oct | Nov | Dec | Avg |
|---|---|---|---|---|---|---|---|---|---|---|---|---|---|
| **Fixed array** | | | | | | | | | | | | | |
| Lat - 15 | 2.8 | 3.5 | 4.3 | 5.2 | 5.7 | 6.0 | 5.9 | 5.7 | 5.0 | 4.3 | 2.9 | 2.4 | 4.5 |
| Latitude | 3.1 | 3.8 | 4.4 | 5.1 | 5.4 | 5.6 | 5.5 | 5.5 | 5.1 | 4.6 | 3.2 | 2.7 | 4.5 |
| Lat + 15 | 3.3 | 3.8 | 4.3 | 4.7 | 4.8 | 4.8 | 4.8 | 5.0 | 4.8 | 4.6 | 3.4 | 2.9 | 4.3 |
| **Single axis tracker** | | | | | | | | | | | | | |
| Lat - 15 | 3.3 | 4.2 | 5.3 | 6.6 | 7.3 | 7.8 | 7.6 | 7.4 | 6.4 | 5.4 | 3.5 | 2.8 | 5.6 |
| Latitude | 3.5 | 4.4 | 5.4 | 6.5 | 7.1 | 7.5 | 7.3 | 7.2 | 6.4 | 5.6 | 3.7 | 3.1 | 5.7 |
| Lat + 15 | 3.7 | 4.5 | 5.3 | 6.2 | 6.7 | 7 | 6.9 | 6.9 | 6.2 | 5.6 | 3.8 | 3.2 | 5.5 |
| **Dual axis tracker** | | | | | | | | | | | | | |
| | 3.7 | 4.5 | 5.4 | 6.6 | 7.4 | 8.0 | 7.7 | 7.4 | 6.4 | 5.7 | 3.9 | 3.2 | 5.8 |

## LOUISVILLE    KY    Latitude: 38.18 degrees    Elevation: 149 meters

|  | Jan | Feb | Mar | Apr | May | Jun | Jul | Aug | Sep | Oct | Nov | Dec | Avg |
|---|---|---|---|---|---|---|---|---|---|---|---|---|---|
| Fixed array |  |  |  |  |  |  |  |  |  |  |  |  |  |
| Lat - 15 | 2.8 | 3.6 | 4.4 | 5.3 | 5.8 | 6.2 | 6.0 | 5.9 | 5.1 | 4.4 | 3.0 | 2.4 | 4.6 |
| Latitude | 3.1 | 3.9 | 4.5 | 5.2 | 5.5 | 5.7 | 5.6 | 5.7 | 5.2 | 4.7 | 3.3 | 2.7 | 4.6 |
| Lat + 15 | 3.3 | 4.0 | 4.4 | 4.8 | 4.9 | 4.9 | 4.9 | 5.1 | 5.0 | 4.7 | 3.4 | 2.9 | 4.4 |
| Single axis tracker |  |  |  |  |  |  |  |  |  |  |  |  |  |
| Lat - 15 | 3.3 | 4.4 | 5.4 | 6.7 | 7.5 | 8.0 | 7.8 | 7.5 | 6.5 | 5.4 | 3.5 | 2.8 | 5.7 |
| Latitude | 3.6 | 4.6 | 5.5 | 6.6 | 7.2 | 7.7 | 7.5 | 7.4 | 6.6 | 5.7 | 3.8 | 3.1 | 5.8 |
| Lat + 15 | 3.7 | 4.7 | 5.4 | 6.3 | 6.8 | 7.2 | 7.1 | 7 | 6.4 | 5.7 | 3.9 | 3.2 | 5.6 |
| Dual axis tracker |  |  |  |  |  |  |  |  |  |  |  |  |  |
|  | 3.8 | 4.7 | 5.5 | 6.7 | 7.6 | 8.2 | 8.0 | 7.6 | 6.6 | 5.7 | 3.9 | 3.2 | 6.0 |

## BATON ROUGE    LA    Latitude: 30.53 degrees    Elevation: 23 meters

|  | Jan | Feb | Mar | Apr | May | Jun | Jul | Aug | Sep | Oct | Nov | Dec | Avg |
|---|---|---|---|---|---|---|---|---|---|---|---|---|---|
| Fixed array |  |  |  |  |  |  |  |  |  |  |  |  |  |
| Lat - 15 | 3.2 | 4.0 | 4.8 | 5.6 | 5.9 | 5.9 | 5.6 | 5.6 | 5.2 | 5.0 | 3.7 | 3.1 | 4.8 |
| Latitude | 3.6 | 4.4 | 5.0 | 5.5 | 5.5 | 5.5 | 5.3 | 5.4 | 5.2 | 5.3 | 4.1 | 3.6 | 4.9 |
| Lat + 15 | 3.8 | 4.5 | 4.9 | 5.1 | 4.9 | 4.7 | 4.6 | 4.9 | 5.0 | 5.4 | 4.4 | 3.8 | 4.7 |
| Single axis tracker |  |  |  |  |  |  |  |  |  |  |  |  |  |
| Lat - 15 | 3.9 | 5.0 | 6.1 | 7.1 | 7.4 | 7.5 | 7.0 | 6.9 | 6.5 | 6.3 | 4.6 | 3.8 | 6.0 |
| Latitude | 4.2 | 5.3 | 6.2 | 7.0 | 7.2 | 7.2 | 6.7 | 6.8 | 6.5 | 6.6 | 4.9 | 4.2 | 6.1 |
| Lat + 15 | 4.4 | 5.4 | 6.1 | 6.7 | 6.8 | 6.7 | 6.3 | 6.4 | 6.4 | 6.7 | 5.1 | 4.4 | 5.9 |
| Dual axis tracker |  |  |  |  |  |  |  |  |  |  |  |  |  |
|  | 4.4 | 5.4 | 6.2 | 7.1 | 7.5 | 7.6 | 7.1 | 6.9 | 6.6 | 6.7 | 5.1 | 4.5 | 6.3 |

## LAKE CHARLES    LA    Latitude: 30.12 degrees    Elevation: 3 meters

|  | Jan | Feb | Mar | Apr | May | Jun | Jul | Aug | Sep | Oct | Nov | Dec | Avg |
|---|---|---|---|---|---|---|---|---|---|---|---|---|---|
| Fixed array |  |  |  |  |  |  |  |  |  |  |  |  |  |
| Lat - 15 | 3.3 | 4.1 | 4.9 | 5.5 | 6.0 | 6.2 | 5.9 | 5.7 | 5.4 | 5.0 | 3.9 | 3.2 | 4.9 |
| Latitude | 3.7 | 4.5 | 5.1 | 5.4 | 5.6 | 5.7 | 5.5 | 5.6 | 5.4 | 5.4 | 4.3 | 3.7 | 5.0 |
| Lat + 15 | 3.9 | 4.6 | 4.9 | 5.1 | 5.0 | 5.0 | 4.9 | 5.1 | 5.2 | 5.4 | 4.6 | 3.9 | 4.8 |
| Single axis tracker |  |  |  |  |  |  |  |  |  |  |  |  |  |
| Lat - 15 | 4.0 | 5.2 | 6.1 | 6.9 | 7.5 | 7.8 | 7.3 | 7.2 | 6.7 | 6.3 | 4.8 | 3.9 | 6.1 |
| Latitude | 4.3 | 5.4 | 6.2 | 6.9 | 7.2 | 7.5 | 7.0 | 7.0 | 6.8 | 6.6 | 5.2 | 4.3 | 6.2 |
| Lat + 15 | 4.5 | 5.5 | 6.1 | 6.6 | 6.8 | 6.9 | 6.6 | 6.7 | 6.6 | 6.6 | 5.3 | 4.5 | 6.1 |
| Dual axis tracker |  |  |  |  |  |  |  |  |  |  |  |  |  |
|  | 4.5 | 5.5 | 6.2 | 7.0 | 7.5 | 7.9 | 7.4 | 7.2 | 6.8 | 6.7 | 5.4 | 4.5 | 6.4 |

## NEW ORLEANS    LA    Latitude: 29.98 degrees    Elevation: 3 meters

|  | Jan | Feb | Mar | Apr | May | Jun | Jul | Aug | Sep | Oct | Nov | Dec | Avg |
|---|---|---|---|---|---|---|---|---|---|---|---|---|---|
| Fixed array |  |  |  |  |  |  |  |  |  |  |  |  |  |
| Lat - 15 | 3.3 | 4.2 | 4.9 | 5.7 | 6.0 | 6.0 | 5.7 | 5.6 | 5.3 | 5.0 | 3.8 | 3.2 | 4.9 |
| Latitude | 3.7 | 4.5 | 5.0 | 5.6 | 5.7 | 5.5 | 5.3 | 5.4 | 5.3 | 5.3 | 4.3 | 3.7 | 5.0 |
| Lat + 15 | 3.9 | 4.6 | 4.9 | 5.3 | 5.1 | 4.8 | 4.7 | 4.9 | 5.1 | 5.4 | 4.5 | 3.9 | 4.8 |
| Single axis tracker |  |  |  |  |  |  |  |  |  |  |  |  |  |
| Lat - 15 | 4.0 | 5.2 | 6.1 | 7.3 | 7.7 | 7.5 | 6.9 | 6.9 | 6.6 | 6.3 | 4.7 | 3.9 | 6.1 |
| Latitude | 4.3 | 5.5 | 6.3 | 7.3 | 7.4 | 7.2 | 6.7 | 6.7 | 6.6 | 6.6 | 5.1 | 4.3 | 6.2 |
| Lat + 15 | 4.5 | 5.6 | 6.2 | 7 | 7 | 6.7 | 6.2 | 6.4 | 6.5 | 6.7 | 5.3 | 4.5 | 6 |
| Dual axis tracker |  |  |  |  |  |  |  |  |  |  |  |  |  |
|  | 4.5 | 5.6 | 6.3 | 7.3 | 7.8 | 7.7 | 7.0 | 6.9 | 6.6 | 6.7 | 5.3 | 4.5 | 6.4 |

## SHREVEPORT LA  Latitude: 32.47 degrees  Elevation: 79 meters

| | Jan | Feb | Mar | Apr | May | Jun | Jul | Aug | Sep | Oct | Nov | Dec | Avg |
|---|---|---|---|---|---|---|---|---|---|---|---|---|---|
| **Fixed array** | | | | | | | | | | | | | |
| Lat - 15 | 3.4 | 4.1 | 4.9 | 5.6 | 6.0 | 6.3 | 6.3 | 6.2 | 5.4 | 4.9 | 3.7 | 3.2 | 5.0 |
| Latitude | 3.8 | 4.5 | 5.1 | 5.5 | 5.7 | 5.8 | 5.9 | 6.0 | 5.5 | 5.2 | 4.2 | 3.7 | 5.1 |
| Lat + 15 | 4.0 | 4.6 | 5.0 | 5.1 | 5.0 | 5.0 | 5.2 | 5.5 | 5.3 | 5.3 | 4.4 | 3.9 | 4.9 |
| **Single axis tracker** | | | | | | | | | | | | | |
| Lat - 15 | 4.1 | 5.2 | 6.2 | 7.1 | 7.6 | 8.1 | 8.2 | 8.0 | 6.9 | 6.2 | 4.6 | 3.9 | 6.3 |
| Latitude | 4.4 | 5.4 | 6.3 | 7.0 | 7.4 | 7.7 | 7.9 | 7.8 | 7.0 | 6.5 | 5.0 | 4.3 | 6.4 |
| Lat + 15 | 4.6 | 5.5 | 6.2 | 6.8 | 7 | 7.2 | 7.3 | 7.4 | 6.8 | 6.7 | 5.3 | 4.5 | 6 |
| **Dual axis tracker** | | | | | | | | | | | | | |
| | 4.7 | 5.5 | 6.3 | 7.1 | 7.7 | 8.2 | 8.3 | 8.0 | 7.0 | 6.6 | 5.2 | 4.6 | 6.6 |

## BOSTON MA  Latitude: 42.37 degrees  Elevation: 5 meters

| | Jan | Feb | Mar | Apr | May | Jun | Jul | Aug | Sep | Oct | Nov | Dec | Avg |
|---|---|---|---|---|---|---|---|---|---|---|---|---|---|
| **Fixed array** | | | | | | | | | | | | | |
| Lat - 15 | 3.0 | 3.8 | 4.6 | 5.2 | 5.7 | 6.0 | 6.0 | 5.7 | 5.0 | 4.1 | 2.8 | 2.5 | 4.5 |
| Latitude | 3.4 | 4.2 | 4.7 | 5.0 | 5.3 | 5.5 | 5.6 | 5.5 | 5.1 | 4.3 | 3.1 | 2.9 | 4.6 |
| Lat + 15 | 3.6 | 4.3 | 4.6 | 4.7 | 4.7 | 4.8 | 4.9 | 5.0 | 4.9 | 4.4 | 3.3 | 3.1 | 4.4 |
| **Single axis tracker** | | | | | | | | | | | | | |
| Lat - 15 | 3.6 | 4.7 | 5.7 | 6.5 | 7.3 | 7.7 | 7.8 | 7.3 | 6.3 | 5.0 | 3.4 | 2.9 | 5.7 |
| Latitude | 3.9 | 5.0 | 5.9 | 6.5 | 7.1 | 7.4 | 7.5 | 7.1 | 6.4 | 5.2 | 3.6 | 3.2 | 5.7 |
| Lat + 15 | 4.1 | 5.1 | 5.8 | 6.2 | 6.6 | 6.9 | 7 | 6.8 | 6.2 | 5.2 | 3.7 | 3.4 | 5.6 |
| **Dual axis tracker** | | | | | | | | | | | | | |
| | 4.1 | 5.1 | 5.9 | 6.6 | 7.4 | 7.9 | 7.9 | 7.3 | 6.4 | 5.3 | 3.8 | 3.4 | 5.9 |

## WORCESTER MA  Latitude: 42.27 degrees  Elevation: 301 meters

| | Jan | Feb | Mar | Apr | May | Jun | Jul | Aug | Sep | Oct | Nov | Dec | Avg |
|---|---|---|---|---|---|---|---|---|---|---|---|---|---|
| **Fixed array** | | | | | | | | | | | | | |
| Lat - 15 | 3.0 | 3.8 | 4.6 | 5.1 | 5.5 | 5.8 | 5.9 | 5.6 | 4.9 | 4.0 | 2.8 | 2.4 | 4.5 |
| Latitude | 3.4 | 4.2 | 4.8 | 5.0 | 5.2 | 5.4 | 5.5 | 5.3 | 5.0 | 4.3 | 3.0 | 2.8 | 4.5 |
| Lat + 15 | 3.6 | 4.4 | 4.7 | 4.6 | 4.6 | 4.6 | 4.8 | 4.8 | 4.7 | 4.3 | 3.2 | 3.0 | 4.3 |
| **Single axis tracker** | | | | | | | | | | | | | |
| Lat - 15 | 3.5 | 4.7 | 5.7 | 6.4 | 7.0 | 7.5 | 7.6 | 7.1 | 6.2 | 4.9 | 3.2 | 2.9 | 5.6 |
| Latitude | 3.9 | 5.0 | 5.9 | 6.4 | 6.8 | 7.2 | 7.3 | 6.9 | 6.2 | 5.1 | 3.5 | 3.1 | 5.6 |
| Lat + 15 | 4.1 | 5.1 | 5.8 | 6.1 | 6.4 | 6.7 | 6.8 | 6.6 | 6 | 5.1 | 3.6 | 3.3 | 5.5 |
| **Dual axis tracker** | | | | | | | | | | | | | |
| | 4.1 | 5.1 | 5.9 | 6.5 | 7.1 | 7.7 | 7.7 | 7.1 | 6.2 | 5.2 | 3.6 | 3.4 | 5.8 |

## BALTIMORE MD  Latitude: 39.18 degrees  Elevation: 47 meters

| | Jan | Feb | Mar | Apr | May | Jun | Jul | Aug | Sep | Oct | Nov | Dec | Avg |
|---|---|---|---|---|---|---|---|---|---|---|---|---|---|
| **Fixed array** | | | | | | | | | | | | | |
| Lat - 15 | 3.1 | 3.8 | 4.6 | 5.3 | 5.7 | 6.0 | 6.0 | 5.6 | 5.0 | 4.3 | 3.2 | 2.7 | 4.6 |
| Latitude | 3.5 | 4.2 | 4.8 | 5.2 | 5.3 | 5.6 | 5.5 | 5.4 | 5.1 | 4.6 | 3.6 | 3.1 | 4.6 |
| Lat + 15 | 3.7 | 4.3 | 4.7 | 4.8 | 4.7 | 4.8 | 4.9 | 4.9 | 4.8 | 4.6 | 3.7 | 3.3 | 4.4 |
| **Single axis tracker** | | | | | | | | | | | | | |
| Lat - 15 | 3.7 | 4.7 | 5.9 | 6.8 | 7.2 | 7.8 | 7.7 | 7.1 | 6.3 | 5.3 | 3.9 | 3.2 | 5.8 |
| Latitude | 4.1 | 5.0 | 6.0 | 6.7 | 7.0 | 7.5 | 7.4 | 7.0 | 6.4 | 5.5 | 4.2 | 3.5 | 5.9 |
| Lat + 15 | 4.3 | 5.1 | 5.9 | 6.4 | 6.6 | 7 | 6.9 | 6.6 | 6.2 | 5.6 | 4.3 | 3.7 | 5.7 |
| **Dual axis tracker** | | | | | | | | | | | | | |
| | 4.3 | 5.1 | 6.0 | 6.8 | 7.3 | 8.0 | 7.8 | 7.2 | 6.4 | 5.6 | 4.3 | 3.7 | 6.0 |

## CARIBOU ME — Latitude: 46.87 degrees — Elevation: 190 meters

| | Jan | Feb | Mar | Apr | May | Jun | Jul | Aug | Sep | Oct | Nov | Dec | Avg |
|---|---|---|---|---|---|---|---|---|---|---|---|---|---|
| Fixed array | | | | | | | | | | | | | |
| Lat - 15 | 2.9 | 3.9 | 5.0 | 5.1 | 5.3 | 5.6 | 5.6 | 5.2 | 4.3 | 3.1 | 2.2 | 2.2 | 4.2 |
| Latitude | 3.3 | 4.3 | 5.2 | 5.0 | 4.9 | 5.1 | 5.2 | 4.9 | 4.3 | 3.3 | 2.4 | 2.5 | 4.2 |
| Lat + 15 | 3.5 | 4.5 | 5.2 | 4.7 | 4.4 | 4.5 | 4.6 | 4.5 | 4.1 | 3.3 | 2.5 | 2.7 | 4.0 |
| Single axis tracker | | | | | | | | | | | | | |
| Lat - 15 | 3.4 | 4.8 | 6.3 | 6.6 | 7.0 | 7.4 | 7.4 | 6.7 | 5.5 | 3.8 | 2.5 | 2.6 | 5.3 |
| Latitude | 3.7 | 5.1 | 6.5 | 6.5 | 6.7 | 7.1 | 7.2 | 6.6 | 5.6 | 3.9 | 2.7 | 2.8 | 5.4 |
| Lat + 15 | 3.9 | 5.3 | 6.5 | 6.3 | 6.3 | 6.6 | 6.7 | 6.3 | 5.4 | 3.9 | 2.8 | 3 | 5.2 |
| Dual axis tracker | | | | | | | | | | | | | |
| | 4.0 | 5.2 | 6.5 | 6.6 | 7.1 | 7.6 | 7.6 | 6.8 | 5.6 | 4.0 | 2.8 | 3.0 | 5.6 |

## PORTLAND ME — Latitude: 43.65 degrees — Elevation: 19 meters

| | Jan | Feb | Mar | Apr | May | Jun | Jul | Aug | Sep | Oct | Nov | Dec | Avg |
|---|---|---|---|---|---|---|---|---|---|---|---|---|---|
| Fixed array | | | | | | | | | | | | | |
| Lat - 15 | 3.1 | 4.1 | 4.8 | 5.2 | 5.7 | 6.0 | 6.0 | 5.8 | 5.1 | 4.0 | 2.8 | 2.6 | 4.6 |
| Latitude | 3.6 | 4.5 | 5.0 | 5.1 | 5.3 | 5.5 | 5.6 | 5.5 | 5.1 | 4.3 | 3.1 | 3.0 | 4.6 |
| Lat + 15 | 3.9 | 4.7 | 5.0 | 4.7 | 4.7 | 4.7 | 4.9 | 5.0 | 4.9 | 4.3 | 3.2 | 3.2 | 4.4 |
| Single axis tracker | | | | | | | | | | | | | |
| Lat - 15 | 3.8 | 5.1 | 6.1 | 6.7 | 7.4 | 7.9 | 7.9 | 7.5 | 6.5 | 5.0 | 3.3 | 3.1 | 5.9 |
| Latitude | 4.2 | 5.4 | 6.3 | 6.6 | 7.1 | 7.5 | 7.6 | 7.4 | 6.6 | 5.2 | 3.6 | 3.4 | 5.9 |
| Lat + 15 | 4.4 | 5.6 | 6.2 | 6.3 | 6.7 | 7 | 7.1 | 7 | 6.4 | 5.2 | 3.7 | 3.6 | 5.8 |
| Dual axis tracker | | | | | | | | | | | | | |
| | 4.5 | 5.6 | 6.3 | 6.7 | 7.5 | 8.1 | 8.1 | 7.6 | 6.6 | 5.3 | 3.7 | 3.6 | 6.1 |

## ALPENA MI — Latitude: 45.07 degrees — Elevation: 210 meters

| | Jan | Feb | Mar | Apr | May | Jun | Jul | Aug | Sep | Oct | Nov | Dec | Avg |
|---|---|---|---|---|---|---|---|---|---|---|---|---|---|
| Fixed array | | | | | | | | | | | | | |
| Lat - 15 | 2.5 | 3.6 | 4.7 | 5.2 | 5.8 | 6.1 | 6.1 | 5.5 | 4.5 | 3.3 | 2.1 | 1.8 | 4.3 |
| Latitude | 2.8 | 3.9 | 4.9 | 5.1 | 5.4 | 5.5 | 5.7 | 5.3 | 4.5 | 3.5 | 2.3 | 2.0 | 4.2 |
| Lat + 15 | 2.9 | 4.0 | 4.9 | 4.7 | 4.8 | 4.8 | 4.9 | 4.8 | 4.3 | 3.5 | 2.3 | 2.1 | 4.0 |
| Single axis tracker | | | | | | | | | | | | | |
| Lat - 15 | 2.9 | 4.3 | 6.0 | 6.8 | 7.8 | 8.2 | 8.4 | 7.4 | 5.7 | 4.0 | 2.4 | 2.1 | 5.5 |
| Latitude | 3.1 | 4.6 | 6.1 | 6.7 | 7.5 | 7.9 | 8.1 | 7.2 | 5.8 | 4.2 | 2.5 | 2.2 | 5.5 |
| Lat + 15 | 3.3 | 4.7 | 6.1 | 6.4 | 7.1 | 7.4 | 7.6 | 6.9 | 5.6 | 4.2 | 2.6 | 2.3 | 5.4 |
| Dual axis tracker | | | | | | | | | | | | | |
| | 3.3 | 4.7 | 6.1 | 6.8 | 7.9 | 8.5 | 8.6 | 7.4 | 5.8 | 4.2 | 2.6 | 2.4 | 5.7 |

## DETROIT MI — Latitude: 42.42 degrees — Elevation: 191 meters

| | Jan | Feb | Mar | Apr | May | Jun | Jul | Aug | Sep | Oct | Nov | Dec | Avg |
|---|---|---|---|---|---|---|---|---|---|---|---|---|---|
| Fixed array | | | | | | | | | | | | | |
| Lat - 15 | 2.4 | 3.3 | 4.1 | 5.0 | 5.7 | 6.1 | 6.1 | 5.6 | 4.8 | 3.7 | 2.4 | 1.9 | 4.3 |
| Latitude | 2.7 | 3.6 | 4.2 | 4.9 | 5.4 | 5.6 | 5.6 | 5.4 | 4.8 | 3.9 | 2.6 | 2.1 | 4.2 |
| Lat + 15 | 2.8 | 3.7 | 4.1 | 4.5 | 4.8 | 4.9 | 4.9 | 4.9 | 4.6 | 3.9 | 2.6 | 2.2 | 4.0 |
| Single axis tracker | | | | | | | | | | | | | |
| Lat - 15 | 2.8 | 4.0 | 5.0 | 6.3 | 7.4 | 8.0 | 8.0 | 7.3 | 6.0 | 4.5 | 2.7 | 2.1 | 5.3 |
| Latitude | 3.0 | 4.2 | 5.1 | 6.2 | 7.1 | 7.7 | 7.7 | 7.1 | 6.1 | 4.7 | 2.9 | 2.3 | 5.3 |
| Lat + 15 | 3.1 | 4.3 | 5 | 5.9 | 6.7 | 7.1 | 7.2 | 6.8 | 5.9 | 4.7 | 2.9 | 2.4 | 5.2 |
| Dual axis tracker | | | | | | | | | | | | | |
| | 3.1 | 4.3 | 5.1 | 6.3 | 7.5 | 8.2 | 8.2 | 7.3 | 6.1 | 4.7 | 3.0 | 2.4 | 5.5 |

## FLINT  MI  Latitude: 42.97 degrees  Elevation: 233 meters

|  | Jan | Feb | Mar | Apr | May | Jun | Jul | Aug | Sep | Oct | Nov | Dec | Avg |
|---|---|---|---|---|---|---|---|---|---|---|---|---|---|
| Fixed array | | | | | | | | | | | | | |
| Lat - 15 | 2.3 | 3.3 | 4.1 | 5.0 | 5.7 | 6.0 | 6.0 | 5.6 | 4.7 | 3.5 | 2.2 | 1.8 | 4.2 |
| Latitude | 2.6 | 3.6 | 4.2 | 4.8 | 5.3 | 5.5 | 5.5 | 5.3 | 4.7 | 3.7 | 2.4 | 2.0 | 4.1 |
| Lat + 15 | 2.7 | 3.7 | 4.1 | 4.5 | 4.7 | 4.7 | 4.8 | 4.8 | 4.5 | 3.7 | 2.5 | 2.1 | 3.9 |
| Single axis tracker | | | | | | | | | | | | | |
| Lat - 15 | 2.7 | 4.0 | 5.0 | 6.3 | 7.3 | 7.8 | 7.9 | 7.2 | 5.9 | 4.3 | 2.5 | 2.0 | 5.2 |
| Latitude | 2.9 | 4.2 | 5.1 | 6.2 | 7.1 | 7.5 | 7.6 | 7.0 | 5.9 | 4.4 | 2.7 | 2.2 | 5.2 |
| Lat + 15 | 3 | 4.3 | 5.1 | 5.9 | 6.6 | 7 | 7.2 | 6.7 | 5.7 | 4.4 | 2.8 | 2.3 | 5.1 |
| Dual axis tracker | | | | | | | | | | | | | |
|  | 3.1 | 4.3 | 5.1 | 6.3 | 7.4 | 8.0 | 8.1 | 7.2 | 6.0 | 4.5 | 2.8 | 2.3 | 5.4 |

## GRAND RAPIDS  MI  Latitude: 42.88 degrees  Elevation: 245 meters

|  | Jan | Feb | Mar | Apr | May | Jun | Jul | Aug | Sep | Oct | Nov | Dec | Avg |
|---|---|---|---|---|---|---|---|---|---|---|---|---|---|
| Fixed array | | | | | | | | | | | | | |
| Lat - 15 | 2.3 | 3.3 | 4.2 | 5.1 | 5.8 | 6.2 | 6.1 | 5.7 | 4.8 | 3.6 | 2.2 | 1.8 | 4.3 |
| Latitude | 2.5 | 3.5 | 4.3 | 5.0 | 5.4 | 5.7 | 5.7 | 5.4 | 4.8 | 3.8 | 2.4 | 2.0 | 4.2 |
| Lat + 15 | 2.6 | 3.6 | 4.2 | 4.6 | 4.8 | 4.9 | 5.0 | 4.9 | 4.6 | 3.8 | 2.5 | 2.0 | 4.0 |
| Single axis tracker | | | | | | | | | | | | | |
| Lat - 15 | 2.6 | 3.9 | 5.1 | 6.5 | 7.6 | 8.2 | 8.2 | 7.4 | 6.0 | 4.3 | 2.5 | 2.0 | 5.4 |
| Latitude | 2.8 | 4.1 | 5.2 | 6.4 | 7.3 | 7.9 | 7.9 | 7.3 | 6.1 | 4.5 | 2.7 | 2.1 | 5.4 |
| Lat + 15 | 2.9 | 4.2 | 5.1 | 6.1 | 6.9 | 7.4 | 7.4 | 6.9 | 5.9 | 4.5 | 2.7 | 2.2 | 5.2 |
| Dual axis tracker | | | | | | | | | | | | | |
|  | 2.9 | 4.2 | 5.2 | 6.5 | 7.7 | 8.4 | 8.3 | 7.5 | 6.1 | 4.5 | 2.8 | 2.2 | 5.5 |

## HOUGHTON  MI  Latitude: 47.17 degrees  Elevation: 329 meters

|  | Jan | Feb | Mar | Apr | May | Jun | Jul | Aug | Sep | Oct | Nov | Dec | Avg |
|---|---|---|---|---|---|---|---|---|---|---|---|---|---|
| Fixed array | | | | | | | | | | | | | |
| Lat - 15 | 2.1 | 3.2 | 4.5 | 5.2 | 5.6 | 5.9 | 6.0 | 5.5 | 4.4 | 3.2 | 1.9 | 1.6 | 4.1 |
| Latitude | 2.3 | 3.5 | 4.7 | 5.2 | 5.3 | 5.4 | 5.6 | 5.2 | 4.4 | 3.4 | 2.1 | 1.8 | 4.1 |
| Lat + 15 | 2.5 | 3.6 | 4.7 | 4.8 | 4.7 | 4.7 | 4.8 | 4.7 | 4.2 | 3.4 | 2.1 | 1.9 | 3.8 |
| Single axis tracker | | | | | | | | | | | | | |
| Lat - 15 | 2.4 | 3.8 | 5.7 | 6.8 | 7.6 | 8.2 | 8.3 | 7.4 | 5.6 | 3.9 | 2.2 | 1.8 | 5.3 |
| Latitude | 2.6 | 4.1 | 5.8 | 6.8 | 7.4 | 7.8 | 8.0 | 7.2 | 5.6 | 4.0 | 2.3 | 2.0 | 5.3 |
| Lat + 15 | 2.7 | 4.1 | 5.8 | 6.6 | 7 | 7.3 | 7.5 | 6.9 | 5.5 | 4 | 2.4 | 2.1 | 5.2 |
| Dual axis tracker | | | | | | | | | | | | | |
|  | 2.7 | 4.1 | 5.8 | 6.9 | 7.7 | 8.4 | 8.4 | 7.4 | 5.7 | 4.1 | 2.4 | 2.1 | 5.5 |

## LANSING  MI  Latitude: 42.78 degrees  Elevation: 256 meters

|  | Jan | Feb | Mar | Apr | May | Jun | Jul | Aug | Sep | Oct | Nov | Dec | Avg |
|---|---|---|---|---|---|---|---|---|---|---|---|---|---|
| Fixed array | | | | | | | | | | | | | |
| Lat - 15 | 2.4 | 3.3 | 4.2 | 5.0 | 5.7 | 6.0 | 6.1 | 5.6 | 4.7 | 3.6 | 2.3 | 1.8 | 4.2 |
| Latitude | 2.6 | 3.6 | 4.3 | 4.9 | 5.3 | 5.6 | 5.6 | 5.4 | 4.8 | 3.8 | 2.4 | 2.0 | 4.2 |
| Lat + 15 | 2.8 | 3.7 | 4.2 | 4.5 | 4.7 | 4.8 | 4.9 | 4.9 | 4.5 | 3.7 | 2.5 | 2.1 | 4.0 |
| Single axis tracker | | | | | | | | | | | | | |
| Lat - 15 | 2.7 | 4.0 | 5.1 | 6.3 | 7.4 | 7.9 | 8.0 | 7.2 | 5.9 | 4.3 | 2.6 | 2.1 | 5.3 |
| Latitude | 2.9 | 4.2 | 5.2 | 6.2 | 7.1 | 7.6 | 7.7 | 7.1 | 6.0 | 4.5 | 2.7 | 2.2 | 5.3 |
| Lat + 15 | 3.1 | 4.3 | 5.1 | 6 | 6.7 | 7.1 | 7.2 | 6.7 | 5.8 | 4.5 | 2.8 | 2.3 | 5.1 |
| Dual axis tracker | | | | | | | | | | | | | |
|  | 3.1 | 4.3 | 5.2 | 6.3 | 7.5 | 8.1 | 8.1 | 7.3 | 6.0 | 4.5 | 2.8 | 2.3 | 5.5 |

## MUSKEGON MI Latitude: 43.17 degrees Elevation: 191 meters

| | Jan | Feb | Mar | Apr | May | Jun | Jul | Aug | Sep | Oct | Nov | Dec | Avg |
|---|---|---|---|---|---|---|---|---|---|---|---|---|---|
| Fixed array | | | | | | | | | | | | | |
| Lat - 15 | 2.0 | 3.1 | 4.2 | 5.2 | 6.0 | 6.3 | 6.4 | 5.8 | 4.8 | 3.5 | 2.1 | 1.6 | 4.3 |
| Latitude | 2.2 | 3.3 | 4.3 | 5.1 | 5.6 | 5.8 | 5.9 | 5.6 | 4.9 | 3.7 | 2.2 | 1.8 | 4.2 |
| Lat + 15 | 2.3 | 3.4 | 4.2 | 4.7 | 5.0 | 5.0 | 5.2 | 5.1 | 4.6 | 3.7 | 2.3 | 1.8 | 3.9 |
| Single axis tracker | | | | | | | | | | | | | |
| Lat - 15 | 2.3 | 3.7 | 5.1 | 6.6 | 7.8 | 8.3 | 8.5 | 7.6 | 6.1 | 4.2 | 2.3 | 1.8 | 5.4 |
| Latitude | 2.4 | 3.8 | 5.2 | 6.6 | 7.6 | 8.0 | 8.2 | 7.5 | 6.1 | 4.4 | 2.5 | 1.9 | 5.3 |
| Lat + 15 | 2.5 | 3.9 | 5.2 | 6.3 | 7.1 | 7.5 | 7.7 | 7.1 | 6 | 4.3 | 2.5 | 2 | 5.2 |
| Dual axis tracker | | | | | | | | | | | | | |
| | 2.5 | 3.9 | 5.3 | 6.7 | 7.9 | 8.5 | 8.7 | 7.7 | 6.2 | 4.4 | 2.5 | 2.0 | 5.5 |

## SAULT STE. MARIE MI Latitude: 46.47 degrees Elevation: 221 meters

| | Jan | Feb | Mar | Apr | May | Jun | Jul | Aug | Sep | Oct | Nov | Dec | Avg |
|---|---|---|---|---|---|---|---|---|---|---|---|---|---|
| Fixed array | | | | | | | | | | | | | |
| Lat - 15 | 2.5 | 3.8 | 5.1 | 5.4 | 5.8 | 5.9 | 6.0 | 5.4 | 4.2 | 3.0 | 1.9 | 1.9 | 4.3 |
| Latitude | 2.8 | 4.2 | 5.3 | 5.3 | 5.4 | 5.4 | 5.6 | 5.2 | 4.2 | 3.1 | 2.1 | 2.2 | 4.2 |
| Lat + 15 | 2.9 | 4.3 | 5.3 | 5.0 | 4.8 | 4.7 | 4.9 | 4.7 | 4.0 | 3.1 | 2.1 | 2.3 | 4.0 |
| Single axis tracker | | | | | | | | | | | | | |
| Lat - 15 | 2.9 | 4.6 | 6.4 | 7.1 | 7.9 | 8.1 | 8.3 | 7.2 | 5.3 | 3.6 | 2.2 | 2.2 | 5.5 |
| Latitude | 3.1 | 4.9 | 6.6 | 7.1 | 7.6 | 7.8 | 8.0 | 7.0 | 5.3 | 3.7 | 2.3 | 2.4 | 5.5 |
| Lat + 15 | 3.2 | 5 | 6.6 | 6.8 | 7.2 | 7.3 | 7.5 | 6.7 | 5.2 | 3.7 | 2.4 | 2.5 | 5.3 |
| Dual axis tracker | | | | | | | | | | | | | |
| | 3.3 | 5.0 | 6.6 | 7.1 | 8.0 | 8.3 | 8.5 | 7.2 | 5.4 | 3.7 | 2.4 | 2.5 | 5.7 |

## TRAVERSE CITY MI Latitude: 44.73 degrees Elevation: 192 meters

| | Jan | Feb | Mar | Apr | May | Jun | Jul | Aug | Sep | Oct | Nov | Dec | Avg |
|---|---|---|---|---|---|---|---|---|---|---|---|---|---|
| Fixed array | | | | | | | | | | | | | |
| Lat - 15 | 2.1 | 3.3 | 4.4 | 5.0 | 5.8 | 6.1 | 6.1 | 5.4 | 4.4 | 3.2 | 2.0 | 1.6 | 4.1 |
| Latitude | 2.3 | 3.5 | 4.6 | 4.9 | 5.4 | 5.6 | 5.6 | 5.2 | 4.4 | 3.4 | 2.1 | 1.8 | 4.1 |
| Lat + 15 | 2.4 | 3.6 | 4.5 | 4.6 | 4.8 | 4.8 | 4.9 | 4.7 | 4.2 | 3.3 | 2.1 | 1.8 | 3.8 |
| Single axis tracker | | | | | | | | | | | | | |
| Lat - 15 | 2.3 | 3.8 | 5.4 | 6.5 | 7.6 | 8.1 | 8.2 | 7.1 | 5.6 | 3.8 | 2.2 | 1.8 | 5.2 |
| Latitude | 2.5 | 4.0 | 5.6 | 6.4 | 7.4 | 7.8 | 7.9 | 7.0 | 5.6 | 3.9 | 2.3 | 1.9 | 5.2 |
| Lat + 15 | 2.6 | 4.1 | 5.5 | 6.2 | 7 | 7.3 | 7.4 | 6.6 | 5.4 | 3.9 | 2.4 | 2 | 5 |
| Dual axis tracker | | | | | | | | | | | | | |
| | 2.6 | 4.1 | 5.6 | 6.5 | 7.7 | 8.3 | 8.3 | 7.2 | 5.6 | 4.0 | 2.4 | 2.0 | 5.4 |

## DULUTH MN Latitude: 46.83 degrees Elevation: 432 meters

| | Jan | Feb | Mar | Apr | May | Jun | Jul | Aug | Sep | Oct | Nov | Dec | Avg |
|---|---|---|---|---|---|---|---|---|---|---|---|---|---|
| Fixed array | | | | | | | | | | | | | |
| Lat - 15 | 2.8 | 4.0 | 5.0 | 5.5 | 5.7 | 5.9 | 6.1 | 5.5 | 4.5 | 3.4 | 2.4 | 2.2 | 4.4 |
| Latitude | 3.2 | 4.4 | 5.2 | 5.4 | 5.4 | 5.4 | 5.6 | 5.3 | 4.5 | 3.6 | 2.6 | 2.5 | 4.4 |
| Lat + 15 | 3.4 | 4.5 | 5.2 | 5.1 | 4.7 | 4.7 | 4.9 | 4.8 | 4.3 | 3.6 | 2.7 | 2.7 | 4.2 |
| Single axis tracker | | | | | | | | | | | | | |
| Lat - 15 | 3.3 | 4.9 | 6.3 | 7.2 | 7.6 | 7.9 | 8.3 | 7.3 | 5.7 | 4.2 | 2.7 | 2.6 | 5.7 |
| Latitude | 3.7 | 5.2 | 6.5 | 7.1 | 7.4 | 7.6 | 8.0 | 7.2 | 5.7 | 4.3 | 2.9 | 2.9 | 5.7 |
| Lat + 15 | 3.8 | 5.3 | 6.5 | 6.9 | 7 | 7.1 | 75 | 6.8 | 5.5 | 4.3 | 3 | 3 | 5.6 |
| Dual axis tracker | | | | | | | | | | | | | |
| | 3.9 | 5.3 | 6.5 | 7.2 | 7.8 | 8.1 | 8.5 | 7.4 | 5.7 | 4.4 | 3.1 | 3.1 | 5.9 |

## INTERNATIONAL FALLS  MN          Latitude: 48.57 degrees          Elevation: 361 meters

|              | Jan | Feb | Mar | Apr | May | Jun | Jul | Aug | Sep | Oct | Nov | Dec | Avg |
|--------------|-----|-----|-----|-----|-----|-----|-----|-----|-----|-----|-----|-----|-----|
| Fixed array  |     |     |     |     |     |     |     |     |     |     |     |     |     |
| Lat - 15     | 2.6 | 3.9 | 4.9 | 5.5 | 5.6 | 5.7 | 5.9 | 5.4 | 4.3 | 3.1 | 2.1 | 2.1 | 4.3 |
| Latitude     | 3.0 | 4.3 | 5.1 | 5.4 | 5.2 | 5.2 | 5.4 | 5.1 | 4.3 | 3.2 | 2.3 | 2.4 | 4.3 |
| Lat + 15     | 3.2 | 4.5 | 5.1 | 5.1 | 4.6 | 4.5 | 4.7 | 4.6 | 4.1 | 3.2 | 2.4 | 2.6 | 4.0 |
| Single axis tracker |  |     |     |     |     |     |     |     |     |     |     |     |     |
| Lat - 15     | 3.1 | 4.8 | 6.3 | 7.3 | 7.6 | 7.7 | 8.1 | 7.3 | 5.5 | 3.7 | 2.4 | 2.4 | 5.5 |
| Latitude     | 3.4 | 5.1 | 6.4 | 7.2 | 7.4 | 7.4 | 7.8 | 7.1 | 5.5 | 3.8 | 2.6 | 2.7 | 5.5 |
| Lat + 15     | 3.6 | 5.2 | 6.4 | 7   | 7   | 7   | 7.4 | 6.8 | 5.3 | 3.8 | 2.7 | 2.9 | 5.4 |
| Dual axis tracker |   |     |     |     |     |     |     |     |     |     |     |     |     |
|              | 3.6 | 5.2 | 6.4 | 7.3 | 7.7 | 8.0 | 8.3 | 7.3 | 5.5 | 3.9 | 2.7 | 2.9 | 5.7 |

## MINNEAPOLIS          MN          Latitude: 44.88 degrees          Elevation: 255 meters

|              | Jan | Feb | Mar | Apr | May | Jun | Jul | Aug | Sep | Oct | Nov | Dec | Avg |
|--------------|-----|-----|-----|-----|-----|-----|-----|-----|-----|-----|-----|-----|-----|
| Fixed array  |     |     |     |     |     |     |     |     |     |     |     |     |     |
| Lat - 15     | 3.1 | 4.1 | 4.8 | 5.3 | 5.8 | 6.1 | 6.4 | 5.8 | 4.9 | 3.9 | 2.6 | 2.3 | 4.6 |
| Latitude     | 3.5 | 4.5 | 5.0 | 5.1 | 5.5 | 5.6 | 5.9 | 5.6 | 5.0 | 4.1 | 2.9 | 2.7 | 4.6 |
| Lat + 15     | 3.8 | 4.7 | 4.9 | 4.8 | 4.8 | 4.9 | 5.1 | 5.1 | 4.7 | 4.2 | 3.0 | 2.9 | 4.4 |
| Single axis tracker |  |     |     |     |     |     |     |     |     |     |     |     |     |
| Lat - 15     | 3.7 | 5.0 | 6.0 | 6.8 | 7.8 | 8.3 | 8.7 | 7.8 | 6.3 | 4.8 | 3.1 | 2.7 | 5.9 |
| Latitude     | 4.0 | 5.4 | 6.2 | 6.7 | 7.5 | 8.0 | 8.4 | 7.7 | 6.4 | 5.0 | 3.3 | 3.0 | 6.0 |
| Lat + 15     | 4.3 | 5.5 | 6.1 | 6.5 | 7.1 | 7.4 | 7.9 | 7.3 | 6.2 | 5   | 3.4 | 3.2 | 5.8 |
| Dual axis tracker |   |     |     |     |     |     |     |     |     |     |     |     |     |
|              | 4.3 | 5.5 | 6.2 | 6.8 | 7.9 | 8.5 | 8.9 | 7.9 | 6.4 | 5.1 | 3.4 | 3.2 | 6.2 |

## ROCHESTER          MN          Latitude: 43.92 degrees          Elevation: 402 meters

|              | Jan | Feb | Mar | Apr | May | Jun | Jul | Aug | Sep | Oct | Nov | Dec | Avg |
|--------------|-----|-----|-----|-----|-----|-----|-----|-----|-----|-----|-----|-----|-----|
| Fixed array  |     |     |     |     |     |     |     |     |     |     |     |     |     |
| Lat - 15     | 2.9 | 3.9 | 4.6 | 5.1 | 5.7 | 6.0 | 6.2 | 5.7 | 4.8 | 3.8 | 2.6 | 2.3 | 4.5 |
| Latitude     | 3.3 | 4.2 | 4.7 | 4.9 | 5.3 | 5.5 | 5.7 | 5.5 | 4.8 | 4.1 | 2.8 | 2.6 | 4.5 |
| Lat + 15     | 3.6 | 4.4 | 4.7 | 4.6 | 4.7 | 4.8 | 5.0 | 4.9 | 4.6 | 4.1 | 2.9 | 2.8 | 4.3 |
| Single axis tracker |  |     |     |     |     |     |     |     |     |     |     |     |     |
| Lat - 15     | 3.5 | 4.8 | 5.7 | 6.5 | 7.5 | 8.0 | 8.3 | 7.5 | 6.1 | 4.7 | 3.0 | 2.7 | 5.7 |
| Latitude     | 3.8 | 5.1 | 5.8 | 6.4 | 7.2 | 7.7 | 8.0 | 7.4 | 6.2 | 4.9 | 3.2 | 2.9 | 5.7 |
| Lat + 15     | 4   | 5.2 | 5.8 | 6.1 | 6.8 | 7.2 | 7.5 | 7   | 6   | 5   | 3.3 | 3.1 | 5.6 |
| Dual axis tracker |   |     |     |     |     |     |     |     |     |     |     |     |     |
|              | 4.1 | 5.2 | 5.8 | 6.5 | 7.6 | 8.2 | 8.5 | 7.6 | 6.2 | 5.0 | 3.3 | 3.1 | 5.9 |

## SAINT CLOUD          MN          Latitude: 45.55 degrees          Elevation: 313 meters

|              | Jan | Feb | Mar | Apr | May | Jun | Jul | Aug | Sep | Oct | Nov | Dec | Avg |
|--------------|-----|-----|-----|-----|-----|-----|-----|-----|-----|-----|-----|-----|-----|
| Fixed array  |     |     |     |     |     |     |     |     |     |     |     |     |     |
| Lat - 15     | 3.0 | 4.1 | 4.8 | 5.3 | 5.8 | 6.1 | 6.3 | 5.8 | 4.8 | 3.8 | 2.6 | 2.3 | 4.6 |
| Latitude     | 3.4 | 4.5 | 5.0 | 5.1 | 5.4 | 5.6 | 5.8 | 5.6 | 4.9 | 4.1 | 2.9 | 2.6 | 4.6 |
| Lat + 15     | 3.7 | 4.7 | 5.0 | 4.7 | 4.8 | 4.8 | 5.1 | 5.0 | 4.6 | 4.1 | 3.0 | 2.8 | 4.4 |
| Single axis tracker |  |     |     |     |     |     |     |     |     |     |     |     |     |
| Lat - 15     | 3.6 | 5.0 | 6.2 | 6.8 | 7.7 | 8.2 | 8.7 | 7.8 | 6.3 | 4.8 | 3.1 | 2.7 | 5.9 |
| Latitude     | 3.9 | 5.4 | 6.3 | 6.8 | 7.4 | 7.9 | 8.4 | 7.7 | 6.3 | 5.0 | 3.3 | 2.9 | 5.9 |
| Lat + 15     | 4.1 | 5.5 | 6.3 | 6.5 | 7   | 7.4 | 7.9 | 7.3 | 6.1 | 5   | 3.4 | 3.1 | 5.8 |
| Dual axis tracker |   |     |     |     |     |     |     |     |     |     |     |     |     |
|              | 4.2 | 5.5 | 6.3 | 6.9 | 7.8 | 8.5 | 8.9 | 7.9 | 6.3 | 5.0 | 3.4 | 3.1 | 6.1 |

## COLUMBIA          MO          Latitude: 38.82 degrees          Elevation: 270 meters

| | Jan | Feb | Mar | Apr | May | Jun | Jul | Aug | Sep | Oct | Nov | Dec | Avg |
|---|---|---|---|---|---|---|---|---|---|---|---|---|---|
| Fixed array | | | | | | | | | | | | | |
| Lat - 15 | 3.3 | 4.0 | 4.7 | 5.6 | 6.0 | 6.4 | 6.6 | 6.2 | 5.3 | 4.5 | 3.2 | 2.8 | 4.9 |
| Latitude | 3.8 | 4.4 | 4.9 | 5.5 | 5.6 | 5.9 | 6.1 | 5.9 | 5.4 | 4.9 | 3.6 | 3.2 | 4.9 |
| Lat + 15 | 4.0 | 4.5 | 4.8 | 5.1 | 5.0 | 5.1 | 5.3 | 5.4 | 5.1 | 4.9 | 3.8 | 3.4 | 4.7 |
| Single axis tracker | | | | | | | | | | | | | |
| Lat - 15 | 4.0 | 4.9 | 5.9 | 7.2 | 7.8 | 8.4 | 8.7 | 8.0 | 6.8 | 5.7 | 3.9 | 3.3 | 6.2 |
| Latitude | 4.4 | 5.2 | 6.0 | 7.1 | 7.6 | 8.1 | 8.4 | 7.9 | 6.9 | 6.0 | 4.2 | 3.7 | 6.3 |
| Lat + 15 | 4.6 | 5.3 | 5.9 | 6.8 | 7.1 | 7.6 | 7.8 | 7.5 | 6.7 | 6 | 4.4 | 3.9 | 6.1 |
| Dual axis tracker | | | | | | | | | | | | | |
| | 4.6 | 5.3 | 6.0 | 7.2 | 7.9 | 8.6 | 8.8 | 8.1 | 6.9 | 6.0 | 4.4 | 3.9 | 6.5 |

## KANSAS CITY          MO          Latitude: 39.30 degrees          Elevation: 315 meters

| | Jan | Feb | Mar | Apr | May | Jun | Jul | Aug | Sep | Oct | Nov | Dec | Avg |
|---|---|---|---|---|---|---|---|---|---|---|---|---|---|
| Fixed array | | | | | | | | | | | | | |
| Lat - 15 | 3.4 | 4.0 | 4.7 | 5.5 | 5.9 | 6.3 | 6.5 | 6.1 | 5.3 | 4.6 | 3.4 | 2.9 | 4.9 |
| Latitude | 3.8 | 4.3 | 4.8 | 5.4 | 5.6 | 5.8 | 6.0 | 5.9 | 5.4 | 5.0 | 3.8 | 3.3 | 4.9 |
| Lat + 15 | 4.1 | 4.4 | 4.7 | 5.0 | 4.9 | 5.0 | 5.3 | 5.3 | 5.2 | 5.0 | 4.0 | 3.5 | 4.7 |
| Single axis tracker | | | | | | | | | | | | | |
| Lat - 15 | 4.1 | 4.9 | 5.9 | 7.1 | 7.8 | 8.4 | 8.8 | 8.1 | 6.9 | 5.9 | 4.1 | 3.5 | 6.3 |
| Latitude | 4.5 | 5.2 | 6.0 | 7.0 | 7.5 | 8.1 | 8.5 | 8.0 | 7.0 | 6.2 | 4.5 | 3.8 | 6.4 |
| Lat + 15 | 4.7 | 5.3 | 5.9 | 6.7 | 7.1 | 7.5 | 7.9 | 7.6 | 6.8 | 6.2 | 4.6 | 4 | 6.2 |
| Dual axis tracker | | | | | | | | | | | | | |
| | 4.8 | 5.3 | 6.0 | 7.1 | 7.9 | 8.6 | 8.9 | 8.2 | 7.0 | 6.2 | 4.7 | 4.1 | 6.6 |

## ST. LOUIS          MO          Latitude: 38.75 degrees          Elevation: 172 meters

| | Jan | Feb | Mar | Apr | May | Jun | Jul | Aug | Sep | Oct | Nov | Dec | Avg |
|---|---|---|---|---|---|---|---|---|---|---|---|---|---|
| Fixed array | | | | | | | | | | | | | |
| Lat - 15 | 3.2 | 3.8 | 4.6 | 5.4 | 5.9 | 6.3 | 6.3 | 6.0 | 5.3 | 4.5 | 3.2 | 2.7 | 4.8 |
| Latitude | 3.6 | 4.2 | 4.7 | 5.3 | 5.6 | 5.8 | 5.9 | 5.7 | 5.3 | 4.8 | 3.5 | 3.1 | 4.8 |
| Lat + 15 | 3.8 | 4.3 | 4.6 | 4.9 | 4.9 | 5.0 | 5.1 | 5.2 | 5.1 | 4.8 | 3.7 | 3.3 | 4.6 |
| Single axis tracker | | | | | | | | | | | | | |
| Lat - 15 | 3.8 | 4.7 | 5.7 | 6.9 | 7.7 | 8.3 | 8.4 | 7.8 | 6.8 | 5.6 | 3.8 | 3.2 | 6.1 |
| Latitude | 4.2 | 5.0 | 5.8 | 6.8 | 7.5 | 7.9 | 8.1 | 7.7 | 6.8 | 5.9 | 4.1 | 3.5 | 6.1 |
| Lat + 15 | 4.4 | 5.1 | 5.7 | 6.5 | 7 | 7.4 | 7.6 | 7.3 | 6.7 | 5.9 | 4.2 | 3.7 | 6 |
| Dual axis tracker | | | | | | | | | | | | | |
| | 4.4 | 5.1 | 5.8 | 6.9 | 7.8 | 8.5 | 8.5 | 7.9 | 6.9 | 5.9 | 4.3 | 3.7 | 6.3 |

## SPRINGFIELD          MO          Latitude: 37.23 degrees          Elevation: 387 meters

| | Jan | Feb | Mar | Apr | May | Jun | Jul | Aug | Sep | Oct | Nov | Dec | Avg |
|---|---|---|---|---|---|---|---|---|---|---|---|---|---|
| Fixed array | | | | | | | | | | | | | |
| Lat - 15 | 3.4 | 3.9 | 4.7 | 5.6 | 5.9 | 6.2 | 6.5 | 6.2 | 5.3 | 4.6 | 3.4 | 2.9 | 4.9 |
| Latitude | 3.8 | 4.3 | 4.9 | 5.4 | 5.6 | 5.7 | 6.0 | 5.9 | 5.4 | 4.9 | 3.8 | 3.3 | 4.9 |
| Lat + 15 | 4.1 | 4.4 | 4.8 | 5.0 | 4.9 | 5.0 | 5.2 | 5.4 | 5.1 | 5.0 | 3.9 | 3.5 | 4.7 |
| Single axis tracker | | | | | | | | | | | | | |
| Lat - 15 | 4.1 | 4.9 | 6.0 | 7.1 | 7.7 | 8.2 | 8.7 | 8.2 | 6.8 | 5.9 | 4.1 | 3.5 | 6.3 |
| Latitude | 4.5 | 5.1 | 6.1 | 7.1 | 7.5 | 7.9 | 8.4 | 8.0 | 6.9 | 6.1 | 4.4 | 3.8 | 6.3 |
| Lat + 15 | 4.2 | 5.7 | 6 | 6.8 | 7 | 7.4 | 7.9 | 7.6 | 6.7 | 6.2 | 4.6 | 4 | 6.2 |
| Dual axis tracker | | | | | | | | | | | | | |
| | 4.7 | 5.2 | 6.1 | 7.2 | 7.8 | 8.4 | 8.9 | 8.2 | 6.9 | 6.2 | 4.6 | 4.1 | 6.5 |

## JACKSON — MS — Latitude: 32.32 degrees — Elevation: 101 meters

| | Jan | Feb | Mar | Apr | May | Jun | Jul | Aug | Sep | Oct | Nov | Dec | Avg |
|---|---|---|---|---|---|---|---|---|---|---|---|---|---|
| Fixed array | | | | | | | | | | | | | |
| Lat - 15 | 3.3 | 4.1 | 5.0 | 5.8 | 6.1 | 6.3 | 6.1 | 6.0 | 5.4 | 5.0 | 3.7 | 3.2 | 5.0 |
| Latitude | 3.7 | 4.5 | 5.2 | 5.7 | 5.8 | 5.8 | 5.7 | 5.8 | 5.5 | 5.3 | 4.2 | 3.6 | 5.1 |
| Lat + 15 | 4.0 | 4.6 | 5.1 | 5.3 | 5.1 | 5.0 | 5.0 | 5.3 | 5.2 | 5.4 | 4.4 | 3.9 | 4.9 |
| Single axis tracker | | | | | | | | | | | | | |
| Lat - 15 | 4.1 | 5.2 | 6.4 | 7.4 | 7.8 | 8.0 | 7.7 | 7.5 | 6.8 | 6.3 | 4.6 | 3.9 | 6.3 |
| Latitude | 4.4 | 5.4 | 6.5 | 7.3 | 7.6 | 7.7 | 7.4 | 7.4 | 6.8 | 6.6 | 5.0 | 4.2 | 6.4 |
| Lat + 15 | 4.6 | 5.5 | 6.4 | 7.1 | 7.1 | 7.1 | 6.9 | 7 | 6.7 | 6.6 | 5.2 | 4.4 | 6.2 |
| Dual axis tracker | | | | | | | | | | | | | |
| | 4.6 | 5.6 | 6.5 | 7.4 | 7.9 | 8.2 | 7.8 | 7.6 | 6.8 | 6.7 | 5.2 | 4.5 | 6.6 |

## MERIDIAN — MS — Latitude: 32.33 degrees — Elevation: 94 meters

| | Jan | Feb | Mar | Apr | May | Jun | Jul | Aug | Sep | Oct | Nov | Dec | Avg |
|---|---|---|---|---|---|---|---|---|---|---|---|---|---|
| Fixed array | | | | | | | | | | | | | |
| Lat - 15 | 3.2 | 4.1 | 4.8 | 5.7 | 5.9 | 6.0 | 5.8 | 5.7 | 5.2 | 4.8 | 3.7 | 3.1 | 4.8 |
| Latitude | 3.6 | 4.4 | 5.0 | 5.6 | 5.6 | 5.6 | 5.4 | 5.5 | 5.2 | 5.2 | 4.1 | 3.5 | 4.9 |
| Lat + 15 | 3.8 | 4.5 | 4.9 | 5.2 | 5.0 | 4.9 | 4.8 | 5.0 | 5.0 | 5.2 | 4.3 | 3.8 | 4.7 |
| Single axis tracker | | | | | | | | | | | | | |
| Lat - 15 | 4.0 | 5.1 | 6.1 | 7.2 | 7.5 | 7.6 | 7.2 | 7.1 | 6.5 | 6.1 | 4.5 | 3.8 | 6.0 |
| Latitude | 4.3 | 5.3 | 6.2 | 7.1 | 7.2 | 7.3 | 6.9 | 7.0 | 6.5 | 6.4 | 4.9 | 4.1 | 6.1 |
| Lat + 15 | 4.4 | 5.4 | 6.1 | 6.8 | 6.8 | 6.8 | 6.5 | 6.6 | 6.3 | 6.4 | 5 | 4.3 | 6 |
| Dual axis tracker | | | | | | | | | | | | | |
| | 4.5 | 5.4 | 6.2 | 7.2 | 7.6 | 7.8 | 7.3 | 7.1 | 6.5 | 6.4 | 5.1 | 4.4 | 6.3 |

## BILLINGS — MT — Latitude: 45.80 degrees — Elevation: 1088 meters

| | Jan | Feb | Mar | Apr | May | Jun | Jul | Aug | Sep | Oct | Nov | Dec | Avg |
|---|---|---|---|---|---|---|---|---|---|---|---|---|---|
| Fixed array | | | | | | | | | | | | | |
| Lat - 15 | 2.9 | 3.9 | 4.9 | 5.6 | 6.0 | 6.6 | 7.1 | 6.7 | 5.6 | 4.5 | 3.2 | 2.7 | 5.0 |
| Latitude | 3.3 | 4.2 | 5.1 | 5.5 | 5.6 | 6.0 | 6.5 | 6.5 | 5.7 | 4.8 | 3.5 | 3.1 | 5.0 |
| Lat + 15 | 3.5 | 4.4 | 5.1 | 5.1 | 5.0 | 5.2 | 5.7 | 5.9 | 5.5 | 4.9 | 3.7 | 3.3 | 4.8 |
| Single axis tracker | | | | | | | | | | | | | |
| Lat - 15 | 3.5 | 4.7 | 6.3 | 7.4 | 8.2 | 9.2 | 10.1 | 9.4 | 7.6 | 5.7 | 3.8 | 3.2 | 6.6 |
| Latitude | 3.8 | 5.0 | 6.4 | 7.3 | 7.9 | 8.8 | 9.7 | 9.2 | 7.6 | 6.0 | 4.1 | 3.5 | 6.6 |
| Lat + 15 | 4 | 5.1 | 6.4 | 7 | 7.5 | 8.3 | 9.1 | 8.8 | 7.5 | 6 | 4.3 | 3.7 | 6.5 |
| Dual axis tracker | | | | | | | | | | | | | |
| | 4.0 | 5.1 | 6.5 | 7.4 | 8.3 | 9.4 | 10.3 | 9.4 | 7.7 | 6.1 | 4.3 | 3.8 | 6.9 |

## CUT BANK — MT — Latitude: 48.60 degrees — Elevation: 1170 meters

| | Jan | Feb | Mar | Apr | May | Jun | Jul | Aug | Sep | Oct | Nov | Dec | Avg |
|---|---|---|---|---|---|---|---|---|---|---|---|---|---|
| Fixed array | | | | | | | | | | | | | |
| Lat - 15 | 2.6 | 3.7 | 4.8 | 5.6 | 6.1 | 6.5 | 7.0 | 6.5 | 5.4 | 4.3 | 2.9 | 2.4 | 4.8 |
| Latitude | 3.0 | 4.0 | 4.9 | 5.5 | 5.7 | 5.9 | 6.5 | 6.2 | 5.4 | 4.6 | 3.3 | 2.7 | 4.8 |
| Lat + 15 | 3.2 | 4.2 | 4.9 | 5.1 | 5.1 | 5.1 | 5.6 | 5.6 | 5.2 | 4.6 | 3.5 | 2.9 | 4.6 |
| Single axis tracker | | | | | | | | | | | | | |
| Lat - 15 | 3.1 | 4.5 | 6.1 | 7.6 | 8.4 | 9.1 | 10.1 | 9.0 | 7.1 | 5.4 | 3.5 | 2.8 | 6.4 |
| Latitude | 3.4 | 4.8 | 6.2 | 7.5 | 8.2 | 8.7 | 9.8 | 8.9 | 7.2 | 5.7 | 3.8 | 3.1 | 6.5 |
| Lat + 15 | 3.6 | 4.9 | 6.2 | 7.2 | 7.7 | 8.2 | 9.2 | 8.4 | 7 | 5.7 | 4 | 3.2 | 6.3 |
| Dual axis tracker | | | | | | | | | | | | | |
| | 3.6 | 4.9 | 6.2 | 7.6 | 8.6 | 9.3 | 10.3 | 9.1 | 7.2 | 5.7 | 4.0 | 3.3 | 6.7 |

## GLASGOW   MT   Latitude: 48.22 degrees   Elevation: 700 meters

| | Jan | Feb | Mar | Apr | May | Jun | Jul | Aug | Sep | Oct | Nov | Dec | Avg |
|---|---|---|---|---|---|---|---|---|---|---|---|---|---|
| Fixed array | | | | | | | | | | | | | |
| Lat - 15 | 2.7 | 3.7 | 4.8 | 5.4 | 5.8 | 6.4 | 6.8 | 6.4 | 5.2 | 4.1 | 2.8 | 2.3 | 4.7 |
| Latitude | 3.1 | 4.1 | 5.0 | 5.3 | 5.5 | 5.9 | 6.3 | 6.1 | 5.3 | 4.3 | 3.1 | 2.6 | 4.7 |
| Lat + 15 | 3.3 | 4.2 | 4.9 | 4.9 | 4.8 | 5.0 | 5.5 | 5.5 | 5.0 | 4.4 | 3.3 | 2.8 | 4.5 |
| Single axis tracker | | | | | | | | | | | | | |
| Lat - 15 | 3.2 | 4.5 | 6.1 | 7.2 | 8.0 | 9.0 | 9.8 | 8.9 | 6.9 | 5.1 | 3.3 | 2.7 | 6.2 |
| Latitude | 3.5 | 4.8 | 6.3 | 7.2 | 7.7 | 8.6 | 9.4 | 8.7 | 6.9 | 5.3 | 3.6 | 3.0 | 6.3 |
| Lat + 15 | 3.7 | 4.9 | 6.2 | 6.9 | 7.3 | 8.1 | 8.9 | 8.3 | 6.8 | 5.3 | 3.7 | 3.1 | 6.1 |
| Dual axis tracker | | | | | | | | | | | | | |
| | 3.7 | 5.0 | 6.3 | 7.3 | 8.1 | 9.2 | 10.0 | 8.9 | 7.0 | 5.4 | 3.7 | 3.2 | 6.5 |

## GREAT FALLS   MT   Latitude: 47.48 degrees   Elevation: 1116 meters

| | Jan | Feb | Mar | Apr | May | Jun | Jul | Aug | Sep | Oct | Nov | Dec | Avg |
|---|---|---|---|---|---|---|---|---|---|---|---|---|---|
| Fixed array | | | | | | | | | | | | | |
| Lat - 15 | 2.6 | 3.7 | 4.8 | 5.5 | 6.0 | 6.6 | 7.2 | 6.5 | 5.4 | 4.3 | 2.9 | 2.3 | 4.8 |
| Latitude | 3.0 | 4.0 | 5.0 | 5.4 | 5.6 | 6.0 | 6.6 | 6.3 | 5.5 | 4.6 | 3.3 | 2.7 | 4.8 |
| Lat + 15 | 3.2 | 4.2 | 5.0 | 5.0 | 4.9 | 5.2 | 5.7 | 5.7 | 5.3 | 4.6 | 3.4 | 2.9 | 4.6 |
| Single axis tracker | | | | | | | | | | | | | |
| Lat - 15 | 3.1 | 4.5 | 6.1 | 7.2 | 8.1 | 9.1 | 10.2 | 9.0 | 7.2 | 5.3 | 3.5 | 2.7 | 6.3 |
| Latitude | 3.4 | 4.8 | 6.3 | 7.2 | 7.8 | 8.7 | 9.8 | 8.9 | 7.3 | 5.6 | 3.7 | 3.0 | 6.4 |
| Lat + 15 | 3.5 | 4.9 | 6.2 | 6.9 | 7.4 | 8.2 | 9.2 | 8.4 | 7.1 | 5.6 | 3.9 | 3.2 | 6.2 |
| Dual axis tracker | | | | | | | | | | | | | |
| | 3.6 | 4.9 | 6.3 | 7.3 | 8.2 | 9.3 | 10.4 | 9.1 | 7.3 | 5.6 | 3.9 | 3.2 | 6.6 |

## HELENA   MT   Latitude: 46.60 degrees   Elevation: 1188 meters

| | Jan | Feb | Mar | Apr | May | Jun | Jul | Aug | Sep | Oct | Nov | Dec | Avg |
|---|---|---|---|---|---|---|---|---|---|---|---|---|---|
| Fixed array | | | | | | | | | | | | | |
| Lat - 15 | 2.5 | 3.6 | 4.6 | 5.4 | 5.9 | 6.4 | 7.1 | 6.5 | 5.5 | 4.3 | 2.8 | 2.2 | 4.7 |
| Latitude | 2.8 | 3.9 | 4.8 | 5.3 | 5.5 | 5.8 | 6.5 | 6.2 | 5.6 | 4.6 | 3.1 | 2.6 | 4.7 |
| Lat + 15 | 3.0 | 4.0 | 4.7 | 4.9 | 4.9 | 5.0 | 5.7 | 5.7 | 5.4 | 4.7 | 3.3 | 2.8 | 4.5 |
| Single axis tracker | | | | | | | | | | | | | |
| Lat - 15 | 3.0 | 4.3 | 5.8 | 7.1 | 8.1 | 8.9 | 10.2 | 9.2 | 7.4 | 5.5 | 3.4 | 2.6 | 6.3 |
| Latitude | 3.2 | 4.6 | 6.0 | 7.0 | 7.8 | 8.6 | 9.9 | 9.0 | 7.5 | 5.7 | 3.6 | 2.9 | 6.3 |
| Lat + 15 | 3.4 | 4.7 | 5.9 | 6.7 | 7.4 | 8 | 9.3 | 8.6 | 7.3 | 5.7 | 3.8 | 3 | 6.2 |
| Dual axis tracker | | | | | | | | | | | | | |
| | 3.4 | 4.7 | 6.0 | 7.1 | 8.2 | 9.2 | 10.4 | 9.2 | 7.5 | 5.8 | 3.8 | 3.1 | 6.5 |

## KALISPELL   MT   Latitude: 48.30 degrees   Elevation: 904 meters

| | Jan | Feb | Mar | Apr | May | Jun | Jul | Aug | Sep | Oct | Nov | Dec | Avg |
|---|---|---|---|---|---|---|---|---|---|---|---|---|---|
| Fixed array | | | | | | | | | | | | | |
| Lat - 15 | 1.9 | 2.9 | 4.0 | 4.9 | 5.5 | 6.0 | 6.7 | 6.2 | 5.1 | 3.6 | 1.9 | 1.5 | 4.2 |
| Latitude | 2.1 | 3.1 | 4.1 | 4.8 | 5.1 | 5.5 | 6.2 | 5.9 | 5.2 | 3.8 | 2.0 | 1.7 | 4.1 |
| Lat + 15 | 2.2 | 3.2 | 4.0 | 4.4 | 4.6 | 4.7 | 5.4 | 5.4 | 5.0 | 3.8 | 2.1 | 1.7 | 3.9 |
| Single axis tracker | | | | | | | | | | | | | |
| Lat - 15 | 2.1 | 3.5 | 4.9 | 6.4 | 7.5 | 8.3 | 9.7 | 8.7 | 6.8 | 4.5 | 2.1 | 1.7 | 5.5 |
| Latitude | 2.3 | 3.6 | 5.0 | 6.3 | 7.2 | 8.0 | 9.4 | 8.5 | 6.9 | 4.7 | 2.3 | 1.8 | 5.5 |
| Lat + 15 | 3.4 | 3.7 | 8 | 6 | 6.8 | 7.5 | 8.9 | 8.1 | 6.7 | 4.7 | 2.3 | 1.9 | 5.3 |
| Dual axis tracker | | | | | | | | | | | | | |
| | 2.4 | 3.7 | 5.1 | 6.4 | 7.6 | 8.6 | 10.0 | 8.7 | 6.9 | 4.7 | 2.3 | 1.9 | 5.7 |

## LEWISTOWN  MT  Latitude: 47.05 degrees  Elevation: 1264 meters

|  | Jan | Feb | Mar | Apr | May | Jun | Jul | Aug | Sep | Oct | Nov | Dec | Avg |
|---|---|---|---|---|---|---|---|---|---|---|---|---|---|
| Fixed array |  |  |  |  |  |  |  |  |  |  |  |  |  |
| Lat - 15 | 2.6 | 3.6 | 4.7 | 5.4 | 5.8 | 6.3 | 6.9 | 6.4 | 5.3 | 4.2 | 2.9 | 2.3 | 4.7 |
| Latitude | 3.0 | 3.9 | 4.9 | 5.3 | 5.4 | 5.8 | 6.4 | 6.1 | 5.4 | 4.5 | 3.2 | 2.7 | 4.7 |
| Lat + 15 | 3.2 | 4.1 | 4.8 | 4.9 | 4.8 | 5.0 | 5.5 | 5.5 | 5.1 | 4.6 | 3.4 | 2.8 | 4.5 |
| Single axis tracker |  |  |  |  |  |  |  |  |  |  |  |  |  |
| Lat - 15 | 3.1 | 4.4 | 5.9 | 7.2 | 7.9 | 8.8 | 9.8 | 8.9 | 7.0 | 5.3 | 3.4 | 2.7 | 6.2 |
| Latitude | 3.4 | 4.7 | 6.1 | 7.1 | 7.7 | 8.5 | 9.5 | 8.7 | 7.1 | 5.6 | 3.7 | 3.0 | 6.3 |
| Lat + 15 | 3.5 | 4.8 | 6 | 6.8 | 7.2 | 7.9 | 8.9 | 8.3 | 6.9 | 5.6 | 3.8 | 3.2 | 6.1 |
| Dual axis tracker |  |  |  |  |  |  |  |  |  |  |  |  |  |
|  | 3.6 | 4.8 | 6.1 | 7.2 | 8.0 | 9.0 | 10.0 | 8.9 | 7.1 | 5.6 | 3.9 | 3.2 | 6.5 |

## MILES CITY  MT  Latitude: 46.43 degrees  Elevation: 803 meters

|  | Jan | Feb | Mar | Apr | May | Jun | Jul | Aug | Sep | Oct | Nov | Dec | Avg |
|---|---|---|---|---|---|---|---|---|---|---|---|---|---|
| Fixed array |  |  |  |  |  |  |  |  |  |  |  |  |  |
| Lat - 15 | 3.0 | 3.9 | 5.0 | 5.6 | 6.0 | 6.7 | 7.1 | 6.7 | 5.6 | 4.4 | 3.0 | 2.6 | 5.0 |
| Latitude | 3.4 | 4.3 | 5.2 | 5.5 | 5.7 | 6.1 | 6.5 | 6.5 | 5.7 | 4.7 | 3.4 | 3.0 | 5.0 |
| Lat + 15 | 3.6 | 4.5 | 5.1 | 5.1 | 5.0 | 5.3 | 5.6 | 5.8 | 5.4 | 4.8 | 3.6 | 3.2 | 4.8 |
| Single axis tracker |  |  |  |  |  |  |  |  |  |  |  |  |  |
| Lat - 15 | 3.5 | 4.8 | 6.3 | 7.5 | 8.3 | 9.4 | 10.1 | 9.4 | 7.5 | 5.6 | 3.6 | 3.0 | 6.6 |
| Latitude | 3.9 | 5.1 | 6.5 | 7.4 | 8.0 | 9.1 | 9.8 | 9.2 | 7.6 | 5.9 | 4.0 | 3.4 | 6.7 |
| Lat + 15 | 4.1 | 5.3 | 6.4 | 7.1 | 7.6 | 8.5 | 9.2 | 8.8 | 7.4 | 5.9 | 4.1 | 3.6 | 6.5 |
| Dual axis tracker |  |  |  |  |  |  |  |  |  |  |  |  |  |
|  | 4.1 | 5.3 | 6.5 | 7.5 | 8.4 | 9.7 | 10.3 | 9.4 | 7.6 | 5.9 | 4.1 | 3.6 | 6.9 |

## MISSOULA  MT  Latitude: 46.92 degrees  Elevation: 972 meters

|  | Jan | Feb | Mar | Apr | May | Jun | Jul | Aug | Sep | Oct | Nov | Dec | Avg |
|---|---|---|---|---|---|---|---|---|---|---|---|---|---|
| Fixed array |  |  |  |  |  |  |  |  |  |  |  |  |  |
| Lat - 15 | 2.0 | 3.0 | 4.0 | 5.0 | 5.6 | 6.1 | 7.0 | 6.4 | 5.3 | 3.9 | 2.1 | 1.7 | 4.4 |
| Latitude | 2.2 | 3.2 | 4.1 | 4.9 | 5.2 | 5.6 | 6.4 | 6.1 | 5.4 | 4.1 | 2.3 | 1.9 | 4.3 |
| Lat + 15 | 2.3 | 3.2 | 4.0 | 4.5 | 4.6 | 4.8 | 5.6 | 5.5 | 5.1 | 4.1 | 2.4 | 2.0 | 4.0 |
| Single axis tracker |  |  |  |  |  |  |  |  |  |  |  |  |  |
| Lat - 15 | 2.3 | 3.5 | 5.0 | 6.5 | 7.5 | 8.5 | 10.2 | 9.0 | 7.1 | 4.9 | 2.5 | 1.9 | 5.7 |
| Latitude | 2.4 | 3.7 | 5.1 | 6.5 | 7.3 | 8.2 | 9.9 | 8.9 | 7.1 | 5.1 | 2.6 | 2.0 | 5.7 |
| Lat + 15 | 2.5 | 3.7 | 5 | 6.2 | 6.9 | 7.7 | 9.3 | 5.8 | 7 | 5.1 | 2.7 | 2.1 | 5.6 |
| Dual axis tracker |  |  |  |  |  |  |  |  |  |  |  |  |  |
|  | 2.6 | 3.8 | 5.1 | 6.6 | 7.6 | 8.7 | 10.4 | 9.1 | 7.2 | 5.1 | 2.7 | 2.1 | 5.9 |

## ASHEVILLE  NC  Latitude: 35.43 degrees  Elevation: 661 meters

|  | Jan | Feb | Mar | Apr | May | Jun | Jul | Aug | Sep | Oct | Nov | Dec | Avg |
|---|---|---|---|---|---|---|---|---|---|---|---|---|---|
| Fixed array |  |  |  |  |  |  |  |  |  |  |  |  |  |
| Lat - 15 | 3.5 | 4.2 | 4.9 | 5.7 | 5.8 | 5.9 | 5.7 | 5.5 | 5.0 | 4.6 | 3.6 | 3.1 | 4.8 |
| Latitude | 3.9 | 4.6 | 5.1 | 5.6 | 5.4 | 5.4 | 5.3 | 5.3 | 5.0 | 5.0 | 4.1 | 3.6 | 4.9 |
| Lat + 15 | 4.2 | 4.7 | 5.0 | 5.2 | 4.8 | 4.7 | 4.7 | 4.8 | 4.8 | 5.0 | 4.3 | 3.9 | 4.7 |
| Single axis tracker |  |  |  |  |  |  |  |  |  |  |  |  |  |
| Lat - 15 | 4.3 | 5.3 | 6.3 | 7.4 | 7.3 | 7.4 | 7.1 | 6.7 | 6.1 | 5.8 | 4.5 | 3.8 | 6.0 |
| Latitude | 4.7 | 5.6 | 6.4 | 7.3 | 7.1 | 7.1 | 6.8 | 6.6 | 6.1 | 6.1 | 4.9 | 4.2 | 6.1 |
| Lat + 15 | 4.9 | 5.7 | 6.4 | 7.1 | 6.7 | 6.6 | 6.4 | 6.2 | 6 | 6.1 | 5 | 4.5 | 6 |
| Dual axis tracker |  |  |  |  |  |  |  |  |  |  |  |  |  |
|  | 4.9 | 5.7 | 6.4 | 7.4 | 7.4 | 7.6 | 7.2 | 6.7 | 6.1 | 6.2 | 5.1 | 4.5 | 6.3 |

## CAPE HATTERAS    NC     Latitude: 35.27 degrees     Elevation: 2 meters

|  | Jan | Feb | Mar | Apr | May | Jun | Jul | Aug | Sep | Oct | Nov | Dec | Avg |
|---|---|---|---|---|---|---|---|---|---|---|---|---|---|
| **Fixed array** | | | | | | | | | | | | | |
| Lat - 15 | 3.3 | 4.1 | 5.1 | 6.0 | 6.1 | 6.2 | 6.1 | 5.8 | 5.3 | 4.6 | 3.7 | 3.2 | 5.0 |
| Latitude | 3.8 | 4.5 | 5.2 | 5.9 | 5.8 | 5.7 | 5.7 | 5.6 | 5.4 | 4.9 | 4.2 | 3.6 | 5.0 |
| Lat + 15 | 4.0 | 4.6 | 5.2 | 5.5 | 5.1 | 5.0 | 5.0 | 5.1 | 5.1 | 4.9 | 4.5 | 3.9 | 4.8 |
| **Single axis tracker** | | | | | | | | | | | | | |
| Lat - 15 | 4.1 | 5.1 | 6.5 | 7.8 | 7.9 | 7.9 | 7.7 | 7.3 | 6.6 | 5.7 | 4.6 | 3.9 | 6.3 |
| Latitude | 4.5 | 5.4 | 6.6 | 7.7 | 7.6 | 7.6 | 7.4 | 7.2 | 6.7 | 6.0 | 5.0 | 4.3 | 6.3 |
| Lat + 15 | 4.7 | 5.5 | 6.5 | 7.4 | 7.2 | 7 | 6.9 | 6.8 | 6.5 | 6 | 5.2 | 4.5 | 6.2 |
| **Dual axis tracker** | | | | | | | | | | | | | |
|  | 4.7 | 5.5 | 6.6 | 7.8 | 7.9 | 8.1 | 7.8 | 7.4 | 6.7 | 6.0 | 5.2 | 4.6 | 6.5 |

## CHARLOTTE    NC     Latitude: 35.22 degrees     Elevation: 234 meters

|  | Jan | Feb | Mar | Apr | May | Jun | Jul | Aug | Sep | Oct | Nov | Dec | Avg |
|---|---|---|---|---|---|---|---|---|---|---|---|---|---|
| **Fixed array** | | | | | | | | | | | | | |
| Lat - 15 | 3.4 | 4.1 | 5.0 | 5.9 | 6.0 | 6.1 | 6.0 | 5.8 | 5.2 | 4.7 | 3.7 | 3.1 | 4.9 |
| Latitude | 3.8 | 4.5 | 5.2 | 5.7 | 5.7 | 5.7 | 5.6 | 5.6 | 5.3 | 5.1 | 4.2 | 3.6 | 5.0 |
| Lat + 15 | 4.1 | 4.6 | 5.1 | 5.3 | 5.0 | 4.9 | 4.9 | 5.1 | 5.0 | 5.2 | 4.4 | 3.9 | 4.8 |
| **Single axis tracker** | | | | | | | | | | | | | |
| Lat - 15 | 4.2 | 5.2 | 6.4 | 7.6 | 7.7 | 7.8 | 7.5 | 7.3 | 6.5 | 6.0 | 4.6 | 3.8 | 6.2 |
| Latitude | 4.5 | 5.5 | 6.6 | 7.5 | 7.4 | 7.5 | 7.3 | 7.1 | 6.6 | 6.3 | 5.0 | 4.2 | 6.3 |
| Lat + 15 | 4.7 | 5.6 | 6.5 | 7.2 | 7 | 6.6 | 6.8 | 6.8 | 6.4 | 6.3 | 5.2 | 4.5 | 6.2 |
| **Dual axis tracker** | | | | | | | | | | | | | |
|  | 4.8 | 5.6 | 6.6 | 7.6 | 7.7 | 8.0 | 7.7 | 7.3 | 6.6 | 6.3 | 5.2 | 4.5 | 6.5 |

## RALEIGH    NC     Latitude: 35.87 degrees     Elevation: 134 meters

|  | Jan | Feb | Mar | Apr | May | Jun | Jul | Aug | Sep | Oct | Nov | Dec | Avg |
|---|---|---|---|---|---|---|---|---|---|---|---|---|---|
| **Fixed array** | | | | | | | | | | | | | |
| Lat - 15 | 3.4 | 4.1 | 5.0 | 5.8 | 6.0 | 6.2 | 6.0 | 5.7 | 5.1 | 4.6 | 3.7 | 3.1 | 4.9 |
| Latitude | 3.8 | 4.5 | 5.2 | 5.7 | 5.7 | 5.7 | 5.6 | 5.5 | 5.2 | 4.9 | 4.1 | 3.6 | 5.0 |
| Lat + 15 | 4.1 | 4.6 | 5.1 | 5.3 | 5.0 | 4.9 | 4.9 | 5.0 | 5.0 | 5.0 | 4.3 | 3.8 | 4.8 |
| **Single axis tracker** | | | | | | | | | | | | | |
| Lat - 15 | 4.1 | 5.2 | 6.4 | 7.5 | 7.6 | 7.8 | 7.5 | 7.0 | 6.4 | 5.8 | 4.5 | 3.8 | 6.1 |
| Latitude | 4.5 | 5.5 | 6.5 | 7.5 | 7.4 | 7.5 | 7.2 | 6.9 | 6.4 | 6.0 | 4.9 | 4.2 | 6.2 |
| Lat + 15 | 4.7 | 5.6 | 6.5 | 7.2 | 7 | 7 | 6.7 | 6.5 | 6.3 | 6.2 | 5 | 4.4 | 6.1 |
| **Dual axis tracker** | | | | | | | | | | | | | |
|  | 4.7 | 5.6 | 6.6 | 7.5 | 7.7 | 8.0 | 7.6 | 7.1 | 6.5 | 6.1 | 5.1 | 4.4 | 6.4 |

## GREENSBORO    NC     Latitude: 36.08 degrees     Elevation: 270 meters

|  | Jan | Feb | Mar | Apr | May | Jun | Jul | Aug | Sep | Oct | Nov | Dec | Avg |
|---|---|---|---|---|---|---|---|---|---|---|---|---|---|
| **Fixed array** | | | | | | | | | | | | | |
| Lat - 15 | 3.3 | 4.1 | 5.0 | 5.8 | 6.0 | 6.1 | 6.0 | 5.7 | 5.2 | 4.6 | 3.6 | 3.1 | 4.9 |
| Latitude | 3.8 | 4.5 | 5.2 | 5.7 | 5.6 | 5.6 | 5.6 | 5.5 | 5.2 | 5.0 | 4.1 | 3.6 | 5.0 |
| Lat + 15 | 4.1 | 4.6 | 5.1 | 5.2 | 5.0 | 4.9 | 4.9 | 5.1 | 5.0 | 5.0 | 4.3 | 3.8 | 4.8 |
| **Single axis tracker** | | | | | | | | | | | | | |
| Lat - 15 | 4.1 | 5.2 | 6.4 | 7.5 | 7.6 | 7.8 | 7.6 | 7.3 | 6.5 | 5.9 | 4.5 | 3.8 | 6.2 |
| Latitude | 4.5 | 5.5 | 6.6 | 7.4 | 7.4 | 7.5 | 7.3 | 7.1 | 6.5 | 6.1 | 4.8 | 4.2 | 6.3 |
| Lat + 15 | 4.7 | 5.6 | 6.5 | 7.1 | 7 | 7 | 6.8 | 6.8 | 6.4 | 6.2 | 5 | 4.4 | 6.1 |
| **Dual axis tracker** | | | | | | | | | | | | | |
|  | 4.8 | 5.6 | 6.6 | 7.5 | 7.7 | 8.0 | 7.7 | 7.3 | 6.6 | 6.2 | 5.0 | 4.5 | 6.5 |

## WILMINGTON NC Latitude: 34.27 degrees Elevation: 9 meters

| | Jan | Feb | Mar | Apr | May | Jun | Jul | Aug | Sep | Oct | Nov | Dec | Avg |
|---|---|---|---|---|---|---|---|---|---|---|---|---|---|
| Fixed array | | | | | | | | | | | | | |
| Lat - 15 | 3.5 | 4.2 | 5.2 | 6.0 | 6.1 | 6.1 | 5.9 | 5.6 | 5.1 | 4.7 | 3.9 | 3.3 | 5.0 |
| Latitude | 4.0 | 4.6 | 5.4 | 5.9 | 5.8 | 5.6 | 5.5 | 5.4 | 5.2 | 5.0 | 4.4 | 3.8 | 5.0 |
| Lat + 15 | 4.2 | 4.7 | 5.3 | 5.5 | 5.1 | 4.9 | 4.8 | 4.9 | 5.0 | 5.1 | 4.6 | 4.1 | 4.9 |
| Single axis tracker | | | | | | | | | | | | | |
| Lat - 15 | 4.3 | 5.3 | 6.6 | 7.8 | 7.8 | 7.8 | 7.4 | 7.0 | 6.4 | 5.9 | 4.8 | 4.0 | 6.3 |
| Latitude | 4.7 | 5.6 | 6.7 | 7.8 | 7.6 | 7.4 | 7.2 | 6.9 | 6.4 | 6.2 | 5.2 | 4.5 | 6.3 |
| Lat + 15 | 4.9 | 5.7 | 6.6 | 7.5 | 7.1 | 6.9 | 6.7 | 6.5 | 6.3 | 6.2 | 5.4 | 4.7 | 6.2 |
| Dual axis tracker | | | | | | | | | | | | | |
| | 4.9 | 5.7 | 6.7 | 7.9 | 7.9 | 7.9 | 7.6 | 7.0 | 6.5 | 6.3 | 5.4 | 4.8 | 6.6 |

## BISMARCK ND Latitude: 46.77 degrees Elevation: 502 meters

| | Jan | Feb | Mar | Apr | May | Jun | Jul | Aug | Sep | Oct | Nov | Dec | Avg |
|---|---|---|---|---|---|---|---|---|---|---|---|---|---|
| Fixed array | | | | | | | | | | | | | |
| Lat - 15 | 3.1 | 4.0 | 5.0 | 5.6 | 6.1 | 6.5 | 6.8 | 6.4 | 5.3 | 4.2 | 2.9 | 2.6 | 4.9 |
| Latitude | 3.5 | 4.4 | 5.2 | 5.5 | 5.7 | 5.9 | 6.3 | 6.1 | 5.4 | 4.5 | 3.2 | 3.0 | 4.9 |
| Lat + 15 | 3.7 | 4.5 | 5.1 | 5.1 | 5.1 | 5.1 | 5.5 | 5.5 | 5.1 | 4.5 | 3.4 | 3.2 | 4.7 |
| Single axis tracker | | | | | | | | | | | | | |
| Lat - 15 | 3.6 | 4.9 | 6.3 | 7.3 | 8.3 | 8.9 | 9.5 | 8.7 | 7.0 | 5.3 | 3.4 | 3.0 | 6.4 |
| Latitude | 4.0 | 5.2 | 6.4 | 7.3 | 8.0 | 8.6 | 9.2 | 8.5 | 7.0 | 5.5 | 3.7 | 3.3 | 6.4 |
| Lat + 15 | 4.2 | 5.3 | 6.4 | 7 | 7.6 | 8 | 8.6 | 8.1 | 6.8 | 5.5 | 3.9 | 3.5 | 6.3 |
| Dual axis tracker | | | | | | | | | | | | | |
| | 4.2 | 5.3 | 6.5 | 7.4 | 8.4 | 9.2 | 9.7 | 8.7 | 7.0 | 5.5 | 3.9 | 3.6 | 6.6 |

## FARGO ND Latitude: 46.90 degrees Elevation: 274 meters

| | Jan | Feb | Mar | Apr | May | Jun | Jul | Aug | Sep | Oct | Nov | Dec | Avg |
|---|---|---|---|---|---|---|---|---|---|---|---|---|---|
| Fixed array | | | | | | | | | | | | | |
| Lat - 15 | 2.9 | 3.9 | 4.8 | 5.3 | 5.9 | 6.1 | 6.5 | 6.1 | 4.9 | 3.9 | 2.6 | 2.3 | 4.6 |
| Latitude | 3.4 | 4.3 | 5.0 | 5.2 | 5.5 | 5.6 | 6.0 | 5.8 | 5.0 | 4.1 | 2.9 | 2.7 | 4.6 |
| Lat + 15 | 3.6 | 4.5 | 5.0 | 4.8 | 4.9 | 4.8 | 5.2 | 5.3 | 4.7 | 4.1 | 3.0 | 2.9 | 4.4 |
| Single axis tracker | | | | | | | | | | | | | |
| Lat - 15 | 3.5 | 4.8 | 6.1 | 7.0 | 7.9 | 8.4 | 9.1 | 8.3 | 6.4 | 4.8 | 3.1 | 2.7 | 6.0 |
| Latitude | 3.8 | 5.1 | 6.2 | 6.9 | 7.7 | 8.1 | 8.8 | 8.1 | 6.5 | 5.0 | 3.3 | 3.0 | 6.1 |
| Lat + 15 | 4.1 | 5.2 | 6.2 | 6.7 | 7.2 | 7.6 | 8.2 | 7.7 | 6.3 | 5 | 3.4 | 3.2 | 5.9 |
| Dual axis tracker | | | | | | | | | | | | | |
| | 4.1 | 5.2 | 6.3 | 7.0 | 8.0 | 8.7 | 9.3 | 8.3 | 6.5 | 5.1 | 3.5 | 3.2 | 6.3 |

## MINOT ND Latitude: 48.27 degrees Elevation: 522 meters

| | Jan | Feb | Mar | Apr | May | Jun | Jul | Aug | Sep | Oct | Nov | Dec | Avg |
|---|---|---|---|---|---|---|---|---|---|---|---|---|---|
| Fixed array | | | | | | | | | | | | | |
| Lat - 15 | 2.9 | 3.8 | 4.9 | 5.6 | 6.0 | 6.3 | 6.6 | 6.2 | 5.0 | 4.1 | 2.8 | 2.4 | 4.7 |
| Latitude | 3.3 | 4.1 | 5.1 | 5.5 | 5.6 | 5.8 | 6.1 | 6.0 | 5.1 | 4.4 | 3.1 | 2.8 | 4.7 |
| Lat + 15 | 3.5 | 4.3 | 5.0 | 5.1 | 5.0 | 5.0 | 5.3 | 5.4 | 4.8 | 4.4 | 3.3 | 3.0 | 4.5 |
| Single axis tracker | | | | | | | | | | | | | |
| Lat - 15 | 3.4 | 4.6 | 6.2 | 7.4 | 8.3 | 8.9 | 9.5 | 8.6 | 6.6 | 5.1 | 3.3 | 2.8 | 6.2 |
| Latitude | 3.8 | 4.9 | 6.4 | 7.4 | 8.1 | 8.5 | 9.2 | 8.5 | 6.7 | 5.3 | 3.6 | 3.1 | 6.3 |
| Lat + 15 | 4 | 5 | 6.3 | 7.1 | 7.6 | 8 | 8.6 | 8.1 | 6.5 | 5.4 | 3.7 | 3.3 | 6.1 |
| Dual axis tracker | | | | | | | | | | | | | |
| | 4.0 | 5.0 | 6.4 | 7.5 | 8.4 | 9.1 | 9.7 | 8.7 | 6.7 | 5.4 | 3.7 | 3.4 | 6.5 |

## GRAND ISLAND  NE  Latitude: 40.97 degrees  Elevation: 566 meters

| | Jan | Feb | Mar | Apr | May | Jun | Jul | Aug | Sep | Oct | Nov | Dec | Avg |
|---|---|---|---|---|---|---|---|---|---|---|---|---|---|
| Fixed array | | | | | | | | | | | | | |
| Lat - 15 | 3.6 | 4.2 | 5.0 | 5.8 | 6.1 | 6.7 | 6.8 | 6.4 | 5.5 | 4.7 | 3.5 | 3.1 | 5.1 |
| Latitude | 4.1 | 4.6 | 5.2 | 5.7 | 5.7 | 6.2 | 6.3 | 6.1 | 5.6 | 5.1 | 4.0 | 3.6 | 5.2 |
| Lat + 15 | 4.4 | 4.8 | 5.2 | 5.3 | 5.1 | 5.3 | 5.5 | 5.6 | 5.4 | 5.2 | 4.2 | 3.9 | 5.0 |
| Single axis tracker | | | | | | | | | | | | | |
| Lat - 15 | 4.4 | 5.3 | 6.5 | 7.7 | 8.2 | 9.2 | 9.3 | 8.6 | 7.3 | 6.1 | 4.3 | 3.8 | 6.7 |
| Latitude | 4.9 | 5.6 | 6.6 | 7.6 | 7.9 | 8.8 | 9.0 | 8.5 | 7.4 | 6.4 | 4.7 | 4.2 | 6.8 |
| Lat + 15 | 5.1 | 5.8 | 6.6 | 7.3 | 7.4 | 8.2 | 8.4 | 8 | 7.2 | 6.4 | 4.9 | 4.5 | 6.7 |
| Dual axis tracker | | | | | | | | | | | | | |
| | 5.2 | 5.8 | 6.6 | 7.7 | 8.3 | 9.4 | 9.5 | 8.6 | 7.4 | 6.4 | 4.9 | 4.5 | 7.0 |

## NORFOLK  NE  Latitude: 41.98 degrees  Elevation: 471 meters

| | Jan | Feb | Mar | Apr | May | Jun | Jul | Aug | Sep | Oct | Nov | Dec | Avg |
|---|---|---|---|---|---|---|---|---|---|---|---|---|---|
| Fixed array | | | | | | | | | | | | | |
| Lat - 15 | 3.3 | 4.0 | 4.9 | 5.6 | 6.1 | 6.6 | 6.7 | 6.2 | 5.4 | 4.5 | 3.3 | 2.9 | 5.0 |
| Latitude | 3.8 | 4.4 | 5.1 | 5.5 | 5.7 | 6.0 | 6.2 | 6.0 | 5.4 | 4.9 | 3.7 | 3.3 | 5.0 |
| Lat + 15 | 4.1 | 4.6 | 5.0 | 5.1 | 5.1 | 5.2 | 5.4 | 5.4 | 5.2 | 4.9 | 3.9 | 3.6 | 4.8 |
| Single axis tracker | | | | | | | | | | | | | |
| Lat - 15 | 4.1 | 5.0 | 6.2 | 7.3 | 8.1 | 8.9 | 9.1 | 8.3 | 7.0 | 5.8 | 4.0 | 3.4 | 6.5 |
| Latitude | 4.5 | 5.3 | 6.4 | 7.3 | 7.9 | 8.6 | 8.8 | 8.2 | 7.1 | 6.0 | 4.3 | 3.8 | 6.5 |
| Lat + 15 | 4.7 | 5.4 | 6.3 | 7 | 7.4 | 8 | 8.2 | 7.8 | 6.9 | 6 | 4.5 | 4 | 6.4 |
| Dual axis tracker | | | | | | | | | | | | | |
| | 4.8 | 5.4 | 6.4 | 7.4 | 8.2 | 9.2 | 9.3 | 8.4 | 7.1 | 6.1 | 4.5 | 4.1 | 6.7 |

## NORTH PLATTE  NE  Latitude: 41.13 degrees  Elevation: 849 meters

| | Jan | Feb | Mar | Apr | May | Jun | Jul | Aug | Sep | Oct | Nov | Dec | Avg |
|---|---|---|---|---|---|---|---|---|---|---|---|---|---|
| Fixed array | | | | | | | | | | | | | |
| Lat - 15 | 3.6 | 4.3 | 5.1 | 5.8 | 6.1 | 6.7 | 6.8 | 6.4 | 5.7 | 4.9 | 3.6 | 3.3 | 5.2 |
| Latitude | 4.1 | 4.7 | 5.3 | 5.7 | 5.7 | 6.1 | 6.3 | 6.1 | 5.7 | 5.3 | 4.1 | 3.8 | 5.3 |
| Lat + 15 | 4.4 | 4.9 | 5.2 | 5.3 | 5.1 | 5.3 | 5.5 | 5.6 | 5.5 | 5.3 | 4.3 | 4.1 | 5.0 |
| Single axis tracker | | | | | | | | | | | | | |
| Lat - 15 | 4.4 | 5.4 | 6.5 | 7.8 | 8.1 | 9.1 | 9.3 | 8.6 | 7.5 | 6.3 | 4.5 | 4.0 | 6.8 |
| Latitude | 4.9 | 5.7 | 6.7 | 7.7 | 7.8 | 8.8 | 9.0 | 8.5 | 7.5 | 6.6 | 4.8 | 4.4 | 6.9 |
| Lat + 15 | 5.1 | 5.9 | 6.6 | 7.4 | 7.4 | 8.2 | 8.4 | 8.1 | 7.4 | 6.6 | 5 | 4.7 | 6.7 |
| Dual axis tracker | | | | | | | | | | | | | |
| | 5.2 | 5.9 | 6.7 | 7.8 | 8.2 | 9.4 | 9.5 | 8.7 | 7.6 | 6.7 | 5.0 | 4.7 | 7.1 |

## OMAHA  NE  Latitude: 41.37 degrees  Elevation: 404 meters

| | Jan | Feb | Mar | Apr | May | Jun | Jul | Aug | Sep | Oct | Nov | Dec | Avg |
|---|---|---|---|---|---|---|---|---|---|---|---|---|---|
| Fixed array | | | | | | | | | | | | | |
| Lat - 15 | 3.3 | 4.0 | 4.7 | 5.5 | 6.0 | 6.5 | 6.5 | 6.1 | 5.2 | 4.4 | 3.2 | 2.7 | 4.9 |
| Latitude | 3.8 | 4.4 | 4.9 | 5.3 | 5.6 | 6.0 | 6.0 | 5.8 | 5.3 | 4.7 | 3.5 | 3.2 | 4.9 |
| Lat + 15 | 4.1 | 4.6 | 4.8 | 5.0 | 5.0 | 5.2 | 5.3 | 5.3 | 5.1 | 4.7 | 3.7 | 3.4 | 4.7 |
| Single axis tracker | | | | | | | | | | | | | |
| Lat - 15 | 4.1 | 5.0 | 6.0 | 7.0 | 7.8 | 8.7 | 8.6 | 8.0 | 6.7 | 5.5 | 3.8 | 3.3 | 6.2 |
| Latitude | 4.5 | 5.3 | 6.1 | 7.0 | 7.5 | 8.3 | 8.3 | 7.8 | 6.7 | 5.7 | 4.1 | 3.6 | 6.3 |
| Lat + 15 | 4.7 | 5.4 | 6 | 6.7 | 7.1 | 7.8 | 7.8 | 7.4 | 6.6 | 5.8 | 4.2 | 3.8 | 6.1 |
| Dual axis tracker | | | | | | | | | | | | | |
| | 4.7 | 5.4 | 6.1 | 7.1 | 7.9 | 8.9 | 8.8 | 8.0 | 6.8 | 5.8 | 4.3 | 3.9 | 6.5 |

## SCOTTSBLUFF NE — Latitude: 41.87 degrees — Elevation: 1206 meters

| | Jan | Feb | Mar | Apr | May | Jun | Jul | Aug | Sep | Oct | Nov | Dec | Avg |
|---|---|---|---|---|---|---|---|---|---|---|---|---|---|
| Fixed array | | | | | | | | | | | | | |
| Lat - 15 | 3.5 | 4.3 | 5.1 | 5.8 | 6.0 | 6.7 | 7.0 | 6.7 | 5.9 | 4.9 | 3.6 | 3.2 | 5.2 |
| Latitude | 4.0 | 4.8 | 5.3 | 5.7 | 5.7 | 6.2 | 6.5 | 6.4 | 6.0 | 5.3 | 4.1 | 3.8 | 5.3 |
| Lat + 15 | 4.3 | 4.9 | 5.2 | 5.3 | 5.0 | 5.3 | 5.6 | 5.8 | 5.7 | 5.3 | 4.3 | 4.1 | 5.1 |
| Single axis tracker | | | | | | | | | | | | | |
| Lat - 15 | 4.3 | 5.5 | 6.5 | 7.6 | 8.1 | 9.2 | 9.6 | 9.1 | 7.9 | 6.3 | 4.4 | 3.9 | 6.9 |
| Latitude | 4.7 | 5.8 | 6.7 | 7.6 | 7.8 | 8.9 | 9.2 | 8.9 | 7.9 | 6.6 | 4.8 | 4.4 | 6.9 |
| Lat + 15 | 4.9 | 5.9 | 6.6 | 7.3 | 7.4 | 8.3 | 8.7 | 8.5 | 7.6 | 6.7 | 5 | 4.6 | 6.8 |
| Dual axis tracker | | | | | | | | | | | | | |
| | 5.0 | 5.9 | 6.7 | 7.7 | 8.2 | 9.4 | 9.8 | 9.1 | 8.0 | 6.7 | 5.0 | 4.7 | 7.2 |

## CONCORD NH — Latitude: 43.20 degrees — Elevation: 105 meters

| | Jan | Feb | Mar | Apr | May | Jun | Jul | Aug | Sep | Oct | Nov | Dec | Avg |
|---|---|---|---|---|---|---|---|---|---|---|---|---|---|
| Fixed array | | | | | | | | | | | | | |
| Lat - 15 | 3.1 | 4.1 | 4.8 | 5.2 | 5.7 | 5.9 | 6.0 | 5.7 | 5.0 | 3.9 | 2.7 | 2.5 | 4.6 |
| Latitude | 3.5 | 4.5 | 5.0 | 5.1 | 5.3 | 5.4 | 5.6 | 5.5 | 5.0 | 4.2 | 3.0 | 2.8 | 4.6 |
| Lat + 15 | 3.8 | 4.7 | 5.0 | 4.7 | 4.7 | 4.7 | 4.9 | 4.9 | 4.8 | 4.2 | 3.1 | 3.1 | 4.4 |
| Single axis tracker | | | | | | | | | | | | | |
| Lat - 15 | 3.7 | 5.0 | 6.1 | 6.7 | 7.4 | 7.9 | 7.9 | 7.4 | 6.3 | 4.8 | 3.2 | 2.9 | 5.8 |
| Latitude | 4.1 | 5.4 | 6.3 | 6.6 | 7.1 | 7.5 | 7.7 | 7.3 | 6.3 | 5.0 | 3.4 | 3.2 | 5.8 |
| Lat + 15 | 4.3 | 5.5 | 6.2 | 6.3 | 6.7 | 7 | 7.2 | 6.9 | 6.1 | 5.1 | 3.5 | 3.4 | 5.7 |
| Dual axis tracker | | | | | | | | | | | | | |
| | 4.3 | 5.5 | 6.3 | 6.7 | 7.5 | 8.1 | 8.1 | 7.4 | 6.3 | 5.1 | 3.5 | 3.5 | 6.0 |

## ATLANTIC CITY NJ — Latitude: 39.45 degrees — Elevation: 20 meters

| | Jan | Feb | Mar | Apr | May | Jun | Jul | Aug | Sep | Oct | Nov | Dec | Avg |
|---|---|---|---|---|---|---|---|---|---|---|---|---|---|
| Fixed array | | | | | | | | | | | | | |
| Lat - 15 | 3.0 | 3.8 | 4.6 | 5.3 | 5.7 | 6.0 | 5.9 | 5.6 | 5.0 | 4.3 | 3.2 | 2.7 | 4.6 |
| Latitude | 3.5 | 4.1 | 4.8 | 5.2 | 5.3 | 5.5 | 5.5 | 5.4 | 5.1 | 4.6 | 3.6 | 3.1 | 4.6 |
| Lat + 15 | 3.7 | 4.3 | 4.7 | 4.8 | 4.7 | 4.8 | 4.8 | 4.9 | 4.9 | 4.6 | 3.8 | 3.3 | 4.4 |
| Single axis tracker | | | | | | | | | | | | | |
| Lat - 15 | 3.7 | 4.7 | 5.8 | 6.7 | 7.2 | 7.7 | 7.5 | 7.2 | 6.4 | 5.4 | 3.9 | 3.2 | 5.8 |
| Latitude | 4.0 | 5.0 | 5.9 | 6.6 | 7.0 | 7.4 | 7.2 | 7.0 | 6.4 | 5.6 | 4.2 | 3.6 | 5.8 |
| Lat + 15 | 4.2 | 5.1 | 5.9 | 6.4 | 6.5 | 6.9 | 6.8 | 6.7 | 6.3 | 5.7 | 4.4 | 3.7 | 5.7 |
| Dual axis tracker | | | | | | | | | | | | | |
| | 4.3 | 5.1 | 6.0 | 6.7 | 7.3 | 7.9 | 7.6 | 7.2 | 6.4 | 5.7 | 4.4 | 3.8 | 6.0 |

## NEWARK NJ — Latitude: 40.70 degrees — Elevation: 9 meters

| | Jan | Feb | Mar | Apr | May | Jun | Jul | Aug | Sep | Oct | Nov | Dec | Avg |
|---|---|---|---|---|---|---|---|---|---|---|---|---|---|
| Fixed array | | | | | | | | | | | | | |
| Lat - 15 | 2.9 | 3.7 | 4.5 | 5.2 | 5.5 | 5.8 | 5.8 | 5.5 | 4.9 | 4.1 | 2.9 | 2.4 | 4.4 |
| Latitude | 3.3 | 4.0 | 4.6 | 5.1 | 5.2 | 5.4 | 5.4 | 5.3 | 5.0 | 4.4 | 3.2 | 2.8 | 4.5 |
| Lat + 15 | 3.5 | 4.1 | 4.5 | 4.7 | 4.6 | 4.7 | 4.7 | 4.8 | 4.7 | 4.4 | 3.3 | 3.0 | 4.3 |
| Single axis tracker | | | | | | | | | | | | | |
| Lat - 15 | 3.5 | 4.5 | 5.6 | 6.5 | 7.0 | 7.4 | 7.4 | 6.9 | 6.1 | 5.1 | 3.4 | 2.9 | 5.5 |
| Latitude | 3.8 | 4.7 | 5.7 | 6.5 | 6.7 | 7.1 | 7.1 | 6.8 | 6.2 | 5.3 | 3.7 | 3.1 | 5.6 |
| Lat + 15 | 4 | 4.8 | 5.6 | 6.2 | 6.3 | 6.6 | 6.6 | 6.5 | 6 | 5.3 | 3.8 | 3.3 | 5.4 |
| Dual axis tracker | | | | | | | | | | | | | |
| | 4.0 | 4.8 | 5.7 | 6.6 | 7.0 | 7.5 | 7.5 | 7.0 | 6.2 | 5.4 | 3.8 | 3.3 | 5.7 |

## ALBUQUERQUE  NM  Latitude: 35.05 degrees  Elevation: 1619 meters

|  | Jan | Feb | Mar | Apr | May | Jun | Jul | Aug | Sep | Oct | Nov | Dec | Avg |
|---|---|---|---|---|---|---|---|---|---|---|---|---|---|
| Fixed array |  |  |  |  |  |  |  |  |  |  |  |  |  |
| Lat - 15 | 4.6 | 5.4 | 6.3 | 7.3 | 7.7 | 7.8 | 7.4 | 7.2 | 6.6 | 5.9 | 4.8 | 4.3 | 6.3 |
| Latitude | 5.3 | 6.0 | 6.5 | 7.2 | 7.2 | 7.1 | 6.9 | 6.9 | 6.8 | 6.5 | 5.5 | 5.0 | 6.4 |
| Lat + 15 | 5.8 | 6.2 | 6.5 | 6.6 | 6.3 | 6.1 | 6.0 | 6.3 | 6.5 | 6.6 | 5.9 | 5.5 | 6.2 |
| Single axis tracker |  |  |  |  |  |  |  |  |  |  |  |  |  |
| Lat - 15 | 5.9 | 7.1 | 8.3 | 10.0 | 10.6 | 10.8 | 9.9 | 9.5 | 8.8 | 7.9 | 6.3 | 5.5 | 8.4 |
| Latitude | 6.5 | 7.5 | 8.6 | 9.9 | 10.3 | 10.4 | 9.5 | 9.3 | 9.0 | 8.3 | 6.8 | 6.1 | 8.5 |
| Lat + 15 | 6.9 | 7.7 | 8.5 | 9.5 | 9.7 | 9.7 | 8.9 | 8.9 | 8.8 | 8.4 | 7.1 | 6.5 | 8.4 |
| Dual axis tracker |  |  |  |  |  |  |  |  |  |  |  |  |  |
|  | 6.9 | 7.7 | 8.6 | 10.0 | 10.8 | 11.1 | 10.0 | 9.5 | 9.0 | 8.4 | 7.2 | 6.6 | 8.8 |

## TUCUMCARI  NM  Latitude: 35.18 degrees  Elevation: 1231 meters

|  | Jan | Feb | Mar | Apr | May | Jun | Jul | Aug | Sep | Oct | Nov | Dec | Avg |
|---|---|---|---|---|---|---|---|---|---|---|---|---|---|
| Fixed array |  |  |  |  |  |  |  |  |  |  |  |  |  |
| Lat - 15 | 4.3 | 5.1 | 5.9 | 6.8 | 7.0 | 7.2 | 7.1 | 6.8 | 6.2 | 5.6 | 4.5 | 4.0 | 5.9 |
| Latitude | 5.0 | 5.6 | 6.2 | 6.7 | 6.6 | 6.6 | 6.6 | 6.5 | 6.3 | 6.1 | 5.2 | 4.8 | 6.0 |
| Lat + 15 | 5.4 | 5.9 | 6.1 | 6.2 | 5.8 | 5.7 | 5.7 | 5.9 | 6.1 | 6.2 | 5.5 | 5.2 | 5.8 |
| Single axis tracker |  |  |  |  |  |  |  |  |  |  |  |  |  |
| Lat - 15 | 5.5 | 6.6 | 7.9 | 9.3 | 9.5 | 9.9 | 9.6 | 9.1 | 8.3 | 7.5 | 5.9 | 5.2 | 7.9 |
| Latitude | 6.0 | 7.1 | 8.1 | 9.2 | 9.3 | 9.5 | 9.3 | 9.0 | 8.4 | 7.9 | 6.4 | 5.7 | 8.0 |
| Lat + 15 | 6.9 | 7.7 | 8.5 | 9.5 | 9.7 | 9.7 | 8.9 | 8.9 | 8.8 | 8.4 | 7.1 | 6.5 | 8.4 |
| Dual axis tracker |  |  |  |  |  |  |  |  |  |  |  |  |  |
|  | 6.4 | 7.2 | 8.1 | 9.3 | 9.6 | 10.1 | 9.8 | 9.2 | 8.4 | 8.0 | 6.7 | 6.2 | 8.3 |

## ELKO  NV  Latitude: 40.83 degrees  Elevation: 1547 meters

|  | Jan | Feb | Mar | Apr | May | Jun | Jul | Aug | Sep | Oct | Nov | Dec | Avg |
|---|---|---|---|---|---|---|---|---|---|---|---|---|---|
| Fixed array |  |  |  |  |  |  |  |  |  |  |  |  |  |
| Lat - 15 | 3.3 | 4.1 | 4.9 | 5.7 | 6.4 | 6.9 | 7.3 | 7.0 | 6.5 | 5.2 | 3.5 | 3.1 | 5.3 |
| Latitude | 3.8 | 4.5 | 5.1 | 5.6 | 6.0 | 6.3 | 6.8 | 6.8 | 6.6 | 5.7 | 4.0 | 3.6 | 5.4 |
| Lat + 15 | 4.1 | 4.7 | 5.0 | 5.2 | 5.3 | 5.4 | 5.9 | 6.1 | 6.4 | 5.8 | 4.2 | 3.9 | 5.2 |
| Single axis tracker |  |  |  |  |  |  |  |  |  |  |  |  |  |
| Lat - 15 | 4.1 | 5.2 | 6.4 | 7.7 | 8.8 | 9.8 | 10.5 | 9.9 | 8.9 | 6.9 | 4.3 | 3.8 | 7.2 |
| Latitude | 4.5 | 5.5 | 6.5 | 7.6 | 8.5 | 9.5 | 10.2 | 9.7 | 9.0 | 7.2 | 4.7 | 4.2 | 7.3 |
| Lat + 15 | 4.7 | 5.6 | 6.4 | 7.3 | 8 | 8.9 | 9.6 | 9.3 | 8.9 | 7.3 | 4.9 | 4.5 | 7.1 |
| Dual axis tracker |  |  |  |  |  |  |  |  |  |  |  |  |  |
|  | 4.8 | 5.6 | 6.5 | 7.7 | 8.9 | 10.1 | 10.7 | 9.9 | 9.1 | 7.3 | 4.9 | 4.5 | 7.5 |

## ELY  NV  Latitude: 39.28 degrees  Elevation: 1906 meters

|  | Jan | Feb | Mar | Apr | May | Jun | Jul | Aug | Sep | Oct | Nov | Dec | Avg |
|---|---|---|---|---|---|---|---|---|---|---|---|---|---|
| Fixed array |  |  |  |  |  |  |  |  |  |  |  |  |  |
| Lat - 15 | 4.0 | 4.7 | 5.5 | 6.3 | 6.6 | 7.2 | 7.2 | 6.9 | 6.6 | 5.5 | 4.1 | 3.6 | 5.7 |
| Latitude | 4.6 | 5.2 | 5.7 | 6.2 | 6.2 | 6.6 | 6.6 | 6.6 | 6.7 | 6.0 | 4.7 | 4.3 | 5.8 |
| Lat + 15 | 5.0 | 5.5 | 5.6 | 5.7 | 5.5 | 5.6 | 5.7 | 6.0 | 6.5 | 6.1 | 5.0 | 4.7 | 5.6 |
| Single axis tracker |  |  |  |  |  |  |  |  |  |  |  |  |  |
| Lat - 15 | 5.0 | 6.1 | 7.2 | 8.5 | 9.2 | 10.3 | 10.3 | 9.6 | 9.1 | 7.3 | 5.2 | 4.6 | 7.7 |
| Latitude | 5.5 | 6.5 | 7.4 | 8.4 | 8.9 | 9.9 | 9.9 | 9.5 | 9.2 | 7.7 | 5.6 | 5.1 | 7.8 |
| Lat + 15 | 5.8 | 6.6 | 7.3 | 8.1 | 8.4 | 9.2 | 9.3 | 9 | 9 | 7.7 | 5.9 | 5.4 | 7.7 |
| Dual axis tracker |  |  |  |  |  |  |  |  |  |  |  |  |  |
|  | 5.9 | 6.6 | 7.4 | 8.5 | 9.3 | 10.5 | 10.5 | 9.7 | 9.2 | 7.8 | 5.9 | 5.5 | 8.1 |

## LAS VEGAS NV Latitude: 36.08 degrees Elevation: 664 meters

| | Jan | Feb | Mar | Apr | May | Jun | Jul | Aug | Sep | Oct | Nov | Dec | Avg |
|---|---|---|---|---|---|---|---|---|---|---|---|---|---|
| Fixed array | | | | | | | | | | | | | |
| Lat - 15 | 4.4 | 5.3 | 6.4 | 7.5 | 7.8 | 8.1 | 7.7 | 7.5 | 7.1 | 6.1 | 4.8 | 4.2 | 6.4 |
| Latitude | 5.1 | 5.9 | 6.7 | 7.4 | 7.3 | 7.4 | 7.1 | 7.2 | 7.2 | 6.6 | 5.5 | 4.9 | 6.5 |
| Lat + 15 | 5.6 | 6.1 | 6.6 | 6.8 | 6.5 | 6.3 | 6.2 | 6.5 | 7.0 | 6.8 | 5.9 | 5.4 | 6.3 |
| Single axis tracker | | | | | | | | | | | | | |
| Lat - 15 | 5.7 | 6.9 | 8.5 | 10.3 | 11.0 | 11.5 | 10.8 | 10.4 | 9.7 | 8.1 | 6.2 | 5.3 | 8.7 |
| Latitude | 6.2 | 7.3 | 8.8 | 10.2 | 10.6 | 11.1 | 10.4 | 10.3 | 9.8 | 8.6 | 6.7 | 5.9 | 8.8 |
| Lat + 15 | 6.5 | 7.5 | 8.7 | 9.8 | 10 | 10.3 | 9.8 | 9.8 | 9.6 | 8.7 | 7 | 6.2 | 8.7 |
| Dual axis tracker | | | | | | | | | | | | | |
| | 6.6 | 7.5 | 8.8 | 10.3 | 11.1 | 11.8 | 11.0 | 10.5 | 9.8 | 8.7 | 7.1 | 6.3 | 9.1 |

## RENO NV Latitude: 39.50 degrees Elevation: 1341 meters

| | Jan | Feb | Mar | Apr | May | Jun | Jul | Aug | Sep | Oct | Nov | Dec | Avg |
|---|---|---|---|---|---|---|---|---|---|---|---|---|---|
| Fixed array | | | | | | | | | | | | | |
| Lat - 15 | 3.6 | 4.4 | 5.5 | 6.5 | 7.1 | 7.4 | 7.7 | 7.4 | 6.8 | 5.6 | 3.9 | 3.3 | 5.8 |
| Latitude | 4.1 | 4.9 | 5.7 | 6.4 | 6.6 | 6.8 | 7.1 | 7.1 | 6.9 | 6.1 | 4.4 | 3.9 | 5.8 |
| Lat + 15 | 4.4 | 5.1 | 5.6 | 5.9 | 5.8 | 5.8 | 6.1 | 6.4 | 6.7 | 6.2 | 4.6 | 4.2 | 5.6 |
| Single axis tracker | | | | | | | | | | | | | |
| Lat - 15 | 4.4 | 5.6 | 7.2 | 8.8 | 9.8 | 10.5 | 11.1 | 10.4 | 9.4 | 7.4 | 4.8 | 4.1 | 7.8 |
| Latitude | 4.8 | 6.0 | 7.4 | 8.7 | 9.5 | 10.1 | 10.8 | 10.3 | 9.5 | 7.8 | 5.2 | 4.5 | 7.9 |
| Lat + 15 | 5.1 | 6.1 | 7.3 | 8.4 | 9 | 9.4 | 10.1 | 9.8 | 9.3 | 7.9 | 5.4 | 4.8 | 7.7 |
| Dual axis tracker | | | | | | | | | | | | | |
| | 5.1 | 6.1 | 7.4 | 8.8 | 10.0 | 10.8 | 11.4 | 10.5 | 9.5 | 7.9 | 5.5 | 4.9 | 8.2 |

## TONOPAH NV Latitude: 38.07 degrees Elevation: 1653 meters

| | Jan | Feb | Mar | Apr | May | Jun | Jul | Aug | Sep | Oct | Nov | Dec | Avg |
|---|---|---|---|---|---|---|---|---|---|---|---|---|---|
| Fixed array | | | | | | | | | | | | | |
| Lat - 15 | 4.1 | 4.9 | 5.8 | 6.7 | 7.1 | 7.6 | 7.7 | 7.3 | 6.9 | 5.9 | 4.4 | 3.8 | 6.0 |
| Latitude | 4.8 | 5.4 | 6.1 | 6.6 | 6.7 | 6.9 | 7.1 | 7.1 | 7.1 | 6.4 | 5.0 | 4.5 | 6.1 |
| Lat + 15 | 5.1 | 5.6 | 6.0 | 6.1 | 5.9 | 5.9 | 6.1 | 6.4 | 6.8 | 6.5 | 5.3 | 4.9 | 5.9 |
| Single axis tracker | | | | | | | | | | | | | |
| Lat - 15 | 5.2 | 6.4 | 7.8 | 9.2 | 10.0 | 10.9 | 11.0 | 10.3 | 9.6 | 7.9 | 5.6 | 4.9 | 8.2 |
| Latitude | 5.7 | 6.8 | 8.0 | 9.2 | 9.7 | 10.4 | 10.6 | 10.2 | 9.8 | 8.3 | 6.1 | 5.4 | 8.4 |
| Lat + 15 | 6 | 6.9 | 7.9 | 8.8 | 9.1 | 9.8 | 9.9 | 9.7 | 9.6 | 8.4 | 6.4 | 5.7 | 8.2 |
| Dual axis tracker | | | | | | | | | | | | | |
| | 6.1 | 6.9 | 8.0 | 9.3 | 10.1 | 11.1 | 11.2 | 10.4 | 9.8 | 8.4 | 6.4 | 5.8 | 8.6 |

## WINNEMUCCA NV Latitude: 40.90 degrees Elevation: 1323 meters

| | Jan | Feb | Mar | Apr | May | Jun | Jul | Aug | Sep | Oct | Nov | Dec | Avg |
|---|---|---|---|---|---|---|---|---|---|---|---|---|---|
| Fixed array | | | | | | | | | | | | | |
| Lat - 15 | 3.3 | 4.1 | 5.0 | 6.0 | 6.7 | 7.1 | 7.6 | 7.2 | 6.6 | 5.3 | 3.5 | 3.0 | 5.5 |
| Latitude | 3.7 | 4.5 | 5.2 | 5.9 | 6.2 | 6.5 | 7.0 | 6.9 | 6.7 | 5.7 | 3.9 | 3.5 | 5.5 |
| Lat + 15 | 4.0 | 4.6 | 5.1 | 5.5 | 5.5 | 5.6 | 6.0 | 6.3 | 6.5 | 5.8 | 4.1 | 3.8 | 5.2 |
| Single axis tracker | | | | | | | | | | | | | |
| Lat - 15 | 4.0 | 5.1 | 6.5 | 8.0 | 9.2 | 10.1 | 10.9 | 10.2 | 9.1 | 7.0 | 4.3 | 3.7 | 7.4 |
| Latitude | 4.4 | 5.4 | 6.7 | 8.0 | 8.9 | 9.7 | 10.6 | 10.0 | 9.2 | 7.3 | 4.7 | 4.1 | 7.4 |
| Lat + 15 | 4.6 | 5.5 | 6.6 | 7.7 | 8.4 | 9.1 | 9.9 | 9.6 | 9 | 7.4 | 4.8 | 4.3 | 7.2 |
| Dual axis tracker | | | | | | | | | | | | | |
| | 4.6 | 5.5 | 6.7 | 8.1 | 9.3 | 10.4 | 11.2 | 10.2 | 9.2 | 7.4 | 4.9 | 4.4 | 7.7 |

## ALBANY NY Latitude: 42.75 degrees Elevation: 89 meters

| | Jan | Feb | Mar | Apr | May | Jun | Jul | Aug | Sep | Oct | Nov | Dec | Avg |
|---|---|---|---|---|---|---|---|---|---|---|---|---|---|
| Fixed array | | | | | | | | | | | | | |
| Lat - 15 | 2.7 | 3.6 | 4.4 | 5.0 | 5.5 | 5.8 | 6.0 | 5.5 | 4.8 | 3.7 | 2.4 | 2.1 | 4.3 |
| Latitude | 3.0 | 3.9 | 4.5 | 4.9 | 5.1 | 5.4 | 5.5 | 5.2 | 4.8 | 3.9 | 2.6 | 2.4 | 4.3 |
| Lat + 15 | 3.2 | 4.1 | 4.4 | 4.5 | 4.6 | 4.6 | 4.8 | 4.8 | 4.6 | 3.9 | 2.7 | 2.5 | 4.1 |
| Single axis tracker | | | | | | | | | | | | | |
| Lat - 15 | 3.2 | 4.4 | 5.4 | 6.4 | 7.0 | 7.5 | 7.7 | 7.0 | 6.0 | 4.5 | 2.7 | 2.4 | 5.4 |
| Latitude | 3.5 | 4.6 | 5.5 | 6.3 | 6.8 | 7.2 | 7.5 | 6.9 | 6.0 | 4.6 | 2.9 | 2.6 | 5.4 |
| Lat + 15 | 3.6 | 4.7 | 5.4 | 6 | 6.4 | 6.7 | 7 | 6.5 | 5.9 | 4.6 | 3 | 2.7 | 5.2 |
| Dual axis tracker | | | | | | | | | | | | | |
| | 3.7 | 4.7 | 5.5 | 6.4 | 7.1 | 7.7 | 7.9 | 7.1 | 6.1 | 4.7 | 3.0 | 2.8 | 5.6 |

## BINGHAMTON NY Latitude: 42.22 degrees Elevation: 499 meters

| | Jan | Feb | Mar | Apr | May | Jun | Jul | Aug | Sep | Oct | Nov | Dec | Avg |
|---|---|---|---|---|---|---|---|---|---|---|---|---|---|
| Fixed array | | | | | | | | | | | | | |
| Lat - 15 | 2.5 | 3.3 | 4.2 | 4.8 | 5.3 | 5.6 | 5.7 | 5.3 | 4.5 | 3.5 | 2.3 | 1.9 | 4.1 |
| Latitude | 2.8 | 3.5 | 4.3 | 4.7 | 5.0 | 5.2 | 5.3 | 5.1 | 4.5 | 3.7 | 2.4 | 2.1 | 4.1 |
| Lat + 15 | 2.9 | 3.6 | 4.2 | 4.4 | 4.4 | 4.5 | 4.6 | 4.6 | 4.3 | 3.7 | 2.5 | 2.2 | 3.8 |
| Single axis tracker | | | | | | | | | | | | | |
| Lat - 15 | 2.9 | 3.9 | 5.1 | 6.1 | 6.8 | 7.3 | 7.5 | 6.8 | 5.7 | 4.3 | 2.6 | 2.2 | 5.1 |
| Latitude | 3.1 | 4.1 | 5.2 | 6.0 | 6.6 | 7.0 | 7.2 | 6.7 | 5.7 | 4.4 | 2.7 | 2.3 | 5.1 |
| Lat + 15 | 3.2 | 4.2 | 5.2 | 5.7 | 6.2 | 6.5 | 6.7 | 6.3 | 5.5 | 4.4 | 2.8 | 2.4 | 4.9 |
| Dual axis tracker | | | | | | | | | | | | | |
| | 3.3 | 4.2 | 5.2 | 6.1 | 6.9 | 7.5 | 7.6 | 6.8 | 5.7 | 4.5 | 2.8 | 2.4 | 5.3 |

## BUFFALO NY Latitude: 42.93 degrees Elevation: 215 meters

| | Jan | Feb | Mar | Apr | May | Jun | Jul | Aug | Sep | Oct | Nov | Dec | Avg |
|---|---|---|---|---|---|---|---|---|---|---|---|---|---|
| Fixed array | | | | | | | | | | | | | |
| Lat - 15 | 2.2 | 3.1 | 4.1 | 5.0 | 5.6 | 6.0 | 6.0 | 5.5 | 4.6 | 3.4 | 2.1 | 1.8 | 4.1 |
| Latitude | 2.4 | 3.3 | 4.2 | 4.8 | 5.2 | 5.5 | 5.5 | 5.3 | 4.6 | 3.6 | 2.3 | 1.9 | 4.1 |
| Lat + 15 | 2.5 | 3.4 | 4.1 | 4.5 | 4.6 | 4.8 | 4.8 | 4.8 | 4.3 | 3.6 | 2.3 | 2.0 | 3.8 |
| Single axis tracker | | | | | | | | | | | | | |
| Lat - 15 | 2.5 | 3.6 | 5.0 | 6.3 | 7.2 | 7.9 | 7.9 | 7.2 | 5.8 | 4.1 | 2.4 | 1.9 | 5.2 |
| Latitude | 2.7 | 3.8 | 5.1 | 6.2 | 7.0 | 7.6 | 7.7 | 7.0 | 5.8 | 4.3 | 2.5 | 2.1 | 5.1 |
| Lat + 15 | 2.8 | 3.8 | 5 | 6 | 6.6 | 7.1 | 7.2 | 6.7 | 5.6 | 4.3 | 2.5 | 2.2 | 5 |
| Dual axis tracker | | | | | | | | | | | | | |
| | 2.8 | 3.8 | 5.1 | 6.3 | 7.3 | 8.1 | 8.1 | 7.2 | 5.8 | 4.3 | 2.6 | 2.2 | 5.3 |

## MASSENA NY Latitude: 44.93 degrees Elevation: 63 meters

| | Jan | Feb | Mar | Apr | May | Jun | Jul | Aug | Sep | Oct | Nov | Dec | Avg |
|---|---|---|---|---|---|---|---|---|---|---|---|---|---|
| Fixed array | | | | | | | | | | | | | |
| Lat - 15 | 2.7 | 3.8 | 4.7 | 5.2 | 5.6 | 5.9 | 6.0 | 5.4 | 4.6 | 3.5 | 2.2 | 2.1 | 4.3 |
| Latitude | 3.0 | 4.2 | 4.9 | 5.0 | 5.2 | 5.4 | 5.6 | 5.2 | 4.7 | 3.7 | 2.4 | 2.3 | 4.3 |
| Lat + 15 | 3.2 | 4.3 | 4.8 | 4.7 | 4.6 | 4.7 | 4.9 | 4.7 | 4.4 | 3.7 | 2.5 | 2.5 | 4.1 |
| Single axis tracker | | | | | | | | | | | | | |
| Lat - 15 | 3.2 | 4.7 | 6.0 | 6.7 | 7.4 | 8.0 | 8.2 | 7.2 | 6.0 | 4.3 | 2.5 | 2.4 | 5.6 |
| Latitude | 3.4 | 5.0 | 6.1 | 6.6 | 7.2 | 7.6 | 7.9 | 7.1 | 6.0 | 4.5 | 2.7 | 2.6 | 5.6 |
| Lat + 15 | 3.6 | 5.1 | 6.1 | 6.4 | 6.8 | 7.2 | 7.4 | 6.7 | 5.8 | 4.5 | 2.8 | 2.8 | 5.4 |
| Dual axis tracker | | | | | | | | | | | | | |
| | 3.6 | 5.1 | 6.1 | 6.7 | 7.6 | 8.2 | 8.4 | 7.3 | 6.0 | 4.5 | 2.8 | 2.8 | 5.8 |

## NEW YORK CITY    NY      Latitude: 40.78 degrees      Elevation: 57 meters

| | Jan | Feb | Mar | Apr | May | Jun | Jul | Aug | Sep | Oct | Nov | Dec | Avg |
|---|---|---|---|---|---|---|---|---|---|---|---|---|---|
| Fixed array | | | | | | | | | | | | | |
| Lat - 15 | 2.9 | 3.7 | 4.6 | 5.3 | 5.8 | 6.0 | 6.0 | 5.7 | 5.0 | 4.1 | 2.9 | 2.4 | 4.5 |
| Latitude | 3.2 | 4.0 | 4.8 | 5.2 | 5.4 | 5.5 | 5.6 | 5.5 | 5.0 | 4.4 | 3.2 | 2.8 | 4.6 |
| Lat + 15 | 3.4 | 4.1 | 4.6 | 4.8 | 4.8 | 4.8 | 4.9 | 5.0 | 4.8 | 4.4 | 3.3 | 3.0 | 4.3 |
| Single axis tracker | | | | | | | | | | | | | |
| Lat - 15 | 3.4 | 4.5 | 5.7 | 6.7 | 7.2 | 7.5 | 7.5 | 7.1 | 6.2 | 5.1 | 3.4 | 2.8 | 5.6 |
| Latitude | 3.7 | 4.7 | 5.8 | 6.6 | 6.9 | 7.2 | 7.2 | 6.9 | 6.2 | 5.3 | 3.7 | 3.1 | 5.6 |
| Lat + 15 | 3.9 | 4.8 | 5.7 | 6.3 | 6.5 | 6.7 | 6.7 | 6.6 | 6 | 5.3 | 3.8 | 3.3 | 5.5 |
| Dual axis tracker | | | | | | | | | | | | | |
| | 3.9 | 4.8 | 5.8 | 6.7 | 7.3 | 7.7 | 7.6 | 7.1 | 6.2 | 5.3 | 3.8 | 3.3 | 5.8 |

## ROCHESTER    NY      Latitude: 43.12 degrees      Elevation: 169 meters

| | Jan | Feb | Mar | Apr | May | Jun | Jul | Aug | Sep | Oct | Nov | Dec | Avg |
|---|---|---|---|---|---|---|---|---|---|---|---|---|---|
| Fixed array | | | | | | | | | | | | | |
| Lat - 15 | 2.3 | 3.1 | 4.1 | 5.0 | 5.6 | 5.9 | 6.0 | 5.5 | 4.6 | 3.4 | 2.1 | 1.8 | 4.1 |
| Latitude | 2.5 | 3.3 | 4.2 | 4.9 | 5.2 | 5.4 | 5.5 | 5.3 | 4.6 | 3.6 | 2.3 | 2.0 | 4.1 |
| Lat + 15 | 2.6 | 3.4 | 4.1 | 4.5 | 4.6 | 4.7 | 4.8 | 4.8 | 4.4 | 3.6 | 2.3 | 2.0 | 3.8 |
| Single axis tracker | | | | | | | | | | | | | |
| Lat - 15 | 2.6 | 3.7 | 5.0 | 6.4 | 7.3 | 7.9 | 8.0 | 7.2 | 5.8 | 4.2 | 2.4 | 2.0 | 5.2 |
| Latitude | 2.8 | 3.9 | 5.1 | 6.3 | 7.1 | 7.6 | 7.7 | 7.0 | 5.9 | 4.3 | 2.5 | 2.1 | 5.2 |
| Lat + 15 | 2.9 | 3.9 | 5 | 6 | 6.6 | 7.1 | 7.3 | 6.7 | 5.7 | 4.3 | 2.6 | 2.2 | 5 |
| Dual axis tracker | | | | | | | | | | | | | |
| | 2.9 | 3.9 | 5.1 | 6.4 | 7.4 | 8.1 | 8.2 | 7.2 | 5.9 | 4.3 | 2.6 | 2.2 | 5.4 |

## SYRACUSE    NY      Latitude: 43.12 degrees      Elevation: 124 meters

| | Jan | Feb | Mar | Apr | May | Jun | Jul | Aug | Sep | Oct | Nov | Dec | Avg |
|---|---|---|---|---|---|---|---|---|---|---|---|---|---|
| Fixed array | | | | | | | | | | | | | |
| Lat - 15 | 2.4 | 3.3 | 4.2 | 5.0 | 5.6 | 5.9 | 6.0 | 5.5 | 4.7 | 3.5 | 2.1 | 1.8 | 4.2 |
| Latitude | 2.7 | 3.5 | 4.3 | 4.9 | 5.2 | 5.4 | 5.6 | 5.3 | 4.7 | 3.7 | 2.3 | 2.0 | 4.1 |
| Lat + 15 | 2.8 | 3.6 | 4.2 | 4.5 | 4.6 | 4.7 | 4.9 | 4.8 | 4.5 | 3.7 | 2.3 | 2.1 | 3.9 |
| Single axis tracker | | | | | | | | | | | | | |
| Lat - 15 | 2.8 | 3.9 | 5.2 | 6.4 | 7.3 | 7.9 | 8.0 | 7.2 | 5.9 | 4.3 | 2.4 | 2.0 | 5.3 |
| Latitude | 3.0 | 4.1 | 5.3 | 6.3 | 7.0 | 7.6 | 7.8 | 7.1 | 5.9 | 4.4 | 2.5 | 2.2 | 5.3 |
| Lat + 15 | 3.1 | 4.1 | 5.2 | 6.1 | 6.6 | 7.1 | 7.3 | 6.7 | 5.8 | 4.4 | 2.6 | 2.3 | 5.1 |
| Dual axis tracker | | | | | | | | | | | | | |
| | 3.1 | 4.1 | 5.3 | 6.4 | 7.4 | 8.1 | 8.2 | 7.2 | 6.0 | 4.5 | 2.6 | 2.3 | 5.4 |

## AKRON    OH      Latitude: 40.92 degrees      Elevation: 377 meters

| | Jan | Feb | Mar | Apr | May | Jun | Jul | Aug | Sep | Oct | Nov | Dec | Avg |
|---|---|---|---|---|---|---|---|---|---|---|---|---|---|
| Fixed array | | | | | | | | | | | | | |
| Lat - 15 | 2.3 | 3.1 | 4.0 | 4.9 | 5.5 | 5.9 | 5.9 | 5.5 | 4.8 | 3.8 | 2.3 | 1.8 | 4.2 |
| Latitude | 2.5 | 3.4 | 4.1 | 4.8 | 5.2 | 5.5 | 5.5 | 5.3 | 4.8 | 4.0 | 2.5 | 2.0 | 4.1 |
| Lat + 15 | 2.7 | 3.4 | 3.9 | 4.4 | 4.6 | 4.7 | 4.8 | 4.8 | 4.6 | 4.0 | 2.6 | 2.1 | 3.9 |
| Single axis tracker | | | | | | | | | | | | | |
| Lat - 15 | 2.6 | 3.7 | 4.8 | 6.1 | 7.0 | 7.6 | 7.7 | 7.0 | 6.0 | 4.5 | 2.7 | 2.0 | 5.2 |
| Latitude | 2.8 | 3.9 | 4.8 | 6.0 | 6.8 | 7.3 | 7.4 | 6.9 | 6.0 | 4.7 | 2.8 | 2.2 | 5.1 |
| Lat + 15 | 2.9 | 3.9 | 4.7 | 5.8 | 6.3 | 6.8 | 6.9 | 6.6 | 5.9 | 4.7 | 2.9 | 2.3 | 5 |
| Dual axis tracker | | | | | | | | | | | | | |
| | 3.0 | 3.9 | 4.8 | 6.1 | 7.1 | 7.8 | 7.8 | 7.1 | 6.1 | 4.8 | 2.9 | 2.3 | 5.3 |

## CLEVELAND     OH     Latitude: 41.40 degrees     Elevation: 245 meters

|  | Jan | Feb | Mar | Apr | May | Jun | Jul | Aug | Sep | Oct | Nov | Dec | Avg |
|---|---|---|---|---|---|---|---|---|---|---|---|---|---|
| Fixed array |  |  |  |  |  |  |  |  |  |  |  |  |  |
| Lat - 15 | 2.2 | 3.1 | 3.9 | 5.0 | 5.6 | 6.0 | 6.1 | 5.6 | 4.7 | 3.6 | 2.2 | 1.7 | 4.2 |
| Latitude | 2.4 | 3.3 | 4.0 | 4.9 | 5.3 | 5.5 | 5.6 | 5.3 | 4.8 | 3.8 | 2.4 | 1.9 | 4.1 |
| Lat + 15 | 2.5 | 3.3 | 3.9 | 4.5 | 4.7 | 4.8 | 4.9 | 4.9 | 4.6 | 3.8 | 2.4 | 2.0 | 3.9 |
| Single axis tracker |  |  |  |  |  |  |  |  |  |  |  |  |  |
| Lat - 15 | 2.5 | 3.6 | 4.7 | 6.2 | 7.2 | 7.8 | 7.9 | 7.1 | 5.9 | 4.3 | 2.5 | 1.9 | 5.1 |
| Latitude | 2.6 | 3.8 | 4.7 | 6.1 | 6.9 | 7.5 | 7.6 | 7.0 | 5.9 | 4.5 | 2.6 | 2.0 | 5.1 |
| Lat + 15 | 2.7 | 3.8 | 4.6 | 5.9 | 6.5 | 7 | 7.1 | 6.6 | 5.8 | 4.5 | 2.7 | 2.1 | 4.9 |
| Dual axis tracker |  |  |  |  |  |  |  |  |  |  |  |  |  |
|  | 2.8 | 3.8 | 4.7 | 6.2 | 7.3 | 8.0 | 8.0 | 7.2 | 6.0 | 4.5 | 2.7 | 2.1 | 5.3 |

## COLUMBUS     OH     Latitude: 40.00 degrees     Elevation: 254 meters

|  | Jan | Feb | Mar | Apr | May | Jun | Jul | Aug | Sep | Oct | Nov | Dec | Avg |
|---|---|---|---|---|---|---|---|---|---|---|---|---|---|
| Fixed array |  |  |  |  |  |  |  |  |  |  |  |  |  |
| Lat - 15 | 2.5 | 3.2 | 4.0 | 4.9 | 5.5 | 5.9 | 5.8 | 5.5 | 4.9 | 4.0 | 2.6 | 2.0 | 4.2 |
| Latitude | 2.7 | 3.4 | 4.1 | 4.8 | 5.2 | 5.4 | 5.4 | 5.3 | 4.9 | 4.3 | 2.8 | 2.2 | 4.2 |
| Lat + 15 | 2.9 | 3.5 | 4.0 | 4.4 | 4.6 | 4.7 | 4.8 | 4.8 | 4.7 | 4.3 | 2.9 | 2.3 | 4.0 |
| Single axis tracker |  |  |  |  |  |  |  |  |  |  |  |  |  |
| Lat - 15 | 2.8 | 3.8 | 4.8 | 6.1 | 7.0 | 7.5 | 7.5 | 7.1 | 6.2 | 4.9 | 2.9 | 2.3 | 5.3 |
| Latitude | 3.1 | 4.0 | 4.9 | 6.0 | 6.8 | 7.2 | 7.2 | 6.9 | 6.2 | 5.1 | 3.1 | 2.5 | 5.3 |
| Lat + 15 | 3.2 | 4 | 4.8 | 5.8 | 6.4 | 6.7 | 6.8 | 6.6 | 6 | 5.1 | 3.2 | 2.5 | 5.1 |
| Dual axis tracker |  |  |  |  |  |  |  |  |  |  |  |  |  |
|  | 3.2 | 4.1 | 4.9 | 6.1 | 7.1 | 7.7 | 7.6 | 7.1 | 6.2 | 5.2 | 3.2 | 2.6 | 5.4 |

## DAYTON     OH     Latitude: 39.90 degrees     Elevation: 306 meters

|  | Jan | Feb | Mar | Apr | May | Jun | Jul | Aug | Sep | Oct | Nov | Dec | Avg |
|---|---|---|---|---|---|---|---|---|---|---|---|---|---|
| Fixed array |  |  |  |  |  |  |  |  |  |  |  |  |  |
| Lat - 15 | 2.7 | 3.4 | 4.2 | 5.1 | 5.7 | 6.1 | 6.0 | 5.7 | 5.1 | 4.1 | 2.7 | 2.1 | 4.4 |
| Latitude | 3.0 | 3.7 | 4.3 | 4.9 | 5.3 | 5.6 | 5.6 | 5.5 | 5.1 | 4.4 | 2.9 | 2.4 | 4.4 |
| Lat + 15 | 3.1 | 3.8 | 4.1 | 4.6 | 4.7 | 4.8 | 4.9 | 5.0 | 4.9 | 4.4 | 3.0 | 2.5 | 4.2 |
| Single axis tracker |  |  |  |  |  |  |  |  |  |  |  |  |  |
| Lat - 15 | 3.1 | 4.1 | 5.1 | 6.4 | 7.3 | 7.9 | 7.8 | 7.4 | 6.5 | 5.1 | 3.1 | 2.4 | 5.5 |
| Latitude | 3.4 | 4.3 | 5.1 | 6.3 | 7.1 | 7.6 | 7.5 | 7.3 | 6.5 | 5.3 | 3.3 | 2.6 | 5.5 |
| Lat + 15 | 3.5 | 4.4 | 5 | 6 | 6.6 | 7.1 | 7 | 6.9 | 6.3 | 5.3 | 3.4 | 2.8 | 5.4 |
| Dual axis tracker |  |  |  |  |  |  |  |  |  |  |  |  |  |
|  | 3.6 | 4.4 | 5.2 | 6.4 | 7.4 | 8.1 | 8.0 | 7.5 | 6.5 | 5.3 | 3.4 | 2.8 | 5.7 |

## MANSFIELD     OH     Latitude: 40.82 degrees     Elevation: 395 meters

|  | Jan | Feb | Mar | Apr | May | Jun | Jul | Aug | Sep | Oct | Nov | Dec | Avg |
|---|---|---|---|---|---|---|---|---|---|---|---|---|---|
| Fixed array |  |  |  |  |  |  |  |  |  |  |  |  |  |
| Lat - 15 | 2.3 | 3.2 | 4.0 | 4.9 | 5.6 | 5.9 | 6.0 | 5.6 | 4.8 | 3.8 | 2.4 | 1.9 | 4.2 |
| Latitude | 2.6 | 3.4 | 4.0 | 4.8 | 5.2 | 5.5 | 5.5 | 5.3 | 4.9 | 4.1 | 2.6 | 2.1 | 4.2 |
| Lat + 15 | 2.7 | 3.5 | 3.9 | 4.4 | 4.6 | 4.7 | 4.8 | 4.9 | 4.6 | 4.1 | 2.7 | 2.2 | 3.9 |
| Single axis tracker |  |  |  |  |  |  |  |  |  |  |  |  |  |
| Lat - 15 | 2.7 | 3.8 | 4.7 | 6.1 | 7.1 | 7.7 | 7.7 | 7.1 | 6.1 | 4.7 | 2.7 | 2.1 | 5.2 |
| Latitude | 2.9 | 3.9 | 4.8 | 6.0 | 6.8 | 7.4 | 7.4 | 7.0 | 6.1 | 4.8 | 2.9 | 2.3 | 5.2 |
| Lat + 15 | 3 | 4 | 4.7 | 5.7 | 6.4 | 6.9 | 7 | 6.6 | 5.9 | 4.8 | 3 | 2.4 | 5 |
| Dual axis tracker |  |  |  |  |  |  |  |  |  |  |  |  |  |
|  | 3.0 | 4.0 | 4.8 | 6.1 | 7.2 | 7.9 | 7.9 | 7.2 | 6.1 | 4.9 | 3.0 | 2.4 | 5.4 |

## TOLEDO      OH      Latitude: 41.60 degrees      Elevation: 211 meters

| | Jan | Feb | Mar | Apr | May | Jun | Jul | Aug | Sep | Oct | Nov | Dec | Avg |
|---|---|---|---|---|---|---|---|---|---|---|---|---|---|
| Fixed array | | | | | | | | | | | | | |
| Lat - 15 | 2.5 | 3.4 | 4.2 | 5.1 | 5.8 | 6.2 | 6.2 | 5.8 | 5.0 | 3.9 | 2.5 | 2.0 | 4.4 |
| Latitude | 2.8 | 3.7 | 4.3 | 5.0 | 5.5 | 5.7 | 5.7 | 5.6 | 5.0 | 4.1 | 2.7 | 2.2 | 4.4 |
| Lat + 15 | 3.0 | 3.8 | 4.2 | 4.6 | 4.9 | 5.0 | 5.0 | 5.0 | 4.8 | 4.1 | 2.8 | 2.3 | 4.1 |
| Single axis tracker | | | | | | | | | | | | | |
| Lat - 15 | 2.9 | 4.1 | 5.1 | 6.5 | 7.5 | 8.1 | 8.1 | 7.5 | 6.2 | 4.7 | 2.8 | 2.2 | 5.5 |
| Latitude | 3.2 | 4.3 | 5.2 | 6.4 | 7.3 | 7.8 | 7.8 | 7.3 | 6.3 | 4.9 | 3.0 | 2.4 | 5.5 |
| Lat + 15 | 3.3 | 4.4 | 5.1 | 6.1 | 6.8 | 7.3 | 7.3 | 7 | 6.1 | 4.9 | 3.1 | 2.5 | 5.3 |
| Dual axis tracker | | | | | | | | | | | | | |
| | 3.3 | 4.4 | 5.2 | 6.5 | 7.6 | 8.3 | 8.2 | 7.5 | 6.3 | 5.0 | 3.1 | 2.5 | 5.7 |

## YOUNGSTOWN      OH      Latitude: 41.27 degrees      Elevation: 361 meters

| | Jan | Feb | Mar | Apr | May | Jun | Jul | Aug | Sep | Oct | Nov | Dec | Avg |
|---|---|---|---|---|---|---|---|---|---|---|---|---|---|
| Fixed array | | | | | | | | | | | | | |
| Lat - 15 | 2.2 | 3.0 | 3.8 | 4.7 | 5.4 | 5.8 | 5.7 | 5.3 | 4.6 | 3.6 | 2.2 | 1.7 | 4.0 |
| Latitude | 2.4 | 3.2 | 3.9 | 4.6 | 5.0 | 5.3 | 5.3 | 5.0 | 4.6 | 3.8 | 2.4 | 1.9 | 3.9 |
| Lat + 15 | 2.5 | 3.2 | 3.8 | 4.3 | 4.5 | 4.6 | 4.7 | 4.6 | 4.3 | 3.7 | 2.4 | 1.9 | 3.7 |
| Single axis tracker | | | | | | | | | | | | | |
| Lat - 15 | 2.4 | 3.5 | 4.6 | 5.9 | 6.7 | 7.4 | 7.4 | 6.7 | 5.7 | 4.3 | 2.5 | 1.9 | 4.9 |
| Latitude | 2.6 | 3.6 | 4.6 | 5.8 | 6.5 | 7.1 | 7.1 | 6.5 | 5.7 | 4.4 | 2.6 | 2.0 | 4.9 |
| Lat + 15 | 2.7 | 3.7 | 4.5 | 5.5 | 6.1 | 6.6 | 6.7 | 6.2 | 5.5 | 4.4 | 2.6 | 2.1 | 4.7 |
| Dual axis tracker | | | | | | | | | | | | | |
| | 2.7 | 3.7 | 4.6 | 5.9 | 6.8 | 7.6 | 7.6 | 6.7 | 5.7 | 4.5 | 2.7 | 2.1 | 5.1 |

## OKLAHOMA CITY      OK      Latitude: 35.40 degrees      Elevation: 397 meters

| | Jan | Feb | Mar | Apr | May | Jun | Jul | Aug | Sep | Oct | Nov | Dec | Avg |
|---|---|---|---|---|---|---|---|---|---|---|---|---|---|
| Fixed array | | | | | | | | | | | | | |
| Lat - 15 | 3.9 | 4.4 | 5.3 | 6.0 | 6.2 | 6.6 | 6.8 | 6.5 | 5.6 | 5.0 | 3.9 | 3.5 | 5.3 |
| Latitude | 4.4 | 4.9 | 5.5 | 5.9 | 5.8 | 6.0 | 6.3 | 6.3 | 5.7 | 5.4 | 4.4 | 4.1 | 5.4 |
| Lat + 15 | 4.7 | 5.0 | 5.4 | 5.5 | 5.2 | 5.2 | 5.5 | 5.7 | 5.5 | 5.5 | 4.7 | 4.4 | 5.2 |
| Single axis tracker | | | | | | | | | | | | | |
| Lat - 15 | 4.8 | 5.6 | 6.8 | 7.9 | 8.1 | 8.8 | 9.3 | 8.7 | 7.4 | 6.5 | 4.9 | 4.3 | 6.9 |
| Latitude | 5.3 | 5.9 | 7.0 | 7.8 | 7.9 | 8.4 | 8.9 | 8.6 | 7.5 | 6.8 | 5.3 | 4.8 | 7.0 |
| Lat + 15 | 5.5 | 6 | 6.9 | 7.5 | 7.4 | 7.9 | 8.4 | 8.1 | 7.3 | 6.8 | 5.5 | 5.1 | 6.9 |
| Dual axis tracker | | | | | | | | | | | | | |
| | 5.6 | 6.1 | 7.0 | 7.9 | 8.2 | 9.0 | 9.4 | 8.7 | 7.5 | 6.8 | 5.6 | 5.1 | 7.3 |

## TULSA      OK      Latitude: 36.20 degrees      Elevation: 206 meters

| | Jan | Feb | Mar | Apr | May | Jun | Jul | Aug | Sep | Oct | Nov | Dec | Avg |
|---|---|---|---|---|---|---|---|---|---|---|---|---|---|
| Fixed array | | | | | | | | | | | | | |
| Lat - 15 | 3.5 | 4.2 | 5.0 | 5.7 | 5.9 | 6.3 | 6.6 | 6.3 | 5.3 | 4.7 | 3.6 | 3.2 | 5.0 |
| Latitude | 4.0 | 4.5 | 5.2 | 5.6 | 5.5 | 5.8 | 6.1 | 6.0 | 5.4 | 5.1 | 4.0 | 3.7 | 5.1 |
| Lat + 15 | 4.3 | 4.7 | 5.1 | 5.2 | 4.9 | 5.0 | 5.3 | 5.5 | 5.2 | 5.1 | 4.2 | 4.0 | 4.9 |
| Single axis tracker | | | | | | | | | | | | | |
| Lat - 15 | 4.4 | 5.2 | 6.3 | 7.3 | 7.6 | 8.3 | 8.8 | 8.3 | 6.9 | 6.0 | 4.4 | 3.9 | 6.5 |
| Latitude | 4.8 | 5.5 | 6.5 | 7.2 | 7.4 | 7.9 | 8.5 | 8.2 | 6.9 | 6.3 | 4.8 | 4.3 | 6.5 |
| Lat + 15 | 5 | 5.6 | 6.4 | 6.9 | 7 | 7.4 | 8 | 7.8 | 6.8 | 6.3 | 5 | 4.5 | 6.4 |
| Dual axis tracker | | | | | | | | | | | | | |
| | 5.0 | 5.6 | 6.5 | 7.3 | 7.7 | 8.5 | 9.0 | 8.4 | 6.9 | 6.4 | 5.0 | 4.6 | 6.7 |

## ASTORIA     OR     Latitude: 46.15 degrees     Elevation: 7 meters

| | Jan | Feb | Mar | Apr | May | Jun | Jul | Aug | Sep | Oct | Nov | Dec | Avg |
|---|---|---|---|---|---|---|---|---|---|---|---|---|---|
| Fixed array | | | | | | | | | | | | | |
| Lat - 15 | 1.7 | 2.4 | 3.4 | 4.3 | 5.0 | 5.2 | 5.4 | 5.1 | 4.7 | 3.3 | 2.0 | 1.5 | 3.7 |
| Latitude | 1.9 | 2.6 | 3.4 | 4.2 | 4.7 | 4.7 | 5.0 | 4.9 | 4.7 | 3.5 | 2.1 | 1.7 | 3.6 |
| Lat + 15 | 2.0 | 2.6 | 3.3 | 3.8 | 4.1 | 4.1 | 4.4 | 4.4 | 4.5 | 3.5 | 2.2 | 1.8 | 3.4 |
| Single axis tracker | | | | | | | | | | | | | |
| Lat - 15 | 1.9 | 2.8 | 4.0 | 5.3 | 6.2 | 6.4 | 6.8 | 6.4 | 5.9 | 4.0 | 2.2 | 1.7 | 4.5 |
| Latitude | 2.1 | 2.9 | 4.1 | 5.2 | 6.0 | 6.1 | 6.6 | 6.3 | 5.9 | 4.1 | 2.3 | 1.8 | 4.5 |
| Lat + 15 | 2.2 | 2.9 | 4 | 4.9 | 5.6 | 5.7 | 6.1 | 6 | 5.8 | 4.1 | 2.4 | 1.9 | 4.3 |
| Dual axis tracker | | | | | | | | | | | | | |
| | 2.2 | 3.0 | 4.1 | 5.3 | 6.3 | 6.6 | 7.0 | 6.5 | 6.0 | 4.2 | 2.4 | 1.9 | 4.6 |

## BURNS     OR     Latitude: 43.58 degrees     Elevation: 1271 meters

| | Jan | Feb | Mar | Apr | May | Jun | Jul | Aug | Sep | Oct | Nov | Dec | Avg |
|---|---|---|---|---|---|---|---|---|---|---|---|---|---|
| Fixed array | | | | | | | | | | | | | |
| Lat - 15 | 2.8 | 3.7 | 4.7 | 5.8 | 6.5 | 6.9 | 7.5 | 7.0 | 6.3 | 4.8 | 2.8 | 2.4 | 5.1 |
| Latitude | 3.1 | 4.0 | 4.9 | 5.7 | 6.1 | 6.3 | 6.9 | 6.8 | 6.4 | 5.2 | 3.1 | 2.7 | 5.1 |
| Lat + 15 | 3.3 | 4.1 | 4.8 | 5.3 | 5.4 | 5.4 | 6.0 | 6.1 | 6.1 | 5.2 | 3.3 | 2.9 | 4.8 |
| Single axis tracker | | | | | | | | | | | | | |
| Lat - 15 | 3.3 | 4.5 | 6.0 | 7.7 | 9.0 | 9.8 | 10.9 | 10.0 | 8.6 | 6.2 | 3.4 | 2.8 | 6.8 |
| Latitude | 3.5 | 4.8 | 6.1 | 7.6 | 8.7 | 9.4 | 10.5 | 9.8 | 8.7 | 6.5 | 3.6 | 3.1 | 6.9 |
| Lat + 15 | 3.7 | 4.8 | 6.1 | 7.3 | 8.2 | 8.8 | 9.9 | 9.3 | 8.5 | 6.5 | 3.7 | 3.2 | 6.7 |
| Dual axis tracker | | | | | | | | | | | | | |
| | 3.7 | 4.9 | 6.2 | 7.7 | 9.1 | 10.0 | 11.1 | 10.0 | 8.7 | 6.5 | 3.7 | 3.3 | 7.1 |

## EUGENE     OR     Latitude: 44.12 degrees     Elevation: 109 meters

| | Jan | Feb | Mar | Apr | May | Jun | Jul | Aug | Sep | Oct | Nov | Dec | Avg |
|---|---|---|---|---|---|---|---|---|---|---|---|---|---|
| Fixed array | | | | | | | | | | | | | |
| Lat - 15 | 1.8 | 2.6 | 3.8 | 4.8 | 5.6 | 6.0 | 6.7 | 6.3 | 5.4 | 3.6 | 1.9 | 1.4 | 4.2 |
| Latitude | 2.0 | 2.7 | 3.8 | 4.7 | 5.3 | 5.5 | 6.2 | 6.0 | 5.5 | 3.8 | 2.1 | 1.6 | 4.1 |
| Lat + 15 | 2.0 | 2.8 | 3.7 | 4.3 | 4.6 | 4.8 | 5.4 | 5.5 | 5.2 | 3.8 | 2.1 | 1.6 | 3.8 |
| Single axis tracker | | | | | | | | | | | | | |
| Lat - 15 | 2.0 | 3.0 | 4.5 | 6.0 | 7.3 | 8.1 | 9.4 | 8.5 | 7.1 | 4.4 | 2.2 | 1.6 | 5.3 |
| Latitude | 2.1 | 3.1 | 4.6 | 5.9 | 7.1 | 7.8 | 9.1 | 8.4 | 7.1 | 4.6 | 2.3 | 1.7 | 5.3 |
| Lat + 15 | 2.2 | 3.1 | 4.5 | 5.6 | 6.7 | 7.3 | 8.5 | 8 | 6.9 | 4.6 | 2.3 | 1.7 | 5.1 |
| Dual axis tracker | | | | | | | | | | | | | |
| | 2.2 | 3.2 | 4.6 | 6.0 | 7.4 | 8.3 | 9.6 | 8.6 | 7.1 | 4.6 | 2.3 | 1.7 | 5.5 |

## MEDFORD     OR     Latitude: 42.37 degrees     Elevation: 396 meters

| | Jan | Feb | Mar | Apr | May | Jun | Jul | Aug | Sep | Oct | Nov | Dec | Avg |
|---|---|---|---|---|---|---|---|---|---|---|---|---|---|
| Fixed array | | | | | | | | | | | | | |
| Lat - 15 | 2.1 | 3.2 | 4.5 | 5.7 | 6.6 | 7.1 | 7.7 | 7.2 | 6.3 | 4.5 | 2.3 | 1.7 | 4.9 |
| Latitude | 2.3 | 3.5 | 4.6 | 5.6 | 6.2 | 6.5 | 7.1 | 6.9 | 6.4 | 4.8 | 2.4 | 1.9 | 4.9 |
| Lat + 15 | 2.4 | 3.5 | 4.5 | 5.2 | 5.5 | 5.6 | 6.2 | 6.3 | 6.2 | 4.8 | 2.5 | 2.0 | 4.5 |
| Single axis tracker | | | | | | | | | | | | | |
| Lat - 15 | 2.4 | 3.8 | 5.5 | 7.4 | 8.9 | 9.9 | 11.1 | 10.0 | 8.5 | 5.6 | 2.5 | 1.9 | 6.5 |
| Latitude | 2.6 | 4.0 | 5.6 | 7.3 | 8.7 | 9.5 | 10.7 | 9.9 | 5.9 | 2.7 | 2.0 | 1.8 | 6.5 |
| Lat + 15 | 2.6 | 4 | 5.5 | 7 | 8.1 | 8.9 | 10.1 | 9.4 | 8.4 | 5.9 | 2.8 | 2.1 | 6.2 |
| Dual axis tracker | | | | | | | | | | | | | |
| | 2.7 | 4.0 | 5.6 | 7.4 | 9.0 | 10.1 | 11.3 | 10.1 | 8.6 | 5.9 | 2.8 | 2.1 | 6.7 |

243

## NORTH BEND     OR     Latitude: 43.42 degrees     Elevation: 5 meters

| | Jan | Feb | Mar | Apr | May | Jun | Jul | Aug | Sep | Oct | Nov | Dec | Avg |
|---|---|---|---|---|---|---|---|---|---|---|---|---|---|
| Fixed array | | | | | | | | | | | | | |
| Lat - 15 | 2.4 | 3.0 | 4.1 | 5.1 | 5.8 | 6.1 | 6.5 | 6.0 | 5.4 | 4.0 | 2.6 | 2.1 | 4.4 |
| Latitude | 2.6 | 3.2 | 4.2 | 5.0 | 5.5 | 5.6 | 6.1 | 5.8 | 5.5 | 4.3 | 2.8 | 2.4 | 4.4 |
| Lat + 15 | 2.7 | 3.3 | 4.1 | 4.6 | 4.8 | 4.8 | 5.3 | 5.3 | 5.2 | 4.3 | 2.9 | 2.5 | 4.2 |
| Single axis tracker | | | | | | | | | | | | | |
| Lat - 15 | 2.7 | 3.5 | 5.0 | 6.4 | 7.4 | 7.8 | 8.5 | 7.7 | 6.9 | 4.9 | 3.0 | 2.4 | 5.5 |
| Latitude | 3.0 | 3.7 | 5.0 | 6.3 | 7.2 | 7.5 | 8.2 | 7.6 | 7.0 | 5.1 | 3.2 | 2.7 | 5.5 |
| Lat + 15 | 3.1 | 3.7 | 4.9 | 6.1 | 6.7 | 6.9 | 7.7 | 7.2 | 6.8 | 5.2 | 3.3 | 2.8 | 5.4 |
| Dual axis tracker | | | | | | | | | | | | | |
| | 3.1 | 3.8 | 5.1 | 6.4 | 7.5 | 8.0 | 8.7 | 7.8 | 7.0 | 5.2 | 3.3 | 2.8 | 5.7 |

## PENDLETON     OR     Latitude: 45.68 degrees     Elevation: 456 meters

| | Jan | Feb | Mar | Apr | May | Jun | Jul | Aug | Sep | Oct | Nov | Dec | Avg |
|---|---|---|---|---|---|---|---|---|---|---|---|---|---|
| Fixed array | | | | | | | | | | | | | |
| Lat - 15 | 2.0 | 3.0 | 4.4 | 5.5 | 6.3 | 6.8 | 7.4 | 6.9 | 6.1 | 4.4 | 2.3 | 1.7 | 4.7 |
| Latitude | 2.2 | 3.2 | 4.5 | 5.4 | 5.9 | 6.2 | 6.8 | 6.7 | 6.2 | 4.7 | 2.5 | 1.9 | 4.7 |
| Lat + 15 | 2.2 | 3.2 | 4.3 | 5.0 | 5.2 | 5.3 | 5.9 | 6.0 | 5.9 | 4.8 | 2.6 | 2.0 | 4.4 |
| Single axis tracker | | | | | | | | | | | | | |
| Lat - 15 | 2.2 | 3.5 | 5.5 | 7.2 | 8.6 | 9.5 | 10.7 | 9.8 | 8.2 | 5.6 | 2.7 | 1.9 | 6.3 |
| Latitude | 2.4 | 3.7 | 5.6 | 7.1 | 8.4 | 9.1 | 10.4 | 9.6 | 8.3 | 5.9 | 2.8 | 2.1 | 6.3 |
| Lat + 15 | 2.4 | 3.7 | 5.5 | 6.9 | 7.9 | 8.6 | 9.8 | 9.2 | 8.1 | 5.9 | 2.9 | 2.2 | 6.1 |
| Dual axis tracker | | | | | | | | | | | | | |
| | 2.5 | 3.7 | 5.6 | 7.3 | 8.8 | 9.7 | 10.9 | 9.8 | 8.3 | 5.9 | 2.9 | 2.2 | 6.5 |

## PORTLAND     OR     Latitude: 45.60 degrees     Elevation: 12 meters

| | Jan | Feb | Mar | Apr | May | Jun | Jul | Aug | Sep | Oct | Nov | Dec | Avg |
|---|---|---|---|---|---|---|---|---|---|---|---|---|---|
| Fixed array | | | | | | | | | | | | | |
| Lat - 15 | 1.7 | 2.5 | 3.6 | 4.6 | 5.4 | 5.8 | 6.3 | 5.9 | 5.1 | 3.4 | 1.9 | 1.4 | 4.0 |
| Latitude | 1.9 | 2.6 | 3.7 | 4.5 | 5.0 | 5.3 | 5.8 | 5.6 | 5.1 | 3.6 | 2.1 | 1.6 | 3.9 |
| Lat + 15 | 1.9 | 2.7 | 3.5 | 4.1 | 4.4 | 4.6 | 5.1 | 5.1 | 4.9 | 3.6 | 2.1 | 1.6 | 3.6 |
| Single axis tracker | | | | | | | | | | | | | |
| Lat - 15 | 1.9 | 2.9 | 4.3 | 5.6 | 6.8 | 7.5 | 8.4 | 7.7 | 6.4 | 4.1 | 2.1 | 1.6 | 4.9 |
| Latitude | 2.0 | 3.0 | 4.3 | 5.5 | 6.5 | 7.2 | 8.1 | 7.5 | 6.5 | 4.3 | 2.2 | 1.7 | 4.9 |
| Lat + 15 | 2.1 | 3 | 4.2 | 5.3 | 6.1 | 6.7 | 7.6 | 7.2 | 6.3 | 4.2 | 2.3 | 1.7 | 4.7 |
| Dual axis tracker | | | | | | | | | | | | | |
| | 2.1 | 3.0 | 4.4 | 5.6 | 6.9 | 7.7 | 8.6 | 7.7 | 6.5 | 4.3 | 2.3 | 1.8 | 5.1 |

## REDMOND     OR     Latitude: 44.27 degrees     Elevation: 940 meters

| | Jan | Feb | Mar | Apr | May | Jun | Jul | Aug | Sep | Oct | Nov | Dec | Avg |
|---|---|---|---|---|---|---|---|---|---|---|---|---|---|
| Fixed array | | | | | | | | | | | | | |
| Lat - 15 | 2.6 | 3.5 | 4.8 | 5.9 | 6.6 | 7.0 | 7.6 | 7.1 | 6.3 | 4.7 | 2.9 | 2.4 | 5.1 |
| Latitude | 3.0 | 3.8 | 4.9 | 5.7 | 6.2 | 6.4 | 7.0 | 6.9 | 6.4 | 5.1 | 3.2 | 2.7 | 5.1 |
| Lat + 15 | 3.1 | 3.9 | 4.8 | 5.3 | 5.5 | 5.5 | 6.1 | 6.2 | 6.1 | 5.1 | 3.3 | 2.9 | 4.8 |
| Single axis tracker | | | | | | | | | | | | | |
| Lat - 15 | 3.1 | 4.2 | 6.0 | 7.7 | 9.1 | 9.9 | 11.0 | 10.1 | 8.5 | 6.1 | 3.4 | 2.7 | 6.8 |
| Latitude | 3.4 | 4.4 | 6.2 | 7.7 | 8.8 | 9.5 | 10.6 | 9.9 | 8.6 | 6.3 | 3.6 | 3.0 | 6.9 |
| Lat + 15 | 3.5 | 4.5 | 6.1 | 7.4 | 8.3 | 8.9 | 10 | 9.5 | 8.4 | 6.4 | 3.7 | 3.2 | 6.7 |
| Dual axis tracker | | | | | | | | | | | | | |
| | 3.6 | 4.5 | 6.2 | 7.8 | 9.2 | 10.2 | 11.2 | 10.1 | 8.6 | 6.4 | 3.8 | 3.2 | 7.1 |

## SALEM     OR     Latitude: 44.92 degrees     Elevation: 61 meters

| | Jan | Feb | Mar | Apr | May | Jun | Jul | Aug | Sep | Oct | Nov | Dec | Avg |
|---|---|---|---|---|---|---|---|---|---|---|---|---|---|
| Fixed array | | | | | | | | | | | | | |
| Lat - 15 | 1.8 | 2.7 | 3.8 | 4.8 | 5.6 | 6.0 | 6.6 | 6.2 | 5.3 | 3.6 | 2.0 | 1.5 | 4.2 |
| Latitude | 2.0 | 2.8 | 3.9 | 4.7 | 5.2 | 5.5 | 6.1 | 6.0 | 5.4 | 3.8 | 2.2 | 1.6 | 4.1 |
| Lat + 15 | 2.1 | 2.8 | 3.7 | 4.3 | 4.6 | 4.7 | 5.3 | 5.4 | 5.1 | 3.8 | 2.2 | 1.7 | 3.8 |
| Single axis tracker | | | | | | | | | | | | | |
| Lat - 15 | 2.0 | 3.0 | 4.6 | 6.0 | 7.2 | 7.9 | 9.1 | 8.3 | 6.9 | 4.4 | 2.2 | 1.6 | 5.3 |
| Latitude | 2.2 | 3.2 | 4.6 | 5.9 | 7.0 | 7.6 | 8.7 | 8.2 | 7.0 | 4.5 | 2.4 | 1.8 | 5.3 |
| Lat + 15 | 2.3 | 3.2 | 4.5 | 5.6 | 6.6 | 7.1 | 8.2 | 7.8 | 6.8 | 4.5 | 2.4 | 1.8 | 5.1 |
| Dual axis tracker | | | | | | | | | | | | | |
| | 2.3 | 3.2 | 4.7 | 6.0 | 7.3 | 8.1 | 9.2 | 8.4 | 7.0 | 4.6 | 2.4 | 1.8 | 5.4 |

## ALLENTOWN     PA     Latitude: 40.65 degrees     Elevation: 117 meters

| | Jan | Feb | Mar | Apr | May | Jun | Jul | Aug | Sep | Oct | Nov | Dec | Avg |
|---|---|---|---|---|---|---|---|---|---|---|---|---|---|
| Fixed array | | | | | | | | | | | | | |
| Lat - 15 | 2.8 | 3.6 | 4.4 | 5.1 | 5.5 | 5.8 | 5.8 | 5.5 | 4.8 | 4.0 | 2.7 | 2.3 | 4.4 |
| Latitude | 3.1 | 3.9 | 4.5 | 5.0 | 5.2 | 5.4 | 5.4 | 5.3 | 4.9 | 4.2 | 3.0 | 2.6 | 4.4 |
| Lat + 15 | 3.3 | 4.0 | 4.4 | 4.6 | 4.6 | 4.7 | 4.8 | 4.8 | 4.6 | 4.2 | 3.1 | 2.8 | 4.2 |
| Single axis tracker | | | | | | | | | | | | | |
| Lat - 15 | 3.3 | 4.4 | 5.4 | 6.4 | 6.9 | 7.4 | 7.4 | 6.9 | 6.0 | 4.8 | 3.2 | 2.7 | 5.4 |
| Latitude | 3.6 | 4.6 | 5.5 | 6.3 | 6.7 | 7.1 | 7.1 | 6.8 | 6.0 | 5.0 | 3.4 | 2.9 | 5.4 |
| Lat + 15 | 3.8 | 4.7 | 5.4 | 6 | 6.2 | 6.6 | 6.6 | 6.5 | 5.9 | 5.1 | 3.5 | 3.1 | 5.3 |
| Dual axis tracker | | | | | | | | | | | | | |
| | 3.8 | 4.7 | 5.6 | 6.4 | 7.0 | 7.5 | 7.5 | 7.0 | 6.0 | 5.1 | 3.6 | 3.1 | 5.6 |

## BRADFORD     PA     Latitude: 41.80 degrees     Elevation: 600 meters

| | Jan | Feb | Mar | Apr | May | Jun | Jul | Aug | Sep | Oct | Nov | Dec | Avg |
|---|---|---|---|---|---|---|---|---|---|---|---|---|---|
| Fixed array | | | | | | | | | | | | | |
| Lat - 15 | 2.4 | 3.4 | 4.3 | 5.0 | 5.4 | 5.8 | 5.8 | 5.3 | 4.5 | 3.6 | 2.3 | 1.9 | 4.2 |
| Latitude | 2.7 | 3.7 | 4.4 | 4.8 | 5.1 | 5.3 | 5.4 | 5.1 | 4.5 | 3.8 | 2.4 | 2.1 | 4.1 |
| Lat + 15 | 2.8 | 3.8 | 4.4 | 4.5 | 4.5 | 4.6 | 4.7 | 4.6 | 4.3 | 3.8 | 2.5 | 2.2 | 3.9 |
| Single axis tracker | | | | | | | | | | | | | |
| Lat - 15 | 2.8 | 4.1 | 5.2 | 6.2 | 6.9 | 7.5 | 7.5 | 6.8 | 5.6 | 4.3 | 2.6 | 2.1 | 5.1 |
| Latitude | 3.0 | 4.3 | 5.4 | 6.1 | 6.7 | 7.2 | 7.2 | 6.6 | 5.6 | 4.5 | 2.7 | 2.3 | 5.1 |
| Lat + 15 | 3.1 | 4.4 | 5.3 | 5.9 | 6.3 | 6.7 | 6.8 | 6.3 | 5.5 | 4.4 | 2.8 | 2.4 | 5 |
| Dual axis tracker | | | | | | | | | | | | | |
| | 3.2 | 4.4 | 5.4 | 6.2 | 7.0 | 7.7 | 7.6 | 6.8 | 5.7 | 4.5 | 2.8 | 2.4 | 5.3 |

## ERIE     PA     Latitude: 42.08 degrees     Elevation: 225 meters

| | Jan | Feb | Mar | Apr | May | Jun | Jul | Aug | Sep | Oct | Nov | Dec | Avg |
|---|---|---|---|---|---|---|---|---|---|---|---|---|---|
| Fixed array | | | | | | | | | | | | | |
| Lat - 15 | 2.1 | 3.1 | 4.0 | 5.0 | 5.7 | 6.1 | 6.2 | 5.6 | 4.7 | 3.5 | 2.1 | 1.6 | 4.2 |
| Latitude | 2.3 | 3.3 | 4.1 | 4.9 | 5.4 | 5.6 | 5.8 | 5.4 | 4.7 | 3.7 | 2.2 | 1.8 | 4.1 |
| Lat + 15 | 2.4 | 3.4 | 4.0 | 4.5 | 4.8 | 4.9 | 5.0 | 4.9 | 4.5 | 3.7 | 2.3 | 1.8 | 3.9 |
| Single axis tracker | | | | | | | | | | | | | |
| Lat - 15 | 2.3 | 3.6 | 4.9 | 6.3 | 7.4 | 8.0 | 8.1 | 7.2 | 5.9 | 4.2 | 2.3 | 1.8 | 5.2 |
| Latitude | 2.5 | 3.8 | 4.9 | 6.2 | 7.1 | 7.7 | 7.8 | 7.1 | 5.9 | 4.3 | 2.5 | 1.9 | 5.2 |
| Lat + 15 | 2.6 | 3.8 | 4.9 | 6 | 6.7 | 7.2 | 7.3 | 6.7 | 5.8 | 4.3 | 2.5 | 1.9 | 5 |
| Dual axis tracker | | | | | | | | | | | | | |
| | 2.6 | 3.8 | 5.0 | 6.3 | 7.5 | 8.2 | 8.3 | 7.3 | 6.0 | 4.4 | 2.5 | 1.9 | 5.3 |

## HARRISBURG    PA    Latitude: 40.22 degrees    Elevation: 106 meters

|  | Jan | Feb | Mar | Apr | May | Jun | Jul | Aug | Sep | Oct | Nov | Dec | Avg |
|---|---|---|---|---|---|---|---|---|---|---|---|---|---|
| Fixed array |  |  |  |  |  |  |  |  |  |  |  |  |  |
| Lat - 15 | 2.9 | 3.7 | 4.5 | 5.2 | 5.6 | 6.0 | 5.9 | 5.5 | 4.9 | 4.1 | 2.9 | 2.4 | 4.5 |
| Latitude | 3.2 | 4.0 | 4.6 | 5.1 | 5.3 | 5.5 | 5.5 | 5.3 | 4.9 | 4.3 | 3.2 | 2.7 | 4.5 |
| Lat + 15 | 3.5 | 4.1 | 4.5 | 4.7 | 4.7 | 4.8 | 4.8 | 4.8 | 4.7 | 4.4 | 3.3 | 2.9 | 4.3 |
| Single axis tracker |  |  |  |  |  |  |  |  |  |  |  |  |  |
| Lat - 15 | 3.4 | 4.5 | 5.6 | 6.6 | 7.1 | 7.6 | 7.5 | 7.0 | 6.1 | 5.0 | 3.4 | 2.8 | 5.6 |
| Latitude | 3.7 | 4.8 | 5.7 | 6.5 | 6.8 | 7.3 | 7.2 | 6.9 | 6.2 | 5.2 | 3.6 | 3.1 | 5.6 |
| Lat + 15 | 3.9 | 4.9 | 5.6 | 6.4 | 6.2 | 6.8 | 6.8 | 6.5 | 6 | 5.2 | 3.7 | 3.2 | 5.4 |
| Dual axis tracker |  |  |  |  |  |  |  |  |  |  |  |  |  |
|  | 3.9 | 4.9 | 5.7 | 6.6 | 7.1 | 7.8 | 7.6 | 7.0 | 6.2 | 5.3 | 3.8 | 3.3 | 5.8 |

## PHILADELPHIA    PA    Latitude: 39.88 degrees    Elevation: 9 meters

|  | Jan | Feb | Mar | Apr | May | Jun | Jul | Aug | Sep | Oct | Nov | Dec | Avg |
|---|---|---|---|---|---|---|---|---|---|---|---|---|---|
| Fixed array |  |  |  |  |  |  |  |  |  |  |  |  |  |
| Lat - 15 | 2.9 | 3.7 | 4.5 | 5.2 | 5.6 | 6.0 | 5.9 | 5.7 | 5.0 | 4.2 | 3.0 | 2.6 | 4.5 |
| Latitude | 3.3 | 4.0 | 4.7 | 5.1 | 5.3 | 5.5 | 5.5 | 5.5 | 5.1 | 4.4 | 3.4 | 2.9 | 4.6 |
| Lat + 15 | 3.5 | 4.1 | 4.6 | 4.7 | 4.7 | 4.8 | 4.8 | 5.0 | 4.8 | 4.5 | 3.5 | 3.1 | 4.3 |
| Single axis tracker |  |  |  |  |  |  |  |  |  |  |  |  |  |
| Lat - 15 | 3.5 | 4.5 | 5.7 | 6.6 | 7.1 | 7.7 | 7.5 | 7.2 | 6.3 | 5.2 | 3.6 | 3.0 | 5.7 |
| Latitude | 3.8 | 4.8 | 5.8 | 6.5 | 6.8 | 7.3 | 7.3 | 7.1 | 6.3 | 5.4 | 3.9 | 3.3 | 5.7 |
| Lat + 15 | 4 | 4.9 | 5.7 | 6.4 | 6.2 | 6.8 | 6.8 | 6.7 | 6.2 | 5.4 | 4.1 | 3.5 | 5.6 |
| Dual axis tracker |  |  |  |  |  |  |  |  |  |  |  |  |  |
|  | 4.0 | 4.9 | 5.8 | 6.6 | 7.2 | 7.8 | 7.7 | 7.3 | 6.4 | 5.4 | 4.1 | 3.5 | 5.9 |

## PITTSBURGH    PA    Latitude: 40.50 degrees    Elevation: 373 meters

|  | Jan | Feb | Mar | Apr | May | Jun | Jul | Aug | Sep | Oct | Nov | Dec | Avg |
|---|---|---|---|---|---|---|---|---|---|---|---|---|---|
| Fixed array |  |  |  |  |  |  |  |  |  |  |  |  |  |
| Lat - 15 | 2.4 | 3.2 | 4.1 | 4.9 | 5.5 | 5.9 | 5.9 | 5.5 | 4.8 | 3.8 | 2.4 | 1.9 | 4.2 |
| Latitude | 2.6 | 3.4 | 4.2 | 4.8 | 5.2 | 5.4 | 5.5 | 5.3 | 4.8 | 4.1 | 2.6 | 2.1 | 4.2 |
| Lat + 15 | 2.7 | 3.5 | 4.1 | 4.4 | 4.6 | 4.7 | 4.8 | 4.8 | 4.6 | 4.1 | 2.7 | 2.2 | 3.9 |
| Single axis tracker |  |  |  |  |  |  |  |  |  |  |  |  |  |
| Lat - 15 | 2.7 | 3.7 | 4.9 | 6.0 | 6.9 | 7.5 | 7.4 | 6.9 | 5.9 | 4.7 | 2.8 | 2.1 | 5.1 |
| Latitude | 2.9 | 3.9 | 5.0 | 6.0 | 6.6 | 7.2 | 7.2 | 6.8 | 6.0 | 4.8 | 2.9 | 2.3 | 5.1 |
| Lat + 15 | 3 | 4 | 4.9 | 5.7 | 6.2 | 6.7 | 6.7 | 6.4 | 5.8 | 4.8 | 3 | 2.3 | 5 |
| Dual axis tracker |  |  |  |  |  |  |  |  |  |  |  |  |  |
|  | 3.0 | 4.0 | 5.0 | 6.1 | 7.0 | 7.7 | 7.6 | 6.9 | 6.0 | 4.9 | 3.0 | 2.4 | 5.3 |

## WILKES-BARRE    PA    Latitude: 41.33 degrees    Elevation: 289 meters

|  | Jan | Feb | Mar | Apr | May | Jun | Jul | Aug | Sep | Oct | Nov | Dec | Avg |
|---|---|---|---|---|---|---|---|---|---|---|---|---|---|
| Fixed array |  |  |  |  |  |  |  |  |  |  |  |  |  |
| Lat - 15 | 2.5 | 3.4 | 4.2 | 5.0 | 5.4 | 5.8 | 5.8 | 5.5 | 4.7 | 3.8 | 2.4 | 2.0 | 4.2 |
| Latitude | 2.8 | 3.6 | 4.3 | 4.8 | 5.1 | 5.4 | 5.4 | 5.3 | 4.7 | 4.0 | 2.6 | 2.2 | 4.2 |
| Lat + 15 | 3.0 | 3.7 | 4.2 | 4.5 | 4.5 | 4.7 | 4.8 | 4.8 | 4.5 | 4.0 | 2.7 | 2.4 | 4.0 |
| Single axis tracker |  |  |  |  |  |  |  |  |  |  |  |  |  |
| Lat - 15 | 2.9 | 4.0 | 5.1 | 6.2 | 6.8 | 7.4 | 7.4 | 6.9 | 5.8 | 4.6 | 2.7 | 2.3 | 5.2 |
| Latitude | 3.2 | 4.2 | 5.2 | 6.1 | 6.6 | 7.1 | 7.1 | 6.8 | 5.8 | 4.8 | 2.9 | 2.5 | 5.2 |
| Lat + 15 | 3.3 | 4.3 | 5.1 | 5.8 | 6.2 | 6.6 | 6.7 | 6.4 | 5.6 | 4.8 | 3 | 2.6 | 5 |
| Dual axis tracker |  |  |  |  |  |  |  |  |  |  |  |  |  |
|  | 3.3 | 4.3 | 5.2 | 6.2 | 6.9 | 7.6 | 7.6 | 7.0 | 5.8 | 4.8 | 3.0 | 2.6 | 5.4 |

## WILLIAMSPORT    PA      Latitude: 41.27 degrees      Elevation: 243 meters

| | Jan | Feb | Mar | Apr | May | Jun | Jul | Aug | Sep | Oct | Nov | Dec | Avg |
|---|---|---|---|---|---|---|---|---|---|---|---|---|---|
| Fixed array | | | | | | | | | | | | | |
| Lat - 15 | 2.6 | 3.4 | 4.3 | 5.0 | 5.5 | 5.9 | 5.9 | 5.4 | 4.6 | 3.7 | 2.4 | 2.1 | 4.2 |
| Latitude | 2.9 | 3.7 | 4.4 | 4.9 | 5.1 | 5.4 | 5.5 | 5.2 | 4.6 | 3.9 | 2.6 | 2.3 | 4.2 |
| Lat + 15 | 3.0 | 3.8 | 4.3 | 4.5 | 4.6 | 4.7 | 4.8 | 4.8 | 4.4 | 3.9 | 2.7 | 2.4 | 4.0 |
| Single axis tracker | | | | | | | | | | | | | |
| Lat - 15 | 3.0 | 4.1 | 5.2 | 6.2 | 6.9 | 7.5 | 7.4 | 6.7 | 5.5 | 4.4 | 2.7 | 2.3 | 5.2 |
| Latitude | 3.2 | 4.3 | 5.3 | 6.1 | 6.6 | 7.2 | 7.1 | 6.6 | 5.6 | 4.5 | 2.9 | 2.5 | 5.2 |
| Lat + 15 | 3.4 | 4.4 | 5.2 | 5.9 | 6.2 | 6.7 | 6.7 | 6.2 | 5.4 | 4.5 | 3 | 2.6 | 5 |
| Dual axis tracker | | | | | | | | | | | | | |
| | 3.4 | 4.4 | 5.3 | 6.2 | 6.9 | 7.6 | 7.5 | 6.8 | 5.6 | 4.6 | 3.0 | 2.6 | 5.3 |

## GUAM    PI      Latitude: 13.55 degrees      Elevation: 110 meters

| | Jan | Feb | Mar | Apr | May | Jun | Jul | Aug | Sep | Oct | Nov | Dec | Avg |
|---|---|---|---|---|---|---|---|---|---|---|---|---|---|
| Fixed array | | | | | | | | | | | | | |
| Lat - 15 | 4.3 | 4.8 | 5.4 | 5.8 | 5.7 | 5.5 | 5.1 | 4.9 | 4.8 | 4.6 | 4.3 | 4.1 | 4.9 |
| Latitude | 5.0 | 5.2 | 5.7 | 5.7 | 5.4 | 5.1 | 4.8 | 4.7 | 4.9 | 4.9 | 4.9 | 4.8 | 5.1 |
| Lat + 15 | 5.3 | 5.4 | 5.6 | 5.3 | 4.9 | 4.5 | 4.3 | 4.4 | 4.7 | 4.9 | 5.1 | 5.1 | 5.0 |
| Single axis tracker | | | | | | | | | | | | | |
| Lat - 15 | 5.6 | 6.1 | 6.9 | 7.4 | 7.3 | 6.9 | 6.2 | 5.8 | 5.9 | 5.7 | 5.6 | 5.3 | 6.2 |
| Latitude | 6.1 | 6.4 | 7.1 | 7.3 | 7.0 | 6.6 | 6.0 | 5.7 | 6.0 | 5.9 | 5.9 | 5.8 | 6.3 |
| Lat + 15 | 6.3 | 6.6 | 7 | 7.1 | 6.6 | 6.1 | 5.6 | 5.5 | 5.8 | 5.9 | 6.1 | 6.1 | 6.2 |
| Dual axis tracker | | | | | | | | | | | | | |
| | 6.4 | 6.6 | 7.1 | 7.4 | 7.3 | 7.0 | 6.3 | 5.9 | 6.0 | 6.0 | 6.2 | 6.2 | 6.5 |

## SAN JUAN    PR      Latitude: 18.43 degrees      Elevation: 19 meters

| | Jan | Feb | Mar | Apr | May | Jun | Jul | Aug | Sep | Oct | Nov | Dec | Avg |
|---|---|---|---|---|---|---|---|---|---|---|---|---|---|
| Fixed array | | | | | | | | | | | | | |
| Lat - 15 | 4.5 | 5.1 | 5.8 | 6.1 | 5.7 | 6.0 | 6.0 | 6.0 | 5.6 | 5.0 | 4.5 | 4.1 | 5.4 |
| Latitude | 5.1 | 5.6 | 6.1 | 6.1 | 5.4 | 5.5 | 5.6 | 5.8 | 5.7 | 5.4 | 5.1 | 4.8 | 5.5 |
| Lat + 15 | 5.5 | 5.8 | 6.0 | 5.7 | 4.9 | 4.8 | 5.0 | 5.3 | 5.5 | 5.5 | 5.4 | 5.2 | 5.4 |
| Single axis tracker | | | | | | | | | | | | | |
| Lat - 15 | 5.8 | 6.5 | 7.5 | 7.8 | 7.2 | 7.5 | 7.6 | 7.5 | 7.0 | 6.4 | 5.7 | 5.2 | 6.8 |
| Latitude | 6.3 | 6.9 | 7.7 | 7.8 | 6.9 | 7.2 | 7.3 | 7.4 | 7.1 | 6.7 | 6.1 | 5.7 | 6.9 |
| Lat + 15 | 6.6 | 7.1 | 7.6 | 7.5 | 6.5 | 6.7 | 6.8 | 7 | 7 | 6.7 | 6.3 | 6 | 6.8 |
| Dual axis tracker | | | | | | | | | | | | | |
| | 6.6 | 7.1 | 7.7 | 7.9 | 7.2 | 7.7 | 7.7 | 7.6 | 7.1 | 6.7 | 6.4 | 6.1 | 7.2 |

## PROVIDENCE    RI      Latitude: 41.73 degrees      Elevation: 19 meters

| | Jan | Feb | Mar | Apr | May | Jun | Jul | Aug | Sep | Oct | Nov | Dec | Avg |
|---|---|---|---|---|---|---|---|---|---|---|---|---|---|
| Fixed array | | | | | | | | | | | | | |
| Lat - 15 | 3.0 | 3.7 | 4.5 | 5.1 | 5.6 | 5.9 | 5.9 | 5.5 | 4.9 | 4.1 | 2.9 | 2.5 | 4.5 |
| Latitude | 3.4 | 4.1 | 4.7 | 5.0 | 5.3 | 5.4 | 5.5 | 5.3 | 5.0 | 4.4 | 3.2 | 2.9 | 4.5 |
| Lat + 15 | 3.6 | 4.2 | 4.6 | 4.6 | 4.7 | 4.7 | 4.8 | 4.8 | 4.7 | 4.4 | 3.3 | 3.1 | 4.3 |
| Single axis tracker | | | | | | | | | | | | | |
| Lat - 15 | 3.5 | 4.6 | 5.6 | 6.4 | 7.1 | 7.5 | 7.5 | 7.0 | 6.2 | 5.0 | 3.4 | 2.9 | 5.6 |
| Latitude | 3.9 | 4.9 | 5.8 | 6.4 | 6.9 | 7.2 | 7.2 | 6.9 | 6.2 | 5.3 | 3.6 | 3.2 | 5.6 |
| Lat + 15 | 4.1 | 5 | 5.7 | 6.1 | 6.5 | 6.7 | 6.7 | 6.5 | 6 | 5.3 | 3.7 | 3.4 | 5.5 |
| Dual axis tracker | | | | | | | | | | | | | |
| | 4.1 | 5.0 | 5.8 | 6.5 | 7.2 | 7.7 | 7.6 | 7.0 | 6.2 | 5.3 | 3.8 | 3.5 | 5.8 |

## CHARLESTON  SC  Latitude: 32.90 degrees  Elevation: 12 meters

| | Jan | Feb | Mar | Apr | May | Jun | Jul | Aug | Sep | Oct | Nov | Dec | Avg |
|---|---|---|---|---|---|---|---|---|---|---|---|---|---|
| Fixed array | | | | | | | | | | | | | |
| Lat - 15 | 3.5 | 4.3 | 5.3 | 6.2 | 6.2 | 6.1 | 6.0 | 5.6 | 5.1 | 4.8 | 3.9 | 3.4 | 5.0 |
| Latitude | 4.0 | 4.7 | 5.5 | 6.1 | 5.8 | 5.6 | 5.6 | 5.4 | 5.2 | 5.2 | 4.5 | 3.9 | 5.1 |
| Lat + 15 | 4.3 | 4.9 | 5.4 | 5.7 | 5.2 | 4.9 | 4.9 | 4.9 | 5.0 | 5.2 | 4.7 | 4.2 | 4.9 |
| Single axis tracker | | | | | | | | | | | | | |
| Lat - 15 | 4.4 | 5.4 | 6.7 | 8.1 | 7.9 | 7.6 | 7.5 | 6.9 | 6.3 | 6.1 | 4.9 | 4.2 | 6.3 |
| Latitude | 4.8 | 5.8 | 6.9 | 8.0 | 7.7 | 7.3 | 7.2 | 6.8 | 6.4 | 6.3 | 5.3 | 4.6 | 6.4 |
| Lat + 15 | 5 | 5.9 | 6.8 | 7.7 | 7.2 | 6.8 | 6.7 | 6.4 | 6.2 | 6.4 | 5.5 | 4.8 | 6.3 |
| Dual axis tracker | | | | | | | | | | | | | |
| | 5.0 | 5.9 | 6.9 | 8.1 | 8.0 | 7.8 | 7.6 | 7.0 | 6.4 | 6.4 | 5.5 | 4.9 | 6.6 |

## COLUMBIA  SC  Latitude: 33.95 degrees  Elevation: 69 meters

| | Jan | Feb | Mar | Apr | May | Jun | Jul | Aug | Sep | Oct | Nov | Dec | Avg |
|---|---|---|---|---|---|---|---|---|---|---|---|---|---|
| Fixed array | | | | | | | | | | | | | |
| Lat - 15 | 3.4 | 4.2 | 5.1 | 6.0 | 6.1 | 6.1 | 6.0 | 5.7 | 5.2 | 4.8 | 3.8 | 3.3 | 5.0 |
| Latitude | 3.9 | 4.6 | 5.3 | 5.9 | 5.7 | 5.7 | 5.6 | 5.5 | 5.3 | 5.2 | 4.3 | 3.8 | 5.1 |
| Lat + 15 | 4.1 | 4.8 | 5.2 | 5.5 | 5.1 | 4.9 | 4.9 | 5.0 | 5.1 | 5.2 | 4.6 | 4.1 | 4.9 |
| Single axis tracker | | | | | | | | | | | | | |
| Lat - 15 | 4.2 | 5.3 | 6.5 | 7.8 | 7.8 | 7.8 | 7.5 | 7.1 | 6.5 | 6.1 | 4.8 | 4.0 | 6.3 |
| Latitude | 4.6 | 5.6 | 6.6 | 7.8 | 7.5 | 7.4 | 7.2 | 6.9 | 6.6 | 6.4 | 5.2 | 4.4 | 6.4 |
| Lat + 15 | 4.8 | 5.7 | 6.5 | 7.5 | 7.1 | 6.9 | 6.7 | 6.6 | 6.4 | 6.4 | 5.3 | 4.7 | 6.2 |
| Dual axis tracker | | | | | | | | | | | | | |
| | 4.8 | 5.8 | 6.6 | 7.8 | 7.8 | 7.9 | 7.6 | 7.1 | 6.6 | 6.4 | 5.4 | 4.7 | 6.6 |

## GREENVILLE  SC  Latitude: 34.90 degrees  Elevation: 296 meters

| | Jan | Feb | Mar | Apr | May | Jun | Jul | Aug | Sep | Oct | Nov | Dec | Avg |
|---|---|---|---|---|---|---|---|---|---|---|---|---|---|
| Fixed array | | | | | | | | | | | | | |
| Lat - 15 | 3.5 | 4.2 | 5.1 | 5.9 | 6.0 | 6.1 | 5.9 | 5.7 | 5.2 | 4.8 | 3.8 | 3.2 | 5.0 |
| Latitude | 4.0 | 4.6 | 5.3 | 5.8 | 5.6 | 5.6 | 5.5 | 5.5 | 5.2 | 5.2 | 4.3 | 3.7 | 5.0 |
| Lat + 15 | 4.2 | 4.8 | 5.2 | 5.4 | 5.0 | 4.9 | 4.8 | 5.1 | 5.0 | 5.2 | 4.5 | 4.0 | 4.8 |
| Single axis tracker | | | | | | | | | | | | | |
| Lat - 15 | 4.3 | 5.4 | 6.5 | 7.7 | 7.7 | 7.8 | 7.5 | 7.3 | 6.5 | 6.1 | 4.7 | 3.9 | 6.3 |
| Latitude | 4.7 | 5.7 | 6.7 | 7.7 | 7.4 | 7.5 | 7.2 | 7.1 | 6.6 | 6.4 | 5.1 | 4.3 | 6.4 |
| Lat + 15 | 5 | 5.8 | 6.6 | 7.4 | 7 | 7 | 6.7 | 6.8 | 6.4 | 6.5 | 5.3 | 4.6 | 6.2 |
| Dual axis tracker | | | | | | | | | | | | | |
| | 5.0 | 5.8 | 6.7 | 7.8 | 7.8 | 8.0 | 7.6 | 7.3 | 6.6 | 6.5 | 5.4 | 4.6 | 6.6 |

## HURON  SD  Latitude: 44.38 degrees  Elevation: 393 meters

| | Jan | Feb | Mar | Apr | May | Jun | Jul | Aug | Sep | Oct | Nov | Dec | Avg |
|---|---|---|---|---|---|---|---|---|---|---|---|---|---|
| Fixed array | | | | | | | | | | | | | |
| Lat - 15 | 3.0 | 3.8 | 4.7 | 5.4 | 6.0 | 6.4 | 6.7 | 6.3 | 5.3 | 4.3 | 3.0 | 2.6 | 4.8 |
| Latitude | 3.4 | 4.2 | 4.8 | 5.3 | 5.6 | 5.9 | 6.2 | 6.1 | 5.4 | 4.6 | 3.3 | 3.0 | 4.8 |
| Lat + 15 | 3.7 | 4.3 | 4.8 | 4.9 | 4.9 | 5.1 | 5.4 | 5.5 | 5.2 | 4.6 | 3.5 | 3.2 | 4.6 |
| Single axis tracker | | | | | | | | | | | | | |
| Lat - 15 | 3.6 | 4.7 | 5.9 | 7.0 | 8.0 | 8.8 | 9.2 | 8.6 | 7.0 | 5.4 | 3.6 | 3.0 | 6.2 |
| Latitude | 4.0 | 5.0 | 6.0 | 6.9 | 7.7 | 8.5 | 8.9 | 8.4 | 7.1 | 5.6 | 3.4 | 6.3 | 6.5 |
| Lat + 15 | 4.2 | 5.1 | 6 | 6.7 | 7.3 | 7.9 | 8.3 | 8 | 6.9 | 5.6 | 4 | 3.6 | 4.1 |
| Dual axis tracker | | | | | | | | | | | | | |
| | 4.2 | 5.1 | 6.0 | 7.0 | 8.1 | 9.0 | 9.4 | 8.6 | 7.1 | 5.7 | 4.0 | 3.6 | 6.5 |

## PIERRE      SD      Latitude: 44.38 degrees      Elevation: 526 meters

| | Jan | Feb | Mar | Apr | May | Jun | Jul | Aug | Sep | Oct | Nov | Dec | Avg |
|---|---|---|---|---|---|---|---|---|---|---|---|---|---|
| Fixed array | | | | | | | | | | | | | |
| Lat - 15 | 3.1 | 3.9 | 4.9 | 5.6 | 6.1 | 6.6 | 6.8 | 6.5 | 5.6 | 4.5 | 3.2 | 2.7 | 4.9 |
| Latitude | 3.6 | 4.3 | 5.1 | 5.4 | 5.7 | 6.0 | 6.3 | 6.3 | 5.6 | 4.8 | 3.5 | 3.1 | 5.0 |
| Lat + 15 | 3.8 | 4.4 | 5.0 | 5.0 | 5.1 | 5.2 | 5.5 | 5.7 | 5.4 | 4.8 | 3.7 | 3.3 | 4.7 |
| Single axis tracker | | | | | | | | | | | | | |
| Lat - 15 | 3.7 | 4.8 | 6.2 | 7.3 | 8.2 | 9.0 | 9.5 | 8.9 | 7.4 | 5.7 | 3.8 | 3.2 | 6.5 |
| Latitude | 4.1 | 5.1 | 6.4 | 7.2 | 8.0 | 8.7 | 9.1 | 8.8 | 7.5 | 5.9 | 4.1 | 3.5 | 6.5 |
| Lat + 15 | 4.3 | 5.2 | 6.3 | 7 | 7.5 | 8.1 | 8.6 | 8.4 | 7.3 | 6 | 4.2 | 3.7 | 6.4 |
| Dual axis tracker | | | | | | | | | | | | | |
| | 4.4 | 5.2 | 6.4 | 7.4 | 8.3 | 9.3 | 9.7 | 9.0 | 7.5 | 6.0 | 4.3 | 3.8 | 6.8 |

## RAPID CITY      SD      Latitude: 44.05 degrees      Elevation: 966 meters

| | Jan | Feb | Mar | Apr | May | Jun | Jul | Aug | Sep | Oct | Nov | Dec | Avg |
|---|---|---|---|---|---|---|---|---|---|---|---|---|---|
| Fixed array | | | | | | | | | | | | | |
| Lat - 15 | 3.2 | 4.1 | 5.1 | 5.7 | 6.1 | 6.6 | 6.8 | 6.6 | 5.8 | 4.7 | 3.4 | 3.0 | 5.1 |
| Latitude | 3.7 | 4.5 | 5.3 | 5.6 | 5.7 | 6.0 | 6.3 | 6.4 | 5.9 | 5.1 | 3.9 | 3.4 | 5.2 |
| Lat + 15 | 4.0 | 4.7 | 5.2 | 5.2 | 5.0 | 5.2 | 5.5 | 5.8 | 5.7 | 5.2 | 4.1 | 3.7 | 4.9 |
| Single axis tracker | | | | | | | | | | | | | |
| Lat - 15 | 3.9 | 5.2 | 6.5 | 7.6 | 8.3 | 9.2 | 9.5 | 9.2 | 7.8 | 6.1 | 4.2 | 3.6 | 6.8 |
| Latitude | 4.3 | 5.5 | 6.7 | 7.5 | 8.0 | 8.8 | 9.2 | 9.1 | 7.9 | 6.4 | 4.6 | 4.0 | 6.8 |
| Lat + 15 | 4.5 | 5.6 | 6.6 | 7.2 | 7.6 | 8.3 | 8.7 | 8.6 | 7.7 | 6.5 | 4.7 | 4.2 | 6.7 |
| Dual axis tracker | | | | | | | | | | | | | |
| | 4.6 | 5.6 | 6.7 | 7.6 | 8.4 | 9.4 | 9.7 | 9.3 | 7.9 | 6.5 | 4.8 | 4.3 | 7.1 |

## SIOUX FALLS      SD      Latitude: 43.57 degrees      Elevation: 435 meters

| | Jan | Feb | Mar | Apr | May | Jun | Jul | Aug | Sep | Oct | Nov | Dec | Avg |
|---|---|---|---|---|---|---|---|---|---|---|---|---|---|
| Fixed array | | | | | | | | | | | | | |
| Lat - 15 | 3.1 | 3.9 | 4.7 | 5.4 | 5.9 | 6.4 | 6.6 | 6.1 | 5.2 | 4.3 | 3.0 | 2.6 | 4.8 |
| Latitude | 3.6 | 4.3 | 4.9 | 5.2 | 5.5 | 5.8 | 6.1 | 5.9 | 5.3 | 4.6 | 3.3 | 3.0 | 4.8 |
| Lat + 15 | 3.8 | 4.5 | 4.8 | 4.8 | 4.9 | 5.1 | 5.3 | 5.3 | 5.1 | 4.6 | 3.5 | 3.2 | 4.6 |
| Single axis tracker | | | | | | | | | | | | | |
| Lat - 15 | 3.8 | 4.8 | 6.0 | 7.0 | 7.9 | 8.7 | 9.0 | 8.2 | 6.9 | 5.4 | 3.6 | 3.1 | 6.2 |
| Latitude | 4.2 | 5.1 | 6.1 | 6.9 | 7.6 | 8.3 | 8.7 | 8.1 | 6.9 | 5.6 | 3.9 | 3.4 | 6.2 |
| Lat + 15 | 4.4 | 5.3 | 6.1 | 6.6 | 7.2 | 7.8 | 8.2 | 7.7 | 7.6 | 5.6 | 4 | 3.6 | 6.1 |
| Dual axis tracker | | | | | | | | | | | | | |
| | 4.4 | 5.3 | 6.1 | 7.0 | 8.0 | 8.9 | 9.2 | 8.3 | 6.9 | 5.7 | 4.0 | 3.7 | 6.5 |

## BRISTOL      TN      Latitude: 36.48 degrees      Elevation: 459 meters

| | Jan | Feb | Mar | Apr | May | Jun | Jul | Aug | Sep | Oct | Nov | Dec | Avg |
|---|---|---|---|---|---|---|---|---|---|---|---|---|---|
| Fixed array | | | | | | | | | | | | | |
| Lat - 15 | 2.9 | 3.6 | 4.6 | 5.4 | 5.7 | 6.0 | 5.8 | 5.6 | 5.0 | 4.5 | 3.2 | 2.7 | 4.6 |
| Latitude | 3.3 | 3.9 | 4.7 | 5.3 | 5.4 | 5.5 | 5.4 | 5.4 | 5.1 | 4.8 | 3.6 | 3.1 | 4.6 |
| Lat + 15 | 3.5 | 4.0 | 4.6 | 4.9 | 4.8 | 4.8 | 4.7 | 5.0 | 4.9 | 4.8 | 3.7 | 3.3 | 4.4 |
| Single axis tracker | | | | | | | | | | | | | |
| Lat - 15 | 3.5 | 4.5 | 5.7 | 6.9 | 7.2 | 7.6 | 7.3 | 7.1 | 6.3 | 5.6 | 3.9 | 3.2 | 5.7 |
| Latitude | 3.8 | 4.7 | 5.8 | 6.8 | 7.0 | 7.3 | 7.0 | 6.9 | 6.3 | 5.8 | 4.2 | 3.5 | 5.8 |
| Lat + 15 | 4 | 4.8 | 5.7 | 6.5 | 6.6 | 6.8 | 6.5 | 6.6 | 6.2 | 5.9 | 4.3 | 3.7 | 5.6 |
| Dual axis tracker | | | | | | | | | | | | | |
| | 4.0 | 4.8 | 5.8 | 6.9 | 7.3 | 7.8 | 7.4 | 7.1 | 6.4 | 5.9 | 4.3 | 3.7 | 6.0 |

## CHATTANOOGA  TN  Latitude: 35.03 degrees  Elevation: 210 meters

|  | Jan | Feb | Mar | Apr | May | Jun | Jul | Aug | Sep | Oct | Nov | Dec | Avg |
|---|---|---|---|---|---|---|---|---|---|---|---|---|---|
| Fixed array | | | | | | | | | | | | | |
| Lat - 15 | 3.1 | 3.8 | 4.6 | 5.6 | 5.8 | 6.0 | 5.8 | 5.7 | 5.0 | 4.6 | 3.4 | 2.8 | 4.7 |
| Latitude | 3.5 | 4.1 | 4.8 | 5.4 | 5.4 | 5.5 | 5.4 | 5.5 | 5.0 | 4.9 | 3.8 | 3.2 | 4.7 |
| Lat + 15 | 3.7 | 4.2 | 4.6 | 5.1 | 4.9 | 4.8 | 4.7 | 5.0 | 4.8 | 4.9 | 4.0 | 3.4 | 4.5 |
| Single axis tracker | | | | | | | | | | | | | |
| Lat - 15 | 3.8 | 4.7 | 5.7 | 7.1 | 7.3 | 7.6 | 7.2 | 7.1 | 6.2 | 5.7 | 4.1 | 3.4 | 5.8 |
| Latitude | 4.1 | 4.9 | 5.9 | 7.0 | 7.1 | 7.2 | 6.9 | 6.9 | 6.2 | 5.9 | 4.4 | 3.7 | 5.9 |
| Lat + 15 | 4.2 | 5 | 5.8 | 6.7 | 6.6 | 6.7 | 6.5 | 6.6 | 6 | 6 | 4.6 | 3.9 | 5.7 |
| Dual axis tracker | | | | | | | | | | | | | |
|  | 4.3 | 5.0 | 5.9 | 7.1 | 7.4 | 7.7 | 7.3 | 7.1 | 6.2 | 6.0 | 4.6 | 3.9 | 6.0 |

## KNOXVILLE  TN  Latitude: 35.82 degrees  Elevation: 299 meters

|  | Jan | Feb | Mar | Apr | May | Jun | Jul | Aug | Sep | Oct | Nov | Dec | Avg |
|---|---|---|---|---|---|---|---|---|---|---|---|---|---|
| Fixed array | | | | | | | | | | | | | |
| Lat - 15 | 3.0 | 3.7 | 4.6 | 5.5 | 5.8 | 6.1 | 5.8 | 5.7 | 5.0 | 4.6 | 3.3 | 2.8 | 4.7 |
| Latitude | 3.4 | 4.0 | 4.7 | 5.4 | 5.5 | 5.6 | 5.4 | 5.5 | 5.1 | 4.9 | 3.7 | 3.1 | 4.7 |
| Lat + 15 | 3.6 | 4.1 | 4.6 | 5.0 | 4.9 | 4.9 | 4.8 | 5.0 | 4.8 | 4.9 | 3.8 | 3.3 | 4.5 |
| Single axis tracker | | | | | | | | | | | | | |
| Lat - 15 | 3.6 | 4.6 | 5.7 | 6.9 | 7.4 | 7.7 | 7.3 | 7.1 | 6.2 | 5.7 | 4.0 | 3.3 | 5.8 |
| Latitude | 3.9 | 4.8 | 5.8 | 6.9 | 7.1 | 7.4 | 7.1 | 7.0 | 6.3 | 5.9 | 4.3 | 3.6 | 5.8 |
| Lat + 15 | 4.1 | 4.9 | 5.7 | 6.6 | 6.7 | 6.9 | 6.6 | 6.7 | 6.1 | 6 | 4.4 | 3.7 | 5.7 |
| Dual axis tracker | | | | | | | | | | | | | |
|  | 4.1 | 4.9 | 5.8 | 7.0 | 7.4 | 7.9 | 7.4 | 7.2 | 6.3 | 6.0 | 4.4 | 3.8 | 6.0 |

## MEMPHIS  TN  Latitude: 35.05 degrees  Elevation: 87 meters

|  | Jan | Feb | Mar | Apr | May | Jun | Jul | Aug | Sep | Oct | Nov | Dec | Avg |
|---|---|---|---|---|---|---|---|---|---|---|---|---|---|
| Fixed array | | | | | | | | | | | | | |
| Lat - 15 | 3.3 | 4.0 | 4.8 | 5.7 | 6.1 | 6.4 | 6.4 | 6.2 | 5.4 | 4.9 | 3.5 | 3.0 | 5.0 |
| Latitude | 3.7 | 4.4 | 5.0 | 5.6 | 5.8 | 5.9 | 6.0 | 6.0 | 5.4 | 5.2 | 3.9 | 3.4 | 5.0 |
| Lat + 15 | 4.0 | 4.5 | 4.9 | 5.2 | 5.1 | 5.1 | 5.2 | 5.5 | 5.2 | 5.3 | 4.1 | 3.6 | 4.8 |
| Single axis tracker | | | | | | | | | | | | | |
| Lat - 15 | 4.0 | 5.0 | 6.1 | 7.4 | 8.0 | 8.5 | 8.4 | 8.1 | 6.8 | 6.2 | 4.3 | 3.6 | 6.4 |
| Latitude | 4.4 | 5.3 | 6.2 | 7.3 | 7.7 | 8.1 | 8.1 | 7.9 | 6.9 | 6.5 | 4.6 | 3.9 | 6.4 |
| Lat + 15 | 4.6 | 5.4 | 6.1 | 7 | 7.3 | 7.6 | 7.6 | 7.5 | 6.7 | 6.5 | 4.8 | 4.1 | 6.3 |
| Dual axis tracker | | | | | | | | | | | | | |
|  | 4.6 | 5.4 | 6.2 | 7.4 | 8.0 | 8.6 | 8.5 | 8.1 | 6.9 | 6.5 | 4.8 | 4.2 | 6.6 |

## NASHVILLE  TN  Latitude: 36.12 degrees  Elevation: 180 meters

|  | Jan | Feb | Mar | Apr | May | Jun | Jul | Aug | Sep | Oct | Nov | Dec | Avg |
|---|---|---|---|---|---|---|---|---|---|---|---|---|---|
| Fixed array | | | | | | | | | | | | | |
| Lat - 15 | 3.1 | 3.9 | 4.7 | 5.7 | 6.0 | 6.4 | 6.2 | 5.9 | 5.2 | 4.6 | 3.3 | 2.8 | 4.8 |
| Latitude | 3.5 | 4.2 | 4.8 | 5.6 | 5.7 | 5.9 | 5.8 | 5.7 | 5.3 | 4.9 | 3.6 | 3.1 | 4.9 |
| Lat + 15 | 3.7 | 4.3 | 4.7 | 5.2 | 5.0 | 5.1 | 5.1 | 5.2 | 5.0 | 5.0 | 3.8 | 3.4 | 4.6 |
| Single axis tracker | | | | | | | | | | | | | |
| Lat - 15 | 3.8 | 4.8 | 5.8 | 7.3 | 7.7 | 8.1 | 7.8 | 7.5 | 6.5 | 5.8 | 4.0 | 3.3 | 6.0 |
| Latitude | 4.1 | 5.0 | 6.0 | 7.2 | 7.5 | 7.8 | 7.5 | 7.3 | 6.5 | 6.0 | 4.2 | 3.6 | 6.1 |
| Lat + 15 | 4.2 | 5.1 | 5.9 | 6.9 | 7 | 7.3 | 7 | 6.9 | 6.3 | 6.1 | 4.4 | 3.8 | 5.9 |
| Dual axis tracker | | | | | | | | | | | | | |
|  | 4.3 | 5.1 | 6.0 | 7.3 | 7.8 | 8.3 | 7.9 | 7.5 | 6.5 | 6.1 | 4.4 | 3.8 | 6.3 |

## ABLIENE TX — Latitude: 32.43 degrees — Elevation: 534 meters

|  | Jan | Feb | Mar | Apr | May | Jun | Jul | Aug | Sep | Oct | Nov | Dec | Avg |
|---|---|---|---|---|---|---|---|---|---|---|---|---|---|
| **Fixed array** | | | | | | | | | | | | | |
| Lat - 15 | 4.1 | 4.8 | 5.8 | 6.4 | 6.5 | 6.8 | 6.8 | 6.5 | 5.7 | 5.3 | 4.3 | 3.9 | 5.6 |
| Latitude | 4.7 | 5.3 | 6.0 | 6.3 | 6.1 | 6.2 | 6.3 | 6.3 | 5.8 | 5.7 | 4.9 | 4.5 | 5.7 |
| Lat + 15 | 5.1 | 5.5 | 6.0 | 5.8 | 5.4 | 5.4 | 5.5 | 5.7 | 5.6 | 5.8 | 5.2 | 4.9 | 5.5 |
| **Single axis tracker** | | | | | | | | | | | | | |
| Lat - 15 | 5.2 | 6.2 | 7.6 | 8.5 | 8.6 | 9.1 | 9.3 | 8.7 | 7.5 | 6.9 | 5.5 | 4.9 | 7.3 |
| Latitude | 5.7 | 6.6 | 7.8 | 8.4 | 8.4 | 8.7 | 8.9 | 8.6 | 7.6 | 7.2 | 6.0 | 5.4 | 7.4 |
| Lat + 15 | 6 | 6.7 | 7.7 | 8.1 | 7.9 | 8.2 | 8.4 | 8.2 | 7.4 | 7.3 | 6.2 | 5.7 | 7.3 |
| **Dual axis tracker** | | | | | | | | | | | | | |
|  | 6.0 | 6.7 | 7.8 | 8.5 | 8.7 | 9.3 | 9.4 | 8.8 | 7.6 | 7.3 | 6.3 | 5.8 | 7.7 |

## AMARILLO TX — Latitude: 35.23 degrees — Elevation: 1098 meters

|  | Jan | Feb | Mar | Apr | May | Jun | Jul | Aug | Sep | Oct | Nov | Dec | Avg |
|---|---|---|---|---|---|---|---|---|---|---|---|---|---|
| **Fixed array** | | | | | | | | | | | | | |
| Lat - 15 | 4.2 | 4.9 | 5.8 | 6.5 | 6.6 | 6.9 | 6.9 | 6.5 | 5.9 | 5.5 | 4.4 | 3.9 | 5.7 |
| Latitude | 4.9 | 5.4 | 6.0 | 6.4 | 6.2 | 6.3 | 6.4 | 6.3 | 6.0 | 5.9 | 5.0 | 4.6 | 5.8 |
| Lat + 15 | 5.3 | 5.7 | 5.9 | 6.0 | 5.5 | 5.5 | 5.5 | 5.7 | 5.8 | 6.0 | 5.3 | 5.0 | 5.6 |
| **Single axis tracker** | | | | | | | | | | | | | |
| Lat - 15 | 5.4 | 6.4 | 7.6 | 8.8 | 8.9 | 9.4 | 9.4 | 8.8 | 7.8 | 7.2 | 5.6 | 5.0 | 7.5 |
| Latitude | 6.0 | 6.8 | 7.8 | 8.7 | 8.6 | 9.0 | 9.0 | 8.7 | 7.9 | 7.6 | 6.1 | 5.5 | 7.6 |
| Lat + 15 | 6.3 | 6.9 | 7.7 | 8.4 | 8.1 | 8.4 | 8.4 | 8.2 | 7.7 | 7.7 | 6.4 | 5.9 | 7.5 |
| **Dual axis tracker** | | | | | | | | | | | | | |
|  | 6.3 | 7.0 | 7.8 | 8.8 | 9.0 | 9.6 | 9.5 | 8.9 | 7.9 | 7.7 | 6.4 | 6.0 | 7.9 |

## AUSTIN TX — Latitude: 30.30 degrees — Elevation: 189 meters

|  | Jan | Feb | Mar | Apr | May | Jun | Jul | Aug | Sep | Oct | Nov | Dec | Avg |
|---|---|---|---|---|---|---|---|---|---|---|---|---|---|
| **Fixed array** | | | | | | | | | | | | | |
| Lat - 15 | 3.7 | 4.4 | 5.2 | 5.6 | 5.8 | 6.4 | 6.7 | 6.5 | 5.7 | 5.0 | 4.1 | 3.5 | 5.2 |
| Latitude | 4.2 | 4.8 | 5.4 | 5.5 | 5.5 | 5.9 | 6.2 | 6.3 | 5.8 | 5.4 | 4.6 | 5.3 | 5.4 |
| Lat + 15 | 4.4 | 5.0 | 5.3 | 5.1 | 4.9 | 5.1 | 5.4 | 5.7 | 5.5 | 5.5 | 4.8 | 4.3 | 5.1 |
| **Single axis tracker** | | | | | | | | | | | | | |
| Lat - 15 | 4.6 | 5.6 | 6.6 | 7.1 | 7.3 | 8.3 | 8.7 | 8.5 | 7.3 | 6.5 | 5.1 | 4.4 | 6.7 |
| Latitude | 5.0 | 5.9 | 6.7 | 7.0 | 7.1 | 8.0 | 8.4 | 8.3 | 7.3 | 6.8 | 5.5 | 4.8 | 6.7 |
| Lat + 15 | 5.2 | 6 | 6.6 | 6.7 | 6.7 | 7.4 | 7.8 | 7.9 | 7.2 | 6.8 | 5.7 | 5 | 6.6 |
| **Dual axis tracker** | | | | | | | | | | | | | |
|  | 5.2 | 6.0 | 6.7 | 7.1 | 7.4 | 8.5 | 8.9 | 8.5 | 7.4 | 6.9 | 5.7 | 5.1 | 7.0 |

## BROWNSVILLE TX — Latitude: 25.90 degrees — Elevation: 6 meters

|  | Jan | Feb | Mar | Apr | May | Jun | Jul | Aug | Sep | Oct | Nov | Dec | Avg |
|---|---|---|---|---|---|---|---|---|---|---|---|---|---|
| **Fixed array** | | | | | | | | | | | | | |
| Lat - 15 | 4.3 | 5.3 | 6 | 6.3 | 6.4 | 7.1 | 7.4 | 7.3 | 6.9 | 6.7 | 5.4 | 4.3 | 6.1 |
| Latitude | 3.6 | 4.3 | 5.0 | 5.3 | 5.4 | 5.7 | 5.9 | 5.8 | 5.5 | 5.3 | 4.4 | 3.6 | 5.0 |
| Lat + 15 | 3.8 | 4.4 | 4.9 | 5.0 | 4.8 | 5.0 | 5.2 | 5.3 | 5.3 | 5.4 | 4.6 | 3.8 | 4.8 |
| **Single axis tracker** | | | | | | | | | | | | | |
| Lat - 15 | 3.9 | 5.0 | 6.0 | 6.6 | 7.1 | 7.9 | 8.2 | 7.8 | 6.9 | 6.3 | 4.9 | 3.8 | 6.2 |
| Latitude | 4.2 | 5.2 | 6.1 | 6.6 | 6.8 | 7.6 | 7.9 | 7.7 | 7.0 | 6.6 | 5.3 | 4.1 | 6.3 |
| Lat + 15 | 3.9 | 5.0 | 6.0 | 6.6 | 7.1 | 7.9 | 8.2 | 7.8 | 6.9 | 6.3 | 4.9 | 3.8 | 6.2 |
| **Dual axis tracker** | | | | | | | | | | | | | |
|  | 4.4 | 5.3 | 6.1 | 6.6 | 7.1 | 8.1 | 8.4 | 7.8 | 7.0 | 6.7 | 5.5 | 4.4 | 6.5 |

## CORPUS CHRISTI    TX        Latitude: 27.77 degrees       Elevation: 13 meters

| | Jan | Feb | Mar | Apr | May | Jun | Jul | Aug | Sep | Oct | Nov | Dec | Avg |
|---|---|---|---|---|---|---|---|---|---|---|---|---|---|
| Fixed array | | | | | | | | | | | | | |
| Lat - 15 | 3.2 | 4.0 | 4.7 | 5.2 | 5.4 | 5.9 | 6.1 | 5.9 | 5.3 | 4.8 | 3.9 | 3.1 | 4.8 |
| Latitude | 3.6 | 4.3 | 4.9 | 5.1 | 5.1 | 5.5 | 5.7 | 5.7 | 5.4 | 5.2 | 4.3 | 3.6 | 4.9 |
| Lat + 15 | 3.8 | 4.5 | 4.8 | 4.7 | 4.6 | 4.8 | 5.0 | 5.2 | 5.2 | 5.3 | 4.6 | 3.8 | 4.7 |
| Single axis tracker | | | | | | | | | | | | | |
| Lat - 15 | 3.9 | 5.0 | 5.9 | 6.4 | 6.8 | 7.6 | 8.0 | 7.7 | 6.9 | 6.3 | 4.9 | 3.8 | 6.1 |
| Latitude | 4.2 | 5.2 | 6.0 | 6.3 | 6.5 | 7.3 | 7.7 | 7.5 | 7.0 | 6.5 | 5.2 | 4.2 | 6.1 |
| Lat + 15 | 4.4 | 5.3 | 5.9 | 6.1 | 6.1 | 6.8 | 7.2 | 7.2 | 6.8 | 6.6 | 5.4 | 4.4 | 6 |
| Dual axis tracker | | | | | | | | | | | | | |
| | 4.4 | 5.3 | 6.0 | 6.4 | 6.8 | 7.7 | 8.1 | 7.7 | 7.0 | 6.6 | 5.5 | 4.4 | 6.3 |

## EL PASO           TX        Latitude: 31.80 degrees       Elevation: 1194 meters

| | Jan | Feb | Mar | Apr | May | Jun | Jul | Aug | Sep | Oct | Nov | Dec | Avg |
|---|---|---|---|---|---|---|---|---|---|---|---|---|---|
| Fixed array | | | | | | | | | | | | | |
| Lat - 15 | 4.6 | 5.6 | 6.6 | 7.4 | 7.8 | 7.7 | 7.2 | 6.9 | 6.4 | 5.9 | 4.9 | 4.4 | 6.3 |
| Latitude | 5.3 | 6.2 | 7.0 | 7.3 | 7.3 | 7.1 | 6.7 | 6.7 | 6.6 | 6.4 | 5.7 | 5.1 | 6.5 |
| Lat + 15 | 5.8 | 6.5 | 6.9 | 6.8 | 6.4 | 6.0 | 5.8 | 6.1 | 6.3 | 6.6 | 6.1 | 5.6 | 6.2 |
| Single axis tracker | | | | | | | | | | | | | |
| Lat - 15 | 5.9 | 7.4 | 8.9 | 10.2 | 10.7 | 10.6 | 9.5 | 9.1 | 8.6 | 7.8 | 6.4 | 5.6 | 8.4 |
| Latitude | 6.5 | 7.9 | 9.2 | 10.1 | 10.4 | 10.2 | 9.2 | 9.0 | 8.7 | 8.2 | 7.0 | 6.3 | 8.6 |
| Lat + 15 | 6.9 | 8.1 | 9.1 | 9.7 | 9.8 | 9.5 | 8.6 | 8.5 | 8.5 | 8.3 | 7.3 | 6.6 | 8.4 |
| Dual axis tracker | | | | | | | | | | | | | |
| | 6.9 | 8.1 | 9.2 | 10.2 | 10.8 | 10.9 | 9.7 | 9.2 | 8.7 | 8.3 | 7.4 | 6.7 | 8.9 |

## FORT WORTH       TX        Latitude: 32.83 degrees       Elevation: 164 meters

| | Jan | Feb | Mar | Apr | May | Jun | Jul | Aug | Sep | Oct | Nov | Dec | Avg |
|---|---|---|---|---|---|---|---|---|---|---|---|---|---|
| Fixed array | | | | | | | | | | | | | |
| Lat - 15 | 3.8 | 4.5 | 5.3 | 5.9 | 6.2 | 6.7 | 6.9 | 6.5 | 5.7 | 5.0 | 4.0 | 3.5 | 5.3 |
| Latitude | 4.3 | 4.9 | 5.5 | 5.7 | 5.8 | 6.2 | 6.4 | 6.3 | 5.8 | 5.4 | 4.5 | 4.1 | 5.4 |
| Lat + 15 | 4.6 | 5.1 | 5.4 | 5.3 | 5.2 | 5.3 | 5.6 | 5.7 | 5.6 | 5.5 | 4.8 | 4.4 | 5.2 |
| Single axis tracker | | | | | | | | | | | | | |
| Lat - 15 | 4.7 | 5.7 | 6.7 | 7.5 | 7.9 | 8.8 | 9.1 | 8.6 | 7.3 | 6.4 | 5.1 | 4.4 | 6.9 |
| Latitude | 5.1 | 6.0 | 6.9 | 7.4 | 7.7 | 8.4 | 8.8 | 8.5 | 7.4 | 6.7 | 5.5 | 4.9 | 6.9 |
| Lat + 15 | 5.3 | 6.1 | 6.8 | 7.1 | 7.2 | 7.8 | 8.2 | 8 | 7.2 | 6.8 | 5.7 | 5.1 | 6.8 |
| Dual axis tracker | | | | | | | | | | | | | |
| | 5.4 | 6.2 | 6.9 | 7.5 | 8.0 | 9.0 | 9.3 | 8.6 | 7.4 | 6.8 | 5.7 | 5.2 | 7.2 |

## HOUSTON        TX        Latitude: 29.98 degrees       Elevation: 33 meters

| | Jan | Feb | Mar | Apr | May | Jun | Jul | Aug | Sep | Oct | Nov | Dec | Avg |
|---|---|---|---|---|---|---|---|---|---|---|---|---|---|
| Fixed array | | | | | | | | | | | | | |
| Lat - 15 | 3.2 | 3.9 | 4.6 | 5.2 | 5.6 | 5.9 | 5.8 | 5.7 | 5.2 | 4.8 | 3.7 | 3.0 | 4.7 |
| Latitude | 3.6 | 4.3 | 4.7 | 5.1 | 5.3 | 5.4 | 5.4 | 5.5 | 5.3 | 5.2 | 4.1 | 3.5 | 4.8 |
| Lat + 15 | 3.8 | 4.4 | 4.6 | 4.7 | 4.7 | 4.7 | 4.8 | 5.0 | 5.1 | 5.2 | 4.3 | 3.7 | 4.6 |
| Single axis tracker | | | | | | | | | | | | | |
| Lat - 15 | 3.9 | 4.9 | 5.7 | 6.4 | 7.0 | 7.5 | 7.4 | 7.3 | 6.7 | 6.1 | 4.6 | 3.7 | 5.9 |
| Latitude | 4.2 | 5.1 | 5.8 | 6.4 | 6.8 | 7.2 | 7.2 | 7.2 | 6.7 | 6.4 | 4.9 | 4.0 | 6.0 |
| Lat + 15 | 4.3 | 5.2 | 5.7 | 6.1 | 6.4 | 6.7 | 6.7 | 6.8 | 6.6 | 6.4 | 5.1 | 4.2 | 5.9 |
| Dual axis tracker | | | | | | | | | | | | | |
| | 4.4 | 5.2 | 5.8 | 6.5 | 7.1 | 7.7 | 7.6 | 7.3 | 6.7 | 6.5 | 5.1 | 4.2 | 6.2 |

## LUBBOCK TX Latitude: 33.65 degrees Elevation: 988 meters

| | Jan | Feb | Mar | Apr | May | Jun | Jul | Aug | Sep | Oct | Nov | Dec | Avg |
|---|---|---|---|---|---|---|---|---|---|---|---|---|---|
| Fixed array | | | | | | | | | | | | | |
| Lat - 15 | 4.2 | 5.0 | 5.8 | 6.6 | 6.7 | 6.9 | 6.8 | 6.5 | 5.8 | 5.4 | 4.4 | 3.9 | 5.7 |
| Latitude | 4.9 | 5.5 | 6.1 | 6.5 | 6.3 | 6.3 | 6.3 | 6.3 | 5.9 | 5.9 | 5.1 | 4.6 | 5.8 |
| Lat + 15 | 5.3 | 5.7 | 6.0 | 6.0 | 5.6 | 5.4 | 5.5 | 5.7 | 5.7 | 6.0 | 5.4 | 5.0 | 5.6 |
| Single axis tracker | | | | | | | | | | | | | |
| Lat - 15 | 5.4 | 6.4 | 7.7 | 8.8 | 9.0 | 9.3 | 9.3 | 8.8 | 7.6 | 7.2 | 5.7 | 5.0 | 7.5 |
| Latitude | 5.9 | 6.8 | 7.9 | 8.7 | 8.7 | 8.9 | 8.9 | 8.6 | 7.7 | 7.5 | 6.2 | 5.6 | 7.6 |
| Lat + 15 | 6.2 | 7 | 7.8 | 8.4 | 8.2 | 8.3 | 8.4 | 8.2 | 7.5 | 7.6 | 6.5 | 5.9 | 7.5 |
| Dual axis tracker | | | | | | | | | | | | | |
| | 6.3 | 7.0 | 7.9 | 8.8 | 9.1 | 9.5 | 9.4 | 8.8 | 7.7 | 7.6 | 6.5 | 6.0 | 7.9 |

## LUFKIN TX Latitude: 31.23 degrees Elevation: 96 meters

| | Jan | Feb | Mar | Apr | May | Jun | Jul | Aug | Sep | Oct | Nov | Dec | Avg |
|---|---|---|---|---|---|---|---|---|---|---|---|---|---|
| Fixed array | | | | | | | | | | | | | |
| Lat - 15 | 3.4 | 4.1 | 4.9 | 5.5 | 5.9 | 6.2 | 6.3 | 6.2 | 5.5 | 5.0 | 3.9 | 3.2 | 5.0 |
| Latitude | 3.8 | 4.5 | 5.1 | 5.4 | 5.6 | 5.8 | 5.9 | 6.0 | 5.6 | 5.4 | 4.3 | 3.7 | 5.1 |
| Lat + 15 | 4.0 | 4.6 | 5.0 | 5.0 | 4.9 | 5.0 | 5.2 | 5.4 | 5.4 | 5.4 | 4.6 | 4.0 | 4.9 |
| Single axis tracker | | | | | | | | | | | | | |
| Lat - 15 | 4.1 | 5.2 | 6.2 | 6.9 | 7.4 | 8.0 | 8.0 | 7.9 | 7.0 | 6.4 | 4.8 | 4.0 | 6.3 |
| Latitude | 4.4 | 5.5 | 6.3 | 6.9 | 7.2 | 7.7 | 7.8 | 7.8 | 7.0 | 6.6 | 5.2 | 4.3 | 6.4 |
| Lat + 15 | 4.6 | 5.5 | 6.2 | 6.6 | 6.7 | 7.1 | 7.2 | 7.4 | 6.9 | 6.7 | 5.3 | 4.5 | 6.2 |
| Dual axis tracker | | | | | | | | | | | | | |
| | 4.6 | 5.6 | 6.3 | 7.0 | 7.5 | 8.2 | 8.2 | 8.0 | 7.1 | 6.7 | 5.4 | 4.6 | 6.6 |

## MIDLAND TX Latitude: 31.93 degrees Elevation: 871 meters

| | Jan | Feb | Mar | Apr | May | Jun | Jul | Aug | Sep | Oct | Nov | Dec | Avg |
|---|---|---|---|---|---|---|---|---|---|---|---|---|---|
| Fixed array | | | | | | | | | | | | | |
| Lat - 15 | 4.3 | 5.1 | 6.2 | 6.8 | 7.0 | 7.1 | 6.9 | 6.6 | 5.9 | 5.5 | 4.6 | 4.1 | 5.8 |
| Latitude | 5.0 | 5.7 | 6.5 | 6.7 | 6.5 | 6.5 | 6.4 | 6.4 | 6.0 | 6.0 | 5.3 | 4.8 | 6.0 |
| Lat + 15 | 5.4 | 5.9 | 6.4 | 6.2 | 5.8 | 5.6 | 5.5 | 5.8 | 5.8 | 6.1 | 5.6 | 5.2 | 5.8 |
| Single axis tracker | | | | | | | | | | | | | |
| Lat - 15 | 5.6 | 6.7 | 8.2 | 9.1 | 9.3 | 9.5 | 9.3 | 8.8 | 7.8 | 7.3 | 6.0 | 5.3 | 7.7 |
| Latitude | 6.1 | 7.1 | 8.4 | 9.0 | 9.0 | 9.1 | 9.0 | 8.7 | 7.8 | 7.6 | 6.5 | 5.8 | 7.9 |
| Lat + 15 | 6.4 | 7.3 | 8.3 | 8.7 | 8.5 | 8.5 | 8.4 | 8.3 | 7.7 | 7.7 | 6.8 | 6.2 | 7.7 |
| Dual axis tracker | | | | | | | | | | | | | |
| | 6.5 | 7.3 | 8.4 | 9.1 | 9.4 | 9.7 | 9.5 | 8.9 | 7.9 | 7.7 | 6.8 | 6.3 | 8.1 |

## PORT ARTHUR TX Latitude: 29.95 degrees Elevation: 7 meters

| | Jan | Feb | Mar | Apr | May | Jun | Jul | Aug | Sep | Oct | Nov | Dec | Avg |
|---|---|---|---|---|---|---|---|---|---|---|---|---|---|
| Fixed array | | | | | | | | | | | | | |
| Lat - 15 | 3.3 | 4.1 | 4.7 | 5.3 | 5.8 | 6.1 | 6.0 | 5.8 | 5.4 | 4.9 | 3.8 | 3.2 | 4.9 |
| Latitude | 3.7 | 4.4 | 4.9 | 5.2 | 5.5 | 5.7 | 5.5 | 5.6 | 5.5 | 5.3 | 4.2 | 3.6 | 4.9 |
| Lat + 15 | 3.9 | 4.5 | 4.8 | 4.9 | 4.9 | 4.9 | 4.9 | 5.2 | 5.3 | 5.4 | 4.4 | 3.9 | 4.7 |
| Single axis tracker | | | | | | | | | | | | | |
| Lat - 15 | 4.0 | 5.1 | 5.9 | 6.7 | 7.3 | 7.9 | 7.6 | 7.4 | 6.8 | 6.3 | 4.7 | 3.9 | 6.1 |
| Latitude | 4.3 | 5.3 | 6.0 | 6.6 | 7.1 | 7.5 | 7.3 | 7.3 | 6.9 | 6.6 | 5.0 | 4.2 | 6.2 |
| Lat + 15 | 4.4 | 5.4 | 5.9 | 6.3 | 6.7 | 7 | 6.8 | 6.9 | 6.7 | 6.7 | 5.2 | 4.4 | 6 |
| Dual axis tracker | | | | | | | | | | | | | |
| | 4.5 | 5.4 | 6.0 | 6.7 | 7.4 | 8.0 | 7.7 | 7.5 | 6.9 | 6.7 | 5.2 | 4.5 | 6.4 |

## SAN ANGELO     TX     Latitude: 31.37 degrees     Elevation: 582 meters

| | Jan | Feb | Mar | Apr | May | Jun | Jul | Aug | Sep | Oct | Nov | Dec | Avg |
|---|---|---|---|---|---|---|---|---|---|---|---|---|---|
| **Fixed array** | | | | | | | | | | | | | |
| Lat - 15 | 4.1 | 4.9 | 5.8 | 6.4 | 6.5 | 6.7 | 6.8 | 6.5 | 5.7 | 5.3 | 4.4 | 3.9 | 5.6 |
| Latitude | 4.7 | 5.4 | 6.1 | 6.3 | 6.1 | 6.2 | 6.3 | 6.3 | 5.8 | 5.7 | 5.0 | 4.6 | 5.7 |
| Lat + 15 | 5.1 | 5.6 | 6.0 | 5.8 | 5.4 | 5.3 | 5.5 | 5.7 | 5.6 | 5.8 | 5.4 | 5.0 | 5.5 |
| **Single axis tracker** | | | | | | | | | | | | | |
| Lat - 15 | 5.2 | 6.3 | 7.6 | 8.3 | 8.4 | 9.0 | 9.2 | 8.7 | 7.5 | 6.9 | 5.7 | 5.0 | 7.3 |
| Latitude | 5.7 | 6.7 | 7.8 | 8.3 | 8.2 | 8.6 | 8.8 | 8.6 | 7.6 | 7.2 | 6.2 | 5.5 | 7.4 |
| Lat + 15 | 6 | 6.8 | 7.8 | 8 | 7.7 | 8 | 8.3 | 8.2 | 7.4 | 7.3 | 6.4 | 5.8 | 7.3 |
| **Dual axis tracker** | | | | | | | | | | | | | |
| | 6.1 | 6.9 | 7.8 | 8.4 | 8.5 | 9.2 | 9.3 | 8.8 | 7.6 | 7.3 | 6.5 | 5.9 | 7.7 |

## SAN ANTONIO     TX     Latitude: 29.53 degrees     Elevation: 242 meters

| | Jan | Feb | Mar | Apr | May | Jun | Jul | Aug | Sep | Oct | Nov | Dec | Avg |
|---|---|---|---|---|---|---|---|---|---|---|---|---|---|
| **Fixed array** | | | | | | | | | | | | | |
| Lat - 15 | 3.7 | 4.5 | 5.2 | 5.7 | 5.9 | 6.5 | 6.7 | 6.6 | 5.8 | 5.1 | 4.1 | 3.5 | 5.3 |
| Latitude | 4.3 | 4.9 | 5.4 | 5.6 | 5.6 | 6.0 | 6.3 | 6.3 | 5.9 | 5.5 | 4.6 | 4.1 | 5.4 |
| Lat + 15 | 4.5 | 5.0 | 5.3 | 5.2 | 5.0 | 5.2 | 5.5 | 5.8 | 5.7 | 5.6 | 4.9 | 4.4 | 5.2 |
| **Single axis tracker** | | | | | | | | | | | | | |
| Lat - 15 | 4.7 | 5.7 | 6.6 | 7.1 | 7.3 | 8.3 | 8.7 | 8.4 | 7.4 | 6.6 | 5.2 | 4.4 | 6.7 |
| Latitude | 5.1 | 6.0 | 6.8 | 7.0 | 7.1 | 7.9 | 8.4 | 8.3 | 7.5 | 6.9 | 5.6 | 4.8 | 6.8 |
| Lat + 15 | 5.3 | 6.1 | 6.7 | 6.7 | 6.7 | 7.4 | 7.8 | 7.9 | 7.3 | 6.9 | 5.8 | 5 | 6.6 |
| **Dual axis tracker** | | | | | | | | | | | | | |
| | 5.3 | 6.1 | 6.8 | 7.1 | 7.4 | 8.4 | 8.8 | 8.4 | 7.5 | 6.9 | 5.8 | 5.1 | 7.0 |

## VICTORIA     TX     Latitude: 28.85 degrees     Elevation: 32 meters

| | Jan | Feb | Mar | Apr | May | Jun | Jul | Aug | Sep | Oct | Nov | Dec | Avg |
|---|---|---|---|---|---|---|---|---|---|---|---|---|---|
| **Fixed array** | | | | | | | | | | | | | |
| Lat - 15 | 3.3 | 4.1 | 4.8 | 5.2 | 5.6 | 6.1 | 6.1 | 5.9 | 5.4 | 4.9 | 3.9 | 3.2 | 4.9 |
| Latitude | 3.7 | 4.5 | 4.9 | 5.1 | 5.3 | 5.6 | 5.7 | 5.7 | 5.4 | 5.3 | 4.3 | 3.6 | 4.9 |
| Lat + 15 | 3.9 | 4.6 | 4.8 | 4.8 | 4.7 | 4.9 | 5.0 | 5.2 | 5.2 | 5.4 | 4.6 | 3.8 | 4.7 |
| **Single axis tracker** | | | | | | | | | | | | | |
| Lat - 15 | 4.1 | 5.1 | 5.9 | 6.4 | 7.0 | 7.7 | 7.8 | 7.6 | 6.9 | 6.3 | 4.8 | 3.9 | 6.1 |
| Latitude | 4.4 | 5.4 | 6.0 | 6.3 | 6.7 | 7.4 | 7.5 | 7.5 | 6.9 | 6.6 | 5.2 | 4.2 | 6.2 |
| Lat + 15 | 4.5 | 5.5 | 5.9 | 6.1 | 6.3 | 6.9 | 7.1 | 7.1 | 6.8 | 6.6 | 5.4 | 4.4 | 6 |
| **Dual axis tracker** | | | | | | | | | | | | | |
| | 4.6 | 5.5 | 6.0 | 6.4 | 7.0 | 7.9 | 7.9 | 7.7 | 6.9 | 6.7 | 5.4 | 4.4 | 6.4 |

## WACO     TX     Latitude: 31.62 degrees     Elevation: 155 meters

| | Jan | Feb | Mar | Apr | May | Jun | Jul | Aug | Sep | Oct | Nov | Dec | Avg |
|---|---|---|---|---|---|---|---|---|---|---|---|---|---|
| **Fixed array** | | | | | | | | | | | | | |
| Lat - 15 | 3.7 | 4.4 | 5.3 | 5.7 | 6.0 | 6.5 | 6.8 | 6.5 | 5.7 | 5.0 | 4.0 | 3.6 | 5.3 |
| Latitude | 4.2 | 4.8 | 5.4 | 5.6 | 5.6 | 6.0 | 6.3 | 6.3 | 5.8 | 5.4 | 4.5 | 4.1 | 5.4 |
| Lat + 15 | 4.5 | 5.0 | 5.3 | 5.2 | 5.0 | 5.2 | 5.5 | 5.7 | 5.6 | 5.5 | 4.8 | 4.4 | 5.1 |
| **Single axis tracker** | | | | | | | | | | | | | |
| Lat - 15 | 4.6 | 5.6 | 6.7 | 7.3 | 7.6 | 8.6 | 9.1 | 8.7 | 7.4 | 6.5 | 5.1 | 4.4 | 6.8 |
| Latitude | 5.0 | 5.9 | 6.8 | 7.2 | 7.4 | 8.2 | 8.7 | 8.5 | 7.5 | 6.8 | 5.5 | 4.9 | 6.9 |
| Lat + 15 | 5.3 | 6 | 6.8 | 6.9 | 7 | 7.6 | 8.2 | 8.1 | 7.3 | 6.9 | 5.7 | 5.1 | 6.7 |
| **Dual axis tracker** | | | | | | | | | | | | | |
| | 5.3 | 6.0 | 6.9 | 7.3 | 7.7 | 8.7 | 9.2 | 8.7 | 7.5 | 6.9 | 5.7 | 5.2 | 7.1 |

## WICHITA FALLS    TX      Latitude: 33.97 degrees      Elevation: 314 meters

| | Jan | Feb | Mar | Apr | May | Jun | Jul | Aug | Sep | Oct | Nov | Dec | Avg |
|---|---|---|---|---|---|---|---|---|---|---|---|---|---|
| Fixed array | | | | | | | | | | | | | |
| Lat - 15 | 3.9 | 4.6 | 5.4 | 6.1 | 6.4 | 6.7 | 6.8 | 6.5 | 5.7 | 5.1 | 4.1 | 3.6 | 5.4 |
| Latitude | 4.5 | 5.0 | 5.6 | 6.0 | 6.0 | 6.2 | 6.3 | 6.3 | 5.8 | 5.5 | 4.6 | 4.2 | 5.5 |
| Lat + 15 | 4.8 | 5.2 | 5.5 | 5.6 | 5.3 | 5.3 | 5.5 | 5.7 | 5.6 | 5.6 | 4.9 | 4.5 | 5.3 |
| Single axis tracker | | | | | | | | | | | | | |
| Lat - 15 | 4.9 | 5.8 | 7.0 | 8.0 | 8.4 | 9.0 | 9.2 | 8.7 | 7.5 | 6.7 | 5.1 | 4.5 | 7.1 |
| Latitude | 5.3 | 6.2 | 7.2 | 7.9 | 8.1 | 8.6 | 8.9 | 8.6 | 7.6 | 7.0 | 5.6 | 5.0 | 7.2 |
| Lat + 15 | 5.6 | 6.3 | 7.1 | 7.6 | 7.6 | 8.1 | 8.3 | 8.2 | 7.4 | 7.1 | 5.8 | 5.2 | 7 |
| Dual axis tracker | | | | | | | | | | | | | |
| | 5.6 | 6.3 | 7.2 | 8.0 | 8.5 | 9.2 | 9.4 | 8.8 | 7.6 | 7.1 | 5.8 | 5.3 | 7.4 |

## CEDAR CITY    UT      Latitude: 37.70 degrees      Elevation: 1712 meters

| | Jan | Feb | Mar | Apr | May | Jun | Jul | Aug | Sep | Oct | Nov | Dec | Avg |
|---|---|---|---|---|---|---|---|---|---|---|---|---|---|
| Fixed array | | | | | | | | | | | | | |
| Lat - 15 | 4.0 | 4.7 | 5.5 | 6.4 | 7.0 | 7.5 | 7.1 | 6.8 | 6.6 | 5.6 | 4.2 | 3.7 | 5.8 |
| Latitude | 4.6 | 5.2 | 5.7 | 6.3 | 6.5 | 6.8 | 6.6 | 6.6 | 6.7 | 6.1 | 4.8 | 4.4 | 5.9 |
| Lat + 15 | 5.0 | 5.4 | 5.7 | 5.9 | 5.7 | 5.9 | 5.7 | 6.0 | 6.5 | 6.2 | 5.1 | 4.8 | 5.7 |
| Single axis tracker | | | | | | | | | | | | | |
| Lat - 15 | 5.1 | 6.1 | 7.3 | 8.8 | 9.7 | 10.8 | 10.1 | 9.5 | 9.1 | 7.5 | 5.4 | 4.7 | 7.9 |
| Latitude | 5.6 | 6.5 | 7.5 | 8.7 | 9.4 | 10.3 | 9.7 | 9.4 | 9.3 | 7.9 | 5.9 | 5.3 | 8.0 |
| Lat + 15 | 5.9 | 6.6 | 7.4 | 8.4 | 8.9 | 9.7 | 9.1 | 9 | 9.1 | 8 | 6.1 | 5.6 | 7.8 |
| Dual axis tracker | | | | | | | | | | | | | |
| | 6.0 | 6.6 | 7.5 | 8.8 | 9.9 | 11.0 | 10.3 | 9.6 | 9.3 | 8.0 | 6.2 | 5.7 | 8.3 |

## SALT LAKE CITY    UT      Latitude: 40.77 degrees      Elevation: 1288 meters

| | Jan | Feb | Mar | Apr | May | Jun | Jul | Aug | Sep | Oct | Nov | Dec | Avg |
|---|---|---|---|---|---|---|---|---|---|---|---|---|---|
| Fixed array | | | | | | | | | | | | | |
| Lat - 15 | 2.9 | 4.0 | 5.0 | 5.9 | 6.6 | 7.2 | 7.3 | 7.0 | 6.3 | 5.0 | 3.3 | 2.5 | 5.2 |
| Latitude | 3.2 | 4.3 | 5.2 | 5.8 | 6.2 | 6.6 | 6.7 | 6.7 | 6.4 | 5.4 | 3.7 | 2.9 | 5.3 |
| Lat + 15 | 3.4 | 4.4 | 5.1 | 5.4 | 5.5 | 5.6 | 5.8 | 6.1 | 6.1 | 5.5 | 3.9 | 3.1 | 5.0 |
| Single axis tracker | | | | | | | | | | | | | |
| Lat - 15 | 3.4 | 4.8 | 6.3 | 7.7 | 8.9 | 10.0 | 10.2 | 9.6 | 8.5 | 6.5 | 4.0 | 3.0 | 6.9 |
| Latitude | 3.7 | 5.1 | 6.5 | 7.7 | 8.7 | 9.6 | 9.8 | 9.4 | 8.6 | 6.8 | 4.3 | 3.3 | 7.0 |
| Lat + 15 | 3.8 | 5.2 | 6.4 | 7.4 | 8.2 | 9 | 9.2 | 9 | 8.4 | 6.9 | 4.5 | 3.4 | 6.8 |
| Dual axis tracker | | | | | | | | | | | | | |
| | 3.9 | 5.2 | 6.5 | 7.8 | 9.1 | 10.3 | 10.4 | 9.6 | 8.6 | 6.9 | 4.5 | 3.5 | 7.2 |

## LYNCHBURG    VA      Latitude: 37.33 degrees      Elevation: 279 meters

| | Jan | Feb | Mar | Apr | May | Jun | Jul | Aug | Sep | Oct | Nov | Dec | Avg |
|---|---|---|---|---|---|---|---|---|---|---|---|---|---|
| Fixed array | | | | | | | | | | | | | |
| Lat - 15 | 3.4 | 4.2 | 5.1 | 5.8 | 6.1 | 6.3 | 6.1 | 5.9 | 5.3 | 4.6 | 3.6 | 3.1 | 5.0 |
| Latitude | 3.9 | 4.6 | 5.3 | 5.7 | 5.7 | 5.8 | 5.7 | 5.7 | 5.3 | 5.0 | 4.0 | 3.6 | 5.0 |
| Lat + 15 | 4.2 | 4.8 | 5.2 | 5.3 | 5.1 | 5.0 | 5.0 | 5.2 | 5.1 | 5.0 | 4.3 | 3.8 | 4.8 |
| Single axis tracker | | | | | | | | | | | | | |
| Lat - 15 | 4.2 | 5.3 | 6.5 | 7.5 | 7.8 | 8.2 | 7.8 | 7.5 | 6.7 | 5.9 | 4.4 | 3.8 | 6.3 |
| Latitude | 4.6 | 5.6 | 6.6 | 7.4 | 7.6 | 7.8 | 7.5 | 7.3 | 6.7 | 6.1 | 4.8 | 4.1 | 6.4 |
| Lat + 15 | 4.8 | 5.7 | 6.6 | 7.1 | 7.1 | 7.3 | 7 | 7 | 6.6 | 6.2 | 5 | 4.4 | 6.2 |
| Dual axis tracker | | | | | | | | | | | | | |
| | 4.9 | 5.7 | 6.7 | 7.5 | 7.9 | 8.3 | 7.9 | 7.5 | 6.8 | 6.2 | 5.0 | 4.4 | 6.6 |

## NORFOLK VA Latitude: 36.90 degrees Elevation: 9 meters

| | Jan | Feb | Mar | Apr | May | Jun | Jul | Aug | Sep | Oct | Nov | Dec | Avg |
|---|---|---|---|---|---|---|---|---|---|---|---|---|---|
| **Fixed array** | | | | | | | | | | | | | |
| Lat - 15 | 3.2 | 3.9 | 4.8 | 5.5 | 5.8 | 6.0 | 5.8 | 5.6 | 5.0 | 4.3 | 3.5 | 2.9 | 4.7 |
| Latitude | 3.6 | 4.3 | 4.9 | 5.4 | 5.5 | 5.6 | 5.4 | 5.4 | 5.1 | 4.6 | 3.9 | 3.4 | 4.8 |
| Lat + 15 | 3.8 | 4.4 | 4.8 | 5.0 | 4.8 | 4.8 | 4.8 | 4.9 | 4.9 | 4.7 | 4.1 | 3.6 | 4.6 |
| **Single axis tracker** | | | | | | | | | | | | | |
| Lat - 15 | 3.9 | 4.9 | 6.0 | 7.1 | 7.4 | 7.7 | 7.4 | 7.1 | 6.3 | 5.4 | 4.3 | 3.5 | 5.9 |
| Latitude | 4.2 | 5.1 | 6.2 | 7.0 | 7.1 | 7.4 | 7.1 | 7.0 | 6.4 | 5.6 | 4.6 | 3.9 | 6.0 |
| Lat + 15 | 4.4 | 5.2 | 6.1 | 6.7 | 6.7 | 6.9 | 6.6 | 6.6 | 6.2 | 5.7 | 4.8 | 4.1 | 5.8 |
| **Dual axis tracker** | | | | | | | | | | | | | |
| | 4.4 | 5.3 | 6.2 | 7.1 | 7.4 | 7.9 | 7.5 | 7.1 | 6.4 | 5.7 | 4.8 | 4.1 | 6.2 |

## RICHMOND VA Latitude: 37.50 degrees Elevation: 50 meters

| | Jan | Feb | Mar | Apr | May | Jun | Jul | Aug | Sep | Oct | Nov | Dec | Avg |
|---|---|---|---|---|---|---|---|---|---|---|---|---|---|
| **Fixed array** | | | | | | | | | | | | | |
| Lat - 15 | 3.2 | 3.9 | 4.8 | 5.5 | 5.8 | 6.1 | 5.9 | 5.7 | 5.1 | 4.4 | 3.5 | 2.9 | 4.7 |
| Latitude | 3.6 | 4.3 | 5.0 | 5.4 | 5.5 | 5.6 | 5.5 | 5.5 | 5.2 | 4.7 | 3.9 | 3.3 | 4.8 |
| Lat + 15 | 3.9 | 4.4 | 4.9 | 5.0 | 4.9 | 4.9 | 4.8 | 5.0 | 4.9 | 4.8 | 4.1 | 3.6 | 4.6 |
| **Single axis tracker** | | | | | | | | | | | | | |
| Lat - 15 | 3.9 | 4.9 | 6.1 | 7.1 | 7.4 | 7.8 | 7.5 | 7.1 | 6.4 | 5.5 | 4.2 | 3.5 | 6.0 |
| Latitude | 4.3 | 5.2 | 6.2 | 7.0 | 7.2 | 7.5 | 7.2 | 7.0 | 6.5 | 5.7 | 4.6 | 3.8 | 6.0 |
| Lat + 15 | 4.5 | 5.3 | 6.1 | 6.7 | 6.8 | 7 | 6.7 | 6.7 | 6.3 | 5.8 | 4.8 | 4.1 | 5.9 |
| **Dual axis tracker** | | | | | | | | | | | | | |
| | 4.5 | 5.3 | 6.2 | 7.1 | 7.5 | 8.0 | 7.6 | 7.2 | 6.5 | 5.8 | 4.8 | 4.1 | 6.2 |

## ROANOKE VA Latitude: 37.32 degrees Elevation: 358 meters

| | Jan | Feb | Mar | Apr | May | Jun | Jul | Aug | Sep | Oct | Nov | Dec | Avg |
|---|---|---|---|---|---|---|---|---|---|---|---|---|---|
| **Fixed array** | | | | | | | | | | | | | |
| Lat - 15 | 3.3 | 4.0 | 4.8 | 5.6 | 5.8 | 6.0 | 5.9 | 5.7 | 5.1 | 4.6 | 3.5 | 2.9 | 4.8 |
| Latitude | 3.7 | 4.3 | 5.0 | 5.5 | 5.5 | 5.6 | 5.5 | 5.5 | 5.1 | 4.9 | 3.9 | 3.4 | 4.8 |
| Lat + 15 | 3.9 | 4.5 | 4.9 | 5.1 | 4.9 | 4.8 | 4.8 | 5.0 | 4.9 | 4.9 | 4.1 | 3.6 | 4.6 |
| **Single axis tracker** | | | | | | | | | | | | | |
| Lat - 15 | 4.0 | 5.0 | 6.1 | 7.2 | 7.4 | 7.7 | 7.4 | 7.2 | 6.4 | 5.7 | 4.2 | 3.6 | 6.0 |
| Latitude | 4.3 | 5.2 | 6.2 | 7.1 | 7.2 | 7.4 | 7.2 | 7.1 | 6.4 | 6.0 | 4.6 | 3.9 | 6.1 |
| Lat + 15 | 4.5 | 5.3 | 6.1 | 6.8 | 6.7 | 6.9 | 6.7 | 6.7 | 6.2 | 6 | 4.7 | 4.1 | 5.9 |
| **Dual axis tracker** | | | | | | | | | | | | | |
| | 4.6 | 5.4 | 6.2 | 7.2 | 7.5 | 7.9 | 7.6 | 7.2 | 6.4 | 6.0 | 4.7 | 4.2 | 6.3 |

## STERLING VA Latitude: 38.95 degrees Elevation: 82 meters

| | Jan | Feb | Mar | Apr | May | Jun | Jul | Aug | Sep | Oct | Nov | Dec | Avg |
|---|---|---|---|---|---|---|---|---|---|---|---|---|---|
| **Fixed array** | | | | | | | | | | | | | |
| Lat - 15 | 3.1 | 3.9 | 4.7 | 5.4 | 5.8 | 6.1 | 6.0 | 5.7 | 5.0 | 4.3 | 3.2 | 2.7 | 4.7 |
| Latitude | 3.5 | 4.2 | 4.8 | 5.3 | 5.5 | 5.7 | 5.6 | 5.5 | 5.1 | 4.6 | 3.6 | 3.1 | 4.7 |
| Lat + 15 | 3.7 | 4.3 | 4.7 | 4.9 | 4.8 | 4.9 | 4.9 | 5.0 | 4.9 | 4.6 | 3.7 | 3.3 | 4.5 |
| **Single axis tracker** | | | | | | | | | | | | | |
| Lat - 15 | 3.7 | 4.8 | 5.9 | 6.9 | 7.3 | 7.8 | 7.5 | 7.0 | 6.3 | 5.3 | 3.9 | 3.2 | 5.8 |
| Latitude | 4.0 | 5.0 | 6.0 | 6.8 | 7.1 | 7.4 | 7.2 | 6.9 | 6.3 | 5.6 | 4.2 | 3.5 | 5.8 |
| Lat + 15 | 4.2 | 5.1 | 5.9 | 6.5 | 6.7 | 6.9 | 6.7 | 6.5 | 6.1 | 5.6 | 4.3 | 3.7 | 5.7 |
| **Dual axis tracker** | | | | | | | | | | | | | |
| | 4.3 | 5.1 | 6.0 | 6.9 | 7.4 | 7.9 | 7.6 | 7.1 | 6.3 | 5.6 | 4.3 | 3.7 | 6.0 |

## BURLINGTON VT — Latitude: 44.47 degrees — Elevation: 104 meters

|  | Jan | Feb | Mar | Apr | May | Jun | Jul | Aug | Sep | Oct | Nov | Dec | Avg |
|---|---|---|---|---|---|---|---|---|---|---|---|---|---|
| Fixed array |  |  |  |  |  |  |  |  |  |  |  |  |  |
| Lat - 15 | 2.6 | 3.6 | 4.5 | 5.0 | 5.6 | 5.9 | 6.1 | 5.6 | 4.7 | 3.5 | 2.2 | 1.9 | 4.3 |
| Latitude | 2.9 | 3.9 | 4.7 | 4.9 | 5.3 | 5.4 | 5.6 | 5.3 | 4.8 | 3.7 | 2.4 | 2.1 | 4.3 |
| Lat + 15 | 3.1 | 4.1 | 4.6 | 4.5 | 4.6 | 4.7 | 4.9 | 4.8 | 4.5 | 3.7 | 2.4 | 2.2 | 4.0 |
| Single axis tracker |  |  |  |  |  |  |  |  |  |  |  |  |  |
| Lat - 15 | 3.0 | 4.4 | 5.7 | 6.4 | 7.3 | 7.8 | 8.0 | 7.2 | 6.0 | 4.2 | 2.5 | 2.2 | 5.4 |
| Latitude | 3.3 | 4.6 | 5.8 | 6.3 | 7.0 | 7.4 | 7.7 | 7.1 | 6.0 | 4.4 | 2.6 | 2.4 | 5.4 |
| Lat + 15 | 3.4 | 4.8 | 5.7 | 6.1 | 6.6 | 6.9 | 7.2 | 6.7 | 5.8 | 4.4 | 2.7 | 2.5 | 5.2 |
| Dual axis tracker |  |  |  |  |  |  |  |  |  |  |  |  |  |
|  | 3.5 | 4.8 | 5.8 | 6.4 | 7.4 | 8.0 | 8.2 | 7.3 | 6.0 | 4.4 | 2.7 | 2.5 | 5.6 |

## OLYMPIA WA — Latitude: 46.97 degrees — Elevation: 61 meters

|  | Jan | Feb | Mar | Apr | May | Jun | Jul | Aug | Sep | Oct | Nov | Dec | Avg |
|---|---|---|---|---|---|---|---|---|---|---|---|---|---|
| Fixed array |  |  |  |  |  |  |  |  |  |  |  |  |  |
| Lat - 15 | 1.4 | 2.3 | 3.4 | 4.4 | 5.1 | 5.5 | 5.9 | 5.5 | 4.6 | 3.0 | 1.6 | 1.2 | 3.7 |
| Latitude | 1.5 | 2.4 | 3.4 | 4.2 | 4.7 | 5.0 | 5.5 | 5.2 | 4.6 | 3.1 | 1.7 | 1.3 | 3.6 |
| Lat + 15 | 1.6 | 2.4 | 3.3 | 3.9 | 4.2 | 4.3 | 4.8 | 4.7 | 4.4 | 3.0 | 1.8 | 1.4 | 3.3 |
| Single axis tracker |  |  |  |  |  |  |  |  |  |  |  |  |  |
| Lat - 15 | 1.6 | 2.5 | 4.0 | 5.3 | 6.5 | 7.1 | 7.9 | 7.1 | 5.8 | 3.5 | 1.8 | 1.3 | 4.5 |
| Latitude | 1.7 | 2.6 | 4.1 | 5.2 | 6.2 | 6.8 | 7.6 | 7.0 | 5.9 | 3.6 | 1.9 | 1.4 | 4.5 |
| Lat + 15 | 1.7 | 2.6 | 4 | 5 | 5.9 | 6.3 | 7.1 | 6.6 | 5.7 | 3.5 | 1.9 | 1.4 | 4.3 |
| Dual axis tracker |  |  |  |  |  |  |  |  |  |  |  |  |  |
|  | 1.7 | 2.7 | 4.1 | 5.3 | 6.6 | 7.3 | 8.0 | 7.2 | 5.9 | 3.6 | 1.9 | 1.5 | 4.7 |

## QUILLAYUTE WA — Latitude: 47.95 degrees — Elevation: 55 meters

|  | Jan | Feb | Mar | Apr | May | Jun | Jul | Aug | Sep | Oct | Nov | Dec | Avg |
|---|---|---|---|---|---|---|---|---|---|---|---|---|---|
| Fixed array |  |  |  |  |  |  |  |  |  |  |  |  |  |
| Lat - 15 | 1.5 | 2.3 | 3.2 | 4.1 | 4.8 | 5.0 | 5.2 | 4.9 | 4.4 | 3.0 | 1.8 | 1.4 | 3.5 |
| Latitude | 1.6 | 2.4 | 3.3 | 4.0 | 4.5 | 4.6 | 4.8 | 4.7 | 4.4 | 3.1 | 1.9 | 1.5 | 3.4 |
| Lat + 15 | 1.7 | 2.4 | 3.2 | 3.7 | 4.0 | 4.0 | 4.2 | 4.2 | 4.2 | 3.1 | 2.0 | 1.6 | 3.2 |
| Single axis tracker |  |  |  |  |  |  |  |  |  |  |  |  |  |
| Lat - 15 | 1.7 | 2.7 | 3.9 | 5.1 | 6.0 | 6.3 | 6.6 | 6.2 | 5.5 | 3.6 | 2.0 | 1.5 | 4.3 |
| Latitude | 1.8 | 2.8 | 4.0 | 5.0 | 5.8 | 6.0 | 6.4 | 6.1 | 5.6 | 3.7 | 2.1 | 1.6 | 4.2 |
| Lat + 15 | 1.8 | 2.8 | 3.9 | 4.7 | 5.5 | 5.6 | 6 | 5.7 | 5.4 | 3.7 | 2.2 | 1.7 | 4.1 |
| Dual axis tracker |  |  |  |  |  |  |  |  |  |  |  |  |  |
|  | 1.9 | 2.8 | 4.0 | 5.1 | 6.1 | 6.5 | 6.8 | 6.2 | 5.6 | 3.7 | 2.2 | 1.7 | 4.4 |

## SEATTLE WA — Latitude: 47.45 degrees — Elevation: 122 meters

|  | Jan | Feb | Mar | Apr | May | Jun | Jul | Aug | Sep | Oct | Nov | Dec | Avg |
|---|---|---|---|---|---|---|---|---|---|---|---|---|---|
| Fixed array |  |  |  |  |  |  |  |  |  |  |  |  |  |
| Lat - 15 | 1.5 | 2.3 | 3.5 | 4.6 | 5.4 | 5.7 | 6.1 | 5.6 | 4.7 | 3.0 | 1.7 | 1.3 | 3.8 |
| Latitude | 1.6 | 2.5 | 3.6 | 4.4 | 5.1 | 5.2 | 5.7 | 5.4 | 4.7 | 3.2 | 1.8 | 1.4 | 3.7 |
| Lat + 15 | 1.7 | 2.5 | 3.5 | 4.1 | 4.5 | 4.5 | 4.9 | 4.9 | 4.5 | 3.2 | 1.8 | 1.4 | 3.5 |
| Single axis tracker |  |  |  |  |  |  |  |  |  |  |  |  |  |
| Lat - 15 | 1.6 | 2.7 | 4.2 | 5.6 | 6.9 | 7.3 | 8.2 | 7.3 | 5.8 | 3.6 | 1.9 | 1.4 | 4.7 |
| Latitude | 1.8 | 2.8 | 4.3 | 5.5 | 6.7 | 7.0 | 7.9 | 7.2 | 5.9 | 3.7 | 2.0 | 1.5 | 4.7 |
| Lat + 15 | 1.8 | 2.8 | 4.2 | 5.3 | 6.3 | 6.6 | 7.4 | 6.8 | 5.7 | 3.7 | 2 | 1.5 | 4.5 |
| Dual axis tracker |  |  |  |  |  |  |  |  |  |  |  |  |  |
|  | 1.8 | 2.9 | 4.3 | 5.6 | 7.0 | 7.5 | 8.3 | 7.4 | 5.9 | 3.7 | 2.0 | 1.5 | 4.9 |

## SPOKANE   WA   Latitude: 47.63 degrees   Elevation: 721 meters

|  | Jan | Feb | Mar | Apr | May | Jun | Jul | Aug | Sep | Oct | Nov | Dec | Avg |
|---|---|---|---|---|---|---|---|---|---|---|---|---|---|
| **Fixed array** | | | | | | | | | | | | | |
| Lat - 15 | 2.1 | 3.0 | 4.2 | 5.3 | 6.0 | 6.4 | 7.0 | 6.6 | 5.7 | 4.0 | 2.1 | 1.7 | 4.5 |
| Latitude | 2.3 | 3.2 | 4.4 | 5.2 | 5.6 | 5.9 | 6.5 | 6.3 | 5.7 | 4.3 | 2.3 | 1.9 | 4.5 |
| Lat + 15 | 2.4 | 3.3 | 4.3 | 4.8 | 4.9 | 5.1 | 5.6 | 5.7 | 5.5 | 4.4 | 2.4 | 2.0 | 4.2 |
| **Single axis tracker** | | | | | | | | | | | | | |
| Lat - 15 | 2.4 | 3.6 | 5.4 | 7.0 | 8.2 | 9.1 | 10.2 | 9.3 | 7.6 | 5.1 | 2.4 | 1.9 | 6.0 |
| Latitude | 2.5 | 3.7 | 5.5 | 6.9 | 8.0 | 8.7 | 9.9 | 9.2 | 7.7 | 5.3 | 2.6 | 2.1 | 6.0 |
| Lat + 15 | 2.6 | 3.8 | 5.4 | 6.6 | 7.5 | 8.2 | 9.3 | 8.7 | 7.5 | 5.3 | 2.6 | 2.2 | 5.8 |
| **Dual axis tracker** | | | | | | | | | | | | | |
|  | 2.7 | 3.8 | 5.5 | 7.0 | 8.3 | 9.3 | 10.4 | 9.4 | 7.7 | 5.4 | 2.7 | 2.2 | 6.2 |

## YAKIMA   WA   Latitude: 46.57 degrees   Elevation: 325 meters

|  | Jan | Feb | Mar | Apr | May | Jun | Jul | Aug | Sep | Oct | Nov | Dec | Avg |
|---|---|---|---|---|---|---|---|---|---|---|---|---|---|
| **Fixed array** | | | | | | | | | | | | | |
| Lat - 15 | 2.2 | 3.3 | 4.7 | 5.7 | 6.4 | 6.8 | 7.3 | 6.8 | 6.0 | 4.4 | 2.5 | 1.9 | 4.8 |
| Latitude | 2.5 | 3.5 | 4.8 | 5.5 | 6.0 | 6.2 | 6.7 | 6.6 | 6.1 | 4.7 | 2.7 | 2.2 | 4.8 |
| Lat + 15 | 2.6 | 3.6 | 4.7 | 5.1 | 5.3 | 5.3 | 5.8 | 5.9 | 5.9 | 4.8 | 2.8 | 2.3 | 4.5 |
| **Single axis tracker** | | | | | | | | | | | | | |
| Lat - 15 | 2.5 | 3.9 | 5.9 | 7.5 | 8.9 | 9.6 | 10.5 | 9.6 | 8.2 | 5.6 | 2.9 | 2.2 | 6.5 |
| Latitude | 2.8 | 4.1 | 6.1 | 7.4 | 8.6 | 9.2 | 10.2 | 9.5 | 8.2 | 5.8 | 3.1 | 2.4 | 6.5 |
| Lat + 15 | 2.9 | 4.2 | 6 | 7.1 | 8.2 | 8.6 | 9.6 | 9 | 8.1 | 5.8 | 3.2 | 2.5 | 6.3 |
| **Dual axis tracker** | | | | | | | | | | | | | |
|  | 2.9 | 4.2 | 6.1 | 7.6 | 9.0 | 9.8 | 10.7 | 9.7 | 8.3 | 5.9 | 3.2 | 2.5 | 6.7 |

## EAU CLAIRE   WI   Latitude: 44.87 degrees   Elevation: 273 meters

|  | Jan | Feb | Mar | Apr | May | Jun | Jul | Aug | Sep | Oct | Nov | Dec | Avg |
|---|---|---|---|---|---|---|---|---|---|---|---|---|---|
| **Fixed array** | | | | | | | | | | | | | |
| Lat - 15 | 2.9 | 4.0 | 4.7 | 5.1 | 5.7 | 6.0 | 6.1 | 5.6 | 4.7 | 3.7 | 2.4 | 2.3 | 4.4 |
| Latitude | 3.3 | 4.4 | 4.9 | 5.0 | 5.3 | 5.5 | 5.6 | 5.4 | 4.7 | 3.9 | 2.7 | 2.6 | 4.4 |
| Lat + 15 | 3.6 | 4.6 | 4.8 | 4.6 | 4.7 | 4.7 | 4.9 | 4.8 | 4.5 | 3.9 | 2.8 | 2.8 | 4.2 |
| **Single axis tracker** | | | | | | | | | | | | | |
| Lat - 15 | 3.5 | 4.9 | 5.9 | 6.6 | 7.5 | 8.0 | 8.2 | 7.4 | 5.9 | 4.5 | 2.8 | 2.6 | 5.7 |
| Latitude | 3.8 | 5.2 | 6.1 | 6.5 | 7.3 | 7.7 | 7.9 | 7.3 | 6.0 | 4.7 | 3.0 | 2.9 | 5.7 |
| Lat + 15 | 4 | 5.4 | 6 | 6.3 | 6.9 | 7.4 | 7.2 | 6.9 | 5.8 | 4.7 | 3.1 | 3.1 | 5.6 |
| **Dual axis tracker** | | | | | | | | | | | | | |
|  | 4.1 | 5.4 | 6.1 | 6.6 | 7.7 | 8.2 | 8.3 | 7.5 | 6.0 | 4.7 | 3.1 | 3.1 | 5.9 |

## GREEN BAY   WI   Latitude: 44.48 degrees   Elevation: 214 meters

|  | Jan | Feb | Mar | Apr | May | Jun | Jul | Aug | Sep | Oct | Nov | Dec | Avg |
|---|---|---|---|---|---|---|---|---|---|---|---|---|---|
| **Fixed array** | | | | | | | | | | | | | |
| Lat - 15 | 2.9 | 3.8 | 4.6 | 5.2 | 5.8 | 6.1 | 6.1 | 5.6 | 4.7 | 3.6 | 2.4 | 2.3 | 4.4 |
| Latitude | 3.3 | 4.2 | 4.8 | 5.0 | 5.4 | 5.6 | 5.7 | 5.4 | 4.7 | 3.8 | 2.6 | 2.6 | 4.4 |
| Lat + 15 | 3.5 | 4.3 | 4.7 | 4.7 | 4.8 | 4.9 | 4.9 | 4.9 | 4.5 | 3.8 | 2.7 | 2.8 | 4.2 |
| **Single axis tracker** | | | | | | | | | | | | | |
| Lat - 15 | 3.4 | 4.7 | 5.8 | 6.6 | 7.7 | 8.2 | 8.2 | 7.4 | 6.0 | 4.4 | 2.8 | 2.6 | 5.7 |
| Latitude | 3.8 | 5.0 | 6.0 | 6.5 | 7.4 | 7.9 | 7.9 | 7.2 | 6.0 | 4.6 | 3.0 | 2.9 | 5.7 |
| Lat + 15 | 3.9 | 5.1 | 5.9 | 6.3 | 7 | 7.4 | 7.4 | 6.9 | 5.8 | 4.6 | 3.1 | 3.1 | 5.5 |
| **Dual axis tracker** | | | | | | | | | | | | | |
|  | 4.0 | 5.1 | 6.0 | 6.6 | 7.8 | 8.4 | 8.4 | 7.4 | 6.0 | 4.6 | 3.1 | 3.1 | 5.9 |

## LA CROSSE     WI     Latitude: 43.87 degrees     Elevation: 205 meters

|  | Jan | Feb | Mar | Apr | May | Jun | Jul | Aug | Sep | Oct | Nov | Dec | Avg |
|---|---|---|---|---|---|---|---|---|---|---|---|---|---|
| **Fixed array** | | | | | | | | | | | | | |
| Lat - 15 | 2.9 | 3.9 | 4.6 | 5.2 | 5.8 | 6.1 | 6.2 | 5.8 | 4.8 | 3.8 | 2.6 | 2.3 | 4.5 |
| Latitude | 3.3 | 4.3 | 4.8 | 5.0 | 5.4 | 5.6 | 5.8 | 5.5 | 4.8 | 4.1 | 2.8 | 2.6 | 4.5 |
| Lat + 15 | 3.6 | 4.4 | 4.7 | 4.7 | 4.8 | 4.9 | 5.0 | 5.0 | 4.6 | 4.1 | 2.9 | 2.8 | 4.3 |
| **Single axis tracker** | | | | | | | | | | | | | |
| Lat - 15 | 3.5 | 4.8 | 5.8 | 6.6 | 7.6 | 8.2 | 8.3 | 7.6 | 6.1 | 4.7 | 3.0 | 2.7 | 5.7 |
| Latitude | 3.9 | 5.1 | 5.9 | 6.5 | 7.4 | 7.9 | 8.0 | 7.4 | 6.1 | 4.9 | 3.2 | 3.0 | 5.8 |
| Lat + 15 | 4.1 | 5.2 | 5.9 | 6.2 | 7 | 7.3 | 7.5 | 7.1 | 5.9 | 4.9 | 3.3 | 3.1 | 5.6 |
| **Dual axis tracker** | | | | | | | | | | | | | |
|  | 4.1 | 5.2 | 6.0 | 6.6 | 7.7 | 8.4 | 8.5 | 7.6 | 6.1 | 4.9 | 3.3 | 3.2 | 6.0 |

## MADISON     WI     Latitude: 43.13 degrees     Elevation: 262 meters

|  | Jan | Feb | Mar | Apr | May | Jun | Jul | Aug | Sep | Oct | Nov | Dec | Avg |
|---|---|---|---|---|---|---|---|---|---|---|---|---|---|
| **Fixed array** | | | | | | | | | | | | | |
| Lat - 15 | 3.0 | 3.9 | 4.5 | 5.1 | 5.8 | 6.2 | 6.2 | 5.7 | 4.8 | 3.8 | 2.5 | 2.3 | 4.5 |
| Latitude | 3.4 | 4.3 | 4.7 | 5.0 | 5.5 | 5.7 | 5.8 | 5.5 | 4.8 | 4.0 | 2.8 | 2.6 | 4.5 |
| Lat + 15 | 3.6 | 4.4 | 4.6 | 4.6 | 4.8 | 4.9 | 5.0 | 5.0 | 4.6 | 4.0 | 2.9 | 2.8 | 4.3 |
| **Single axis tracker** | | | | | | | | | | | | | |
| Lat - 15 | 3.5 | 4.8 | 5.6 | 6.5 | 7.5 | 8.1 | 8.0 | 7.3 | 6.0 | 4.6 | 2.9 | 2.7 | 5.6 |
| Latitude | 3.9 | 5.0 | 5.8 | 6.4 | 7.3 | 7.8 | 7.7 | 7.1 | 6.0 | 4.8 | 3.2 | 3.0 | 5.7 |
| Lat + 15 | 4 | 5.2 | 5.7 | 6.1 | 6.8 | 7.2 | 7.2 | 6.8 | 5.8 | 4.8 | 3.2 | 3.1 | 5.5 |
| **Dual axis tracker** | | | | | | | | | | | | | |
|  | 4.1 | 5.2 | 5.8 | 6.5 | 7.6 | 8.3 | 8.2 | 7.3 | 6.0 | 4.8 | 3.3 | 3.2 | 5.9 |

## MILWAUKEE     WI     Latitude: 42.95 degrees     Elevation: 211 meters

|  | Jan | Feb | Mar | Apr | May | Jun | Jul | Aug | Sep | Oct | Nov | Dec | Avg |
|---|---|---|---|---|---|---|---|---|---|---|---|---|---|
| **Fixed array** | | | | | | | | | | | | | |
| Lat - 15 | 2.8 | 3.6 | 4.3 | 5.1 | 5.9 | 6.2 | 6.3 | 5.8 | 4.9 | 3.8 | 2.5 | 2.2 | 4.5 |
| Latitude | 3.2 | 3.9 | 4.4 | 4.9 | 5.5 | 5.7 | 5.8 | 5.6 | 4.9 | 4.0 | 2.8 | 2.5 | 4.5 |
| Lat + 15 | 3.4 | 4.1 | 4.4 | 4.6 | 4.9 | 5.0 | 5.1 | 5.0 | 4.7 | 4.0 | 2.9 | 2.7 | 4.2 |
| **Single axis tracker** | | | | | | | | | | | | | |
| Lat - 15 | 3.3 | 4.4 | 5.3 | 6.4 | 7.6 | 8.3 | 8.4 | 7.6 | 6.2 | 4.7 | 2.6 | 5.6 | 5.9 |
| Latitude | 3.7 | 4.6 | 5.4 | 6.3 | 7.4 | 7.9 | 8.1 | 7.4 | 6.2 | 4.9 | 3.2 | 2.8 | 5.7 |
| Lat + 15 | 3.8 | 4.7 | 54 | 6 | 6.9 | 7.4 | 7.6 | 7.1 | 6.1 | 4.9 | 3.2 | 3 | 5.5 |
| **Dual axis tracker** | | | | | | | | | | | | | |
|  | 3.9 | 4.7 | 5.5 | 6.4 | 7.7 | 8.5 | 8.5 | 7.6 | 6.3 | 4.9 | 3.3 | 3.0 | 5.9 |

## CHARLESTON     WV     Latitude: 38.37 degrees     Elevation: 290 meters

|  | Jan | Feb | Mar | Apr | May | Jun | Jul | Aug | Sep | Oct | Nov | Dec | Avg |
|---|---|---|---|---|---|---|---|---|---|---|---|---|---|
| **Fixed array** | | | | | | | | | | | | | |
| Lat - 15 | 2.7 | 3.3 | 4.3 | 5.1 | 5.6 | 5.9 | 5.7 | 5.5 | 4.9 | 4.1 | 2.9 | 2.3 | 4.4 |
| Latitude | 2.9 | 3.6 | 4.4 | 5.0 | 5.3 | 5.4 | 5.3 | 5.3 | 4.9 | 4.4 | 3.1 | 2.6 | 4.4 |
| Lat + 15 | 3.1 | 3.7 | 4.3 | 4.6 | 4.7 | 4.7 | 4.7 | 4.9 | 4.7 | 4.4 | 3.3 | 2.7 | 4.1 |
| **Single axis tracker** | | | | | | | | | | | | | |
| Lat - 15 | 3.1 | 4.0 | 5.3 | 6.4 | 7.1 | 7.5 | 7.2 | 6.9 | 6.0 | 5.1 | 3.4 | 2.7 | 5.4 |
| Latitude | 3.3 | 4.2 | 5.3 | 6.3 | 6.9 | 7.2 | 6.9 | 6.7 | 6.1 | 5.3 | 3.6 | 2.9 | 5.4 |
| Lat + 15 | 3.5 | 4.2 | 5.3 | 6 | 6.5 | 6.7 | 6.5 | 6.4 | 5.9 | 5.3 | 3.7 | 3 | 5.2 |
| **Dual axis tracker** | | | | | | | | | | | | | |
|  | 3.5 | 4.3 | 5.4 | 6.4 | 7.2 | 7.6 | 7.3 | 6.9 | 6.1 | 5.3 | 3.7 | 3.0 | 5.6 |

## ELKINS     WV     Latitude: 38.88 degrees     Elevation: 594 meters

| | Jan | Feb | Mar | Apr | May | Jun | Jul | Aug | Sep | Oct | Nov | Dec | Avg |
|---|---|---|---|---|---|---|---|---|---|---|---|---|---|
| Fixed array | | | | | | | | | | | | | |
| Lat - 15 | 2.6 | 3.3 | 4.1 | 4.8 | 5.3 | 5.6 | 5.5 | 5.2 | 4.6 | 3.9 | 2.7 | 2.2 | 4.2 |
| Latitude | 2.9 | 3.5 | 4.2 | 4.7 | 5.0 | 5.2 | 5.1 | 5.1 | 4.7 | 4.1 | 2.9 | 2.4 | 4.2 |
| Lat + 15 | 3.0 | 3.6 | 4.1 | 4.4 | 4.4 | 4.5 | 4.5 | 4.6 | 4.5 | 4.1 | 3.0 | 2.6 | 3.9 |
| Single axis tracker | | | | | | | | | | | | | |
| Lat - 15 | 3.0 | 3.9 | 5.0 | 6.0 | 6.6 | 6.9 | 6.7 | 6.4 | 5.6 | 4.6 | 3.1 | 2.5 | 5.0 |
| Latitude | 3.2 | 4.0 | 5.0 | 5.9 | 6.4 | 6.6 | 6.4 | 6.2 | 5.6 | 4.8 | 3.3 | 2.7 | 5.0 |
| Lat + 15 | 3.3 | 4.1 | 5 | 5.6 | 6 | 6.1 | 6 | 5.9 | 5.4 | 4.8 | 3.4 | 2.8 | 4.9 |
| Dual axis tracker | | | | | | | | | | | | | |
| | 3.4 | 4.1 | 5.1 | 6.0 | 6.7 | 7.1 | 6.8 | 6.4 | 5.6 | 4.9 | 3.4 | 2.9 | 5.2 |

## HUNTINGTON     WV     Latitude: 38.37 degrees     Elevation: 255 meters

| | Jan | Feb | Mar | Apr | May | Jun | Jul | Aug | Sep | Oct | Nov | Dec | Avg |
|---|---|---|---|---|---|---|---|---|---|---|---|---|---|
| Fixed array | | | | | | | | | | | | | |
| Lat - 15 | 2.6 | 3.4 | 4.3 | 5.1 | 5.6 | 5.9 | 5.7 | 5.4 | 4.9 | 4.1 | 2.8 | 2.3 | 4.4 |
| Latitude | 2.9 | 3.6 | 4.4 | 4.9 | 5.3 | 5.4 | 5.3 | 5.2 | 4.9 | 4.4 | 3.1 | 2.5 | 4.3 |
| Lat + 15 | 3.1 | 3.7 | 4.3 | 4.6 | 4.7 | 4.7 | 4.7 | 4.8 | 4.7 | 4.4 | 3.2 | 2.7 | 4.1 |
| Single axis tracker | | | | | | | | | | | | | |
| Lat - 15 | 3.1 | 4.1 | 5.2 | 6.4 | 7.1 | 7.5 | 7.3 | 6.8 | 6.1 | 5.1 | 3.3 | 2.6 | 5.4 |
| Latitude | 3.3 | 4.3 | 5.3 | 6.3 | 6.9 | 7.2 | 7.0 | 6.7 | 6.1 | 5.3 | 3.6 | 2.8 | 5.4 |
| Lat + 15 | 3.4 | 4.3 | 5.2 | 6 | 6.5 | 6.7 | 6.6 | 6.4 | 5.9 | 5.3 | 3.7 | 2.9 | 5.3 |
| Dual axis tracker | | | | | | | | | | | | | |
| | 3.5 | 4.3 | 5.4 | 6.4 | 7.2 | 7.7 | 7.4 | 6.9 | 6.1 | 5.4 | 3.7 | 3.0 | 5.6 |

## CASPER     WY     Latitude: 42.92 degrees     Elevation: 1612 meters

| | Jan | Feb | Mar | Apr | May | Jun | Jul | Aug | Sep | Oct | Nov | Dec | Avg |
|---|---|---|---|---|---|---|---|---|---|---|---|---|---|
| Fixed array | | | | | | | | | | | | | |
| Lat - 15 | 3.4 | 4.3 | 5.2 | 5.8 | 6.2 | 6.8 | 7.0 | 6.8 | 6.0 | 4.8 | 3.6 | 3.1 | 5.2 |
| Latitude | 3.9 | 4.7 | 5.4 | 5.7 | 5.8 | 6.2 | 6.5 | 6.5 | 6.1 | 5.2 | 4.1 | 3.6 | 5.3 |
| Lat + 15 | 4.3 | 4.9 | 5.3 | 5.3 | 5.1 | 5.3 | 5.6 | 5.9 | 5.9 | 5.3 | 4.3 | 3.9 | 5.1 |
| Single axis tracker | | | | | | | | | | | | | |
| Lat - 15 | 4.2 | 5.4 | 6.7 | 7.7 | 8.4 | 9.6 | 9.9 | 9.5 | 8.1 | 6.3 | 4.5 | 3.8 | 7.0 |
| Latitude | 4.6 | 5.8 | 6.9 | 7.6 | 8.2 | 9.2 | 9.5 | 9.3 | 8.2 | 6.6 | 4.9 | 4.2 | 7.1 |
| Lat + 15 | 4.9 | 5.9 | 6.9 | 7.3 | 7.7 | 8.6 | 9 | 8.9 | 8 | 6.6 | 5.1 | 4.4 | 7 |
| Dual axis tracker | | | | | | | | | | | | | |
| | 4.9 | 5.9 | 6.9 | 7.7 | 8.5 | 9.8 | 10.1 | 9.5 | 8.2 | 6.7 | 5.1 | 4.5 | 7.3 |

## CHEYENNE     WY     Latitude: 41.15 degrees     Elevation: 1872 meters

| | Jan | Feb | Mar | Apr | May | Jun | Jul | Aug | Sep | Oct | Nov | Dec | Avg |
|---|---|---|---|---|---|---|---|---|---|---|---|---|---|
| Fixed array | | | | | | | | | | | | | |
| Lat - 15 | 3.6 | 4.4 | 5.3 | 5.9 | 6.0 | 6.5 | 6.6 | 6.3 | 5.8 | 5.0 | 3.8 | 3.3 | 5.2 |
| Latitude | 4.1 | 4.9 | 5.5 | 5.8 | 5.6 | 6.0 | 6.1 | 6.1 | 6.0 | 5.4 | 4.3 | 3.9 | 5.3 |
| Lat + 15 | 4.5 | 5.1 | 5.5 | 5.4 | 5.0 | 5.1 | 5.3 | 5.5 | 5.7 | 5.5 | 4.6 | 4.2 | 5.1 |
| Single axis tracker | | | | | | | | | | | | | |
| Lat - 15 | 4.5 | 5.7 | 6.9 | 7.9 | 8.1 | 9.0 | 9.2 | 8.7 | 7.8 | 6.6 | 4.8 | 4.1 | 6.9 |
| Latitude | 4.9 | 6.0 | 7.1 | 7.8 | 7.9 | 8.7 | 8.9 | 8.6 | 7.9 | 6.9 | 5.2 | 4.6 | 7.0 |
| Lat + 15 | 5.2 | 6.2 | 7 | 7.5 | 7.4 | 8.1 | 8.3 | 8.2 | 7.7 | 6.9 | 5.4 | 4.9 | 6.9 |
| Dual axis tracker | | | | | | | | | | | | | |
| | 5.2 | 6.2 | 7.1 | 7.9 | 8.2 | 9.3 | 9.3 | 8.8 | 7.9 | 6.9 | 5.4 | 4.9 | 7.3 |

## LANDER WY Latitude: 42.82 degrees Elevation: 1696 meters

| | Jan | Feb | Mar | Apr | May | Jun | Jul | Aug | Sep | Oct | Nov | Dec | Avg |
|---|---|---|---|---|---|---|---|---|---|---|---|---|---|
| Fixed array | | | | | | | | | | | | | |
| Lat - 15 | 3.7 | 4.8 | 5.7 | 6.2 | 6.5 | 7.0 | 7.0 | 6.8 | 6.1 | 5.1 | 3.8 | 3.4 | 5.5 |
| Latitude | 4.3 | 5.3 | 6.0 | 6.1 | 6.1 | 6.4 | 6.5 | 6.5 | 6.2 | 5.5 | 4.3 | 4.0 | 5.6 |
| Lat + 15 | 4.6 | 5.6 | 5.9 | 5.7 | 5.4 | 5.5 | 5.6 | 5.9 | 6.0 | 5.6 | 4.6 | 4.3 | 5.4 |
| Single axis tracker | | | | | | | | | | | | | |
| Lat - 15 | 4.6 | 6.1 | 7.4 | 8.3 | 8.7 | 9.8 | 9.9 | 9.4 | 8.2 | 6.6 | 4.8 | 4.1 | 7.3 |
| Latitude | 5.0 | 6.5 | 7.6 | 8.3 | 8.5 | 9.4 | 9.5 | 9.3 | 8.3 | 6.9 | 5.2 | 4.6 | 7.4 |
| Lat + 15 | 5.3 | 6.7 | 7.6 | 8 | 8 | 8.8 | 9 | 8.8 | 8.1 | 7 | 5.4 | 4.9 | 7.3 |
| Dual axis tracker | | | | | | | | | | | | | |
| | 5.4 | 6.7 | 7.7 | 8.3 | 8.9 | 10.0 | 10.1 | 9.5 | 8.4 | 7.0 | 5.4 | 5.0 | 7.7 |

## ROCK SPRINGS WY Latitude: 41.60 degrees Elevation: 2056 meters

| | Jan | Feb | Mar | Apr | May | Jun | Jul | Aug | Sep | Oct | Nov | Dec | Avg |
|---|---|---|---|---|---|---|---|---|---|---|---|---|---|
| Fixed array | | | | | | | | | | | | | |
| Lat - 15 | 3.5 | 4.4 | 5.3 | 6.0 | 6.5 | 7.0 | 7.1 | 6.9 | 6.3 | 5.2 | 3.7 | 3.2 | 5.4 |
| Latitude | 4.0 | 4.8 | 5.5 | 5.9 | 6.1 | 6.4 | 6.6 | 6.6 | 6.4 | 5.6 | 4.1 | 3.7 | 5.5 |
| Lat + 15 | 4.3 | 5.1 | 5.5 | 5.5 | 5.4 | 5.5 | 5.7 | 6.0 | 6.1 | 5.7 | 4.4 | 4.1 | 5.3 |
| Single axis tracker | | | | | | | | | | | | | |
| Lat - 15 | 4.3 | 5.6 | 6.9 | 8.0 | 9.0 | 10.0 | 10.1 | 9.6 | 8.5 | 6.8 | 4.6 | 3.9 | 7.3 |
| Latitude | 4.7 | 6.0 | 7.1 | 8.0 | 8.7 | 9.6 | 9.8 | 9.4 | 8.6 | 7.1 | 4.9 | 4.4 | 7.4 |
| Lat + 15 | 5 | 6.1 | 7 | 7.7 | 8.2 | 9 | 9.2 | 9 | 8.5 | 7.2 | 5.1 | 4.6 | 7.2 |
| Dual axis tracker | | | | | | | | | | | | | |
| | 5.0 | 6.1 | 7.1 | 8.1 | 9.1 | 10.3 | 10.3 | 9.6 | 8.7 | 7.2 | 5.2 | 4.7 | 7.6 |

## SHERIDAN WY Latitude: 44.77 degrees Elevation: 1209 meters

| | Jan | Feb | Mar | Apr | May | Jun | Jul | Aug | Sep | Oct | Nov | Dec | Avg |
|---|---|---|---|---|---|---|---|---|---|---|---|---|---|
| Fixed array | | | | | | | | | | | | | |
| Lat - 15 | 3.1 | 4.0 | 5.0 | 5.6 | 5.9 | 6.5 | 6.9 | 6.6 | 5.7 | 4.5 | 3.2 | 2.8 | 5.0 |
| Latitude | 3.5 | 4.4 | 5.2 | 5.5 | 5.6 | 6.0 | 6.4 | 6.3 | 5.8 | 4.8 | 3.6 | 3.2 | 5.0 |
| Lat + 15 | 3.7 | 4.6 | 5.1 | 5.1 | 4.9 | 5.1 | 5.5 | 5.7 | 5.5 | 4.9 | 3.8 | 3.5 | 4.8 |
| Single axis tracker | | | | | | | | | | | | | |
| Lat - 15 | 3.7 | 5.0 | 6.4 | 7.4 | 8.0 | 9.1 | 9.8 | 9.2 | 7.6 | 5.7 | 3.9 | 3.3 | 6.6 |
| Latitude | 4.0 | 5.3 | 6.5 | 7.4 | 7.8 | 8.8 | 9.5 | 9.0 | 7.7 | 6.0 | 4.3 | 3.7 | 6.7 |
| Lat + 15 | 4.2 | 5.4 | 6.5 | 7.1 | 7.3 | 8.2 | 8.9 | 8.6 | 7.5 | 6 | 4.4 | 3.9 | 6.5 |
| Dual axis tracker | | | | | | | | | | | | | |
| | 4.3 | 5.4 | 6.5 | 7.5 | 8.1 | 9.4 | 10.0 | 9.2 | 7.7 | 6.0 | 4.4 | 4.0 | 6.9 |

# World Daily Insolation Data (KWh/m²)

## BISKRA, ALGERIA — Location: 34.85° N, 5.73° W, 124 Meters

| | Jan | Feb | Mar | Apr | May | Jun | Jul | Aug | Sep | Oct | Nov | Dec | Yr |
|---|---|---|---|---|---|---|---|---|---|---|---|---|---|
| Latitude Tilt -15° | | | | | | | | | | | | | |
| Fixed Array | 4.21 | 4.98 | 5.64 | 6.01 | 6.36 | 6.63 | 6.79 | 6.60 | 5.69 | 4.86 | 4.05 | 3.79 | 5.47 |
| Tracking Array | 4.87 | 6.29 | 7.05 | 7.97 | 8.74 | 9.21 | 9.36 | 8.83 | 7.25 | 6.07 | 4.69 | 4.20 | 7.04 |
| Latitude Tilt° | | | | | | | | | | | | | |
| Fixed Array | 4.86 | 5.47 | 5.82 | 5.84 | 5.93 | 6.07 | 6.27 | 6.30 | 5.74 | 5.22 | 4.60 | 4.43 | 5.55 |
| Tracking Array | 5.64 | 6.97 | 7.40 | 7.93 | 8.33 | 8.60 | 8.82 | 8.63 | 7.46 | 6.61 | 5.37 | 4.94 | 7.22 |
| Latitude Tilt +15° | | | | | | | | | | | | | |
| Fixed Array | 5.23 | 5.66 | 5.70 | 5.39 | 5.26 | 5.28 | 5.49 | 5.72 | 5.49 | 5.30 | 4.89 | 4.82 | 5.35 |
| Tracking Array | 6.02 | 7.18 | 7.26 | 7.35 | 7.35 | 7.40 | 7.69 | 7.85 | 7.16 | 6.70 | 5.68 | 5.33 | 6.91 |
| Two Axis Tracking | 6.07 | 7.19 | 7.43 | 8.03 | 8.76 | 9.29 | 9.40 | 8.85 | 7.47 | 6.73 | 5.70 | 5.40 | 7.35 |

## LUANDA, ANGOLA — Location: 8.82° N, 13.22° W, 42 Meters

| | Jan | Feb | Mar | Apr | May | Jun | Jul | Aug | Sep | Oct | Nov | Dec | Yr |
|---|---|---|---|---|---|---|---|---|---|---|---|---|---|
| Latitude Tilt -15° | | | | | | | | | | | | | |
| Fixed Array | 5.92 | 6.07 | 5.43 | 4.89 | 4.60 | 4.18 | 3.36 | 3.70 | 4.57 | 5.06 | 5.60 | 6.16 | 4.96 |
| Tracking Array | 7.62 | 7.83 | 7.02 | 6.19 | 5.61 | 5.01 | 4.17 | 4.75 | 5.96 | 6.66 | 7.27 | 7.87 | 6.33 |
| Latitude Tilt° | | | | | | | | | | | | | |
| Fixed Array | 5.56 | 5.87 | 5.49 | 5.19 | 5.11 | 4.75 | 3.71 | 3.95 | 4.68 | 4.97 | 5.31 | 5.72 | 5.03 |
| Tracking Array | 7.20 | 7.66 | 7.19 | 6.68 | 6.34 | 5.80 | 4.78 | 5.21 | 6.21 | 6.60 | 6.93 | 7.36 | 6.50 |
| Latitude Tilt +15° | | | | | | | | | | | | | |
| Fixed Array | 4.94 | 5.40 | 5.30 | 5.27 | 5.42 | 5.14 | 3.93 | 4.04 | 4.60 | 4.66 | 4.77 | 5.02 | 4.87 |
| Tracking Array | 6.28 | 6.96 | 6.89 | 6.76 | 6.70 | 6.27 | 5.11 | 5.36 | 6.08 | 6.11 | 6.11 | 6.33 | 6.25 |
| Two Axis Tracking | 7.67 | 7.84 | 7.20 | 6.79 | 6.73 | 6.34 | 5.14 | 5.37 | 6.23 | 6.69 | 7.30 | 7.95 | 6.77 |

## BUENOS AIRES, ARGENTINA — Location: 34° 58' S, 58° 48'W, 25 Meters

| | Jan | Feb | Mar | Apr | May | Jun | Jul | Aug | Sep | Oct | Nov | Dec | Yr |
|---|---|---|---|---|---|---|---|---|---|---|---|---|---|
| Latitude Tilt -15° | | | | | | | | | | | | | |
| Fixed Array | 7.13 | 6.49 | 5.45 | 4.46 | 3.57 | 2.93 | 3.24 | 4.11 | 5.07 | 5.90 | 6.47 | 7.12 | 5.16 |
| Tracking Array | 9.80 | 8.72 | 7.02 | 5.50 | 4.07 | 3.13 | 3.57 | 4.98 | 6.38 | 7.86 | 8.90 | 9.85 | 6.65 |
| Latitude Tilt° | | | | | | | | | | | | | |
| Fixed Array | 6.58 | 6.19 | 5.47 | 4.75 | 4.02 | 3.39 | 3.70 | 4.48 | 5.19 | 5.71 | 6.02 | 6.51 | 5.17 |
| Tracking Array | 9.24 | 8.52 | 7.20 | 5.97 | 4.64 | 3.67 | 4.14 | 5.51 | 6.68 | 7.80 | 8.46 | 9.18 | 6.75 |
| Latitude Tilt +15° | | | | | | | | | | | | | |
| Fixed Array | 5.77 | 5.62 | 5.21 | 4.80 | 4.25 | 3.65 | 3.95 | 4.60 | 5.06 | 5.27 | 5.33 | 5.65 | 4.93 |
| Tracking Array | 8.05 | 7.74 | 6.89 | 6.03 | 4.90 | 3.96 | 4.42 | 5.66 | 6.52 | 7.21 | 7.44 | 7.88 | 6.39 |
| Two Axis Tracking | 9.85 | 8.74 | 7.22 | 6.07 | 4.91 | 4.01 | 4.45 | 5.67 | 6.70 | 7.91 | 8.92 | 9.94 | 7.03 |

## CORRIENTES, ARGENTINA — Location: 27.47° S, 58.82° W, 52 Meters

| | Jan | Feb | Mar | Apr | May | Jun | Jul | Aug | Sep | Oct | Nov | Dec | Yr |
|---|---|---|---|---|---|---|---|---|---|---|---|---|---|
| Latitude Tilt -15° | | | | | | | | | | | | | |
| Fixed Array | 6.75 | 6.36 | 5.68 | 4.71 | 3.90 | 3.49 | 3.61 | 4.40 | 5.30 | 5.97 | 6.65 | 6.62 | 5.29 |
| Tracking Array | 9.08 | 8.44 | 7.29 | 5.75 | 4.64 | 3.98 | 4.19 | 5.34 | 6.69 | 7.87 | 8.92 | 8.95 | 6.76 |
| Latitude Tilt° | | | | | | | | | | | | | |
| Fixed Array | 6.26 | 6.09 | 5.71 | 5.02 | 4.38 | 4.03 | 4.11 | 4.79 | 5.44 | 5.80 | 6.21 | 6.08 | 5.32 |
| Tracking Array | 8.56 | 8.24 | 7.47 | 6.24 | 5.29 | 4.67 | 4.85 | 5.90 | 7.00 | 7.81 | 8.47 | 8.33 | 6.90 |
| Latitude Tilt +15° | | | | | | | | | | | | | |
| Fixed Array | 5.51 | 5.54 | 5.45 | 5.07 | 4.63 | 4.35 | 4.39 | 4.93 | 5.31 | 5.36 | 5.51 | 5.30 | 5.11 |
| Tracking Array | 7.45 | 7.48 | 7.15 | 6.30 | 5.58 | 5.03 | 5.18 | 6.07 | 6.83 | 7.21 | 7.45 | 7.15 | 6.57 |
| Two Axis Tracking | 9.13 | 8.45 | 7.49 | 6.34 | 5.60 | 5.09 | 5.21 | 6.08 | 7.02 | 7.92 | 8.94 | 9.03 | 7.19 |

## PATAUGNES, ARGENTINA        Location:  40.80° S,  62.98° W,  34 Meters

|  | Jan | Feb | Mar | Apr | May | Jun | Jul | Aug | Sep | Oct | Nov | Dec | Yr |
|---|---|---|---|---|---|---|---|---|---|---|---|---|---|
| Latitude Tilt -15° | | | | | | | | | | | | | |
| Fixed Array | 6.88 | 6.46 | 5.45 | 4.43 | 3.36 | 2.96 | 2.82 | 4.05 | 4.91 | 5.60 | 6.53 | 6.93 | 5.03 |
| Tracking Array | 9.44 | 8.57 | 7.02 | 5.40 | 3.64 | 3.01 | 2.91 | 4.74 | 6.24 | 7.50 | 8.93 | 9.60 | 6.42 |
| Latitude Tilt° | | | | | | | | | | | | | |
| Fixed Array | 6.69 | 6.44 | 5.57 | 4.81 | 3.85 | 3.48 | 3.27 | 4.49 | 5.12 | 5.53 | 6.39 | 6.69 | 5.19 |
| Tracking Array | 8.91 | 8.38 | 7.21 | 5.87 | 4.15 | 3.53 | 3.37 | 5.25 | 6.54 | 7.45 | 8.50 | 8.96 | 6.51 |
| Latitude Tilt +15° | | | | | | | | | | | | | |
| Fixed Array | 6.37 | 6.27 | 5.45 | 4.96 | 4.13 | 3.83 | 3.54 | 4.71 | 5.10 | 5.26 | 6.12 | 6.34 | 5.17 |
| Tracking Array | 7.77 | 7.62 | 6.91 | 5.93 | 4.39 | 3.83 | 3.61 | 5.40 | 6.39 | 6.89 | 7.49 | 7.70 | 6.16 |
| Two Axis Tracking | 9.48 | 8.59 | 7.23 | 5.97 | 4.40 | 3.86 | 3.63 | 5.41 | 6.56 | 7.56 | 8.95 | 9.68 | 6.78 |

## SAN CARLOS DE BARILOCHE, ARGENTINA  Location:  41.2° S,  71.3° W,  825 Meters

|  | Jan | Feb | Mar | Apr | May | Jun | Jul | Aug | Sep | Oct | Nov | Dec | Yr |
|---|---|---|---|---|---|---|---|---|---|---|---|---|---|
| Latitude Tilt -15° | | | | | | | | | | | | | |
| Fixed Array | 6.99 | 6.78 | 5.48 | 3.95 | 2.78 | 1.81 | 2.39 | 3.39 | 4.82 | 5.94 | 6.96 | 6.62 | 4.83 |
| Tracking Array | 9.59 | 8.95 | 7.07 | 4.69 | 2.92 | 1.81 | 2.42 | 3.83 | 6.13 | 7.91 | 9.47 | 9.21 | 6.17 |
| Latitude Tilt° | | | | | | | | | | | | | |
| Fixed Array | 6.79 | 6.76 | 5.58 | 4.26 | 3.15 | 2.08 | 2.74 | 3.73 | 5.02 | 5.86 | 6.80 | 6.39 | 4.93 |
| Tracking Array | 9.02 | 8.73 | 7.24 | 5.07 | 3.32 | 2.08 | 2.79 | 4.23 | 6.41 | 7.83 | 8.98 | 8.56 | 6.19 |
| Latitude Tilt +15° | | | | | | | | | | | | | |
| Fixed Array | 6.47 | 6.57 | 5.46 | 4.37 | 3.36 | 2.25 | 2.95 | 3.89 | 4.99 | 5.57 | 6.51 | 6.07 | 4.87 |
| Tracking Array | 7.84 | 7.91 | 6.92 | 5.11 | 3.50 | 2.25 | 2.99 | 4.34 | 6.25 | 7.22 | 7.89 | 7.33 | 5.80 |
| Two Axis Tracking | 9.64 | 8.97 | 7.26 | 5.15 | 3.51 | 2.26 | 2.99 | 4.35 | 6.42 | 7.95 | 9.49 | 9.30 | 6.44 |

## SANTIAGO DEL ESTERO, ARGENTINA  Location:  27.8° S,  64.3° W

|  | Jan | Feb | Mar | Apr | May | Jun | Jul | Aug | Sep | Oct | Nov | Dec | Yr |
|---|---|---|---|---|---|---|---|---|---|---|---|---|---|
| Latitude Tilt -15° | | | | | | | | | | | | | |
| Fixed Array | 6.42 | 6.02 | 5.44 | 4.44 | 3.62 | 3.20 | 3.77 | 4.55 | 5.17 | 6.22 | 6.54 | 6.52 | 5.16 |
| Tracking Array | 8.68 | 8.03 | 7.02 | 5.46 | 4.21 | 3.57 | 4.43 | 5.53 | 6.54 | 8.17 | 8.79 | 8.83 | 6.60 |
| Latitude Tilt° | | | | | | | | | | | | | |
| Fixed Array | 5.96 | 5.77 | 5.47 | 4.73 | 4.06 | 3.68 | 4.31 | 4.97 | 5.31 | 6.05 | 6.11 | 5.99 | 5.20 |
| Tracking Array | 8.19 | 7.85 | 7.21 | 5.93 | 4.81 | 4.19 | 5.13 | 6.13 | 6.85 | 8.11 | 8.37 | 8.23 | 6.75 |
| Latitude Tilt +15° | | | | | | | | | | | | | |
| Fixed Array | 5.27 | 5.26 | 5.23 | 4.77 | 4.28 | 3.97 | 4.62 | 5.12 | 5.18 | 5.60 | 5.43 | 5.24 | 5.00 |
| Tracking Array | 7.14 | 7.14 | 6.91 | 6.00 | 5.09 | 4.53 | 5.48 | 6.31 | 6.70 | 7.51 | 7.37 | 7.08 | 6.44 |
| Two Axis Tracking | 8.71 | 8.04 | 7.22 | 6.03 | 5.10 | 4.58 | 5.52 | 6.32 | 6.87 | 8.23 | 8.81 | 8.90 | 7.03 |

## DARWIN, AUSTRALIA        Location:  12.43° S, 30.87° W, 27 Meters

|  | Jan | Feb | Mar | Apr | May | Jun | Jul | Aug | Sep | Oct | Nov | Dec | Yr |
|---|---|---|---|---|---|---|---|---|---|---|---|---|---|
| Latitude Tilt -15° | | | | | | | | | | | | | |
| Fixed Array | 5.17 | 5.33 | 5.57 | 5.05 | 5.14 | 4.96 | 5.25 | 6.14 | 6.41 | 6.52 | 6.22 | 5.68 | 5.62 |
| Tracking Array | 6.83 | 7.02 | 7.18 | 6.34 | 6.15 | 5.79 | 6.19 | 7.46 | 8.09 | 8.38 | 8.06 | 7.41 | 7.08 |
| Latitude Tilt° | | | | | | | | | | | | | |
| Fixed Array | 4.87 | 5.15 | 5.61 | 5.35 | 5.75 | 5.71 | 5.98 | 6.69 | 6.60 | 6.37 | 5.86 | 5.28 | 5.77 |
| Tracking Array | 6.45 | 6.85 | 7.34 | 6.82 | 6.93 | 6.69 | 7.07 | 8.17 | 8.42 | 8.30 | 7.67 | 6.91 | 7.30 |
| Latitude Tilt +15° | | | | | | | | | | | | | |
| Fixed Array | 4.36 | 4.75 | 5.40 | 5.42 | 6.12 | 6.23 | 6.46 | 6.95 | 6.48 | 5.92 | 5.23 | 4.65 | 5.66 |
| Tracking Array | 5.61 | 6.22 | 7.02 | 6.89 | 7.31 | 7.22 | 7.54 | 8.40 | 8.22 | 7.67 | 6.74 | 5.93 | 7.06 |
| Two Axis Tracking | 6.88 | 7.03 | 7.36 | 6.93 | 7.34 | 7.30 | 7.59 | 8.41 | 8.44 | 8.42 | 8.09 | 7.49 | 7.61 |

263

## MELBOURNE, AUSTRALIA          Location: 37.82° S, 44.97° W, 35 Meters

| | Jan | Feb | Mar | Apr | May | Jun | Jul | Aug | Sep | Oct | Nov | Dec | Yr |
|---|---|---|---|---|---|---|---|---|---|---|---|---|---|
| Latitude Tilt -15° | | | | | | | | | | | | | |
| Fixed Array | 7.15 | 6.37 | 3.96 | 4.14 | 3.51 | 3.13 | 3.31 | 3.72 | 4.61 | 5.36 | 5.37 | 5.93 | 4.71 |
| Tracking Array | 9.95 | 8.63 | 5.38 | 5.06 | 3.93 | 3.32 | 3.61 | 4.37 | 5.89 | 7.27 | 7.62 | 8.45 | 6.21 |
| Latitude Tilt° | | | | | | | | | | | | | |
| Fixed Array | 6.60 | 6.07 | 3.94 | 4.41 | 3.96 | 3.65 | 3.80 | 4.05 | 4.72 | 5.18 | 5.01 | 5.45 | 4.74 |
| Tracking Array | 9.39 | 8.44 | 5.53 | 5.49 | 4.49 | 3.90 | 4.19 | 4.85 | 6.17 | 7.22 | 7.25 | 7.88 | 6.23 |
| Latitude Tilt +15° | | | | | | | | | | | | | |
| Fixed Array | 5.78 | 5.51 | 3.74 | 4.45 | 4.20 | 3.96 | 4.08 | 4.17 | 4.59 | 4.77 | 4.45 | 4.77 | 4.54 |
| Tracking Array | 8.19 | 7.68 | 5.30 | 5.55 | 4.74 | 4.22 | 4.48 | 4.99 | 6.04 | 6.68 | 6.39 | 6.78 | 5.92 |
| Two Axis Tracking | 9.99 | 8.65 | 5.54 | 5.58 | 4.76 | 4.27 | 4.51 | 4.99 | 6.19 | 7.32 | 7.63 | 8.51 | 6.50 |

## LA PAZ, BOLIVIA          Location: 16.5° S; 69.6° W, 3658 Meters

| | Jan | Feb | Mar | Apr | May | Jun | Jul | Aug | Sep | Oct | Nov | Dec | Yr |
|---|---|---|---|---|---|---|---|---|---|---|---|---|---|
| Latitude Tilt -15° | | | | | | | | | | | | | |
| Fixed Array | 4.80 | 5.02 | 5.30 | 5.03 | 5.19 | 5.49 | 4.59 | 4.61 | 5.93 | 6.53 | 6.16 | 5.85 | 5.37 |
| Tracking Array | 6.47 | 6.69 | 6.86 | 6.24 | 6.08 | 6.21 | 5.35 | 5.66 | 7.48 | 8.42 | 8.07 | 7.70 | 6.77 |
| Latitude Tilt° | | | | | | | | | | | | | |
| Fixed Array | 4.52 | 4.84 | 5.34 | 5.36 | 5.88 | 6.46 | 5.26 | 5.01 | 6.12 | 6.38 | 5.79 | 5.41 | 5.51 |
| Tracking Array | 6.10 | 6.53 | 7.04 | 6.77 | 6.93 | 7.28 | 6.19 | 6.26 | 7.83 | 8.36 | 7.67 | 7.17 | 7.01 |
| Latitude Tilt +15° | | | | | | | | | | | | | |
| Fixed Array | 4.06 | 4.46 | 5.13 | 5.43 | 6.27 | 7.08 | 5.66 | 5.16 | 6.00 | 5.92 | 5.17 | 4.77 | 5.42 |
| Tracking Array | 5.31 | 5.93 | 6.74 | 6.84 | 7.32 | 7.86 | 6.61 | 6.43 | 7.65 | 7.73 | 6.75 | 6.16 | 6.77 |
| Two Axis Tracking | 6.50 | 6.70 | 7.05 | 6.88 | 7.34 | 7.95 | 6.65 | 6.44 | 7.85 | 8.48 | 8.09 | 7.77 | 7.30 |

## BELEM, BRAZIL          Location: 1.47° S, 48.48° W

| | Jan | Feb | Mar | Apr | May | Jun | Jul | Aug | Sep | Oct | Nov | Dec | Yr |
|---|---|---|---|---|---|---|---|---|---|---|---|---|---|
| Latitude Tilt -15° | | | | | | | | | | | | | |
| Fixed Array | 5.55 | 4.89 | 4.70 | 4.71 | 4.93 | 5.20 | 5.57 | 6.13 | 6.66 | 7.09 | 7.01 | 6.91 | 5.78 |
| Tracking Array | 7.06 | 6.41 | 6.18 | 6.06 | 6.13 | 6.35 | 6.84 | 7.68 | 8.45 | 8.93 | 8.69 | 8.49 | 7.27 |
| Latitude Tilt° | | | | | | | | | | | | | |
| Fixed Array | 5.22 | 4.76 | 4.76 | 4.99 | 5.48 | 5.96 | 6.33 | 6.68 | 6.88 | 6.95 | 6.59 | 6.37 | 5.91 |
| Tracking Array | 6.66 | 6.26 | 6.32 | 6.52 | 6.91 | 7.34 | 7.82 | 8.42 | 8.80 | 8.85 | 8.27 | 7.92 | 7.51 |
| Latitude Tilt +15° | | | | | | | | | | | | | |
| Fixed Array | 4.65 | 4.42 | 4.61 | 5.07 | 5.82 | 6.51 | 6.85 | 6.95 | 6.78 | 6.46 | 5.84 | 5.52 | 5.79 |
| Tracking Array | 5.80 | 5.68 | 6.05 | 6.59 | 7.29 | 7.91 | 8.35 | 8.65 | 8.59 | 8.17 | 7.28 | 6.80 | 7.26 |
| Two Axis Tracking | 7.10 | 6.42 | 6.34 | 6.63 | 7.32 | 8.00 | 8.40 | 8.67 | 8.82 | 8.97 | 8.72 | 8.58 | 7.83 |

## CUIABA, BRAZIL          Location: 15.60° S, 56.10° W

| | Jan | Feb | Mar | Apr | May | Jun | Jul | Aug | Sep | Oct | Nov | Dec | Yr |
|---|---|---|---|---|---|---|---|---|---|---|---|---|---|
| Latitude Tilt -15° | | | | | | | | | | | | | |
| Fixed Array | 4.47 | 4.29 | 4.11 | 4.44 | 4.17 | 3.78 | 4.09 | 4.13 | 4.33 | 5.15 | 5.29 | 4.77 | 4.42 |
| Tracking Array | 6.06 | 5.83 | 5.49 | 5.60 | 5.01 | 4.44 | 4.83 | 5.15 | 5.65 | 6.79 | 7.02 | 6.42 | 5.69 |
| Latitude Tilt° | | | | | | | | | | | | | |
| Fixed Array | 4.21 | 4.15 | 4.13 | 4.71 | 4.66 | 4.34 | 4.64 | 4.46 | 4.43 | 5.03 | 4.99 | 4.46 | 4.52 |
| Tracking Array | 5.72 | 5.70 | 5.64 | 6.08 | 5.72 | 5.21 | 5.60 | 5.70 | 5.92 | 6.74 | 6.67 | 5.98 | 5.89 |
| Latitude Tilt +15° | | | | | | | | | | | | | |
| Fixed Array | 3.80 | 3.84 | 3.97 | 4.76 | 4.92 | 4.67 | 4.97 | 4.58 | 4.33 | 4.70 | 4.49 | 3.98 | 4.42 |
| Tracking Array | 4.98 | 5.18 | 5.40 | 6.15 | 6.04 | 5.62 | 5.98 | 5.86 | 5.79 | 6.23 | 5.87 | 5.14 | 5.69 |
| Two Axis Tracking | 6.09 | 5.85 | 5.65 | 6.18 | 6.06 | 5.68 | 6.02 | 5.87 | 5.94 | 6.84 | 7.03 | 6.47 | 6.14 |

## MACEIO, BRAZIL — Location: 9.57° S, 35.78° W, 15 Meters

| | Jan | Feb | Mar | Apr | May | Jun | Jul | Aug | Sep | Oct | Nov | Dec | Yr |
|---|---|---|---|---|---|---|---|---|---|---|---|---|---|
| **Latitude Tilt -15°** | | | | | | | | | | | | | |
| Fixed Array | 6.67 | 6.33 | 5.86 | 5.29 | 4.64 | 4.28 | 4.43 | 4.97 | 5.64 | 6.33 | 6.52 | 6.58 | 5.63 |
| Tracking Array | 8.50 | 8.14 | 7.53 | 6.64 | 5.65 | 5.11 | 5.35 | 6.18 | 7.20 | 8.13 | 8.35 | 8.37 | 7.10 |
| **Latitude Tilt°** | | | | | | | | | | | | | |
| Fixed Array | 6.22 | 6.11 | 5.92 | 5.61 | 5.16 | 4.87 | 4.99 | 5.38 | 5.80 | 6.19 | 6.14 | 6.08 | 5.71 |
| Tracking Array | 8.03 | 7.95 | 7.70 | 7.15 | 6.37 | 5.91 | 6.11 | 6.78 | 7.50 | 8.05 | 7.94 | 7.82 | 7.28 |
| **Latitude Tilt +15°** | | | | | | | | | | | | | |
| Fixed Array | 5.49 | 5.60 | 5.71 | 5.70 | 5.47 | 5.27 | 5.35 | 5.55 | 5.70 | 5.77 | 5.47 | 5.31 | 5.53 |
| Tracking Array | 6.99 | 7.22 | 7.37 | 7.22 | 6.73 | 6.38 | 6.53 | 6.97 | 7.33 | 7.45 | 6.99 | 6.71 | 6.99 |
| Two Axis Tracking | 8.55 | 8.16 | 7.72 | 7.26 | 6.75 | 6.45 | 6.57 | 6.98 | 7.52 | 8.17 | 8.37 | 8.46 | 7.58 |

## MANAUS, BRAZIL — Location: 3.13° S, 60.03° W

| | Jan | Feb | Mar | Apr | May | Jun | Jul | Aug | Sep | Oct | Nov | Dec | Yr |
|---|---|---|---|---|---|---|---|---|---|---|---|---|---|
| **Latitude Tilt -15°** | | | | | | | | | | | | | |
| Fixed Array | 4.48 | 4.25 | 4.13 | 3.89 | 3.98 | 4.22 | 4.57 | 5.15 | 5.35 | 5.40 | 5.06 | 4.76 | 4.60 |
| Tracking Array | 5.91 | 5.71 | 5.54 | 5.10 | 5.00 | 5.17 | 5.63 | 6.49 | 6.90 | 7.01 | 6.57 | 6.19 | 5.93 |
| **Latitude Tilt°** | | | | | | | | | | | | | |
| Fixed Array | 4.24 | 4.15 | 4.17 | 4.09 | 4.36 | 4.77 | 5.13 | 5.57 | 5.50 | 5.31 | 4.80 | 4.46 | 4.71 |
| Tracking Array | 5.57 | 5.56 | 5.65 | 5.48 | 5.63 | 5.96 | 6.42 | 7.11 | 7.17 | 6.93 | 6.24 | 5.77 | 6.13 |
| **Latitude Tilt +15°** | | | | | | | | | | | | | |
| Fixed Array | 3.83 | 3.86 | 4.04 | 4.13 | 4.59 | 5.14 | 5.50 | 5.76 | 5.41 | 4.97 | 4.33 | 3.97 | 4.63 |
| Tracking Array | 4.84 | 5.04 | 5.40 | 5.53 | 5.93 | 6.42 | 6.85 | 7.29 | 6.99 | 6.40 | 5.48 | 4.94 | 5.93 |
| Two Axis Tracking | 5.95 | 5.71 | 5.66 | 5.56 | 5.95 | 6.49 | 6.89 | 7.31 | 7.19 | 7.04 | 6.59 | 6.27 | 6.38 |

## PORTO NACIONAL, BRAZIL — Location: 10.70° S, 48.42° W

| | Jan | Feb | Mar | Apr | May | Jun | Jul | Aug | Sep | Oct | Nov | Dec | Yr |
|---|---|---|---|---|---|---|---|---|---|---|---|---|---|
| **Latitude Tilt -15°** | | | | | | | | | | | | | |
| Fixed Array | 5.66 | 5.26 | 5.08 | 5.00 | 5.03 | 4.90 | 5.05 | 5.71 | 5.68 | 5.62 | 5.41 | 5.44 | 5.32 |
| Tracking Array | 7.37 | 6.91 | 6.61 | 6.29 | 6.06 | 5.77 | 6.01 | 7.00 | 7.24 | 7.31 | 7.08 | 7.09 | 6.73 |
| **Latitude Tilt°** | | | | | | | | | | | | | |
| Fixed Array | 5.32 | 5.09 | 5.12 | 5.30 | 5.63 | 5.64 | 5.74 | 6.21 | 5.84 | 5.50 | 5.12 | 5.07 | 5.47 |
| Tracking Array | 6.96 | 6.75 | 6.77 | 6.78 | 6.85 | 6.67 | 6.87 | 7.68 | 7.54 | 7.25 | 6.75 | 6.63 | 6.96 |
| **Latitude Tilt +15°** | | | | | | | | | | | | | |
| Fixed Array | 4.74 | 4.70 | 4.94 | 5.38 | 5.99 | 6.15 | 6.20 | 6.45 | 5.74 | 5.14 | 4.62 | 4.49 | 5.38 |
| Tracking Array | 6.06 | 6.14 | 6.48 | 6.86 | 7.23 | 7.21 | 7.35 | 7.91 | 7.37 | 6.70 | 5.94 | 5.69 | 6.75 |
| Two Axis Tracking | 7.41 | 6.92 | 6.78 | 6.89 | 7.26 | 7.29 | 7.40 | 7.92 | 7.56 | 7.35 | 7.11 | 7.16 | 7.26 |

## SAO PAULO, BRAZIL — Location: 23.6° S, 46.6° W, 60 Meters

| | Jan | Feb | Mar | Apr | May | Jun | Jul | Aug | Sep | Oct | Nov | Dec | Yr |
|---|---|---|---|---|---|---|---|---|---|---|---|---|---|
| **Latitude Tilt -15°** | | | | | | | | | | | | | |
| Fixed Array | 5.34 | 5.31 | 4.63 | 4.28 | 3.54 | 3.51 | 3.37 | 3.79 | 4.65 | 5.20 | 5.42 | 5.34 | 4.53 |
| Tracking Array | 7.25 | 7.12 | 6.09 | 5.35 | 4.15 | 4.06 | 3.86 | 4.69 | 5.99 | 6.93 | 7.33 | 7.26 | 5.84 |
| **Latitude Tilt°** | | | | | | | | | | | | | |
| Fixed Array | 4.99 | 5.10 | 4.64 | 4.54 | 3.93 | 4.04 | 3.81 | 4.09 | 4.75 | 5.07 | 5.10 | 4.95 | 4.58 |
| Tracking Array | 6.83 | 6.96 | 6.25 | 5.80 | 4.74 | 4.76 | 4.47 | 5.19 | 6.27 | 6.87 | 6.97 | 6.76 | 5.99 |
| **Latitude Tilt +15°** | | | | | | | | | | | | | |
| Fixed Array | 4.45 | 4.67 | 4.43 | 4.57 | 4.13 | 4.36 | 4.05 | 4.18 | 4.63 | 4.70 | 4.57 | 4.38 | 4.43 |
| Tracking Array | 5.95 | 6.32 | 5.99 | 5.86 | 5.00 | 5.14 | 4.78 | 5.33 | 6.12 | 6.35 | 6.13 | 5.80 | 5.73 |
| Two Axis Tracking | 7.28 | 7.14 | 6.27 | 5.90 | 5.02 | 5.19 | 4.81 | 5.34 | 6.29 | 6.97 | 7.34 | 7.32 | 6.24 |

## PRAIA, CABO VERDE    Location:  14.90° N,  23.52° W,  27 Meters

| | Jan | Feb | Mar | Apr | May | Jun | Jul | Aug | Sep | Oct | Nov | Dec | Yr |
|---|---|---|---|---|---|---|---|---|---|---|---|---|---|
| Latitude Tilt -15° | | | | | | | | | | | | | |
| Fixed Array | 5.45 | 6.19 | 7.31 | 7.81 | 7.46 | 7.23 | 6.22 | 5.73 | 5.93 | 5.87 | 5.12 | 4.53 | 6.24 |
| Tracking Array | 6.34 | 7.46 | 9.10 | 9.93 | 9.56 | 9.27 | 8.08 | 7.49 | 7.58 | 7.21 | 6.06 | 5.26 | 7.78 |
| Latitude Tilt° | | | | | | | | | | | | | |
| Fixed Array | 6.20 | 6.75 | 7.56 | 7.64 | 6.99 | 6.65 | 5.80 | 5.53 | 5.99 | 6.27 | 5.74 | 5.18 | 6.36 |
| Tracking Array | 7.25 | 8.19 | 9.51 | 9.87 | 9.13 | 8.67 | 7.64 | 7.33 | 7.77 | 7.79 | 6.86 | 6.10 | 8.01 |
| Latitude Tilt +15° | | | | | | | | | | | | | |
| Fixed Array | 6.70 | 7.02 | 7.46 | 7.10 | 6.20 | 5.76 | 5.13 | 5.08 | 5.78 | 6.40 | 6.12 | 5.62 | 6.20 |
| Tracking Array | 7.75 | 8.44 | 9.32 | 9.16 | 8.06 | 7.46 | 6.65 | 6.67 | 7.47 | 7.90 | 7.26 | 6.60 | 7.73 |
| Two Axis Tracking | 7.80 | 8.45 | 9.54 | 9.99 | 9.59 | 9.36 | 8.13 | 7.50 | 7.79 | 7.94 | 7.29 | 6.68 | 8.34 |

## SANTIAGO, CHILE    Location:  33.45° S,  70.67° W,  520 Meters

| | Jan | Feb | Mar | Apr | May | Jun | Jul | Aug | Sep | Oct | Nov | Dec | Yr |
|---|---|---|---|---|---|---|---|---|---|---|---|---|---|
| Latitude Tilt -15° | | | | | | | | | | | | | |
| Fixed Array | 7.57 | 7.70 | 6.40 | 5.07 | 3.23 | 2.63 | 2.92 | 3.58 | 4.32 | 6.34 | 6.65 | 7.73 | 5.35 |
| Tracking Array | 10.33 | 10.18 | 8.08 | 6.35 | 3.62 | 2.76 | 3.17 | 4.21 | 5.59 | 8.37 | 9.09 | 10.61 | 6.86 |
| Latitude Tilt° | | | | | | | | | | | | | |
| Fixed Array | 6.98 | 7.35 | 6.44 | 5.43 | 3.61 | 3.01 | 3.30 | 3.87 | 4.41 | 6.15 | 6.19 | 7.06 | 5.32 |
| Tracking Array | 9.73 | 9.94 | 8.28 | 6.89 | 4.12 | 3.24 | 3.66 | 4.65 | 5.85 | 8.31 | 8.64 | 9.88 | 6.93 |
| Latitude Tilt +15° | | | | | | | | | | | | | |
| Fixed Array | 6.11 | 6.66 | 6.15 | 5.50 | 3.80 | 3.23 | 3.51 | 3.96 | 4.27 | 5.67 | 5.48 | 6.10 | 5.04 |
| Tracking Array | 8.47 | 9.02 | 7.93 | 6.96 | 4.35 | 3.49 | 3.91 | 4.78 | 5.71 | 7.67 | 7.60 | 8.48 | 6.53 |
| Two Axis Tracking | 10.38 | 10.20 | 8.30 | 7.00 | 4.36 | 3.53 | 3.93 | 4.79 | 5.86 | 8.43 | 9.11 | 10.71 | 7.22 |

## SHANGHAI, CHINA    Location:  31.28° N, 21.47° W, 3 Meters

| | Jan | Feb | Mar | Apr | May | Jun | Jul | Aug | Sep | Oct | Nov | Dec | Yr |
|---|---|---|---|---|---|---|---|---|---|---|---|---|---|
| Latitude Tilt -15° | | | | | | | | | | | | | |
| Fixed Array | 3.38 | 3.07 | 4.27 | 4.85 | 5.34 | 4.69 | 5.82 | 5.99 | 5.20 | 4.38 | 3.47 | 3.11 | 4.46 |
| Tracking Array | 3.74 | 3.55 | 5.54 | 6.58 | 7.38 | 6.63 | 8.01 | 8.04 | 6.72 | 5.37 | 3.90 | 3.35 | 5.73 |
| Latitude Tilt° | | | | | | | | | | | | | |
| Fixed Array | 3.82 | 3.28 | 4.35 | 4.70 | 4.99 | 4.33 | 5.38 | 5.72 | 5.22 | 4.66 | 3.88 | 3.57 | 4.49 |
| Tracking Array | 4.31 | 3.92 | 5.80 | 6.53 | 7.02 | 6.17 | 7.53 | 7.84 | 6.90 | 5.83 | 4.45 | 3.92 | 5.85 |
| Latitude Tilt +15° | | | | | | | | | | | | | |
| Fixed Array | 4.06 | 3.33 | 4.23 | 4.34 | 4.45 | 3.83 | 4.74 | 5.20 | 4.98 | 4.71 | 4.08 | 3.84 | 4.32 |
| Tracking Array | 4.59 | 4.02 | 5.67 | 6.04 | 6.17 | 5.29 | 6.54 | 7.11 | 6.61 | 5.89 | 4.70 | 4.22 | 5.57 |
| Two Axis Tracking | 4.62 | 4.03 | 5.82 | 6.62 | 7.40 | 6.69 | 8.06 | 8.05 | 6.91 | 5.93 | 4.72 | 4.27 | 6.09 |

## BOGOTA, COLOMBIA    Location:  4.6° N,  74.1° W,  2,560 Meters

| | Jan | Feb | Mar | Apr | May | Jun | Jul | Aug | Sep | Oct | Nov | Dec | Yr |
|---|---|---|---|---|---|---|---|---|---|---|---|---|---|
| Latitude Tilt -15° | | | | | | | | | | | | | |
| Fixed Array | 4.81 | 4.87 | 4.73 | 4.38 | 4.53 | 4.80 | 4.96 | 4.91 | 4.76 | 3.93 | 4.12 | 4.23 | 4.59 |
| Tracking Array | 5.88 | 6.13 | 6.16 | 5.85 | 5.98 | 6.23 | 6.42 | 6.44 | 6.24 | 5.12 | 5.14 | 5.15 | 5.90 |
| Latitude Tilt° | | | | | | | | | | | | | |
| Fixed Array | 5.40 | 5.25 | 4.87 | 4.32 | 4.32 | 4.49 | 4.67 | 4.77 | 4.82 | 4.14 | 4.54 | 4.78 | 4.70 |
| Tracking Array | 6.72 | 6.73 | 6.44 | 5.81 | 5.70 | 5.82 | 6.06 | 6.29 | 6.39 | 5.52 | 5.81 | 5.98 | 6.10 |
| Latitude Tilt +15° | | | | | | | | | | | | | |
| Fixed Array | 5.79 | 5.42 | 4.80 | 4.08 | 3.92 | 3.98 | 4.17 | 4.41 | 4.67 | 4.18 | 4.78 | 5.15 | 4.61 |
| Tracking Array | 7.17 | 6.93 | 6.30 | 5.38 | 5.02 | 5.00 | 5.27 | 5.72 | 6.14 | 5.60 | 6.14 | 6.44 | 5.93 |
| Two Axis Tracking | 7.22 | 6.94 | 6.46 | 5.88 | 5.99 | 6.30 | 6.46 | 6.45 | 6.40 | 5.62 | 6.17 | 6.51 | 6.37 |

## QUITO, ECUADOR          Location: 0° 28' S, 78° 53' W, 2851 Meters

|  | Jan | Feb | Mar | Apr | May | Jun | Jul | Aug | Sep | Oct | Nov | Dec | Yr |
|---|---|---|---|---|---|---|---|---|---|---|---|---|---|
| Latitude Tilt -15° | | | | | | | | | | | | | |
| Fixed Array | 5.38 | 5.21 | 4.09 | 4.10 | 3.82 | 3.91 | 4.23 | 5.30 | 4.47 | 4.88 | 5.12 | 5.14 | 4.64 |
| Tracking Array | 6.84 | 6.76 | 5.49 | 5.37 | 4.85 | 4.84 | 5.27 | 6.71 | 5.89 | 6.40 | 6.59 | 6.55 | 5.96 |
| Latitude Tilt° | | | | | | | | | | | | | |
| Fixed Array | 5.06 | 5.06 | 4.14 | 4.33 | 4.18 | 4.38 | 4.71 | 5.74 | 4.60 | 4.81 | 4.87 | 4.81 | 4.72 |
| Tracking Array | 6.45 | 6.59 | 5.61 | 5.77 | 5.46 | 5.59 | 6.01 | 7.34 | 6.13 | 6.33 | 6.26 | 6.11 | 6.13 |
| Latitude Tilt +15° | | | | | | | | | | | | | |
| Fixed Array | 4.51 | 4.68 | 4.02 | 4.38 | 4.39 | 4.71 | 5.03 | 5.95 | 4.53 | 4.52 | 4.39 | 4.24 | 4.61 |
| Tracking Array | 5.61 | 5.98 | 5.36 | 5.83 | 5.76 | 6.03 | 6.42 | 7.54 | 5.98 | 5.85 | 5.50 | 5.23 | 5.92 |
| Two Axis Tracking | 6.89 | 6.77 | 5.62 | 5.86 | 5.78 | 6.09 | 6.46 | 7.56 | 6.14 | 6.43 | 6.61 | 6.62 | 6.40 |

## SAN SALVADOR, EL SALVADOR   Location: 13.6° N, 89.2° W, 698 Meters

|  | Jan | Feb | Mar | Apr | May | Jun | Jul | Aug | Sep | Oct | Nov | Dec | Yr |
|---|---|---|---|---|---|---|---|---|---|---|---|---|---|
| Latitude Tilt -15° | | | | | | | | | | | | | |
| Fixed Array | 5.75 | 6.29 | 6.49 | 6.37 | 6.15 | 5.24 | 6.19 | 6.93 | 5.23 | 5.78 | 5.80 | 5.79 | 6.00 |
| Tracking Array | 6.72 | 7.61 | 8.16 | 8.21 | 7.98 | 6.90 | 8.01 | 8.90 | 6.77 | 7.14 | 6.84 | 6.66 | 7.50 |
| Latitude Tilt° | | | | | | | | | | | | | |
| Fixed Array | 6.56 | 6.86 | 6.70 | 6.24 | 5.79 | 4.88 | 5.77 | 6.67 | 5.27 | 6.16 | 6.53 | 6.71 | 6.18 |
| Tracking Array | 7.67 | 8.35 | 8.52 | 8.15 | 7.61 | 6.45 | 7.56 | 8.69 | 6.94 | 7.70 | 7.73 | 7.71 | 7.76 |
| Latitude Tilt +15° | | | | | | | | | | | | | |
| Fixed Array | 7.08 | 7.12 | 6.59 | 5.81 | 5.17 | 4.31 | 5.09 | 6.09 | 5.08 | 6.28 | 6.98 | 7.35 | 6.08 |
| Tracking Array | 8.18 | 8.59 | 8.34 | 7.55 | 6.71 | 5.54 | 6.57 | 7.89 | 6.66 | 7.79 | 8.17 | 8.32 | 7.52 |
| Two Axis Tracking | 8.24 | 8.60 | 8.54 | 8.26 | 8.01 | 6.98 | 8.07 | 8.91 | 6.95 | 7.83 | 8.20 | 8.42 | 8.08 |

## PARIS-ST. MAUR, FRANCE      Location: 48.82° N, 2.50° W, 50 Meters

|  | Jan | Feb | Mar | Apr | May | Jun | Jul | Aug | Sep | Oct | Nov | Dec | Yr |
|---|---|---|---|---|---|---|---|---|---|---|---|---|---|
| Latitude Tilt -15° | | | | | | | | | | | | | |
| Fixed Array | 1.77 | 2.47 | 3.75 | 4.32 | 5.01 | 5.37 | 5.14 | 4.59 | 3.95 | 2.74 | 1.71 | 1.56 | 3.53 |
| Tracking Array | 1.77 | 2.54 | 4.56 | 6.02 | 7.39 | 8.04 | 7.66 | 6.60 | 5.04 | 3.01 | 1.71 | 1.56 | 4.66 |
| Latitude Tilt° | | | | | | | | | | | | | |
| Fixed Array | 2.06 | 2.75 | 3.90 | 4.25 | 4.78 | 5.05 | 4.87 | 4.45 | 4.02 | 2.95 | 1.95 | 1.83 | 3.57 |
| Tracking Array | 2.06 | 2.82 | 4.79 | 5.99 | 7.05 | 7.50 | 7.21 | 6.46 | 5.19 | 3.27 | 1.95 | 1.83 | 4.68 |
| Latitude Tilt +15° | | | | | | | | | | | | | |
| Fixed Array | 2.24 | 2.91 | 3.88 | 4.04 | 4.41 | 4.61 | 4.47 | 4.18 | 3.93 | 3.02 | 2.11 | 2.02 | 3.49 |
| Tracking Array | 2.24 | 2.94 | 4.69 | 5.54 | 6.22 | 6.45 | 6.28 | 5.87 | 4.98 | 3.31 | 2.11 | 2.02 | 4.39 |
| Two Axis Tracking | 2.24 | 2.94 | 4.81 | 6.06 | 7.41 | 8.10 | 7.69 | 6.62 | 5.20 | 3.33 | 2.11 | 2.02 | 4.88 |

## GEORGETOWN, GUYANA      Location: 7.8° N, 58.1° W

|  | Jan | Feb | Mar | Apr | May | Jun | Jul | Aug | Sep | Oct | Nov | Dec | Yr |
|---|---|---|---|---|---|---|---|---|---|---|---|---|---|
| Latitude Tilt -15° | | | | | | | | | | | | | |
| Fixed Array | 4.25 | 4.77 | 5.00 | 5.25 | 4.89 | 4.82 | 5.29 | 5.55 | 5.61 | 5.10 | 4.53 | 3.99 | 4.92 |
| Tracking Array | 5.18 | 5.98 | 6.46 | 6.86 | 6.42 | 6.31 | 6.85 | 7.20 | 7.22 | 6.43 | 5.56 | 4.82 | 6.27 |
| Latitude Tilt° | | | | | | | | | | | | | |
| Fixed Array | 4.74 | 5.14 | 5.15 | 5.17 | 4.65 | 4.51 | 4.97 | 5.37 | 5.68 | 5.42 | 5.03 | 4.49 | 5.03 |
| Tracking Array | 5.92 | 6.56 | 6.75 | 6.81 | 6.13 | 5.90 | 6.47 | 7.04 | 7.40 | 6.95 | 6.28 | 5.59 | 6.48 |
| Latitude Tilt +15° | | | | | | | | | | | | | |
| Fixed Array | 5.04 | 5.30 | 5.07 | 4.84 | 4.20 | 4.00 | 4.43 | 4.95 | 5.49 | 5.51 | 5.32 | 4.83 | 4.92 |
| Tracking Array | 6.32 | 6.76 | 6.61 | 6.32 | 5.41 | 5.07 | 5.63 | 6.40 | 7.11 | 7.04 | 6.65 | 6.03 | 6.28 |
| Two Axis Tracking | 6.36 | 6.76 | 6.77 | 6.90 | 6.44 | 6.37 | 6.90 | 7.21 | 7.42 | 7.07 | 6.68 | 6.11 | 6.75 |

## NEW DELHI, INDIA — Location: 28.58° N, 77.20° W, 210 Meters

| | Jan | Feb | Mar | Apr | May | Jun | Jul | Aug | Sep | Oct | Nov | Dec | Yr |
|---|---|---|---|---|---|---|---|---|---|---|---|---|---|
| **Latitude Tilt -15°** | | | | | | | | | | | | | |
| Fixed Array | 5.04 | 6.37 | 7.05 | 7.12 | 7.38 | 6.76 | 4.50 | 5.53 | 5.66 | 6.09 | 5.62 | 4.87 | 6.00 |
| Tracking Array | 6.38 | 8.09 | 8.60 | 9.23 | 9.83 | 9.15 | 6.31 | 7.44 | 7.23 | 7.34 | 7.49 | 6.06 | 7.76 |
| **Latitude Tilt°** | | | | | | | | | | | | | |
| Fixed Array | 5.83 | 7.04 | 7.31 | 6.94 | 6.87 | 6.19 | 4.20 | 5.30 | 5.70 | 6.57 | 6.43 | 5.73 | 6.18 |
| Tracking Array | 7.38 | 8.97 | 9.02 | 9.17 | 9.36 | 8.53 | 5.94 | 7.27 | 7.44 | 7.99 | 8.56 | 7.11 | 8.06 |
| **Latitude Tilt +15°** | | | | | | | | | | | | | |
| Fixed Array | 6.28 | 7.31 | 7.18 | 6.42 | 6.08 | 5.38 | 3.75 | 4.83 | 5.46 | 6.69 | 6.88 | 6.26 | 6.04 |
| Tracking Array | 7.87 | 9.23 | 8.83 | 8.50 | 8.25 | 7.32 | 5.17 | 6.60 | 7.13 | 8.09 | 9.05 | 7.68 | 7.81 |
| Two Axis Tracking | 7.92 | 9.24 | 9.05 | 9.30 | 9.86 | 9.23 | 6.34 | 7.46 | 7.45 | 8.13 | 9.08 | 7.77 | 8.40 |

## TOKYO, JAPAN — Location: 35.68° N, 39.77° W, 4 Meters

| | Jan | Feb | Mar | Apr | May | Jun | Jul | Aug | Sep | Oct | Nov | Dec | Yr |
|---|---|---|---|---|---|---|---|---|---|---|---|---|---|
| **Latitude Tilt -15°** | | | | | | | | | | | | | |
| Fixed Array | 2.95 | 3.22 | 3.42 | 3.63 | 3.81 | 3.32 | 3.68 | 3.80 | 2.99 | 2.56 | 2.63 | 2.68 | 3.22 |
| Tracking Array | 3.14 | 3.64 | 4.52 | 5.21 | 5.61 | 5.03 | 5.47 | 5.49 | 4.28 | 2.98 | 2.79 | 2.76 | 4.24 |
| **Latitude Tilt°** | | | | | | | | | | | | | |
| Fixed Array | 3.34 | 3.47 | 3.47 | 3.50 | 3.58 | 3.09 | 3.43 | 3.62 | 2.96 | 2.67 | 2.92 | 3.08 | 3.26 |
| Tracking Array | 3.63 | 4.03 | 4.74 | 5.18 | 5.34 | 4.69 | 5.15 | 5.37 | 4.40 | 3.24 | 3.19 | 3.24 | 4.35 |
| **Latitude Tilt +15°** | | | | | | | | | | | | | |
| Fixed Array | 3.55 | 3.53 | 3.35 | 3.23 | 3.21 | 2.76 | 3.07 | 3.30 | 2.80 | 2.65 | 3.06 | 3.31 | 3.15 |
| Tracking Array | 3.87 | 4.14 | 4.64 | 4.80 | 4.71 | 4.03 | 4.48 | 4.88 | 4.23 | 3.27 | 3.37 | 3.50 | 4.16 |
| Two Axis Tracking | 3.90 | 4.15 | 4.76 | 5.25 | 5.62 | 5.08 | 5.49 | 5.50 | 4.41 | 3.29 | 3.39 | 3.54 | 4.53 |

## NAIROBI, KENYA — Location: 1.30° N, 36.75° W, 1799 Meters

| | Jan | Feb | Mar | Apr | May | Jun | Jul | Aug | Sep | Oct | Nov | Dec | Yr |
|---|---|---|---|---|---|---|---|---|---|---|---|---|---|
| **Latitude Tilt -15°** | | | | | | | | | | | | | |
| Fixed Array | 6.93 | 7.14 | 6.41 | 5.32 | 4.40 | 4.13 | 3.46 | 4.02 | 5.26 | 5.80 | 5.93 | 6.52 | 5.44 |
| Tracking Array | 8.57 | 8.95 | 8.17 | 6.78 | 5.51 | 5.09 | 4.37 | 5.19 | 6.80 | 7.44 | 7.49 | 8.06 | 6.87 |
| **Latitude Tilt°** | | | | | | | | | | | | | |
| Fixed Array | 6.46 | 6.89 | 6.49 | 5.65 | 4.86 | 4.66 | 3.81 | 4.30 | 5.42 | 5.69 | 5.60 | 6.03 | 5.49 |
| Tracking Array | 8.08 | 8.73 | 8.35 | 7.29 | 6.21 | 5.88 | 4.98 | 5.68 | 7.08 | 7.37 | 7.12 | 7.52 | 7.02 |
| **Latitude Tilt +15°** | | | | | | | | | | | | | |
| Fixed Array | 5.67 | 6.29 | 6.26 | 5.75 | 5.13 | 5.02 | 4.02 | 4.42 | 5.33 | 5.32 | 5.01 | 5.24 | 5.29 |
| Tracking Array | 7.02 | 7.92 | 7.98 | 7.36 | 6.55 | 6.34 | 5.32 | 5.83 | 6.91 | 6.81 | 6.26 | 6.44 | 6.73 |
| Two Axis Tracking | 8.62 | 8.96 | 8.37 | 7.40 | 6.57 | 6.41 | 5.35 | 5.84 | 7.09 | 7.48 | 7.52 | 8.15 | 7.31 |

## PUERTO STANLEY, MALVINAS — Location: 51.7° S, 57.9° W, 23 Meters

| | Jan | Feb | Mar | Apr | May | Jun | Jul | Aug | Sep | Oct | Nov | Dec | Yr |
|---|---|---|---|---|---|---|---|---|---|---|---|---|---|
| **Latitude Tilt -15°** | | | | | | | | | | | | | |
| Fixed Array | 5.37 | 4.56 | 4.11 | 3.00 | 2.29 | 1.76 | 1.99 | 2.90 | 4.13 | 5.07 | 5.64 | 5.58 | 3.87 |
| Tracking Array | 7.88 | 6.45 | 5.28 | 3.31 | 2.30 | 1.76 | 1.99 | 3.04 | 5.14 | 7.06 | 8.11 | 8.25 | 5.05 |
| **Latitude Tilt°** | | | | | | | | | | | | | |
| Fixed Array | 5.28 | 4.56 | 4.18 | 3.24 | 2.62 | 2.07 | 2.31 | 3.20 | 4.31 | 5.00 | 5.56 | 5.45 | 3.98 |
| Tracking Array | 7.43 | 6.30 | 5.42 | 3.59 | 2.62 | 2.07 | 2.31 | 3.37 | 5.38 | 7.01 | 7.71 | 7.69 | 5.08 |
| **Latitude Tilt +15°** | | | | | | | | | | | | | |
| Fixed Array | 5.09 | 4.46 | 4.07 | 3.32 | 2.81 | 2.26 | 2.50 | 3.34 | 4.28 | 4.74 | 5.37 | 5.25 | 3.96 |
| Tracking Array | 6.48 | 5.73 | 5.19 | 3.63 | 2.81 | 2.26 | 2.50 | 3.48 | 5.25 | 6.48 | 6.79 | 6.61 | 4.77 |
| Two Axis Tracking | 7.92 | 6.47 | 5.43 | 3.65 | 2.81 | 2.26 | 2.50 | 3.48 | 5.39 | 7.11 | 8.13 | 8.32 | 5.29 |

## CHIHUAHUA, MEXICO — Location: 7.8° N, 58.1° W

| | Jan | Feb | Mar | Apr | May | Jun | Jul | Aug | Sep | Oct | Nov | Dec | Yr |
|---|---|---|---|---|---|---|---|---|---|---|---|---|---|
| Latitude Tilt -15° | | | | | | | | | | | | | |
| Fixed Array | 5.05 | 5.83 | 6.58 | 7.08 | 7.38 | 7.04 | 6.97 | 6.79 | 6.71 | 6.32 | 5.23 | 4.51 | 6.29 |
| Tracking Array | 6.43 | 7.17 | 8.10 | 9.17 | 9.82 | 9.47 | 9.34 | 8.93 | 8.43 | 7.51 | 6.79 | 5.47 | 8.05 |
| Latitude Tilt | | | | | | | | | | | | | |
| Fixed Array | 5.83 | 6.42 | 6.81 | 6.90 | 6.88 | 6.44 | 6.43 | 6.50 | 6.78 | 6.82 | 5.97 | 5.27 | 6.42 |
| Tracking Array | 7.42 | 7.94 | 8.49 | 9.11 | 9.34 | 8.82 | 8.79 | 8.73 | 8.67 | 8.16 | 7.76 | 6.41 | 8.30 |
| Latitude Tilt +15° | | | | | | | | | | | | | |
| Fixed Array | 6.29 | 6.65 | 6.68 | 6.38 | 6.08 | 5.58 | 5.63 | 5.90 | 6.51 | 6.96 | 6.37 | 5.74 | 6.23 |
| Tracking Array | 7.92 | 8.17 | 8.30 | 8.44 | 8.23 | 7.57 | 7.64 | 7.92 | 8.31 | 8.26 | 8.20 | 6.92 | 7.99 |
| Two Axis Tracking | 7.97 | 8.19 | 8.51 | 9.24 | 9.84 | 9.55 | 9.39 | 8.95 | 8.69 | 8.30 | 8.23 | 7.00 | 8.65 |

## GUAYMAS, MEXICO — Location: 27.50° N, 110.0° W

| | Jan | Feb | Mar | Apr | May | Jun | Jul | Aug | Sep | Oct | Nov | Dec | Yr |
|---|---|---|---|---|---|---|---|---|---|---|---|---|---|
| Latitude Tilt -15° | | | | | | | | | | | | | |
| Fixed Array | 5.05 | 5.83 | 6.58 | 7.08 | 7.38 | 7.37 | 6.97 | 6.79 | 6.71 | 6.71 | 5.23 | 4.51 | 6.35 |
| Tracking Array | 6.43 | 7.17 | 8.10 | 9.17 | 9.82 | 9.88 | 9.34 | 8.93 | 8.43 | 7.95 | 6.79 | 5.47 | 8.12 |
| Latitude Tilt | | | | | | | | | | | | | |
| Fixed Array | 5.83 | 6.42 | 6.81 | 6.90 | 6.88 | 6.73 | 6.43 | 6.50 | 6.78 | 7.26 | 5.97 | 5.27 | 6.48 |
| Tracking Array | 7.42 | 7.94 | 8.49 | 9.11 | 9.34 | 9.21 | 8.79 | 8.73 | 8.67 | 8.65 | 7.76 | 6.41 | 8.38 |
| Latitude Tilt +15° | | | | | | | | | | | | | |
| Fixed Array | 6.29 | 6.65 | 6.68 | 6.38 | 6.08 | 5.83 | 5.63 | 5.90 | 6.51 | 7.41 | 6.37 | 5.74 | 6.29 |
| Tracking Array | 7.92 | 8.17 | 8.30 | 8.44 | 8.23 | 7.90 | 7.64 | 7.92 | 8.31 | 8.75 | 8.20 | 6.92 | 8.06 |
| Two Axis Tracking | 7.97 | 8.19 | 8.51 | 9.24 | 9.84 | 9.97 | 9.39 | 8.95 | 8.69 | 8.80 | 8.23 | 7.00 | 8.73 |

## MEXICO D.F., MEXICO — Location: 19.3° N, 99.2° W, 2,268 Meters

| | Jan | Feb | Mar | Apr | May | Jun | Jul | Aug | Sep | Oct | Nov | Dec | Yr |
|---|---|---|---|---|---|---|---|---|---|---|---|---|---|
| Latitude Tilt -15° | | | | | | | | | | | | | |
| Fixed Array | 4.32 | 6.24 | 7.71 | 6.22 | 5.93 | 4.94 | 4.92 | 5.43 | 5.00 | 4.454 | 4.50 | 4.51 | 5.36 |
| Tracking Array | 5.06 | 7.39 | 9.51 | 8.07 | 7.84 | 6.66 | 6.64 | 7.19 | 6.51 | 5.67 | 5.29 | 5.54 | 6.78 |
| Latitude Tilt° | | | | | | | | | | | | | |
| Fixed Array | 4.90 | 6.86 | 7.99 | 6.07 | 5.57 | 4.58 | 4.60 | 5.22 | 5.04 | 4.82 | 5.06 | 5.23 | 5.50 |
| Tracking Array | 5.85 | 8.17 | 9.96 | 8.02 | 7.45 | 6.20 | 6.24 | 7.02 | 6.69 | 6.15 | 6.04 | 6.49 | 7.04 |
| Latitude Tilt +15° | | | | | | | | | | | | | |
| Fixed Array | 5.23 | 7.11 | 7.86 | 5.64 | 4.97 | 4.06 | 4.10 | 4.78 | 4.84 | 4.87 | 5.36 | 5.68 | 5.38 |
| Tracking Array | 6.23 | 8.40 | 9.74 | 7.41 | 6.56 | 5.32 | 5.42 | 6.37 | 6.41 | 6.22 | 6.38 | 6.99 | 6.79 |
| Two Axis Tracking | 6.27 | 8.41 | 9.99 | 8.13 | 7.86 | 6.72 | 6.67 | 7.20 | 6.70 | 6.26 | 6.40 | 7.07 | 7.31 |

## NAVAJOA, MEXICO — Location: 25.0° N, 109.0° W

| | Jan | Feb | Mar | Apr | May | Jun | Jul | Aug | Sep | Oct | Nov | Dec | Yr |
|---|---|---|---|---|---|---|---|---|---|---|---|---|---|
| Latitude Tilt -15° | | | | | | | | | | | | | |
| Fixed Array | 4.28 | 5.00 | 5.60 | 6.09 | 6.26 | 6.30 | 5.65 | 5.75 | 5.43 | 5.42 | 4.37 | 3.74 | 5.32 |
| Tracking Array | 5.14 | 5.96 | 7.04 | 7.98 | 8.37 | 8.46 | 7.65 | 7.65 | 7.00 | 6.52 | 5.25 | 4.29 | 6.78 |
| Latitude Tilt | | | | | | | | | | | | | |
| Fixed Array | 4.87 | 5.44 | 5.76 | 5.92 | 5.85 | 5.77 | 5.25 | 5.50 | 5.46 | 5.80 | 4.91 | 4.30 | 5.40 |
| Tracking Array | 5.92 | 6.58 | 7.36 | 7.91 | 7.95 | 7.86 | 7.18 | 7.46 | 7.18 | 7.07 | 5.99 | 5.01 | 6.96 |
| Latitude Tilt +15° | | | | | | | | | | | | | |
| Fixed Array | 5.20 | 5.60 | 5.63 | 5.48 | 5.20 | 5.03 | 4.63 | 5.01 | 5.23 | 5.88 | 5.20 | 4.63 | 5.23 |
| Tracking Array | 6.30 | 6.75 | 7.19 | 7.30 | 6.98 | 6.73 | 6.22 | 6.75 | 6.87 | 7.14 | 6.31 | 5.39 | 6.66 |
| Two Axis Tracking | 6.33 | 6.76 | 7.38 | 8.03 | 8.39 | 8.54 | 7.69 | 7.67 | 7.19 | 7.18 | 6.33 | 5.45 | 7.25 |

## PUERTO VALLARTA, MEXICO  Location: 20.0° N, 106.0° W

|  | Jan | Feb | Mar | Apr | May | Jun | Jul | Aug | Sep | Oct | Nov | Dec | Yr |
|---|---|---|---|---|---|---|---|---|---|---|---|---|---|
| **Latitude Tilt -15°** | | | | | | | | | | | | | |
| Fixed Array | 4.60 | 5.23 | 5.81 | 5.93 | 6.09 | 5.94 | 6.06 | 5.99 | 5.47 | 5.20 | 4.59 | 4.10 | 5.42 |
| Tracking Array | 5.45 | 6.29 | 7.31 | 7.75 | 8.05 | 7.88 | 8.02 | 7.87 | 7.05 | 6.39 | 5.38 | 4.88 | 6.86 |
| **Latitude Tilt** | | | | | | | | | | | | | |
| Fixed Array | 5.24 | 5.70 | 5.98 | 5.78 | 5.70 | 5.46 | 5.62 | 5.75 | 5.51 | 5.56 | 5.16 | 4.71 | 5.51 |
| Tracking Array | 6.28 | 6.94 | 7.65 | 7.68 | 7.64 | 7.32 | 7.53 | 7.67 | 7.23 | 6.93 | 6.13 | 5.70 | 7.06 |
| **Latitude Tilt +15°** | | | | | | | | | | | | | |
| Fixed Array | 5.60 | 5.87 | 5.86 | 5.37 | 5.08 | 4.78 | 4.95 | 5.24 | 5.29 | 5.63 | 5.46 | 5.08 | 5.35 |
| Tracking Array | 6.68 | 7.12 | 7.47 | 7.09 | 6.71 | 6.26 | 6.52 | 6.95 | 6.92 | 6.99 | 6.46 | 6.14 | 6.78 |
| Two Axis Tracking | 6.72 | 7.13 | 7.67 | 7.79 | 8.07 | 7.95 | 8.07 | 7.88 | 7.25 | 7.03 | 6.48 | 6.20 | 7.35 |

## TACUBAYA, MEXICO  Location: 19.40° N, 99.10° W, 2,300 Meters

|  | Jan | Feb | Mar | Apr | May | Jun | Jul | Aug | Sep | Oct | Nov | Dec | Yr |
|---|---|---|---|---|---|---|---|---|---|---|---|---|---|
| **Latitude Tilt -15°** | | | | | | | | | | | | | |
| Fixed Array | 4.70 | 5.49 | 6.19 | 6.04 | 5.68 | 5.50 | 4.86 | 5.06 | 4.46 | 4.53 | 4.61 | 4.49 | 5.13 |
| Tracking Array | 5.50 | 6.57 | 7.75 | 7.86 | 7.53 | 7.34 | 6.56 | 6.76 | 5.89 | 5.65 | 5.39 | 5.52 | 6.53 |
| **Latitude Tilt°** | | | | | | | | | | | | | |
| Fixed Array | 5.37 | 6.00 | 6.40 | 5.90 | 5.34 | 5.09 | 4.54 | 4.87 | 4.49 | 4.81 | 5.19 | 5.21 | 5.27 |
| Tracking Array | 6.36 | 7.27 | 8.13 | 7.81 | 7.17 | 6.83 | 6.17 | 6.60 | 6.05 | 6.14 | 6.16 | 6.47 | 6.76 |
| **Latitude Tilt +15°** | | | | | | | | | | | | | |
| Fixed Array | 5.75 | 6.21 | 6.28 | 5.48 | 4.77 | 4.48 | 4.06 | 4.47 | 4.31 | 4.86 | 5.50 | 5.65 | 5.15 |
| Tracking Array | 6.78 | 7.47 | 7.95 | 7.23 | 6.31 | 5.86 | 5.36 | 5.99 | 5.80 | 6.21 | 6.51 | 6.97 | 6.54 |
| Two Axis Tracking | 6.82 | 7.48 | 8.15 | 7.92 | 7.55 | 7.40 | 6.60 | 6.77 | 6.06 | 6.24 | 6.53 | 7.05 | 7.05 |

## TODOS SANTOS, MEXICO  Location: 23.0° N, 110.0° W

|  | Jan | Feb | Mar | Apr | May | Jun | Jul | Aug | Sep | Oct | Nov | Dec | Yr |
|---|---|---|---|---|---|---|---|---|---|---|---|---|---|
| **Latitude Tilt -15°** | | | | | | | | | | | | | |
| Fixed Array | 4.41 | 5.01 | 5.66 | 6.07 | 6.73 | 6.67 | 6.23 | 6.44 | 5.75 | 5.45 | 4.64 | 3.89 | 5.58 |
| Tracking Array | 5.32 | 5.99 | 7.12 | 7.94 | 8.89 | 8.86 | 8.31 | 8.45 | 7.37 | 6.60 | 5.57 | 4.53 | 7.08 |
| **Latitude Tilt** | | | | | | | | | | | | | |
| Fixed Array | 5.02 | 5.45 | 5.82 | 5.91 | 6.28 | 6.10 | 5.77 | 6.16 | 5.80 | 5.83 | 5.23 | 4.47 | 5.65 |
| Tracking Array | 6.13 | 6.61 | 7.45 | 7.87 | 8.45 | 8.23 | 7.80 | 8.23 | 7.56 | 7.15 | 6.35 | 5.29 | 7.26 |
| **Latitude Tilt +15°** | | | | | | | | | | | | | |
| Fixed Array | 5.36 | 5.61 | 5.69 | 5.48 | 5.57 | 5.30 | 5.07 | 5.60 | 5.55 | 5.91 | 5.54 | 4.81 | 5.46 |
| Tracking Array | 6.52 | 6.78 | 7.27 | 7.27 | 7.42 | 7.04 | 6.76 | 7.45 | 7.23 | 7.22 | 6.70 | 5.69 | 6.95 |
| Two Axis Tracking | 6.55 | 6.79 | 7.47 | 7.99 | 8.92 | 8.95 | 8.35 | 8.46 | 7.57 | 7.27 | 6.72 | 5.75 | 7.57 |

## TUXTLA GUTIERREZ, MEXICO  Location: 17.50° N, 93.0° O

|  | Jan | Feb | Mar | Apr | May | Jun | Jul | Aug | Sep | Oct | Nov | Dec | Yr |
|---|---|---|---|---|---|---|---|---|---|---|---|---|---|
| **Latitude Tilt -15°** | | | | | | | | | | | | | |
| Fixed Array | 3.94 | 4.74 | 4.79 | 4.65 | 4.74 | 4.39 | 4.74 | 4.62 | 4.04 | 4.05 | 3.97 | 3.69 | 4.36 |
| Tracking Array | 4.67 | 5.79 | 6.17 | 6.22 | 6.39 | 5.98 | 6.40 | 6.22 | 5.41 | 5.15 | 4.77 | 4.33 | 5.62 |
| **Latitude Tilt** | | | | | | | | | | | | | |
| Fixed Array | 4.44 | 5.14 | 4.92 | 4.55 | 4.48 | 4.10 | 4.45 | 4.46 | 4.06 | 4.28 | 4.43 | 4.21 | 4.46 |
| Tracking Array | 5.40 | 6.41 | 6.47 | 6.19 | 6.08 | 5.57 | 6.02 | 6.08 | 5.56 | 5.60 | 5.45 | 5.08 | 5.82 |
| **Latitude Tilt +15°** | | | | | | | | | | | | | |
| Fixed Array | 4.72 | 5.29 | 4.81 | 4.25 | 4.05 | 3.66 | 3.98 | 4.11 | 3.90 | 4.32 | 4.66 | 4.53 | 4.36 |
| Tracking Array | 5.76 | 6.59 | 6.33 | 5.73 | 5.36 | 4.78 | 5.23 | 5.52 | 5.33 | 5.66 | 5.76 | 5.48 | 5.63 |
| Two Axis Tracking | 5.80 | 6.60 | 6.49 | 6.27 | 6.41 | 6.03 | 6.43 | 6.23 | 5.57 | 5.69 | 5.78 | 5.54 | 6.07 |

## VERACRUZ, MEXICO          Location: 19.20° N, 96.13° W, 12 Meters

|                    | Jan  | Feb  | Mar  | Apr  | May  | Jun  | Jul  | Aug  | Sep  | Oct  | Nov  | Dec  | Yr   |
|--------------------|------|------|------|------|------|------|------|------|------|------|------|------|------|
| Latitude Tilt -15° |      |      |      |      |      |      |      |      |      |      |      |      |      |
| Fixed Array        | 3.98 | 5.24 | 5.42 | 5.83 | 6.86 | 7.09 | 5.76 | 6.45 | 6.14 | 5.94 | 5.12 | 5.02 | 5.74 |
| Tracking Array     | 4.68 | 6.31 | 7.05 | 7.63 | 8.96 | 9.24 | 7.64 | 8.40 | 7.82 | 7.21 | 5.93 | 6.38 | 7.31 |
| Latitude Tilt°     |      |      |      |      |      |      |      |      |      |      |      |      |      |
| Fixed Array        | 4.49 | 5.72 | 6.12 | 5.78 | 6.41 | 6.48 | 5.35 | 6.19 | 6.20 | 6.38 | 5.79 | 5.85 | 5.90 |
| Tracking Array     | 5.40 | 6.98 | 7.80 | 7.23 | 8.51 | 8.60 | 7.18 | 8.20 | 8.03 | 7.83 | 6.77 | 7.46 | 7.51 |
| Latitude Tilt +15° |      |      |      |      |      |      |      |      |      |      |      |      |      |
| Fixed Array        | 4.77 | 5.90 | 6.21 | 5.71 | 5.69 | 5.61 | 4.74 | 5.64 | 5.96 | 6.49 | 6.16 | 6.37 | 5.77 |
| Tracking Array     | 5.75 | 7.16 | 7.78 | 7.23 | 7.49 | 7.37 | 6.23 | 7.43 | 7.69 | 7.91 | 7.14 | 8.04 | 7.28 |
| Two Axis Tracking  | 5.78 | 7.17 | 8.15 | 7.92 | 8.98 | 9.33 | 7.68 | 8.42 | 8.04 | 7.96 | 7.17 | 8.12 | 7.89 |

## ULAN-BATOR, MONGOLIA          Location: 47.85° N, 6.75° W

|                    | Jan  | Feb  | Mar  | Apr  | May  | Jun  | Jul  | Aug  | Sep  | Oct  | Nov  | Dec  | Yr   |
|--------------------|------|------|------|------|------|------|------|------|------|------|------|------|------|
| Latitude Tilt -15° |      |      |      |      |      |      |      |      |      |      |      |      |      |
| Fixed Array        | 4.06 | 4.97 | 5.81 | 5.61 | 6.65 | 6.06 | 5.74 | 5.57 | 4.98 | 4.50 | 3.44 | 3.21 | 5.05 |
| Tracking Array     | 4.12 | 5.68 | 7.83 | 7.63 | 9.41 | 8.90 | 8.39 | 7.74 | 6.59 | 5.32 | 3.48 | 3.21 | 6.53 |
| Latitude Tilt°     |      |      |      |      |      |      |      |      |      |      |      |      |      |
| Fixed Array        | 4.81 | 5.64 | 6.12 | 5.55 | 6.33 | 5.69 | 5.43 | 5.42 | 5.10 | 4.92 | 4.01 | 3.85 | 5.24 |
| Tracking Array     | 4.83 | 6.29 | 8.22 | 7.60 | 8.97 | 8.30 | 7.91 | 7.57 | 6.78 | 5.79 | 4.03 | 3.85 | 6.68 |
| Latitude Tilt +15° |      |      |      |      |      |      |      |      |      |      |      |      |      |
| Fixed Array        | 5.31 | 6.04 | 6.13 | 5.27 | 5.82 | 5.18 | 4.97 | 5.08 | 4.99 | 5.10 | 4.38 | 4.28 | 4.77 |
| Tracking Array     | 5.31 | 6.47 | 8.05 | 7.04 | 7.92 | 7.14 | 6.88 | 6.88 | 6.51 | 5.86 | 4.38 | 4.28 | 6.39 |
| Two Axis Tracking  | 5.31 | 6.48 | 8.24 | 7.69 | 9.43 | 8.97 | 8.43 | 7.75 | 6.80 | 5.89 | 4.38 | 4.28 | 6.97 |

## HUANCAYO, PERU          Location: 12.0° S, 75.3° W, 3,313 Meters

|                    | Jan   | Feb  | Mar  | Apr  | May  | Jun  | Jul  | Aug  | Sep  | Oct  | Nov  | Dec  | Yr   |
|--------------------|-------|------|------|------|------|------|------|------|------|------|------|------|------|
| Latitude Tilt -15° |       |      |      |      |      |      |      |      |      |      |      |      |      |
| Fixed Array        | 7.85  | 6.03 | 6.64 | 6.19 | 5.68 | 5.58 | 5.95 | 6.58 | 6.96 | 7.41 | 7.38 | 7.00 | 6.55 |
| Tracking Array     | 9.97  | 7.83 | 8.45 | 7.65 | 6.77 | 6.49 | 6.98 | 7.98 | 8.73 | 9.44 | 9.42 | 8.94 | 8.22 |
| Latitude Tilt°     |       |      |      |      |      |      |      |      |      |      |      |      |      |
| Fixed Array        | 7.27  | 5.81 | 6.70 | 6.60 | 6.38 | 6.46 | 6.80 | 7.18 | 7.16 | 7.23 | 6.90 | 6.44 | 6.74 |
| Tracking Array     | 9.39  | 7.63 | 8.62 | 8.22 | 7.61 | 7.48 | 7.95 | 8.73 | 9.08 | 9.34 | 8.94 | 8.33 | 8.44 |
| Latitude Tilt +15° |       |      |      |      |      |      |      |      |      |      |      |      |      |
| Fixed Array        | 6.35  | 5.32 | 6.43 | 6.71 | 6.80 | 7.06 | 7.36 | 7.46 | 7.03 | 6.69 | 6.11 | 5.60 | 6.58 |
| Tracking Array     | 8.15  | 6.91 | 8.23 | 8.29 | 8.02 | 8.05 | 8.47 | 8.96 | 8.84 | 8.61 | 7.85 | 7.13 | 8.12 |
| Two Axis Tracking  | 10.04 | 7.84 | 8.64 | 8.34 | 8.04 | 8.14 | 8.52 | 8.97 | 9.10 | 9.49 | 9.46 | 9.06 | 9.59 |

## SAN JUAN, PUERTO RICO          Location: 18.47° N, 66.10° W, 6 Meters

|                    | Jan  | Feb  | Mar  | Apr  | May  | Jun  | Jul  | Aug  | Sep  | Oct  | Nov  | Dec  | Yr   |
|--------------------|------|------|------|------|------|------|------|------|------|------|------|------|------|
| Latitude Tilt -15° |      |      |      |      |      |      |      |      |      |      |      |      |      |
| Fixed Array        | 5.22 | 5.85 | 6.84 | 7.06 | 6.41 | 7.03 | 7.39 | 6.62 | 6.29 | 5.78 | 5.07 | 5.19 | 6.23 |
| Tracking Array     | 5.97 | 6.99 | 8.50 | 9.06 | 8.38 | 9.15 | 9.58 | 8.59 | 7.98 | 7.04 | 5.90 | 6.24 | 7.78 |
| Latitude Tilt°     |      |      |      |      |      |      |      |      |      |      |      |      |      |
| Fixed Array        | 6.00 | 6.42 | 7.08 | 6.89 | 6.00 | 6.44 | 6.82 | 6.36 | 6.36 | 6.21 | 5.74 | 6.08 | 6.37 |
| Tracking Array     | 6.90 | 7.73 | 8.92 | 9.00 | 7.98 | 8.52 | 9.01 | 8.39 | 8.21 | 7.66 | 6.74 | 7.32 | 8.03 |
| Latitude Tilt +15° |      |      |      |      |      |      |      |      |      |      |      |      |      |
| Fixed Array        | 6.46 | 6.65 | 6.96 | 6.40 | 5.35 | 5.58 | 5.96 | 5.80 | 6.12 | 6.32 | 6.11 | 6.64 | 6.19 |
| Tracking Array     | 7.36 | 7.95 | 8.72 | 8.33 | 7.03 | 7.31 | 7.83 | 7.62 | 7.87 | 7.75 | 7.12 | 7.90 | 7.73 |
| Two Axis Tracking  | 7.40 | 7.96 | 8.94 | 9.12 | 8.40 | 9.23 | 9.63 | 8.61 | 8.22 | 7.79 | 7.15 | 7.99 | 8.37 |

## STOCKHOLM, SWEDEN — Location: 59.35° N, 17.95° W, 43 Meters

| | Jan | Feb | Mar | Apr | May | Jun | Jul | Aug | Sep | Oct | Nov | Dec | Yr |
|---|---|---|---|---|---|---|---|---|---|---|---|---|---|
| **Latitude Tilt -15°** | | | | | | | | | | | | | |
| Fixed Array | 1.43 | 2.46 | 3.85 | 4.12 | 5.17 | 5.45 | 5.27 | 4.57 | 3.46 | 2.09 | 1.09 | 1.05 | 3.33 |
| Tracking Array | 1.43 | 2.47 | 4.63 | 5.82 | 8.16 | 8.94 | 8.51 | 6.79 | 4.42 | 2.20 | 1.09 | 1.05 | 4.63 |
| **Latitude Tilt°** | | | | | | | | | | | | | |
| Fixed Array | 1.67 | 2.76 | 4.02 | 4.05 | 4.91 | 5.12 | 4.98 | 4.42 | 3.52 | 2.25 | 1.25 | 1.24 | 3.35 |
| Tracking Array | 1.67 | 2.76 | 4.85 | 5.77 | 7.76 | 8.33 | 8.00 | 6.62 | 4.53 | 2.38 | 1.25 | 1.24 | 4.60 |
| **Latitude Tilt +15°** | | | | | | | | | | | | | |
| Fixed Array | 1.81 | 2.91 | 3.99 | 3.82 | 4.52 | 4.67 | 4.56 | 4.13 | 3.42 | 2.30 | 1.34 | 1.35 | 3.24 |
| Tracking Array | 1.81 | 2.91 | 4.74 | 5.34 | 6.83 | 7.14 | 6.95 | 6.00 | 4.34 | 2.41 | 1.34 | 1.35 | 4.26 |
| Two Axis Tracking | 1.81 | 2.91 | 4.86 | 5.86 | 8.18 | 9.03 | 8.56 | 6.80 | 4.54 | 2.43 | 1.34 | 1.35 | 4.81 |

## BANGKOK, THAILAND — Location: 13.73° N, 0.50° W, 20 Meters

| | Jan | Feb | Mar | Apr | May | Jun | Jul | Aug | Sep | Oct | Nov | Dec | Yr |
|---|---|---|---|---|---|---|---|---|---|---|---|---|---|
| **Latitude Tilt -15°** | | | | | | | | | | | | | |
| Fixed Array | 4.95 | 5.62 | 5.23 | 5.62 | 5.63 | 5.30 | 4.53 | 4.67 | 4.32 | 4.36 | 5.21 | 4.99 | 5.04 |
| Tracking Array | 5.84 | 6.85 | 6.69 | 7.33 | 7.37 | 6.97 | 6.09 | 6.24 | 5.73 | 5.54 | 6.19 | 5.79 | 6.39 |
| **Latitude Tilt°** | | | | | | | | | | | | | |
| Fixed Array | 5.60 | 6.10 | 5.37 | 5.51 | 5.32 | 4.93 | 4.27 | 4.52 | 4.35 | 4.61 | 5.84 | 5.74 | 5.18 |
| Tracking Array | 6.67 | 7.51 | 6.99 | 7.28 | 7.03 | 6.52 | 5.74 | 6.10 | 5.87 | 5.98 | 7.00 | 6.71 | 6.62 |
| **Latitude Tilt +15°** | | | | | | | | | | | | | |
| Fixed Array | 6.01 | 6.32 | 5.27 | 5.14 | 4.77 | 4.36 | 3.84 | 4.18 | 4.19 | 4.66 | 6.23 | 6.25 | 5.10 |
| Tracking Array | 7.12 | 7.73 | 6.85 | 6.75 | 6.20 | 5.60 | 5.00 | 5.55 | 5.64 | 6.06 | 7.41 | 7.24 | 6.43 |
| Two Axis Tracking | 7.17 | 7.74 | 7.01 | 7.37 | 7.39 | 7.05 | 6.12 | 6.25 | 5.88 | 6.09 | 7.44 | 7.33 | 6.90 |

## PUERTO DE ESPANA, TRINIDAD — Location: 10.63° N, 61.40° W

| | Jan | Feb | Mar | Apr | May | Jun | Jul | Aug | Sep | Oct | Nov | Dec | Yr |
|---|---|---|---|---|---|---|---|---|---|---|---|---|---|
| **Latitude Tilt -15°** | | | | | | | | | | | | | |
| Fixed Array | 4.93 | 5.94 | 6.73 | 5.87 | 5.44 | 5.23 | 5.28 | 6.11 | 6.19 | 5.35 | 4.75 | 4.76 | 5.55 |
| Tracking Array | 5.90 | 7.28 | 8.46 | 7.60 | 7.10 | 6.83 | 6.90 | 7.88 | 7.89 | 6.70 | 5.76 | 5.63 | 6.99 |
| **Latitude Tilt°** | | | | | | | | | | | | | |
| Fixed Array | 5.57 | 6.46 | 6.95 | 5.76 | 5.15 | 4.87 | 4.95 | 5.89 | 6.26 | 5.70 | 5.29 | 5.45 | 5.69 |
| Tracking Array | 6.73 | 7.98 | 8.83 | 7.55 | 6.76 | 6.38 | 6.50 | 7.70 | 8.09 | 7.23 | 6.51 | 6.51 | 7.23 |
| **Latitude Tilt +15°** | | | | | | | | | | | | | |
| Fixed Array | 5.97 | 6.70 | 6.85 | 5.38 | 4.63 | 4.31 | 4.41 | 5.41 | 6.04 | 5.80 | 5.61 | 5.91 | 5.58 |
| Tracking Array | 7.19 | 8.22 | 8.65 | 6.99 | 5.97 | 5.48 | 5.66 | 7.00 | 7.76 | 7.32 | 6.88 | 7.03 | 7.01 |
| Two Axis Tracking | 7.23 | 8.23 | 8.86 | 7.64 | 7.12 | 6.90 | 6.94 | 7.90 | 8.10 | 7.35 | 6.91 | 7.11 | 7.52 |

## MONTEVIDEO, URUGUAY — Location: 34.87° S, 56.17° W, 15 Meters

| | Jan | Feb | Mar | Apr | May | Jun | Jul | Aug | Sep | Oct | Nov | Dec | Yr |
|---|---|---|---|---|---|---|---|---|---|---|---|---|---|
| **Latitude Tilt -15°** | | | | | | | | | | | | | |
| Fixed Array | 7.47 | 7.14 | 6.20 | 5.01 | 4.01 | 3.28 | 3.39 | 4.16 | 5.23 | 6.13 | 7.11 | 7.41 | 5.54 |
| Tracking Array | 10.24 | 9.51 | 7.84 | 6.31 | 4.71 | 3.58 | 3.77 | 5.05 | 6.56 | 8.13 | 9.71 | 10.24 | 7.14 |
| **Latitude Tilt°** | | | | | | | | | | | | | |
| Fixed Array | 6.89 | 6.82 | 6.24 | 5.37 | 4.55 | 3.82 | 3.88 | 4.54 | 5.37 | 5.94 | 6.62 | 6.77 | 5.57 |
| Tracking Array | 9.67 | 9.30 | 8.05 | 6.87 | 5.38 | 4.21 | 4.37 | 5.60 | 6.88 | 8.08 | 9.25 | 9.56 | 7.27 |
| **Latitude Tilt +15°** | | | | | | | | | | | | | |
| Fixed Array | 6.04 | 6.19 | 5.97 | 5.45 | 4.84 | 4.14 | 4.16 | 4.68 | 5.24 | 5.48 | 5.85 | 5.88 | 5.33 |
| Tracking Array | 8.43 | 8.46 | 7.72 | 6.95 | 5.69 | 4.55 | 4.68 | 5.76 | 6.73 | 7.48 | 8.15 | 8.22 | 6.90 |
| Two Axis Tracking | 10.28 | 9.53 | 8.07 | 6.98 | 5.71 | 4.60 | 4.71 | 5.77 | 6.90 | 8.19 | 9.73 | 10.32 | 7.57 |

## BARCELONA, VENEZUELA          Location:  10.12° N,  64.68° W,  7 Meters

| | Jan | Feb | Mar | Apr | May | Jun | Jul | Aug | Sep | Oct | Nov | Dec | Yr |
|---|---|---|---|---|---|---|---|---|---|---|---|---|---|
| Latitude Tilt -15° | | | | | | | | | | | | | |
| Fixed Array | 5.05 | 5.57 | 5.89 | 5.70 | 5.52 | 5.18 | 5.65 | 5.83 | 6.02 | 5.35 | 4.98 | 4.85 | 5.47 |
| Tracking Array | 6.04 | 6.87 | 7.49 | 7.41 | 7.18 | 6.76 | 7.31 | 7.56 | 7.70 | 6.71 | 6.02 | 5.74 | 6.90 |
| Latitude Tilt° | | | | | | | | | | | | | |
| Fixed Array | 5.69 | 6.03 | 6.06 | 5.59 | 5.21 | 4.82 | 5.28 | 5.62 | 6.08 | 5.68 | 5.54 | 5.53 | 5.60 |
| Tracking Array | 6.88 | 7.52 | 7.80 | 7.34 | 6.83 | 6.30 | 6.88 | 7.37 | 7.87 | 7.22 | 6.79 | 6.62 | 7.12 |
| Latitude Tilt +15° | | | | | | | | | | | | | |
| Fixed Array | 6.10 | 6.24 | 5.96 | 5.21 | 4.67 | 4.25 | 4.67 | 5.16 | 5.85 | 5.77 | 5.87 | 6.00 | 5.48 |
| Tracking Array | 7.33 | 7.72 | 7.62 | 6.78 | 6.01 | 5.40 | 5.97 | 6.68 | 7.54 | 7.29 | 7.16 | 7.13 | 6.89 |
| Two Axis Tracking | 7.37 | 7.73 | 7.82 | 7.44 | 7.21 | 6.85 | 7.37 | 7.57 | 7.89 | 7.34 | 7.18 | 7.21 | 7.41 |

## CARACAS, VENEZUELA          Location:  10.5° N,  66.9° W,  862 Meters

| | Jan | Feb | Mar | Apr | May | Jun | Jul | Aug | Sep | Oct | Nov | Dec | Yr |
|---|---|---|---|---|---|---|---|---|---|---|---|---|---|
| Latitude Tilt -15° | | | | | | | | | | | | | |
| Fixed Array | 5.00 | 5.95 | 6.12 | 5.99 | 5.02 | 5.23 | 5.58 | 5.84 | 5.70 | 4.92 | 4.54 | 4.69 | 5.38 |
| Tracking Array | 5.98 | 7.29 | 7.75 | 7.73 | 6.62 | 6.82 | 7.24 | 7.56 | 7.32 | 6.21 | 5.53 | 5.55 | 6.80 |
| Latitude Tilt° | | | | | | | | | | | | | |
| Fixed Array | 5.64 | 6.47 | 6.31 | 5.87 | 4.76 | 4.87 | 5.22 | 5.68 | 5.76 | 5.22 | 5.04 | 5.35 | 5.51 |
| Tracking Array | 6.82 | 8.00 | 8.08 | 7.68 | 6.30 | 6.37 | 6.82 | 7.39 | 7.50 | 6.69 | 6.24 | 6.42 | 7.03 |
| Latitude Tilt +15° | | | | | | | | | | | | | |
| Fixed Array | 6.06 | 6.71 | 6.21 | 5.48 | 4.30 | 4.30 | 4.63 | 5.17 | 5.55 | 5.30 | 5.33 | 5.80 | 5.40 |
| Tracking Array | 7.27 | 8.22 | 7.91 | 7.11 | 5.56 | 5.47 | 5.93 | 6.71 | 7.19 | 6.78 | 6.60 | 6.92 | 6.81 |
| Two Axis Tracking | 7.32 | 8.23 | 8.11 | 7.78 | 6.64 | 6.90 | 7.29 | 7.58 | 7.51 | 6.81 | 6.62 | 7.00 | 7.32 |

# Appendix C: Sun Charts

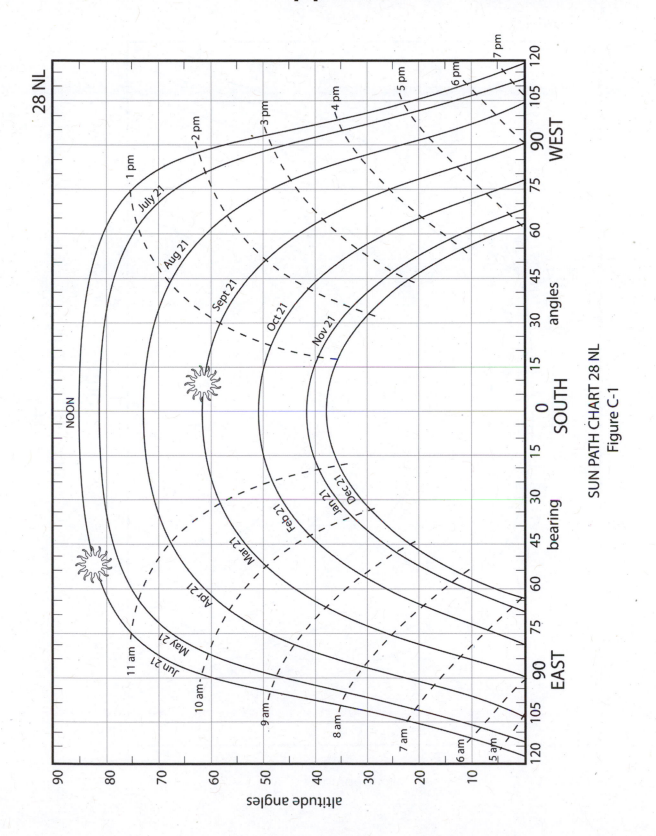

SUN PATH CHART 28 NL
Figure C-1

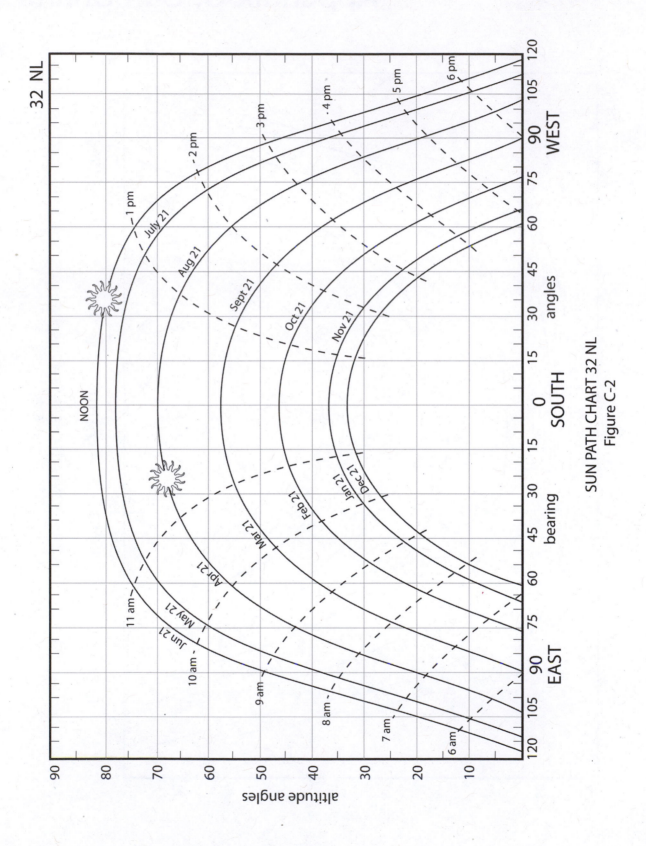

SUN PATH CHART 32 NL

Figure C-2

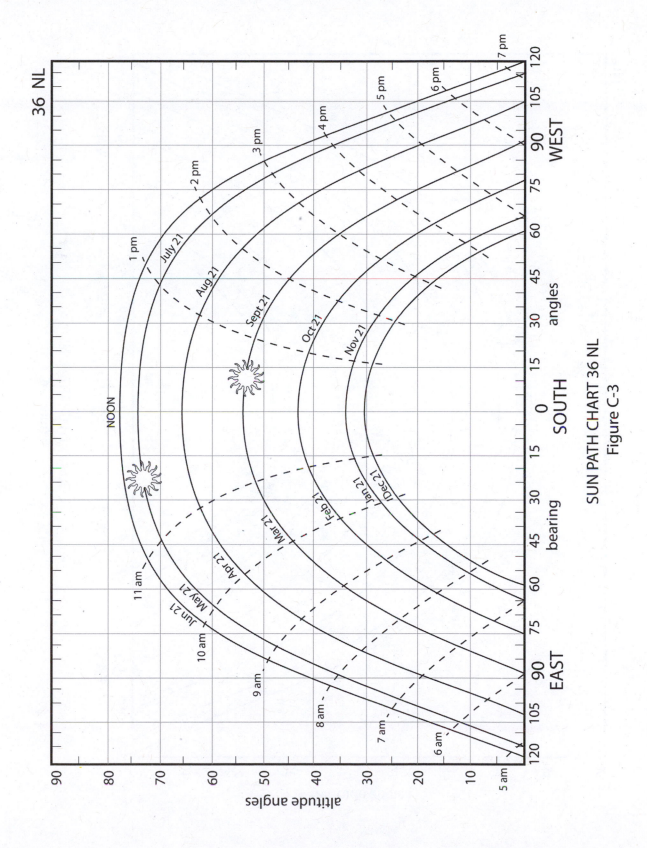

SUN PATH CHART 36 NL
Figure C-3

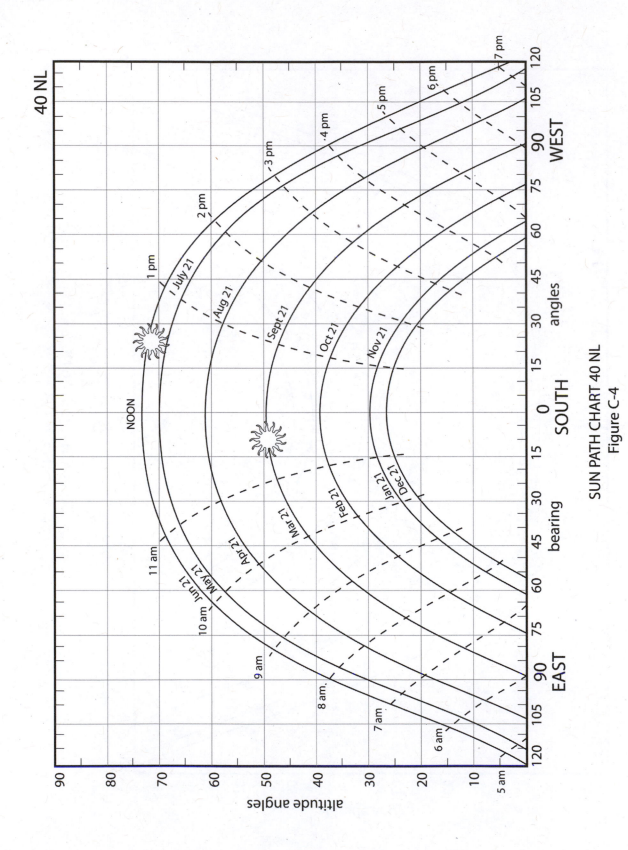

SUN PATH CHART 40 NL

Figure C–4

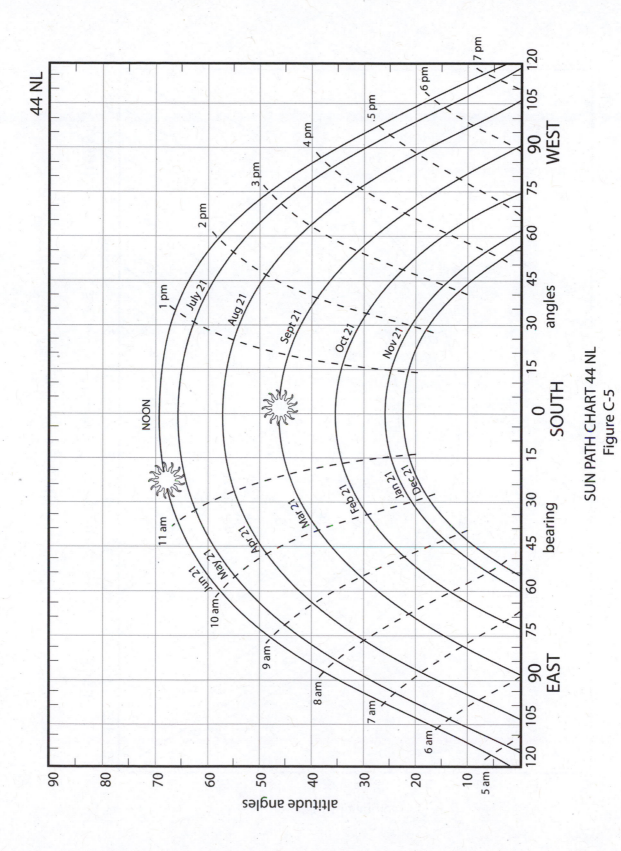

SUN PATH CHART 44 NL

Figure C-5

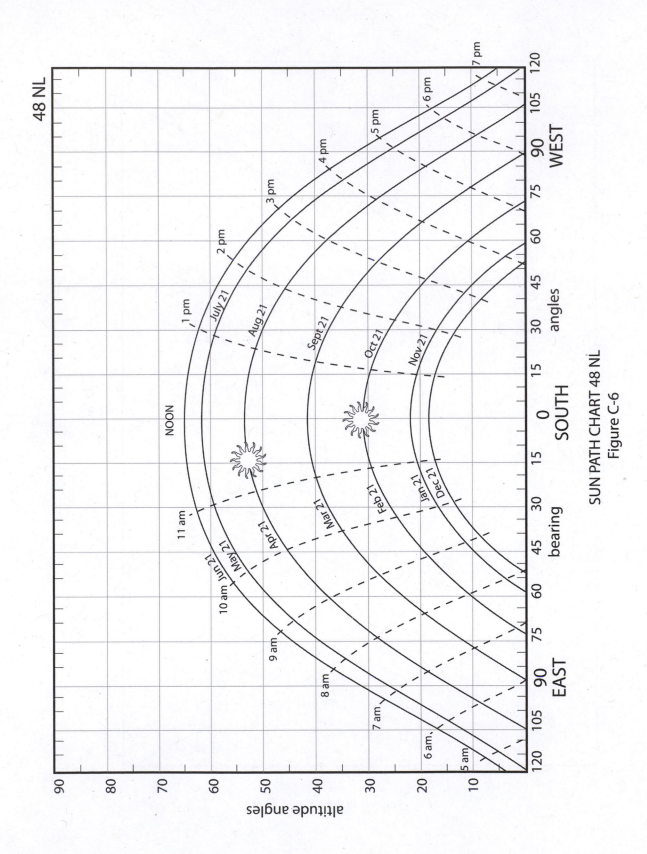

SUN PATH CHART 48 NL

Figure C-6

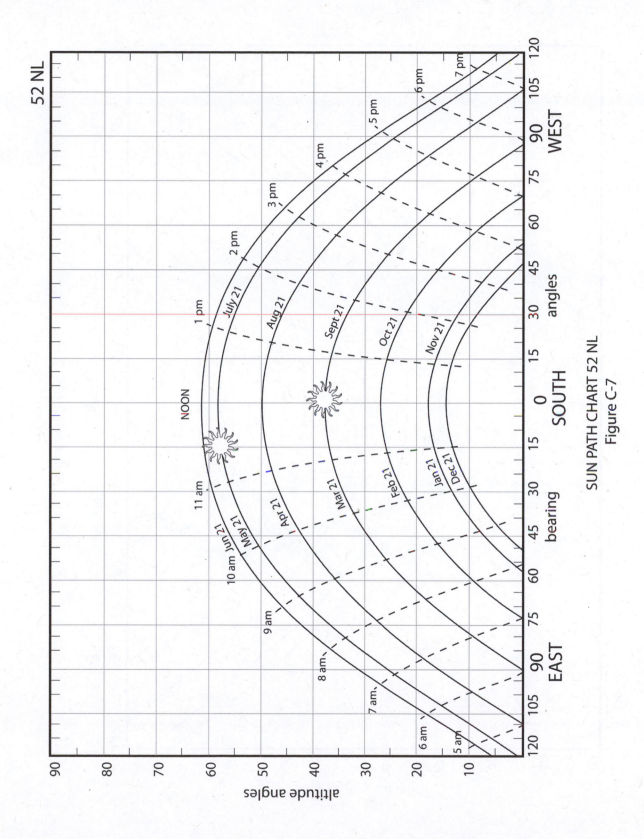

52 NL

SUN PATH CHART 52 NL
Figure C-7

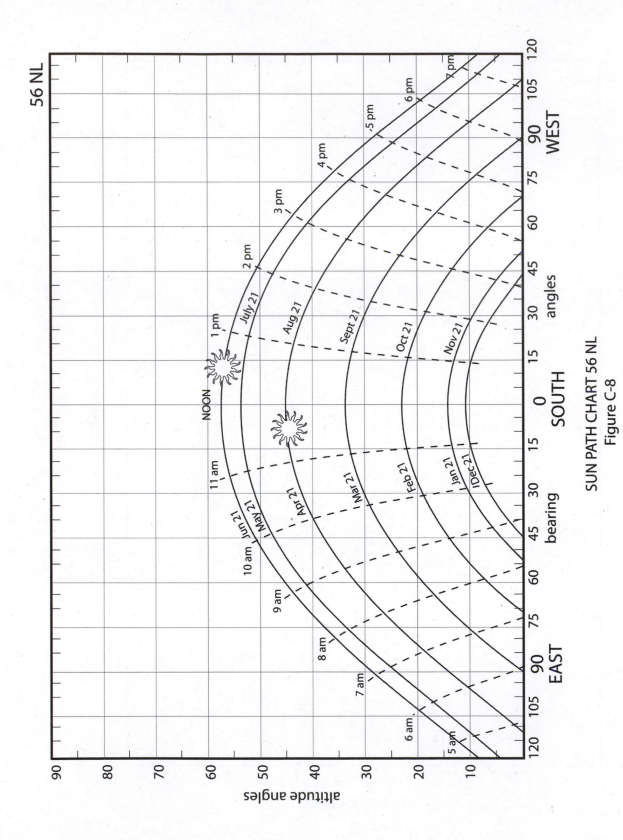

56 NL

SUN PATH CHART 56 NL
Figure C-8

# Appendix D: System Sizing Worksheets

## Stand-Alone Sizing Worksheet

### Load Estimation Worksheet (abbreviated)

| Individual Loads | Qty | X Volts | X Amps | = Watts AC | DC | X Use hrs/day | X Use days/wk | ÷ 7 days | = Watt Hours AC | DC |
|---|---|---|---|---|---|---|---|---|---|---|
| | | | | | | | | 7 | | |
| | | | | | | | | 7 | | |
| | | | | | | | | 7 | | |
| | | | | | | | | 7 | | |
| | | | | | | | | 7 | | |
| | | | | | | | | 7 | | |
| | | | | | | | | 7 | | |
| | | | | | | | | 7 | | |
| | | | | | | | | 7 | | |
| | | | | | | | | | | |

**AC Total Connected Watts:** _____     **AC Average Daily Load:** _____

**DC Total Connected Watts:** _____     **DC Average Daily Load:** _____

### Battery Sizing Worksheet

| AC Average Daily Load (w-hr/day) | ÷ | Inverter Efficiency | + | DC Average Daily Load (w-hr/day) | ÷ | DC System Voltage | = | Average Amp-hours/ Day |
|---|---|---|---|---|---|---|---|---|
| [( | ÷ | | ) + | | ] ÷ | | = | |

| Average Amp-hours/day | X | Days of Autonomy | ÷ | Discharge Limit | ÷ | Battery AH Capacity | = | Batteries in Parallel |
|---|---|---|---|---|---|---|---|---|
| | X | | ÷ | | ÷ | | = | |

| DC System Voltage | ÷ | Battery Voltage | = | Batteries in Series | X | Batteries in Parallel | = | Total Batteries |
|---|---|---|---|---|---|---|---|---|
| | ÷ | | = | | X | | = | |

**Battery Specification**     Make:     Model:

## Stand-Alone Sizing Worksheet (continued)

### Array Sizing Worksheet

| Average Amp-hrs/day | ÷ | Battery Efficiency | ÷ | Peak Sun Hrs/day | = | Array Peak Amps | |
|---|---|---|---|---|---|---|---|
| | ÷ | | ÷ | | = | | |

| Array Peak Amps | ÷ | Peak Amps/module | = | Modules in Parallel | | Module Short Circuit Current | |
|---|---|---|---|---|---|---|---|
| | ÷ | | = | | | | |

| DC System Voltage | ÷ | Nominal Module Voltage | = | Modules in Series | X | Modules in Parallel | = | Total Modules |
|---|---|---|---|---|---|---|---|---|
| | ÷ | | = | | X | | = | |

**Panel Specification**  Make:  Model:

### Controller Sizing Worksheet

| Module Short Circuit Current | X | Modules in Parallel | X | 1.25 | = | Array Short Circuit Amps | Controller Array Amps | Listed Desired Features |
|---|---|---|---|---|---|---|---|---|
| | X | | X | 1.25 | = | | | |

| DC Total Connected Watts | ÷ | DC System Voltage | = | Maximum DC Load Amps | Controller Load Amps | |
|---|---|---|---|---|---|---|
| | ÷ | | = | | | |

**Controller Specification**  Make:  Model:

### Inverter Sizing Worksheet

| AC Total Connected Watts | DC System Voltage | Estimated Surge Watts | Listed Desired Features |
|---|---|---|---|
| | | | |

**Inverter Specification**  Make:  Model:

## Stand-Alone Sizing Worksheet

### Load Estimation Worksheet (abbreviated)

| Individual Loads | Qty | X Volts | X Amps | = Watts AC | Watts DC | X Use hrs/day | X Use days/wk | ÷ 7 days | Watt Hours AC | Watt Hours DC |
|---|---|---|---|---|---|---|---|---|---|---|
| | | | | | | | | 7 | | |
| | | | | | | | | 7 | | |
| | | | | | | | | 7 | | |
| | | | | | | | | 7 | | |
| | | | | | | | | 7 | | |
| | | | | | | | | 7 | | |
| | | | | | | | | 7 | | |
| | | | | | | | | 7 | | |
| | | | | | | | | 7 | | |
| | | | | | | | | | | |

**AC Total Connected Watts:** _____    **AC Average Daily Load:** _____

**DC Total Connected Watts:** _____    **DC Average Daily Load:** _____

### Battery Sizing Worksheet

| AC Average Daily Load (w-hr/day) | ÷ | Inverter Efficiency | + | DC Average Daily Load (w-hr/day) | ÷ | DC System Voltage | = | Average Amp-hours/ Day |
|---|---|---|---|---|---|---|---|---|
| [( | ÷ | | ) + | | ] ÷ | | = | |

| Average Amp-hours/day | X | Days of Autonomy | ÷ | Discharge Limit | ÷ | Battery AH Capacity | = | Batteries in Parallel |
|---|---|---|---|---|---|---|---|---|
| | X | | ÷ | | ÷ | | = | |

| DC System Voltage | ÷ | Battery Voltage | = | Batteries in Series | X | Batteries in Parallel | = | Total Batteries |
|---|---|---|---|---|---|---|---|---|
| | ÷ | | = | | X | | = | |

**Battery Specification**    Make:    Model:

**Stand-Alone Sizing Worksheet (continued)**

## Array Sizing Worksheet

| Average Amp-hrs/day | ÷ | Battery Efficiency | ÷ | Peak Sun Hrs/day | = | Array Peak Amps | |
|---|---|---|---|---|---|---|---|
| | ÷ | | ÷ | | = | | |
| Array Peak Amps | ÷ | Peak Amps/module | = | Modules in Parallel | | Module Short Circuit Current | |
| | ÷ | | = | | | | |
| DC System Voltage | ÷ | Nominal Module Voltage | = | Modules in Series | X | Modules in Parallel | = Total Modules |
| | ÷ | | = | | X | | = |

| **Panel Specification** | Make: | Model: |
|---|---|---|

## Controller Sizing Worksheet

| Module Short Circuit Current | X | Modules in Parallel | X | 1.25 | = | Array Short Circuit Amps | Controller Array Amps | Listed Desired Features |
|---|---|---|---|---|---|---|---|---|
| | X | | X | 1.25 | = | | | |
| DC Total Connected Watts | ÷ | DC System Voltage | = | Maximum DC Load Amps | | Controller Load Amps | | |
| | ÷ | | = | | | | | |

| **Controller Specification** | Make: | Model: |
|---|---|---|

## Inverter Sizing Worksheet

| AC Total Connected Watts | DC System Voltage | Estimated Surge Watts | Listed Desired Features |
|---|---|---|---|
| | | | |

| **Inverter Specification** | Make: | Model: |
|---|---|---|

## Stand-Alone Sizing Worksheet

### Load Estimation Worksheet (abbreviated)

| Individual Loads | Qty | X Volts | X Amps | = Watts AC | Watts DC | X Use hrs/day | X Use days/wk | ÷ 7 days | = Watt Hours AC | Watt Hours DC |
|---|---|---|---|---|---|---|---|---|---|---|
| | | | | | | | | 7 | | |
| | | | | | | | | 7 | | |
| | | | | | | | | 7 | | |
| | | | | | | | | 7 | | |
| | | | | | | | | 7 | | |
| | | | | | | | | 7 | | |
| | | | | | | | | 7 | | |
| | | | | | | | | 7 | | |
| | | | | | | | | 7 | | |

**AC Total Connected Watts:** _____     **AC Average Daily Load:** _____

**DC Total Connected Watts:** _____     **DC Average Daily Load:** _____

### Battery Sizing Worksheet

| AC Average Daily Load (w-hr/day) | ÷ | Inverter Efficiency | + | DC Average Daily Load (w-hr/day) | ÷ | DC System Voltage | = | Average Amp-hours/ Day |
|---|---|---|---|---|---|---|---|---|
| [( | ÷ | | ) + | | ] ÷ | | = | |

| Average Amp-hours/day | X | Days of Autonomy | ÷ | Discharge Limit | ÷ | Battery AH Capacity | = | Batteries in Parallel |
|---|---|---|---|---|---|---|---|---|
| | X | | ÷ | | ÷ | | = | |

| DC System Voltage | ÷ | Battery Voltage | = | Batteries in Series | X | Batteries in Parallel | = | Total Batteries |
|---|---|---|---|---|---|---|---|---|
| | ÷ | | = | | X | | = | |

**Battery Specification**     **Make:**     **Model:**

## Stand-Alone Sizing Worksheet (continued)

### Array Sizing Worksheet

| Average Amp-hrs/day | ÷ | Battery Efficiency | ÷ | Peak Sun Hrs/day | = | Array Peak Amps | | |
|---|---|---|---|---|---|---|---|---|
| | ÷ | | ÷ | | = | | | |
| Array Peak Amps | ÷ | Peak Amps/module | = | Modules in Parallel | | Module Short Circuit Current | | |
| | ÷ | | = | | | | | |
| DC System Voltage | ÷ | Nominal Module Voltage | = | Modules in Series | X | Modules in Parallel | = | Total Modules |
| | ÷ | | = | | X | | = | |

| Panel Specification | Make: | | Model: | |
|---|---|---|---|---|

### Controller Sizing Worksheet

| Module Short Circuit Current | X | Modules in Parallel | X | 1.25 | = | Array Short Circuit Amps | Controller Array Amps | Listed Desired Features |
|---|---|---|---|---|---|---|---|---|
| | X | | X | 1.25 | = | | | |
| DC Total Connected Watts | ÷ | DC System Voltage | = | Maximum DC Load Amps | | Controller Load Amps | | |
| | ÷ | | = | | | | | |

| Controller Specification | Make: | | Model: | |
|---|---|---|---|---|

### Inverter Sizing Worksheet

| AC Total Connected Watts | DC System Voltage | Estimated Surge Watts | Listed Desired Features |
|---|---|---|---|
| | | | |

| Inverter Specification | Make: | | Model: | |
|---|---|---|---|---|

## System Wire Sizing Worksheet

Use the following worksheet to determine system wire sizes.

**PV Combiner Box to Battery**

You can size this section as one wire run from PV to Battery, due to the fact that the controller is basically a pass through device. You can also break this wire run into two sections, PV to Controller and Controller to Battery (see wire sizing worksheet below).

### A. *NEC®* Requirement

$$\frac{\text{Isc of}}{\text{modules}} \ \mathbf{X} \ \frac{\text{\# of modules}}{\text{in parallel}} = \text{Total Amps} \ \mathbf{X} \ 1.25 \ \mathbf{X} \ 1.25 = \textit{NEC}^{®} \text{ required amps}$$

_____ **X** _____ = _____ **X** 1.25 **X** 1.25 = _____

Amperage satisfying *NEC®* = _____        Wire Size from Table 9-4 = _____

### B. Voltage Drop Requirements

System Voltage: _____        Total Amps: \_\_\_\_\_

One Way Distance: \_\_\_\_\_        Voltage Drop(%): \_\_\_\_\_

Wire Size from voltage drop tables (Tables 9-5 through 9-7): \_\_\_\_\_

Is this equal to or greater than the size wire needed for safety? \_\_\_\_\_

- If yes, this is your answer.        • If no, use the wire size from A.

**PV Combiner Box to Controller**

At times, it can be advantageous to break up the PV to Battery wire run into two separate wire runs, PV to Controller and Controller to Battery. Since the Controller is usually very close to the battery, you can usually size this section with wire smaller than the PV to Controller section as long as it passes the *NEC®* required ampacity from the PV array.

### A. *NEC®* Requirement

$$\frac{\text{Isc of}}{\text{modules}} \ \mathbf{X} \ \frac{\text{\# of modules}}{\text{in parallel}} = \text{Total Amps} \ \mathbf{X} \ 1.25 \ \mathbf{X} \ 1.25 = \textit{NEC}^{®} \text{ required amps}$$

_____ **X** _____ = _____ **X** 1.25 **X** 1.25 = _____

Amperage satisfying *NEC®* = _____        Wire Size from Table 9-4 = _____

### B. Voltage Drop Requirements

System Voltage: _____        Total Amps: \_\_\_\_\_

One Way Distance: \_\_\_\_\_        Voltage Drop(%): \_\_\_\_\_

Wire Size from voltage drop tables (Tables 9-5 through 9-7): \_\_\_\_\_

Is this equal to or greater than the size wire needed for safety? \_\_\_\_\_

- If yes, this is your answer.        • If no, use the wire size from A.

## Controller to Battery

### A. *NEC®* Requirement

$$\underset{\text{modules}}{\text{Isc of}} \; \mathbf{X} \; \underset{\text{in parallel}}{\text{\# of modules}} = \text{Total Amps} \; \mathbf{X} \; 1.25 \; \mathbf{X} \; 1.25 = NEC® \text{ required amps}$$

_____ **X** _____ = _____ **X** 1.25 **X** 1.25 = _____

Amperage satisfying *NEC®* = _____          Wire Size from Table 9-4 = _____

### B. Voltage Drop Requirements

System Voltage: _____          Total Amps: _____

One Way Distance: _____          Voltage Drop(%): _____

Wire Size from voltage drop tables (Tables 9-5 through 9-7): _____

Is this equal to or greater than the size wire needed for safety? _____

- If yes, this is your answer.          • If no, use the wire size from A.

## Battery to DC Load Center

### A. *NEC®* Requirement

$$\text{DC load watts} \; \div \; \text{DC voltage} = \text{DC total amps} \; \mathbf{X} \; 1.25 = NEC® \text{ required amps}$$

_____ ÷ _____ = _____ **X** 1.25 = _____

Amperage satisfying *NEC®* = _____          Wire Size from Table 9-4 = _____

### B. Voltage Drop Requirements:

System Voltage: _____          Total Amps: _____

One Way Distance: _____          Voltage Drop(%): _____

Wire Size from voltage drop tables (Tables 9-5 through 9-7): _____

Is this equal to or greater than the size wire needed for safety? _____

- If yes, this is your answer.          • If no, use the wire size from A.

## Battery to Inverter

### A. *NEC®* Requirement

$$\underset{\text{Rated Watts}}{\text{Inverter}} \; \div \; \underset{\text{Efficiency}}{\text{Inverter}} \; \div \; \underset{\text{Voltage}}{\underset{\text{(lowest operating)}}{\text{DC System}}} = \underset{\text{Total Amps}}{\text{Inverter}} \; \mathbf{X} \; 1.25 = \underset{\text{required amps}}{NEC®}$$

_____ ÷ _____ ÷ _____ = _____ **X** 1.25 = _____

Amperage satisfying *NEC®* = _____          Wire Size from Table 9-4 = _____

Verify battery conductor size with inverter manufacturer.

### B. Voltage Drop Requirements

System Voltage: _____          Total Amps: _____

One Way Distance: _____.          Voltage Drop(%): _____

Wire Size from voltage drop tables (Tables 9-5 through 9-7): _____

Is this equal to or greater than the size wire needed for safety? _____

- If yes, this is your answer.          • If no, use the wire size from A.

Temperature deration, conduit fill deration, and device terminal compatibility, are not included in the wire sizing worksheets.

## System Wire Sizing Worksheet

Use the following worksheet to determine system wire sizes.

**PV Combiner Box to Battery**

You can size this section as one wire run from PV to Battery, due to the fact that the controller is basically a pass through device. You can also break this wire run into two sections, PV to Controller and Controller to Battery (see wire sizing worksheet below).

    **A. *NEC*® Requirement**

$$\underset{\text{modules}}{\text{Isc of}} \; \textbf{X} \; \underset{\text{in parallel}}{\text{\# of modules}} = \text{Total Amps} \; \textbf{X} \; 1.25 \; \textbf{X} \; 1.25 = NEC® \text{ required amps}$$

_____ **X** _____ = _____ **X** 1.25 **X** 1.25 = _____

Amperage satisfying *NEC*® = _____            Wire Size from Table 9-4 = _____

    **B. Voltage Drop Requirements**

System Voltage: _____            Total Amps: \_\_\_\_\_

One Way Distance: \_\_\_\_\_           Voltage Drop(%): \_\_\_\_\_

Wire Size from voltage drop tables (Tables 9-5 through 9-7): \_\_\_\_\_

Is this equal to or greater than the size wire needed for safety? \_\_\_\_\_

  • If yes, this is your answer.        • If no, use the wire size from A.

**PV Combiner Box to Controller**

At times, it can be advantageous to break up the PV to Battery wire run into two separate wire runs, PV to Controller and Controller to Battery. Since the Controller is usually very close to the battery, you can usually size this section with wire smaller than the PV to Controller section as long as it passes the *NEC*® required ampacity from the PV array.

    **A. *NEC*® Requirement**

$$\underset{\text{modules}}{\text{Isc of}} \; \textbf{X} \; \underset{\text{in parallel}}{\text{\# of modules}} = \text{Total Amps} \; \textbf{X} \; 1.25 \; \textbf{X} \; 1.25 = NEC® \text{ required amps}$$

_____ **X** _____ = _____ **X** 1.25 **X** 1.25 = _____

Amperage satisfying *NEC*® = _____            Wire Size from Table 9-4 = _____

    **B. Voltage Drop Requirements**

System Voltage: _____            Total Amps: \_\_\_\_\_

One Way Distance: \_\_\_\_\_           Voltage Drop(%): \_\_\_\_\_

Wire Size from voltage drop tables (Tables 9-5 through 9-7): \_\_\_\_\_

Is this equal to or greater than the size wire needed for safety? \_\_\_\_\_

  • If yes, this is your answer.        • If no, use the wire size from A.

**Controller to Battery**

   **A.  *NEC*® Requirement**

$$\frac{\text{Isc of}}{\text{modules}} \quad \textbf{X} \quad \frac{\text{\# of modules}}{\text{in parallel}} = \text{Total Amps} \quad \textbf{X} \ 1.25 \ \textbf{X} \ 1.25 = \textit{NEC}^® \text{ required amps}$$

_____  **X**  _____  = _____  **X** 1.25 **X** 1.25 = _____

Amperage satisfying *NEC*® = _____        Wire Size from Table 9-4 = _____

   **B.  Voltage Drop Requirements**

System Voltage: _____        Total Amps: _____
One Way Distance: _____        Voltage Drop(%): _____

Wire Size from voltage drop tables (Tables 9-5 through 9-7): _____
Is this equal to or greater than the size wire needed for safety? _____

- If yes, this is your answer.        • If no, use the wire size from A.

**Battery to DC Load Center**

   **A.  *NEC*® Requirement**

$$\text{DC load watts} \div \text{DC voltage} = \text{DC total amps} \quad \textbf{X} \ 1.25 = \textit{NEC}^® \text{ required amps}$$

_____  ÷  _____  = _____  **X** 1.25 = _____

Amperage satisfying *NEC*® = _____        Wire Size from Table 9-4 = _____

   **B.  Voltage Drop Requirements:**

System Voltage: _____        Total Amps: _____

One Way Distance: _____        Voltage Drop(%): _____

Wire Size from voltage drop tables (Tables 9-5 through 9-7): _____

Is this equal to or greater than the size wire needed for safety? _____

- If yes, this is your answer.        • If no, use the wire size from A.

**Battery to Inverter**

   **A.  *NEC*® Requirement**

$$\frac{\text{Inverter}}{\text{Rated Watts}} \div \frac{\text{Inverter}}{\text{Efficiency}} \div \frac{\text{DC System}}{\text{(lowest operating)}}_{\text{Voltage}} = \frac{\text{Inverter}}{\text{Total Amps}} \quad \textbf{X} \ 1.25 = \frac{\textit{NEC}^®}{\text{required amps}}$$

_____  ÷ _____  ÷ _____  = _____  **X** 1.25 = _____

Amperage satisfying *NEC*® = _____        Wire Size from Table 9-4 = _____
Verify battery conductor size with inverter manufacturer.

   **B.  Voltage Drop Requirements**

System Voltage: _____        Total Amps: _____

One Way Distance: _____        Voltage Drop(%): _____

Wire Size from voltage drop tables (Tables 9-5 through 9-7): _____

Is this equal to or greater than the size wire needed for safety? _____

- If yes, this is your answer.        • If no, use the wire size from A.

Temperature deration, conduit fill deration, and device terminal compatibility, are not included in the wire sizing worksheets.

## System Wire Sizing Worksheet

Use the following worksheet to determine system wire sizes.

### PV Combiner Box to Battery

You can size this section as one wire run from PV to Battery, due to the fact that the controller is basically a pass through device. You can also break this wire run into two sections, PV to Controller and Controller to Battery (see wire sizing worksheet below).

### A. *NEC* Requirement

$$\frac{\text{Isc of}}{\text{modules}} \times \frac{\text{\# of modules}}{\text{in parallel}} = \text{Total Amps} \times 1.25 \times 1.25 = NEC \text{ required amps}$$

_____ X _____ = _____ X 1.25 X 1.25 = _____

Amperage satisfying *NEC* = _____       Wire Size from Table 9-4 = _____

### B. Voltage Drop Requirements

System Voltage: _____       Total Amps: _____

One Way Distance: _____       Voltage Drop(%): _____

Wire Size from voltage drop tables (Tables 9-5 through 9-7): _____
Is this equal to or greater than the size wire needed for safety? _____

- If yes, this is your answer.       • If no, use the wire size from A.

### PV Combiner Box to Controller

At times, it can be advantageous to break up the PV to Battery wire run into two separate wire runs, PV to Controller and Controller to Battery. Since the Controller is usually very close to the battery, you can usually size this section with wire smaller than the PV to Controller section as long as it passes the *NEC* required ampacity from the PV array.

### A. *NEC* Requirement

$$\frac{\text{Isc of}}{\text{modules}} \times \frac{\text{\# of modules}}{\text{in parallel}} = \text{Total Amps} \times 1.25 \times 1.25 = NEC \text{ required amps}$$

_____ X _____ = _____ X 1.25 X 1.25 = _____

Amperage satisfying *NEC* = _____       Wire Size from Table 9-4 = _____

### B. Voltage Drop Requirements

System Voltage: _____       Total Amps: _____

One Way Distance: _____       Voltage Drop(%): _____

Wire Size from voltage drop tables (Tables 9-5 through 9-7): _____
Is this equal to or greater than the size wire needed for safety? _____

- If yes, this is your answer.       • If no, use the wire size from A.

page 1 of 2

### Controller to Battery
#### A. *NEC*® Requirement

$$\frac{\text{Isc of}}{\text{modules}} \ \text{X} \ \frac{\text{\# of modules}}{\text{in parallel}} = \text{Total Amps X 1.25 X 1.25} = \textit{NEC}\text{® required amps}$$

_____ X _____ = _____ X 1.25 X 1.25 = _____

Amperage satisfying *NEC*® = _____          Wire Size from Table 9-4 = _____

#### B. Voltage Drop Requirements

System Voltage: _____          Total Amps: _____
One Way Distance: _____          Voltage Drop(%): _____

Wire Size from voltage drop tables (Tables 9-5 through 9-7): _____
Is this equal to or greater than the size wire needed for safety? _____

- • If yes, this is your answer.          • If no, use the wire size from A.

### Battery to DC Load Center
#### A. *NEC*® Requirement

$$\text{DC load watts} \div \text{DC voltage} = \text{DC total amps X 1.25} = \textit{NEC}\text{® required amps}$$

_____ ÷ _____ = _____ X 1.25 = _____

Amperage satisfying *NEC*® = _____          Wire Size from Table 9-4 = _____

#### B. Voltage Drop Requirements:

System Voltage: _____          Total Amps: _____

One Way Distance: _____          Voltage Drop(%): _____

Wire Size from voltage drop tables (Tables 9-5 through 9-7): _____

Is this equal to or greater than the size wire needed for safety? _____

- • If yes, this is your answer.          • If no, use the wire size from A.

### Battery to Inverter
#### A. *NEC*® Requirement

$$\frac{\text{Inverter}}{\text{Rated Watts}} \div \frac{\text{Inverter}}{\text{Efficiency}} \div \frac{\text{DC System}}{\text{(lowest operating)}}{\text{Voltage}} = \frac{\text{Inverter}}{\text{Total Amps}} \ \text{X 1.25} = \frac{\textit{NEC}\text{®}}{\text{required amps}}$$

_____ ÷ _____ ÷ _____ = _____ X 1.25 = _____

Amperage satisfying *NEC*® = _____          Wire Size from Table 9-4 = _____
Verify battery conductor size with inverter manufacturer.

#### B. Voltage Drop Requirements

System Voltage: _____          Total Amps: _____

One Way Distance: _____          Voltage Drop(%): _____

Wire Size from voltage drop tables (Tables 9-5 through 9-7): _____

Is this equal to or greater than the size wire needed for safety? _____

- • If yes, this is your answer.          • If no, use the wire size from A.

Temperature deration, conduit fill deration, and device terminal compatibility, are not included in the wire sizing worksheets.

## Grid-tied Sizing Worksheet

**Electric Load Estimation**

1) Figure out the approximate daily average energy usage and PV System kWh/day:

Yearly average energy consumption: _____ kilowatt-hrs/year

_____ kilowatt-hrs/yr ÷ 365 days/yr = _____ average kilowatt-hrs/day
(This is our average daily load.)

_____ % of power to be generated from PV system

_____ Avg. kWh/Day X _____ % of power to be from PV = _____ PV system kilowatt-hrs/day

**Array Sizing**

2) Figure out the PV system kilowatts needed (including derate factors for temperature losses, miscellaneous system losses, and inverter losses):

Average peak sun hours per day: _____

_____ PV System kWh/day ÷ _____ avg. sun hours per day ÷ 0.88 PV Temp Losses (see Notes*)

÷ 0.84 Derate Factor (see Notes**) ÷ _____ inverter efficiency (see Notes***)

= PV array kW needed _____

_____ PV array kW X (1000 watts/kilowatt) = _____ PV array watts

3) Choose a PV module:

Make: _____ Model: _____

STC watt rating: _____ Voc: _____ Vmax: _____

Isc: _____ Imp: _____

_____ PV array watts ÷ _____ STC watt rating _____ # of modules needed

## Grid-tied Sizing Worksheet - Continued

**Inverter Sizing**

4) Choose a specific inverter (or a combination of inverters) that has an appropriate continuous wattage rating:

With grid-tied PV systems an inverter model is chosen based on the maximum amount of watts passing through it from the array (unlike stand-alone PV systems where the inverter size is based on the AC total connected load).

_____ # of modules needed x _____ STC watt rating = _____ max watts inverter(s) must pass

Manufacturer: _____ Model: _____

DC (STC) Continuous Power rating: _____

DC input Voltage Range: _____

5) Calculate how many of these inverters the system will require, and how many modules will be wired into each inverter:

_____ max watts inverter must pass ÷ _____ inverter watt rating = _____ # of inverters

_____ # of PV modules needed ÷ _____ # of inverters = _____ # of modules per inverter

6) Find out how many of our modules the chosen inverter requires in series?
Check with the inverter manufacture to see how many modules this inverter requires in series for it's DC input voltage window.   Most grid-tied inverters have a string sizing program on their website to calculate how many modules in series are needed.

Array Location (City, State): _____

Record Low Temp: _____

Average High Temp: _____

_____

Using the PV modules and the site temperature info,
how many modules does the inverter need in series? _____

Does this work with the number of panels needed per inverter? _____

Remember if using more than one inverter, break up the PV array into subarrays that will feed each inverter. Each inverter must have the appropriate number of modules in series to match the inverter's DC input voltage range. If not, either round the number of modules in the array up or down.  This will affect the percentage of power to be generated by the PV system. Another option is to choose a different module or inverter to be used in the system.

## Notes for Grid-tied Sizing Worksheet

**\*Note:** Standard Test Condition ratings where cell temperature = 25°Celsius is not very realistic when solar modules are in the sun. To account for temperature losses in more realistic situations the sizing sheets use a temperature derate value of 0.88. This assumes an average daytime ambient temperature of 20 degree C. Each module has a slightly different temperature coefficient which is not taken into consideration here.

**\*\*Note:** The Derate Factor of 0.84 accounts for other system losses (including module production tolerance, module mismatch, wiring losses, dust/soiling losses, etc.). This value assumes no shading. See the table below for a summary of how this derate factor is calculated:

| Derate Values | Range of Acceptable Values | Chosen values |
|---|---|---|
| PV module nameplate DC rating | 0.80 - 1.05 | .95 |
| Mismatch Modules | 0.97 - 0.995 | .98 |
| Diodes and connections | 0.99 - 0.997 | .995 |
| DC wiring | 0.97 - 0.99 | .98 |
| AC wiring | 0.98 - 0.993 | .99 |
| Soiling | 0.30 - 0.995 | .95 |
| System availability | 0.00 - 0.995 | .98 |
| Shading | 0.00 - 1.00 | 1.00 |
| Age | 0.70 - 1.00 | 1.00 |
| **DERATE Factor** | | **= 0.84** |

Note: You can adjust this value to reflect conditions for your specific site

Source: http://rredc.nrel.gov/solar/codes_algs/PVWATTS/

**\*\*\*Note:** Typically an inverter efficiency of 0.9 is used as a conservative estimate. A specific inverter's average efficiency number can be implemented from the California Consumer Energy Center's List of Eligible Inverters webpage: http://www.consumerenergycenter.org/cgi-bin/eligible_inverters.cgi. This website provides specific inverter's "average" efficiency. (For more information on how the inverter efficiency values are evaluated see the Emerging Renewable Program Guidebook, Appendix 3)

### Additional Notes:

The National Renewable Energy Laboratory (NREL) has an online Grid-Connected PV system performance calculator called "PVWatts" which has many features to further customize a PV array.

For example, perhaps the only solar access at a particular site is on an east facing roof. This grid-tied sizing method assumes the array orientation is true south. The NREL PVWatts program allows flexibility to calculate the system production for an east facing array and to adjust the tilt angle of the roof pitch. See PVWatts website: http://rredc.nrel.gov/solar/codes_algs/PVWATTS/version1/

Also realize that the results from this grid-tied sizing sheet will differ slightly from the PVWatts performance calculator primarily due to the fact that it uses a simple temperature derate value, while PVWatts utilizes monthly weather data for each location.

Finally it must be realized that annual power production of a PV system is largely dependent on how much available sunlight there is and **weather patterns vary year to year**. This means that even though this grid-tied sizing method (and PVWatts) is utilizing long term solar data, any year could have more or less sunshine available, which means the actual annual power production of a PV system could exceed or fall short of the expectations.

## Grid-tied Sizing Worksheet

**Electric Load Estimation**

1) Figure out the approximate daily average energy usage and PV System kWh/day:

Yearly average energy consumption: _____ kilowatt-hrs/year

_____ kilowatt-hrs/yr ÷ 365 days/yr = _____ average kilowatt-hrs/day
(This is our average daily load.)

_____ % of power to be generated from PV system

_____ Avg. kWh/Day X _____ % of power to be from PV = _____ PV system kilowatt-hrs/day

**Array Sizing**

2) Figure out the PV system kilowatts needed (including derate factors for temperature losses, miscellaneous system losses, and inverter losses):

Average peak sun hours per day: _____

_____ PV System kWh/day ÷ _____ avg. sun hours per day ÷ 0.88 PV Temp Losses (see Notes*)

÷ 0.84 Derate Factor (see Notes**) ÷ _____ inverter efficiency (see Notes***)

= PV array kW needed _____

_____ PV array kW X (1000 watts/kilowatt) = _____ PV array watts

3) Choose a PV module:

Make: _____ Model: _____

STC watt rating: _____ Voc: _____ Vmax: _____

Isc: _____          Imp: _____

_____ PV array watts ÷ _____ STC watt rating _____ # of modules needed

## Grid-tied Sizing Worksheet - Continued

**Inverter Sizing**

4) Choose a specific inverter (or a combination of inverters) that has an appropriate continuous wattage rating:

With grid-tied PV systems an inverter model is chosen based on the maximum amount of watts passing through it from the array (unlike stand-alone PV systems where the inverter size is based on the AC total connected load).

_____ # of modules needed x _____ STC watt rating = _____ max watts inverter(s) must pass

Manufacturer: _____ Model: _____

DC (STC) Continuous Power rating: _____

DC input Voltage Range: _____

5) Calculate how many of these inverters the system will require, and how many modules will be wired into each inverter:

_____ max watts inverter must pass ÷ _____ inverter watt rating = _____ #of inverters

_____ # of PV modules needed ÷ _____ # of inverters = _____ # of modules per inverter

6) Find out how many of our modules the chosen inverter requires in series?
Check with the inverter manufacture to see how many modules this inverter requires in series for it's DC input voltage window.   Most grid-tied inverters have a string sizing program on their website to calculate how many modules in series are needed.

Array Location (City, State): _____

Record Low Temp: _____

Average High Temp: _____

_____

Using the PV modules and the site temperature info,
how many modules does the inverter need in series? _____

Does this work with the number of panels needed per inverter? _____

Remember if using more than one inverter, break up the PV array into subarrays that will feed each inverter. Each inverter must have the appropriate number of modules in series to match the inverter's DC input voltage range. If not, either round the number of modules in the array up or down.  This will affect the percentage of power to be generated by the PV system. Another option is to choose a different module or inverter to be used in the system.

## Notes for Grid-tied Sizing Worksheet

**\*Note:** Standard Test Condition ratings where cell temperature = 25°Celsius is not very realistic when solar modules are in the sun. To account for temperature losses in more realistic situations the sizing sheets use a temperature derate value of 0.88. This assumes an average daytime ambient temperature of 20 degree C. Each module has a slightly different temperature coefficient which is not taken into consideration here.

**\*\*Note:** The Derate Factor of 0.84 accounts for other system losses (including module production tolerance, module mismatch, wiring losses, dust/soiling losses, etc.). This value assumes no shading. See the table below for a summary of how this derate factor is calculated:

| Derate Values | Range of Acceptable Values | Chosen values |
|---|---|---|
| PV module nameplate DC rating | 0.80 - 1.05 | .95 |
| Mismatch Modules | 0.97 - 0.995 | .98 |
| Diodes and connections | 0.99 - 0.997 | .995 |
| DC wiring | 0.97 - 0.99 | .98 |
| AC wiring | 0.98 - 0.993 | .99 |
| Soiling | 0.30 - 0.995 | .95 |
| System availability | 0.00 - 0.995 | .98 |
| Shading | 0.00 - 1.00 | 1.00 |
| Age | 0.70 - 1.00 | 1.00 |
| **DERATE Factor** | | = **0.84** |

Note: You can adjust this value to reflect conditions for your specific site

Source: http://rredc.nrel.gov/solar/codes_algs/PVWATTS/

**\*\*\*Note:** Typically an inverter efficiency of 0.9 is used as a conservative estimate. A specific inverter's average efficiency number can be implemented from the California Consumer Energy Center's List of Eligible Inverters webpage: http://www.consumerenergycenter.org/cgi-bin/eligible_inverters.cgi. This website provides specific inverter's "average" efficiency. (For more information on how the inverter efficiency values are evaluated see the Emerging Renewable Program Guidebook, Appendix 3)

**Additional Notes:**
The National Renewable Energy Laboratory (NREL) has an online Grid-Connected PV system performance calculator called "PVWatts" which has many features to further customize a PV array.

For example, perhaps the only solar access at a particular site is on an east facing roof. This grid-tied sizing method assumes the array orientation is true south. The NREL PVWatts program allows flexibility to calculate the system production for an east facing array and to adjust the tilt angle of the roof pitch. See PVWatts website: http://rredc.nrel.gov/solar/codes_algs/PVWATTS/version1/

Also realize that the results from this grid-tied sizing sheet will differ slightly from the PVWatts performance calculator primarily due to the fact that it uses a simple temperature derate value, while PVWatts utilizes monthly weather data for each location.

Finally it must be realized that annual power production of a PV system is largely dependent on how much available sunlight there is and **weather patterns vary year to year**. This means that even though this grid-tied sizing method (and PVWatts) is utilizing long term solar data, any year could have more or less sunshine available, which means the actual annual power production of a PV system could exceed or fall short of the expectations.

## Grid-tied Sizing Worksheet

**Electric Load Estimation**

1) Figure out the approximate daily average energy usage and PV System kWh/day:

Yearly average energy consumption: _____ kilowatt-hrs/year

_____ kilowatt-hrs/yr ÷ 365 days/yr = _____ average kilowatt-hrs/day
(This is our average daily load.)

_____ % of power to be generated from PV system

_____ Avg. kWh/Day X _____ % of power to be from PV = _____ PV system kilowatt-hrs/day

**Array Sizing**

2) Figure out the PV system kilowatts needed (including derate factors for temperature losses, miscellaneous system losses, and inverter losses):

Average peak sun hours per day: _____

_____ PV System kWh/day ÷ _____ avg. sun hours per day ÷ 0.88 PV Temp Losses (see Notes*)

÷ 0.84 Derate Factor (see Notes**) ÷ _____ inverter efficiency (see Notes***)

= PV array kW needed _____

_____ PV array kW X (1000 watts/kilowatt) = _____ PV array watts

3) Choose a PV module:

Make: _____ Model: _____

STC watt rating: _____ Voc: _____ Vmax: _____

Isc: _____ Imp: _____

_____ PV array watts ÷ _____ STC watt rating _____ # of modules needed

## Grid-tied Sizing Worksheet - Continued

**Inverter Sizing**

4) Choose a specific inverter (or a combination of inverters) that has an appropriate continuous wattage rating:

With grid-tied PV systems an inverter model is chosen based on the maximum amount of watts passing through it from the array (unlike stand-alone PV systems where the inverter size is based on the AC total connected load).

_____ # of modules needed x _____ STC watt rating = _____ max watts inverter(s) must pass

Manufacturer: _____ Model: _____

DC (STC) Continuous Power rating: _____

DC input Voltage Range: _____

5) Calculate how many of these inverters the system will require, and how many modules will be wired into each inverter:

_____ max watts inverter must pass ÷ _____ inverter watt rating = _____ #of inverters

_____ # of PV modules needed ÷ _____ # of inverters = _____ # of modules per inverter

6) Find out how many of our modules the chosen inverter requires in series?
Check with the inverter manufacture to see how many modules this inverter requires in series for it's DC input voltage window.   Most grid-tied inverters have a string sizing program on their website to calculate how many modules in series are needed.

Array Location (City, State): _____

Record Low Temp: _____

Average High Temp: _____

_____

Using the PV modules and the site temperature info,
how many modules does the inverter need in series? _____

Does this work with the number of panels needed per inverter? _____

Remember if using more than one inverter, break up the PV array into subarrays that will feed each inverter. Each inverter must have the appropriate number of modules in series to match the inverter's DC input voltage range. If not, either round the number of modules in the array up or down.  This will affect the percentage of power to be generated by the PV system. Another option is to choose a different module or inverter to be used in the system.

## Notes for Grid-tied Sizing Worksheet

**\*Note:** Standard Test Condition ratings where cell temperature = 25°Celsius is not very realistic when solar modules are in the sun. To account for temperature losses in more realistic situations the sizing sheets use a temperature derate value of 0.88. This assumes an average daytime ambient temperature of 20 degree C. Each module has a slightly different temperature coefficient which is not taken into consideration here.

**\*\*Note:** The Derate Factor of 0.84 accounts for other system losses (including module production tolerance, module mismatch, wiring losses, dust/soiling losses, etc.). This value assumes no shading. See the table below for a summary of how this derate factor is calculated:

| Derate Values | Range of Acceptable Values | Chosen values |
|---|---|---|
| PV module nameplate DC rating | 0.80 - 1.05 | .95 |
| Mismatch Modules | 0.97 - 0.995 | .98 |
| Diodes and connections | 0.99 - 0.997 | .995 |
| DC wiring | 0.97 - 0.99 | .98 |
| AC wiring | 0.98 - 0.993 | .99 |
| Soiling | 0.30 - 0.995 | .95 |
| System availability | 0.00 - 0.995 | .98 |
| Shading | 0.00 - 1.00 | 1.00 |
| Age | 0.70 - 1.00 | <u>1.00</u> |
| **DERATE Factor** | | = **0.84** |

Note: You can adjust this value to reflect conditions for your specific site

Source: http://rredc.nrel.gov/solar/codes_algs/PVWATTS/

**\*\*\*Note:** Typically an inverter efficiency of 0.9 is used as a conservative estimate. A specific inverter's average efficiency number can be implemented from the California Consumer Energy Center's List of Eligible Inverters webpage: http://www.consumerenergycenter.org/cgi-bin/eligible_inverters.cgi. This website provides specific inverter's "average" efficiency. (For more information on how the inverter efficiency values are evaluated see the Emerging Renewable Program Guidebook, Appendix 3)

**Additional Notes:**

The National Renewable Energy Laboratory (NREL) has an online Grid-Connected PV system performance calculator called "PVWatts" which has many features to further customize a PV array.

For example, perhaps the only solar access at a particular site is on an east facing roof. This grid-tied sizing method assumes the array orientation is true south. The NREL PVWatts program allows flexibility to calculate the system production for an east facing array and to adjust the tilt angle of the roof pitch. See PVWatts website: http://rredc.nrel.gov/solar/codes_algs/PVWATTS/version1/

Also realize that the results from this grid-tied sizing sheet will differ slightly from the PVWatts performance calculator primarily due to the fact that it uses a simple temperature derate value, while PVWatts utilizes monthly weather data for each location.

Finally it must be realized that annual power production of a PV system is largely dependent on how much available sunlight there is and **weather patterns vary year to year**. This means that even though this grid-tied sizing method (and PVWatts) is utilizing long term solar data, any year could have more or less sunshine available, which means the actual annual power production of a PV system could exceed or fall short of the expectations.

# Resource Guide

This guide provides additional sources that can be used to further explore photovoltaics. You will find selected organizations, publications, and websites. If you know of other appropriate resources that could prove valuable to future solar students, please contact: Solar Energy International (SEI), PV Manual Department, Email: sei@solarenergy.org, Phone: 970-963-8855, Fax: 970-963-8866, Website: www.solarenergy.org

## Selected Organizations

**American Solar Energy Society (ASES):** A national membership-based organization of solar professionals dedicated to advancing of the use of solar energy. The website and magazine contain the history of renewables, economics, and both residential and commercial applications. The ASES magazine, *Solar Today*, covers PV, passive solar, solar thermal, wind energy, and solar building case studies.
Website: www.ases.org
Phone: 303-443-3130

**Censolar:** An established solar center located in Spain, exclusively dedicated to the dissemination of solar energy information. Censolar is SEI's European affiliate and an SEI INVEST partner. Censolar offers workshops, distance courses, and numerous high quality Spanish solar related publications.
Website: www.censolar.org

**Enersol Associates:** A private, humanitarian non-profit organization using solar energy to improve people's lives in rural Latin America.
Website: www.enersol.org

**European Photovoltaic Industry Association:** Falling beneath the umbrella of the European Renewable Energy Council, this website ties you in with what's new in the photovoltaic industry in Europe. A easy site to use with many useful links leading one to many places of unknown curiousity.
Website: www.epia.org

**Florida Solar Energy Center (FSEC):** An organization that assists individuals in making the right decisions when choosing to use solar energy systems. FSEC offers educational workshops, training, and technical research reports on topics including: photovoltaics, solar thermal, energy efficiency, moisture issues, and cooling strategies.
Website: www.fsec.ucf.edu

*Home Power* **Magazine:** A bi-monthly magazine that is "information central" for details on working and living with PV and other renewable and sustainable technologies. The website has many links to databases, events, non-profits, businesses, and has the current magazine issue available for free viewing. The site also contains an archive of *Home Power* Magazines. Books, videos and CD-ROMs are also available. P.O. BOX 520, Ashland, OR, 97520, USA.
Website: www.homepower.com
Email: hp@homepower.com
Phone: 916-475-3179

**The Institute for Solar Living (Solar Living Center):** This multi-acre institute is a non-profit environmental education/demonstration center. The SLC offers a variety of short workshops that promote energy efficiency, renewable energy technologies and sustainable building. The Solar Living Center, Hopland, CA 95449.
Website: www.solarliving.org

**International Solar Energy Society (ISES):** A UN-accredited NGO that supports the advancement of renewable energy technology, implementation, and education all over the world. ISES is a professional membership based organization.
Website: www.ises.org

**Institute for Sustainable Power (ISP):** A non-profit umbrella organization working worldwide, that strives to improve the quality of renewable energy projects and the development of sustainable, local jobs. An infrastructure providing accreditation to qualified educational organizations and trainers that provide high quality PV training.
Website: www.ispq.org

**Midwest Renewable Energy Association (MREA):** Non-profit network for sharing ideas, resources, and information to promote a sustainable future through renewable energy and energy efficiency. MREA offers workshops and sponsors the annual Midwest Renewable Energy Fair (MREF).
Website: www.the-mrea.org

**North American Board of Certified Energy Practitioners (NABCEP):** Organization developing voluntary standards for renewable energy professionals, beginning with certification for solar electric installers.
Website: www.nabcep.org

**Northern California Solar Energy Association:** A website containing many links to educational organizations. The organization puts out a monthly newsletter of the latest happenings in the California solar world as well as sponsoring a variety of events that promote the use of solar energy.
Website: www.norcalsolar.org

**National Renewable Energy Laboratory (NREL):** Affiliated with U.S. Department of Energy, and is America's national solar R&D laboratory. NREL is working toward securing an energy future that is environmentally and economically sustainable. Portions of the vast website are interactive, and a large photo library is online. Resources, images, and extensive links are available for the novice student as well as the solar professional.
Website: www.nrel.gov

**Rahus Institute:** A non-profit, research and educational organization focusing on resource efficiency and renewable energies in California.
Email: info@rahus.org
Website: www.rahus.org

**SEI INVEST Program:** International Volunteers in Environmentally Sustainable Technologies. Graduates of SEI's courses are connected with organizations working in rural development. Within these organizations alumni can volunteer to help electrify rural communities with renewable energy, build sustainable houses, and train local users and technicians.
Website: www.solarenergy.org/programs/INVEST/

**Solar Electric Light Fund (SELF):** A non-profit charitable organization founded in 1990 to promote, develop, and facilitate solar rural electrification and energy self-sufficiency in developing countries. SELF is an SEI affiliate, and an INVEST partner, and has pioneered rural PV Solar Home Systems worldwide.
Website: www.self.org

## State and Territory Energy Offices

**National Association of State Energy Officials:** The website contains links to the energy offices of all 50 states. A variety of information exists on each state and territory website.
Website: www.naseo.org/members/states.htm

## Books

### General Photovoltaics (PV)

Anderson, Teresa, Alison Doig, Dai Rees, Smail Khennas. *Rural Energy Services: A Handbook for Sustainable Energy Development.* London, UK: Intermediate Technology Publications Ltd., 1999.

Archer, Mary, Robert Hill. *Photoconversion of Solar Energy, Vol 1: Clean Electricity from Photovoltaics.* Colorado, USA: Imperial College Press, 2001.

Canadian Photovoltaic Industries Association. *Photovoltaic Systems Design Manual.* Canada: CANMET - Energy, Mines & Resources, 1991.

Censolar, Centro de Estudios de la Energia Solar. *La Energia Solar: Aplicaciones Practicas.* Spain: Promotora General de Esudios, S.A. 1999.

Cole, Nancy. Sderrett, J.P. *Renewables are Ready: People Creating Renewable Energy Solutions.* Vermont, USA: Chelsea Green Publishing Company, 1995.

Davidson, J. *The New Solar Electric Home: The Photovoltaics How-to Handbook.* Michigan, USA: 1987.

Derrick, Anthony, Catherine Francis, Varis Bokalders. *Solar PV Products: A Guide for Development Workers.* London, UK: Intermediate Technology Publications Ltd, 1991.

Duffie and Beckman. *Solar Engineering of Thermal Processes, 2nd edition.* New York, USA: John Wiley & Sons, 1991.

Ewing, and Ewing. *Power with Nature: Solar and Wind Energy Demystified.* Colorado, USA: PixyJack Press, 2003.

Green. *Solar Cells: Operating Principles, Technology & System Applications.* New York, USA: Prentice-Hall, 1982.

Hackleman, Michael. *Better Use of...Your Electric Lights, Home Appliances, Shop Tools-Everything that Uses Electricity.* California, USA: Peace Press, 1981.

Halacy, Dan. *Home Energy.* Pennsylvania, USA: Rodale Press, 1984.

Hankins, Mark. *Solar Electric Systems for Africa: A Guide for Planning and Installing Solar Electric Systems in Rural Africa.* London, UK: Commonwealth Science Council, 1995.

Imamura, Helm, Palz, Stephens, and Associates. *Photovoltaic System Technology - A European Handbook.* Bedford, UK: Commission of the European Communities, 1992.

Johannsson, (Ed.). *Renewable Energy, Sources for Fuels and Electricity.* Washington DC, USA: Island Press, 1993.

Komp, R. *Practical Photovoltaics: Electricity from Solar Cells,* Third Edition. Michigan, USA: Aatec Publications, 1995.

Landolt-Burnstein. *Numerical Data & Functional Relationships in Science and Technology,* Vol. 4c, Climatology, Part 2. Berlin, Germany: New Series, 1989.

Lasnier, GanAng, and Hilger. *Photovoltaic Engineering Handbook.* Pennsylvania, USA: IOP Publishing, 1990.

Lorenzo, E. Zilles, R. Caamano-Martin, E. *Photovoltaic Rural Electrification: A Fieldwork Picture Book.* Spain: Promotora General de Estudios, S.A. 2001.

Lorenzo, Eduardo, et al. *Solar Electricity: Engineering of Photovoltaic Systems.* Sevilla, Spain: 1994. (Published in Spanish as Electricidad Solar).

Markvart and Tomas, (Ed.). *Solar Electricity.* Southampton, UK: John Wiley & Sons Ltd., 1994.

Maycock and Stirewalt. *A Guide to the Photovoltaic Revolution - Sunlight to Electricity in One Step.* Pennsylvania, USA: Rodale Press, 1985.

Perlin, John. *From Space to Earth: The Story of Solar Electricity.* Michigan, USA: Aatec Publishing, 2001.

Parker, (Ed). *Solar Energy In Agriculture*. Amsterdam, The Netherlands: 1991.

Roberts. *Solar Electricity: A Practical Guide to Designing and Installing Small Photovoltaic Systems*. New Jersey, USA: Prentice Hall, 1991.

Roth, and Schmidt, (Ed.). *Photovoltaik-Anlagen*. Freiburg, Germany: Photovoltaik-Anlagen, Begleitbuch zum Seminar, 1994.

Shepperd, Lisa, W. Richards, H. Elizabeth. *Solar Photovoltaics for Development Applications*. Sandia National Laboratories, 1993.

Sherman, Robin. *Renewables are Ready: A Guide to Teaching Renewable Energy in Junior and Senior High School Classrooms*. Massachusettes, USA: Union of Concerned Scientists, 2003.

Solar Energy International (SEI). *Photovoltaics: Design and Installation Manual*. Co-published with New Society Publishing, 2004.

Solar Energy Research Institute (SERI). *Photovoltaic Fundamentals*. Colorado, USA: National Renewable Energy Laboratories (NREL), 1991.

Spring, Cario, Lisa Stage. *When the Light Goes On: Understanding Energy*. Arizona, USA: Emerald Resource Solutions 2001.

Stamenic and Ingham. *A Power for the World: Solar Photovoltaics Revolution*. North Vancouver, BC. Canada: Sunology International, Inc., 1995.

Strong, S.J. *The Solar Electric House*. Vermont, USA: Chelsea Green Publishing Company, 1991.

Zweibel and Hersch. *Basic Photovoltaic Principles and Methods*. New York, USA: Van Nostrand Reinhold Company, 1984.

**System Design**

Asociacion De La Industria Potovoltaica. *Sistemas de Energia Fotovoltaica: Manual del Instalador*. Spain: Promotora General de Estudios, S.A., 2002.

Brotherton, Miner. *The 12 Volt Bible for Boats*. Maine, USA: Seven Seas Press/International Marine, 1985.

Canadian Photovoltaic Industries Association. *Photovoltaic Systems Design Manual*. Canada: CANMET-Energy, Mines & Resources Canada, 1991.

Chapman, R.N. *Sizing Handbook for Stand-alone Photovoltaic/Storage Systems*. Albuquerque, USA: Sandia National Laboratories, 1987.

Davidson, J. *The New Solar Electric Home: The Photovoltaics How-to Handbook*. Michigan, USA: Aatec Publishers, 1987.

Kiskorski, A.S. *Power Tracking Methods in Photovoltaic Applications*, Proceedings PCIM '93. Nurnberg, Germany: 1993.

Messenger, Roger, Jerry A. Ventre, Gerard, G. Ventre. *Photovoltaic Systems Engineering 2nd Edition*. Florida, USA: CRC Press, 2003.

Paul, Terrance, D. *How to Design an Independent Power System*. Wisconsin, USA: Best Energy Systems for Tomorrow, Inc., 1981.

Sandia National Laboratories, Science Applications, Inc. *Design Handbook for Photovoltaic Power Systems*. McLean, USA: 1981.

Sandia National Laboratories, Photovoltaic System Design Assistance Center (DAC) *The Design of Residential Photovoltaic Systems*, (10 volumes). New Mexico, USA: 1988.

Solar Energy International (SEI). *Photovoltaics: Design and Installation Manual*. Canada: Co-published with New Society Publishing, Canada, 2004.

Wills, R. *The Interconnection of PV Power Systems with the Utility Grid: An Overview for Utility Engineers.* New Mexico, USA: Sandia National Laboratories, 1994.

Yago, Jeffrey, R. *Achieving Energy Independence- One Step at a Time.* Virginia, USA: Dunimis Technology, Inc. 1999.


## Components

Donepudi, Pell, and Royer. *Storage Module Survey: Task 16 - Photovoltaic in Buildings.* Ottawa, Canada: IEA-Solar Heating and Cooling Program, 1993.

Dunselman, Weiden, Zolingen, Heide. *Design Specification for AC Modules.* Holland, The Netherlands: Ecofys report nr. E265, Utrecht, 1993.

Hill and McCarthy. *PV Battery Handbook.* Ireland: Hyperion Energy Systems Ltd., 1992.

Linden, (Ed.). *Handbook of Batteries and Fuel Cells.* USA: McGraw-Hill Inc., 1984.

Panhuber-Fronius and Edelmoser. *Resonant Concept for the Power-Section of a Grid-Coupled Inverter.* Amsterdam, The Netherlands: Proc. of 12th European Solar Energy Conference and Exhibition, 1994.

Perez, Richard, A. *The Complete Battery Book.* Pennsylvania, USA: TAB Books, Inc., 1985.

Rapp, D. *Solar Energy.* NJ, USA: Prentice-Hall, Englewood Cliffs, 1981.

Russell, M.C. *Residential Photovoltaic System Design Handbook.* Massachusetts, USA: MIT 1984.

Schaeffer, J. *Alternative Energy Sourcebook.* California, USA: Real Goods Trading Corporation, 1992.

Wilk, H. *40 kW - Photovoltaic System with IGBT Inverter on the Sound barriers of Motorway A1.* Montreux, Switzerland: 11th European Photovoltaic Solar Energy Conference, 1992.

Wilk H. *200 kW Photovoltaic Rooftop Programme in Austria.* Budapest, Hungary: ISES World Congress, 1993.

Wilk, H. *200 kW PV Rooftop Programme in Austria, First Results.* Amsterdam, The Netherlands: Proc. of 12th European Solar Energy Conference and Exhibition, 1994.


## Architectural Integration

Eiffert and Kiss. *Building-Integrated Photovoltaic Designs for Commercial and Institutional Structures: A Sourcebook for Architects.* Colorado, USA: National Renewable Energy Laboratories (NREL), 2000.

*Independent Energy Guide.* Vermont, USA: Chelsea Green Publishing Company, 1996.

Kiss, G. et al. *Building-Integrated Photovoltaics.* Colorado, USA: National Renewable Energy Laboratory, 1993.

Kiss, G. et al. *Building-Integrated Photovoltaics: A Case Study.* Colorado, USA: National Renewable Energy Laboratory (NREL), 1994.

Kiss, G. et al. *Optimal BIPV Applications.* Colorado, USA: National Renewable Energy Laboratory (NREL), 1995.

Jones, D.L. *Architecture and the Environment: Contemporary Green Buildings.* New York, USA: The Overlook Press, 1998.

NREL, *Photovoltaics in the Built Environment: A Design Guide for Architects and Engineers.* Colorado, USA: National Renewable Energy Laboratory, 1997.

Roaf, S. *Ecohouse: A Design Guide.* USA: Architectural Press, 2001.

Sick and Erge. *Photovoltaics in Buildings: A Design Handbook for Architects and Engineers.* London, UK: James & James Ltd. 1996.

### Installation and Maintenance

Cauldwell, Rex. *Wiring a House*. Connecticut, USA: The Taunton Press, Inc. 2002.

Kardon, Redwood, Hansen, Douglas, and Casey. *Code Check Electrical: A Field Guide to Wiring a Safe House*. Connecticut, USA: Taunton Press, Inc., 2002.

*Maintenance and Operation of Stand-Alone Photovoltaic Systems*. New Mexico, USA: Sandia National Laboratories, 1991.

Wiles, J. C. *Photovoltaic Power Systems and the National Electric Code*. New Mexico, USA: NM State University, 1991.

### Related Reading

Berger. *Charging Ahead: The Business of Renewable Energy and What it Means for America* New York, USA: Henry Holt and Company, 1997.

Butti, K., J. Perlin. *A Golden Thread: 2500 Years of Solar Architecture and Technology*. New York, USA: Cheshire Books, 1980.

Eiffert, P. *The Borrowers Guide to Financing Solar Energy Systems: A Federal Overview, 2nd Edition*. Colorado, USA: National Renewable Energy Laboratories, 1998.

Eiffert, P., G. Leonard, A. Thompson. *Guidelines for the Economic Analysis of Building Integrated Photovoltaic Systems*. Colorado, USA: National Renewable Energy Laboratories (NREL), 2001.

Haas, R. *The Value of Photovoltaic Electricity for Society*. USA: Solar Energy, Vol. 54, No.1, 1995.

Leggett, Jeremy, (Ed.). *Climate Change and the Financial Sector: The Emerging Threat - The Solar Solution*. Munich, Germany: Gerling Akademie Verlag, 1996.

## Publishers

**Chelsea Green Publishing Company:** A publisher of books on a variety of subjects related to sustainable living. P.O Box 428, White River Junction, Vermont 05001 USA.
Phone: 800-639-4099
Website: www.chelseagreen.com

**Intermediate Technology Publications Ltd. (ITDG):** A publisher that builds on the skills and capabilities of people in developing countries through the dissemination of information in many forms. They are an offshoot of Intermediate Technology Development Group, ITDG. ITDG aims to reduce poverty in countries through the use of sustainable technologies.
Phone: 01206 796351
Fax: 01206 799331
Email: sales@portlandpress.com
Website: www.itdg.org/

**Maya Books:** Environmental Publisher with sustainable technology titles. P.O. Box 379, Twickenham TW1 2SU. UK.
Phone/Fax: +44-(0)-181-287-9068
Email: sales@mayabooks.ndirect.co.uk
Website: www.mayabooks.ndirect.co.uk

**New Society Publishers (NSP):** Publishes books about how to build in a sustainable manner, and how to further a just society. All NSP books are printed in an environmental manner. P.O. Box 189, Gabriola Island, BC, V0R IX0, Canada.
Website: www.newsociety.com

## Newsletters & Journals

*Energy* – Monthly international multi-disciplinary resource journal for activities relating to the development, assessment, and management of energy-related topics. Elsevier Science Ltd., The Boulevard, Langford Lane, Kidlington, Oxford OX5 1GB, UK; 655 Ave. of the Americas, New York, NY 10010, USA.
Phone: 212-633-3730
Fax: 212-633-3680

*Photon* – das Solarstrom-Magazin: A bi-monthly magazine (in German) on PV and the PV industry, concentrating on Europe. Editor: Ms. Annegret Kreutzmann, Solar Verlag GmbH, Wilhelmstrasse 34, 52070, Aachen, Germany.
Phone: 0241-47055-0
Fax: 47055-9

*Photon* – The International Photovoltaic Magazine: English-language bi-monthly magazine covering the PV industry worldwide. Editor: Michael Schmela, Solar Verlag GmbH, Wilhelmstrasse 34, D-52070 Aachen, Germany.
Email: michael.schmela@photon-magazine.com
Website: www.photon-magazine.com

*Photovoltaics Bulletin* – Editor: Roberta Thomson, Elsevier Advanced Technology, P.O. BOX 150, Kidlington, Oxford 0X5 1AS, UK.
Phone: +44 1865 843 194
Fax: +44 1865 943 971
Email: R.Thomson@elsevier.co.uk

*Photovoltaic Insider's Report* – Monthly newsletter on the PV industry. Editor: Richard Curry, 1011 W. Colorado Blvd., Dallas, TX 75208, USA.
Phone/Fax: 214-942-5248
Email: rcurry@pvinsider.com
Website: www.pvinsider.com

*PV News* – Monthly newsletter on the PV industry. Editor: Paul Maycock, PV Energy Systems, 4539 Old Auburn Road, Warrenton, VA 20187, USA.
Phone/Fax: 540-349-4497
Email: pves@pvenergy.com
Website: www.pvenergy.com

*Renewable and Sustainable Energy Reviews* – An international journal of RE research. Editor-in-Chief: Lawrence L. Kazmerski, Elsevier Science, The Boulevard, Langford Lane, Kidlington, Oxford OX5 1GB, UK, 655 Ave. of the Americas, New York, NY 10010, USA.
Phone: 212-633-3730
Fax: 212-633-3680

*Renewable Energy* – Monthly international journal to promote and disseminate knowledge of renewable energy. Editor: Ali Sayigh, Elsevier Science Ltd., Langford Lane, Kidlington, Oxford OX5 1GB, UK.
Fax: +44 (0) 1865 843952;
or Elsevier Science, 655 Ave. of the Americas, New York, NY 10010, USA.
Phone: 212-633-3730
Fax: 212-633-3680

*Renewable Energy Bulletin* – Bi-monthly collection from a wide range of journals. Multi-Science Publishing Co., Ltd., 107 High St., Brentwood Esses CM14 4RX, UK.
Phone: +44 1277 223453;
or P.O. BOX 176, Avenel, NJ 07001, USA

*Renewable Energy World* – James & James (Science Publishers) Ltd 35-37 William Road, London NW1 3ER, United Kingdom.
Email: james@jxj.com
Website: www.jxj.com

*Solar Energy* – International journal for scientists, engineers, and technologists published by International Solar Energy Society (ISES). Editor: K.G. Terry Hollands, Elsevier Science Ltd., The Boulevard, Langford Lane, Kidlington, Oxford OX5 1GB, UK.
Fax +44 (0) 1865 843952;
or Elsevier Science, 655 Ave. of the Americas, New York, NY 10010, USA

*Solar Industry Journal* – Quarterly magazine including news, projects, and solar issues of the Solar Energy Industries Association (SEIA). SEIA, 122 C Street NW, Washington, DC 20001, USA.
Phone: 202-383-2600
Fax: 202-383-2670
Website: www.seia.org.

*Solar Today Magazine* – Official magazine of the American Solar Energy Society (ASES). Bi-monthly magazine covering all renewable energy applications, new products, and events. ASES, 2400 Central Avenue, Suite G-1, Boulder, CO 80301, USA.
Phone: 303-443-3130
Fax: 303-443-3212
Website: www.ases.org

*Sun World* – Quarterly magazine of the International Solar Energy Society (ISES). Editor: Leslie F. Jesch, The Franklin Company Consultants Ltd., 192 Franklin Road, Birmingham, B30 2HE, UK.
Email: sunworld@tfc-bham.demon.co.uk
Website: www.demon.co.uk/tfc/sunworld.html.

*The Solar Letter* – Bi-weekly newsletter on all aspects of RE. Editor: Allan L. Frank, ALFA Publishing, 9124 Bradford Rd., Silver Spring, MD 2090, USA.
Phone: 301-565-2532
Fax: 301-565-3298

## Videos

*Residential Solar Electricity with Johnny Weiss.* Practical answers given to the most often asked questions about designing and installing residential photovoltaic systems. Andrews, Scott, S. Renewable Energy with the Experts, 1997.

*Solar Water Pumping with Windy Dankoff.* Topics include watering livestock and crop irrigation. Answers given to the details of solar water pumping. Andrews, Scott, S. Renewable Energy with the Experts, 1998.

*Storage Batteries for Renewable Energy System with Richard Perez.* Considered the heart of a stand-alone systems, battery operation remains a mystery to many RE users. The role of batteries, their limitations and maintenance issues are presented in a clear and concise manner. Andrew, Scott S. Renewable Energy with the Experts, 1998.

## Selected PV Websites

**Australian National University**. Centre for Sustainable Energy Systems: Activities in the area of photovoltaics and concentrating solar thermal.
Website: www.engn.anu.edu.au/solar

**CANMET Energy Diversification Research Laboratory (Canada)**. R&D programs designed to help reduce greenhouse emissions, promote energy efficiency, and deploy renewable energy sources.
Website: www.cedrl.mets.nrcan.gc.ca/eng/programmes_retd.html

**Database of State Incentives for Renewable Energy**. Established in 1995, DSIRE is an ongoing project of the Interstate Renewable Energy Council (IREC), funded by the U.S. Department of Energy's Office of Power Technologies and managed by the North Carolina Solar Center. Comprehensive information on state, local, and utility incentives that promote renewable energy.
Website: www.dsireusa.org

**DOE PV Program**. Information on how a solar cell works complete with animation, links to other pages and documents.
Website: www.eren.doe.gov/pv/

**Energy Efficiency & Renewable Energy Network**. Comprehensive resource of the DOE's renewable energy and energy efficiency information, including 600 links and access to over 80,000 documents.
Website: www.eren.doe.gov

**Energy Star**. Introduced by the US Environmental Protection Agency in 1992 as a voluntary labeling program designed to identify and promote energy-efficient products, in order to reduce carbon dioxide emissions. EPA partnered with the US Department of Energy in 1996 to promote the ENERGY STAR label.
Website: www.energystar.gov

**Florida Solar Energy Center (FSEC)**. Research institute of the University of Florida. Site includes information on solar energy and photovoltaics, equipment testing, education and training, hydrogen energy and teacher resources.
Website: www.fsec.ucf.edu

**Georgia Tech, Univ. Center of Excellence for PV Research and Education**. DOE funded research facility working on fabrication of high efficiency PV cells and providing educational experiences.
Website: www.ece.gatech.edu/research/UCEP

**International Solar Center**. German based organization promoting renewable energies throughout the world.
Website: www.emsolar.ee.tu-berlin.de/iscb/home.html

**Million Solar Roofs, USA**. DOE program working to remove barriers to solar technologies. Information on financing and resources.
Website: www.eere.energy.gov

**National Renewable Energy Laboratory (NREL)**. DOE laboratory for renewable energy research & development. Located in Golden, CO USA. Website includes information on RE basics, national programs, image library, and links to many RE documents.
Website: www.nrel.gov

**Office of Scientific and Technical Information**. DOE's Science and Technology Information and Resources. Links to DOE's research and publications.
Website: www.osti.gov

**PV GAP (Global Approval Program for PV)**. A not-for-profit organization, registered in Switzerland, that certifies the quality of PV components.
Website: www.pvgap.org

**PV Power**. A site for the coordination and dissemination of information of global PV technologies, applications, history, and resources. Site includes a listing of PV manufacturers worldwide.
Website: www.pvpower.com

**PV Portal**. A link that contains breaking news in the photovoltaic industry about the globe, updated routinely the information is pertinant and cutting edge. Many links to manufacterers, designers and installers.
Website: www.pvportal.com

**Sacramento Municipal Utility District (SMUD) PV Program**. Innovative utility program, promoting conservation and PV for its customers.
Website: www.smud.org

**Sandia National Laboratories**. Sandia's PV program goals are to reduce the life-cycle cost of PV systems, reduce barriers to systems acceptance, provide systems best practices and guidelines, performance and reliability testing, standardization, and validation. Site includes basic information on PV systems and balance of system components. Several publications are available for free.
Website: www.sandia.gov/pv

**Solar Energy International (SEI)**. Authors of this text - *Photovoltaics: Design and Installation Manual*, teach SEI's Renewable Energy Education Program (REEP). SEI provides hands-on and online training in the practical use of solar, wind, and water power and in environmental building technologies. Website has training schedule.
Website: www.solarenergy.org

**Univ. of New South Wales-Center for PV Devices & Measurements**. Conducting solar cell research and PV education in Australia.
Website: www.pv.unsw.edu.au

**Utility Photovoltaic Group (UPVG)**. A collaboration of the photovoltaic industry working to create and encourage commercial use of new solar electric power business models. Site includes info on PV events, and utility industry news.
Website: www.solarelectricpower.org

## Technology Education Resources

**Arizona Solar Center**. Source for solar information in Arizona. Site contains maps, data, and AZ solar information.
Website: www.azsolarcenter.com/welcome.html

**Ecological Footprint Quiz**. A great place to go to figure out your rate of consumption within a world context. Fifteen questions are asked, from the answers it is determined how many Earths would be needed to support the entire human race consumed as you do.
Website: www.earthday.net/footprint/index.asp

**Rainbow Power Company Ltd.** Australian RE company. Designs, manufactures, sells, and installs renewable energy equipment.
Website: www.rpc.com.au

**San Juan College**. New Mexico community college offering an Associate Degree in Renewable Energy.
Website: www.sanjuancollege.edu

**Solar Energy International** (SEI). Renewable Energy Education Program teaches the practical use of solar, wind and water power through hands-on workshops and on-line education.
Website: www.solarenergy.org

**Sol Energy**. An interactive website that explains the basics of the solar energy to how a photovoltaic system utilizes such energy to produce electricity.
Website: www.projectsol.aps.com

**U.S. Department of Energy** (DOE). A website devoted to current energy practices within the USA.
Website: www.energy.gov

**U.S. Solar Radiation Resource Maps**. An insolation resource for technical solar radiation data including the extremes in the USA and nearby territories.
Website: http://rredc.nrel.gov/solar/old_data/nsrdb/redbook/atlas

## Associations

**American Council for an Energy-Efficient Economy (ACEEE).** A non-profit organization promoting energy efficiency as a means of fostering both economic prosperity and environmental protection. Offers a list of energy appliances, cars, and trucks. Website: www.aceee.org

**American Solar Energy Society (ASES).** A national organization dedicated to advancing the use of solar energy for the benefit of U.S. citizens and the global environment. Publishes *Solar Today Magazine*. Website: www.ases.org/index.html

**Institute for Sustainable Power (ISP).** Accreditation/Certification: Providing a globally recognized accreditation infrastructure of content modules, training guidelines, testing standards, and third-party qualification. Website: www.ispq-central.com

**Interstate Renewable Energy Council (IREC).** Non-profit organization whose mission is to accelerate the sustainable utilization of renewable energy sources and technologies in and through state and local government and community activities. Includes "Schools Going Solar" program. Website: www.irecusa.org

**North Carolina Solar Center.** Programs and resources for North Carolina and beyond. Services available to the public include a toll-free hotline, a professional referral service, technical assistance and design reviews, free publications, curriculum materials for teachers, training programs. Website: www.ncsc.ncsu.edu

**Center for Renewable Energy & Sustainable Technology (CREST).** Information on RE policy issues and RE in general. Website: www.solstice.crest.org/index.html

**International Solar Energy Society (ISES).** Worldwide membership organization with links and international solar information. Website: www.ises.org

**Solar Energy Industries Association (SEIA).** National trade association of solar energy manufacturers, dealers, distributors, contractors and installer for both PV and solar thermal. Site includes list of members, solar information, national energy policy, legislation and related news. Website: www.seia.org

## Solar Educational Resources

### ORGANIZATION:

| | Internet Courses | Hands-on Training | Children/Teachers | ISP Accredited | Women's Workshops | Degree programs |
|---|:---:|:---:|:---:|:---:|:---:|:---:|
| **Alternative Energy Institute**<br>West Texas A&M University<br>Box 60248; 2402 N. 3rd. Ave.<br>Canyon, TX 79016-0001<br>Phone: (806) 651-2295<br>Fax: (806) 651-2733<br>Email: aeimail@mail.wtamu.edu<br>Web: www.windenergy.org | ✔ | ✔ | ✔ | | | |
| **Appalachian State University**<br>Department of Technology<br>Boone, NC 28608<br>Phone: (828) 262-6361<br>Fax: (828) 265-8696<br>Email: scanlindm@appstate.edu<br>Web: www.appstate.edu | | ✔ | ✔ | | | ✔ |
| **Bronx Community College**<br>Center for Sustainable Energy<br>West 181st. Street & University Ave., GML-104<br>Bronx, New York 10453<br>Phone: (718) 289-5332<br>Fax: (718) 289-6443<br>Email: mail@csebcc.org<br>Web: www.csebcc.org | | ✔ | | | | |
| **Colorado State University**<br>Solar Energy Applications Laboratory (SEAL)<br>College of Engineering<br>Fort Collins, CO 80523<br>Phone: (970) 491-8617<br>Fax: (970) 491-8544<br>Email: seal@lamar.colostate.edu<br>Web: www.colostate.edu/Orgs/SEAL | | | | | ✔ | |
| **Farmingdale State University of New York**<br>Solar Energy Center<br>2350 Broadhollow Road<br>Farmingdale, NY 11735<br>Phone: 631-420-2000<br>Email: dathatyn@farmingdale.edu<br>Web: http://info.lu.farmingdale.edu/depts/<br>met/solar/fsec.html | | ✔ | | ✔ | | |

# Solar Educational Resources

## ORGANIZATION:

| | Internet Courses | Hands-on Training | Children/Teachers | ISP Accredited | Women's Workshops | Degree programs |
|---|---|---|---|---|---|---|
| **Florida Solar Energy Center (FSEC)**<br>1679 Clearlake Road<br>Cocoa, FL 32922-5703<br>Phone: (321) 638-1000<br>Fax: (321) 638-1010<br>Email: info@fsec.ucf.edu<br>Web: www.fsec.ucf.edu | | ✔ | ✔ | ✔ | | |
| **Georgia Institute of Technology**<br>University Center of Excellence for Photovoltaics Research and Education<br>School of Electrical and Computer Engineering<br>777 Atlantic Drive<br>Atlanta, GA 30332-0250<br>Phone: (404) 894-4658<br>Fax: (404) 894-4832<br>Email: ucep@ee.gatech.edu<br>Web: www.ece.gatech.edu | | | | | | ✔ |
| **Humboldt State University**<br>College of Natural Resources and Sciences<br>House 18, 1 Harpst St<br>Arcata, CA 95521<br>Phone: (707) 826-3619<br>Fax: (707) 826-3616<br>Email: ere_dept@Humboldt.edu<br>Web: www.humboldt.edu | | | | | | ✔ |
| **Iowa Renewable Energy Association**<br>PO Box 3405<br>Iowa City, IA 52244-3405<br>Phone: (319) 341-4372<br>Email: irenew@irenew.org<br>Web: www.irenew.org/workshops.html | ✔ | | | | | |
| **Lane Community College**<br>Science Division<br>Building 16, Room 156<br>4000 East 30th Ave.,<br>Eugene, OR 97405<br>Phone: (541) 463-5446<br>Fax (541) 463-3961<br>Web: http://lanecc.edu | | | | | | ✔ |

## Solar Educational Resources

ORGANIZATION:

| Organization | Internet Courses | Hands-on Training | Children/Teachers | ISP Accredited | Women's Workshops | Degree programs |
|---|---|---|---|---|---|---|
| **Midwest Renewable Energy Association (MREA)** | | ✔ | ✔ | ✔ | ✔ | |
| **North Carolina Solar Center** | ✔ | ✔ | | | | |
| **Northeast Sustainable Energy Association** | | ✔ | | | | |
| **Royal Institute of Technology** | | | | | | ✔ |
| **San Juan College** | ✔ | | | | | ✔ |

**Midwest Renewable Energy Association (MREA)**
7558 Deer Road, Custer, WI 54423
Phone: (715) 592-6595
Fax: (715) 592-6596
Email: info@the-mrea.org
Web: www.the-mrea.org

**North Carolina Solar Center**
North Carolina State University
Box 7401
Raleigh, NC 27695-7401
Phone: (919) 515-5666
Fax: (919) 515-5778
Email: ncsun@ncsu.edu
Web: www.ncsc.ncsu.edu

**Northeast Sustainable Energy Association**
50 Miles Street
Greenfield, MA 01301
Phone: (413) 774-6051
Email: nesea@nesea.org
Web: www.nesea.org/

**Royal Institute of Technology**
Department of Energy Technology
SE-100 44 Stockholm, Sweden
Web: www.kth.se.eng

**San Juan College**
4601 College Blvd
Farmington, NM 87402
Phone: (505) 326-3311
Email: bickfordc@sanjuancollege.edu
Web: www.sanjuancollege.edu

# Solar Educational Resources

## ORGANIZATION:

| | Internet Courses | Hands-on Training | Children/Teachers | ISP Accredited | Women's Workshops | Degree programs |
|---|---|---|---|---|---|---|
| **Solar Energy International (SEI)**<br>P.O. Box 715<br>Carbondale, CO 81623<br>Phone: (970) 963-8855<br>Fax: (970) 963-8866<br>Email: sei@solarenergy.org<br>Web: www.solarenergy.org | ✔ | ✔ | ✔ | ✔ | ✔ | |
| **Solar Living Institute**<br>PO Box 836<br>13771 S. Highway 101<br>Hopland, CA 95449<br>Ph: (707) 744-2017<br>Fax: (707) 744-1682<br>Email: sli@solarliving.org<br>Web: www.solarliving.org | | ✔ | ✔ | | ✔ | |
| **SoLEnergy**<br>P.O. Box 217<br>Carbondale, CO 81623<br>Fax: (559) 751-2001<br>Email: SoL@SoLenergy.org<br>Web: www.solenergy.org | ✔ | ✔ | | | | |
| **Southwest Technology Development Institute**<br>New Mexico State University<br>P.O. Box 30001<br>MSC 3 Solar<br>Las Cruces, NM 88003-8001<br>Phone: (505) 646-1049<br>Fax: (505) 646-3841<br>Email: tdi@nmsu.edu<br>Web: www.NMSU.Edu/~tdi/ | | ✔ | | | | |
| **Sunnyside Solar**<br>1014 Green River Road<br>Guilford, VT 05301<br>Phone: (802) 257-1482<br>Fax: (802) 254-4670<br>Email: info@sunnysidesolar.com<br>Web: www.sunnysidesolar.com | | ✔ | | | | |

## Solar Educational Resources

### ORGANIZATION:

| | Internet Courses | Hands-on Training | Children/Teachers | ISP Accredited | Women's Workshops | Degree programs |
|---|---|---|---|---|---|---|
| **University of Massachusetts-Amherst** | | | | | | ✔ |

Mechanical and Industrial Engineering Department
Box 2210
Amherst, MA 01003-2210
Phone: (413) 545-2505
Fax: (413) 545-1027
Email: mie@ecs.umass.edu
Web: http://energy.caeds.eng.uml.edu/

| | | | | | | |
|---|---|---|---|---|---|---|
| **University of Massachusetts-Lowell** | | | | | | ✔ |

Mechanical Engineering Department
One University Avenue
Lowell, MA 01854
Phone: (978) 934-2968
Fax: (978) 934-3048
Email: john-duffy@uml.edu
Web: www.eng.uml.edu/Dept/Energy

| | | | | | | |
|---|---|---|---|---|---|---|
| **University of South Wales** | | | | | | ✔ |

The Centre for Photovoltaic Engineering
Electrical Engineering Building
Sydney, N.S.W, 2052 Australia
Phone: (02) 9385 4018
Email: pv.labs@unsw.edu.au
Web: www.pv.unsw.edu.au/

| | | | | | | |
|---|---|---|---|---|---|---|
| **University of Wisconsin at Madison** | | | | | | ✔ |

Solar Engineering Program
1303 Engineering Research Bldg.
1500 Engineering Drive Madison, WI 53706-1687
Email: beckman@engr.wisc.edu
Web: http://sel.me.wisc.edu/

**FOR OTHER INTERNATIONAL TRAINING OPPORTUNITIES CONTACT:**
**International Solar Energy Society (ISES)**
Villa Tannheim
Wiesentalstr. 50
79115 Freiburg, Germany
Tel.: +49 - 761 - 45906-0
Fax: +49 - 761 - 45906-99
Email: hq@ises.org
Web: www.ises.org/

# Index

absorbed glass mat (AGM) batteries . . . . . . . . . 61

alkaline batteries . . . . . . . . . . . . . . . . . . . . . . . 62

alternating current (AC), defined . . . . . . . . 10, 84

altitude (solar), defined . . . . . . . . . . . . . . . . . 33

American Wire Gauge (AWG) . . . . . . . . . . . . 94

amp-hour, defined . . . . . . . . . . . . . . . . . . . . . 10

ampere, defined . . . . . . . . . . . . . . . . . . . . . . . 10

appliances

   see also:

      loads

   electric resistance . . . . . . . . . . . . . . . . . . . . 38

   matching to the system . . . . . . . . . . . . . . . 11

   refrigeration and health care . . . . . . . . . . . 147

   refrigeration options . . . . . . . . . . . . . . . . . 40

   TV and radio in the developing world . . . 147

   typical wattage requirements . . . . . . . . . . . 46

array (photovoltaic) . . . . . . . see photovoltaic array

automotive batteries, problems with . . . . . . . . 146

azimuth, defined . . . . . . . . . . . . . . . . . . . . . . . 31

batteries

   automotive batteries . . . . . . . . . . . . . . 60, 146

   battery capacity . . . . . . . . . . . . . . . . . . . . . 63

   building a battery enclosure . . . . . . . . . . . 159

   captive electrolyte . . . . . . . . . . . . . . . . . . . 61

   charge termination voltage . . . . . . . . . . . . . 62

   chemical burns, first aid . . . . . . . . . . . . . . 187

   chemical hazards . . . . . . . . . . . . . . . . . . . . 184

   days of autonomy, factors of . . . . . . . . . . . 62

   environmental conditions, effects of . . . . . . 65

   installation checklist . . . . . . . . . . . . . . . . . 163

   installation considerations . . . . . . . . . . . . . 158

   life expectancy . . . . . . . . . . . . . . . . . . . . . 65

   liquid electrolyte . . . . . . . . . . . . . . . . . 61, 67

   maintenance . . . . . . . . . . . . . . . . . . . . . . . 168

      determining state-of charge by measuring

      specific gravity . . . . . . . . . . . . . . . . . . 169

   measuring state of charge . . . . . . . . . . . . . 67

   NEC® regulations . . . . . . . . . . . . . . . . . . . 163

   rate and depth of discharge . . . . . . . . . . . . 64

   reverse polarity . . . . . . . . . . . . . . . . . . . . . 64

   safety rules . . . . . . . . . . . . . . . . . . . . . . . . 68

   sizing exercise . . . . . . . . . . . . . . . . . . . . . . 71

   sizing for stand-alone . . . . . . . . . . . . . . . . 117

   temperature, effects of . . . . . . . . . . . . . . . . 65

types of . . . . . . . . . . . . . . . . . . . . . . . . . . . . . . 60

voltage set points . . . . . . . . . . . . . . . . . . . . . . 63

wiring configuration . . . . . . . . . . . . . . . . . . . . 69

bracket mounting . . . . . . . . . . . . . . . . . . . . . . 141

building integrated photovoltaics . . . . . . . . . . . 143

   facade walls and awnings . . . . . . . . . . . . . 143

   financial considerations . . . . . . . . . . . . . . . 144

   roofing . . . . . . . . . . . . . . . . . . . . . . . . . . . 143

   skylights and windows . . . . . . . . . . . . . . . 143

cell (photovoltaic), defined . . . . . . . . . . . . . . . 50

cell temperature, effects of . . . . . . . . . . . . . . . 55

charge controllers

   features of . . . . . . . . . . . . . . . . . . . . . . . . . 75

   installation checklist . . . . . . . . . . . . . . . . . 163

   installation considerations . . . . . . . . . . . . . 159

   maximum power point tracking . . . . . . . . . 75

   optional features . . . . . . . . . . . . . . . . . . . . 77

   recommended features . . . . . . . . . . . . . . . . 76

   sizing exercise . . . . . . . . . . . . . . . . . . . . . . 79

   sizing for stand-alone . . . . . . . . . . . . . . . . 118

   specifying a controller . . . . . . . . . . . . . . . . 78

   temperature compensation . . . . . . . . . . . . . 77

   types of . . . . . . . . . . . . . . . . . . . . . . . . . . . 74

   voltage step down . . . . . . . . . . . . . . . . . . . 76

checklists

   batteries . . . . . . . . . . . . . . . . . . . . . . . . . . 163

   before testing the system . . . . . . . . . . . . . . 162

   charge controllers . . . . . . . . . . . . . . . . . . . 163

   disconnects . . . . . . . . . . . . . . . . . . . . . . . . 163

   grounding . . . . . . . . . . . . . . . . . . . . . . . . . 165

   installation tools and materials . . . . . . . . . 155

   inverters in battery-based systems . . . . . . . 164

   inverters in grid-tied systems . . . . . . . . . . . 164

   overcurrent protection . . . . . . . . . . . . . . . . 162

   PV array . . . . . . . . . . . . . . . . . . . . . . . . . . 162

   safety signs . . . . . . . . . . . . . . . . . . . . . . . . 165

   troubleshooting . . . . . . . . . . . . . . . . . . . . . 173

   wiring . . . . . . . . . . . . . . . . . . . . . . . . . . . . 162

circuit breakers . . . . . . . . . . . . . . . . . . . . . . . . 107

closed circuit, defined . . . . . . . . . . . . . . . . . . . 11

compact fluorescent lamps . . . . . . . . . . . . . . . 42

321

concentrator modules, defined . . . . . . . . . . . . . 51
conduit, defined . . . . . . . . . . . . . . . . . . . . 94
continuity, how to check . . . . . . . . . . . . . . 171
CPR . . . . . . . . . . . . . . . . . . . . . . . . 187
crystalline silicon . . . . . . . . . . . . . . . . . . 51
current, how to measure . . . . . . . . . . . . . . 172
cycling loads, defined . . . . . . . . . . . . . . . . 38
day use systems, overview . . . . . . . . . . . . . . 4
deep cycle batteries . . . . . . . see lead acid batteries
deration factors, modules . . . . . . . . . . . . . . 136
design month, defined . . . . . . . . . . . . . . . . 35
direct current (DC), defined . . . . . . . . . . . 10
direct mount . . . . . . . . . . . . . . . . . . . . 141

disconnects
    see also:
        overcurrent protection
        wiring
    AC/DC disconnects in inverters . . . . . . . . 87
    alternating current rated switches . . . . . . . 161
    installation checklist . . . . . . . . . . . . . . 163
    placement and specification of . . . . . . . . 108
    safety requirements . . . . . . . . . . . . . . . 181

diversion controllers . . . . . . . . . . . . . . . . . 74
duty cycle, defined . . . . . . . . . . . . . . . . . . 38
electric shock, defined . . . . . . . . . . . . . . . 186

electrical
    boxes . . . . . . . . . . . . . . . . . . . . . . 160
    circuits . . . . . . . . . . . . . . . . . . . . . 11
    common connectors . . . . . . . . . . . . . . 160
    hazards . . . . . . . . . . . . . . . . . . . . . 183
    injuries . . . . . . . . . . . . . . . . . . . . . 186
    terminology . . . . . . . . . . . . . . . . . . . 10
    types of current . . . . . . . . . . . . . . . . . 10

electrical circuit, defined . . . . . . . . . . . . . . 11
electricity, defined . . . . . . . . . . . . . . . . . . 10
equipment grounding . . . . . . . . . . . . . . . . 109

first aid . . . . . . . . . . . . . . . . . . . . . . . 186
    burns . . . . . . . . . . . . . . . . . . . . . . 187
    CPR . . . . . . . . . . . . . . . . . . . . . . . 187
    electrical injuries . . . . . . . . . . . . . . . . 186
    non-electrical injuries . . . . . . . . . . . . . 188

fuses . . . . . . . . . . . . . . . . . . . . . . . . . 107
gel cell batteries . . . . . . . . . . . . . . . . . . . 61
grid-tied systems
    advantages of . . . . . . . . . . . . . . . . . . 126
    calculating annual energy production . . . . . 34
    economics of . . . . . . . . . . . . . . . . . . 130
    NEC® requirements . . . . . . . . . . . . . . 131
    net metering . . . . . . . . . . . . . . . . . . 131
    obtaining an interconnection agreement . . 131
    overview . . . . . . . . . . . . . . . . . . . . . 7
    PURPA regulations . . . . . . . . . . . . . . . 126
    system sizing exercise . . . . . . . . . . . . . 133
    system sizing instructions . . . . . . . . . . . 130
    system sizing worksheet . . . . . . . . . . . . 134
    types of . . . . . . . . . . . . . . . . . . . . . 126
    uninterruptible power supply . . . . . . . . . 127

ground mounting . . . . . . . . . . . . . . . . . . 141
ground-fault protection . . . . . . . . . . . . . 87, 109

grounding
    equipment grounding . . . . . . . . . . . . . 109
    ground-fault protection . . . . . . . . . . . . 109
    grounding electrodes . . . . . . . . . . . . . . 113
    installation checklist . . . . . . . . . . . . . . 165
    safety requirements . . . . . . . . . . . . . . . 181
    sizing conductors for . . . . . . . . . . . . . . 109
    system grounding . . . . . . . . . . . . . . . . 109
    terminology . . . . . . . . . . . . . . . . . . . 108

hazards . . . . . . . . . . . . . . . . . . . . . . . . 182
high voltage PV arrays . . . . . . . . . . . . . . . 14
high-frequency transformers . . . . . . . . . . . . 84

hybrid systems
    advantages of . . . . . . . . . . . . . . . . . . 124
    hybrid systems, overview . . . . . . . . . . . . 7
    PV-generator systems . . . . . . . . . . . . . . 124

hydrogen gas . . . . . . . . . . . . . . . . . . . . . 61
hydrometer, defined . . . . . . . . . . . . . . . . . 169
I-V curve, defined . . . . . . . . . . . . . . . . . . 52
Imp, defined . . . . . . . . . . . . . . . . . . . . . 52
incandescent lamps . . . . . . . . . . . . . . . . . 42
insolation, defined . . . . . . . . . . . . . . . . . . 30

installation
    battery . . . . . . . . . . . . . . . . . . . . . . . . . 158
    controller and inverters . . . . . . . . . . . . . 159
    photovoltaic array . . . . . . . . . . . . . . . . . 158
    site safety . . . . . . . . . . . . . . . . . . . . . . . 185
    system checklist . . . . . . . . . . . . . . . . . . . 162
    tools and materials . . . . . . . . . . . . . . . . 155
    wiring . . . . . . . . . . . . . . . . . . . . . . . . . . 160

interconnection agreement, defined . . . . . . . . 131

inverters
    AC coupled system . . . . . . . . . . . . . . . . . 89
    AC/DC disconnects . . . . . . . . . . . . . . . . . 87
    as charge controller . . . . . . . . . . . . . . . . 88
    batteryless grid-tied inverters . . . . . . . . . . . 86
    features of . . . . . . . . . . . . . . . . . . . . . . . 85
    grid-tied with battery back-up inverters . . . 87
    ground fault protection (GFP) . . . . . . . . . . 87
    installation checklist . . . . . . . . . . . . . . . . 164
    installation considerations . . . . . . . . . . . . 159
    low-voltage disconnect (LVD) . . . . . . . . . . 88
    maximum power point tracking (MPPT) . . 87
    operating principles . . . . . . . . . . . . . . . . . 84
    sizing exercise . . . . . . . . . . . . . . . . . . . . . 90
    sizing for stand-alone . . . . . . . . . . . . . . . 118
    stand-alone inverters . . . . . . . . . . . . . . . . 88
    surge capacity . . . . . . . . . . . . . . . . . . . . . 88
    types of . . . . . . . . . . . . . . . . . . . . . . . . . 84

Isc, defined . . . . . . . . . . . see short circuit current
kilowatt-hour, defined . . . . . . . . . . . . . . . . . 10
lead-acid batteries . . . . . . . . . . . . . . . . . . . 60
LED lamps . . . . . . . . . . . . . . . . . . . . . . . . 42

lighting
    kerosene lanterns . . . . . . . . . . . . . . . . . . 146
    lamp types . . . . . . . . . . . . . . . . . . . . . . . 42
    lighting controls . . . . . . . . . . . . . . . . . . . 41
    lighting, developing world applications . . . 146

liquid vented lead-acid batteries . . . . . . . . . . . 61
load shifting, defined . . . . . . . . . . . . . . . . . . 38

loads
    developing world needs . . . . . . . . . . . . . . 146
    duty cycle . . . . . . . . . . . . . . . . . . . . . . . . 38
    estimating loads, calculations . . . . . . . . . . 43
    estimating loads, considerations . . . . . . . . 43
    estimating surge requirements . . . . . . . . . . 39
    health care and refrigeration . . . . . . . . . . 147
    inductive load, defined . . . . . . . . . . . . . . . 54
    lighting . . . . . . . . . . . . . . . . . . . . . . . . . . 41
    load calculation exercises . . . . . . . . . . . . . 44
    matching loads to the system . . . . . . . . . . 54
    refrigeration load estimating . . . . . . . . . . . 40
    resistive heating . . . . . . . . . . . . . . . . . . . . 38
    typical wattage requirements . . . . . . . . . . . 46
    using energy efficiently . . . . . . . . . . . . . . . 38

low voltage disconnect . . . . . . . . . . . . . . . . . 61
magnetic declination, defined . . . . . . . . . . . . 31

maintenance
    batteries . . . . . . . . . . . . . . . . . . . . . . . . 168
    determining state-of charge . . . . . . . . . . . 169
    PV array . . . . . . . . . . . . . . . . . . . . . . . . 168

material lists
    basic tools . . . . . . . . . . . . . . . . . . . . . . . 155
    for maintenance . . . . . . . . . . . . . . . . . . . 168
    initial site visit tools . . . . . . . . . . . . . . . . 155
    motor transported installation tools . . . . . 155
    non-motor transported installation tools . . 155
    safety equipment . . . . . . . . . . . . . . . . . . 184
    sample installation materials checklist . . . . 156

maximum power point tracking (MPPT) . . 75, 87
maximum power point, defined . . . . . . . . . . . 52
modified square-wave inverters . . . . . . . . . . . 85
module (photovoltaic) . . . see photovoltaic module

mounting
    installation and safety . . . . . . . . . . . . . . . 158
    types and considerations . . . . . . . . . . . . . 140

National Electrical Code® 2005 (NEC®)
    batteries . . . . . . . . . . . . . . . . . . . . . . . . . 163
    disconnects . . . . . . . . . . . . . . . . . . . 108, 165
    ground-fault protection . . . . . . . . . . . . . 109
    grounding . . . . . . . . . . . . 109, 110, 113, 165
    overcurrent protection . . . . . . . . . . . 107, 109
    safety signs and labels . . . . . . . . 113, 162, 165
    wiring . . . . . . . . . . . . . . . . 92-96, 110, 162

National Renewable Energy Laboratory (NREL) 136

net metering . . . . . . . . . . . . . . . . . . . . . . . . 131
    benefits and costs . . . . . . . . . . . . . . . . . . 132
    using the existing meter . . . . . . . . . . . . . 133

open circuit voltage (Voc), defined . . . . . . . . . 53
open circuit, defined . . . . . . . . . . . . . . . . . . . 11

overcurrent protection
    installation checklist . . . . . . . . . . . . . . . . 162
    placement of . . . . . . . . . . . . . . . . . . . . . 107
    sizing exercise . . . . . . . . . . . . . . . . . . . . 107
    sizing of . . . . . . . . . . . . . . . . . . . . . . . . 107
    standard ampacity ratings . . . . . . . . . . . . 107

p-n junction, defined . . . . . . . . . . . . . . . . . . 51
panel (photvoltaic), defined . . . . . . . . . . . . . 50

parallel wiring
    batteries in parallel . . . . . . . . . . . . . . 13, 69
    inverters in parallel . . . . . . . . . . . . . . . . . 86
    loads in parallel . . . . . . . . . . . . . . . . . . . 15
    parallel circuits, defined . . . . . . . . . . . . . . 12
    wiring exercise . . . . . . . . . . . . . . . . . . . . 16

peak load demand . . . . . . . . . . . . . . . . . . . 132
peak sun hours, defined . . . . . . . . . . . . . . . . 30
phantom loads, defined . . . . . . . . . . . . . . . . 39

photovoltaic (PV)
    principles of . . . . . . . . . . . . . . . . . . . . . . 50
    system components . . . . . . . . . . . . . . . . . . 4
    system types . . . . . . . . . . . . . . . . . . . . . . 4

photovoltaic array
    array sizing . . . . . . . . . . . . . . . . . . . . . 117
    defined . . . . . . . . . . . . . . . . . . . . . . . 4, 50
    installation checklist . . . . . . . . . . . . . . . . 162
    installation considerations . . . . . . . . . . . . 158
    maintenance . . . . . . . . . . . . . . . . . . . . . 168
    measuring circuit current . . . . . . 53, 172, 182
    measuring circuit voltage . . . . . . 53, 171, 182
    mounting options . . . . . . . . . . . . . . . . . 140
photovoltaic module
    characteristics of . . . . . . . . . . . . . . . . . . . 51
    defined . . . . . . . . . . . . . . . . . . . . . . . 4, 50
    deration factors . . . . . . . . . . . . . . . . . . . 136
    factors of performance . . . . . . . . . . . . . . . 54
    maximum power point . . . . . . . . . . . . . . . 52
    open circuit voltage . . . . . . . . . . . . . . . . . 53
    shading, effects of . . . . . . . . . . . . . . . . . . 56
    short circuit current . . . . . . . . . . . . . . . . . 53
    Standard Test Conditions . . . . . . . . . . . . . 53

photovoltaic technology . . . . . . . . . . . . . . . . . 2
    advantages of . . . . . . . . . . . . . . . . . . . . . . 3
    development of . . . . . . . . . . . . . . . . . . . . . 2
    disadvantages of . . . . . . . . . . . . . . . . . . . . 3
    environmental, health, and safety Issues . . . . 3
    photovoltaic reaction . . . . . . . . . . . . . . . . 51

polarity . . . . . . . . . . . . . . . . . . . . . . . . . . 172
    how to check . . . . . . . . . . . . . . . . . . . . 172
pole mounting . . . . . . . . . . . . . . . . . . . . . . 141
polyvinyl chloride (PVC) pipe . . . . . . . . . . . . 94
power center, defined . . . . . . . . . . . . . . . . . . 77
PTC ratings, defined . . . . . . . . . . . . . . . . . . 55
Public Utility Regulatory Policies Act (PURPA) 126
pulse width modulation controllers . . . . . . . . . 75
PV Watts, calculation program . . . . . . . . 34, 136
rack mount . . . . . . . . . . . . . . . . . . . . . . . . 141
recombinator cell caps . . . . . . . . . . . . . . . . . 61
refrigeration . . . . . . . . . . . . . . . . . . . . . . . . 39
reverse polarity . . . . . . . . . . . . . . . . . . . . . 172
roof mounting . . . . . . . . . . . . . . . . . . . . . . 141
rural electrification projects . . . . . . . . . . . . . 146

safety
    basic safety . . . . . . . . . . . . . . . . . . . . . . . 180
    first aid . . . . . . . . . . . . . . . . . . . . . . 186
    hazards . . . . . . . . . . . . . . . . . . . . . . 182
    safety equipment . . . . . . . . . . . . . . . . . 184
    safety precautions . . . . . . . . . . . . . . . . 170
    safety signs . . . . . . . . . . . . . . . . . . . . 165
    site safety . . . . . . . . . . . . . . . . . . . . . . 185
    testing high voltage . . . . . . . . . . . . . . . 182

sealed lead-acid batteries . . . . . . . . . . . . . . . . 61

series wiring
    batteries in series . . . . . . . . . . . . . . . 13, 69
    inverters in series . . . . . . . . . . . . . . . . . 86
    loads in series . . . . . . . . . . . . . . . . . . . 15
    series circuits, defined . . . . . . . . . . . . . 12
    wiring exercises . . . . . . . . . . . . . . . . . . 16

shading, effects of . . . . . . . . . . . . . . . . . . . 56
short circuit current (Isc), defined . . . . . . . . . . 53
shunt controllers . . . . . . . . . . . . . . . . . . . . 74
sine-wave Inverters . . . . . . . . . . . . . . . . . . 85
single-stage series controllers . . . . . . . . . . . . 74

site analysis
    determining solar access . . . . . . . . . . . . . 36
    determining solar access with a sun chart . 152
    gathering site data . . . . . . . . . . . . . . . . 35
    identifying shading obstacles . . . . . . . . . . 36
    site evaluation . . . . . . . . . . . . . . . . . . 158
    solar orientation . . . . . . . . . . . . . . . . . 31
    sun chart . . . . . . . . . . . . . . . . . . . . . . 36
    sun's path . . . . . . . . . . . . . . . . . . . . . 30
    tilt angle . . . . . . . . . . . . . . . . . . . . . . 33
    world magnetic declination chart . . . . . . . 32

solar noon, defined . . . . . . . . . . . . . . . . . . 33
solar window, defined . . . . . . . . . . . . . . . . . 36
square wave inverters . . . . . . . . . . . . . . . . . 84

stand-alone systems
    design penalties . . . . . . . . . . . . . . . . . 116
    developing world applications . . . . . . . . . 145
    hybrid systems . . . . . . . . . . . . . . . . . . 124
    system sizing . . . . . . . . . . . . . . . . . . . 117
        array sizing . . . . . . . . . . . . . . . . . 117
        battery sizing . . . . . . . . . . . . . . . . 117
        controller specification . . . . . . . . . . . 118
        electric load estimation . . . . . . . . . . 117
        inverter specification . . . . . . . . . . . . 118
    system sizing exercise . . . . . . . . . . . . . . 119
    system sizing worksheet . . . . . . . . . . . . 121

stand-off mount . . . . . . . . . . . . . . . . . . . . 142

STC ratings . . . . . . . . . . . . . . . . . . . . . . . 55
    factors of module performance . . . . . . . . . 136
surge capacity . . . . . . . . . . . . . . . . . . . . . 89
surge loads, defined . . . . . . . . . . . . . . . . . . 39
surge suppression . . . . . . . . . . . . . . . . . . . 113
system grounding . . . . . . . . . . . . . . . . . . . 109
temperature compensation . . . . . . . . . . . . . . 77
temperature deration . . . . . . . . . . . . . . . . . 136
tools . . . . . . . . . . . . . . . . . . . see materials
tracking mounting . . . . . . . . . . . . . . . . . . 142
transformer, defined . . . . . . . . . . . . . . . . . 84

troubleshooting
    checking for continuity . . . . . . . . . . . . . 171
    checking polarity . . . . . . . . . . . . . . . . . 172
    common system faults . . . . . . . . . . . . . 170
    measuring voltage . . . . . . . . . . . . . . . . 171
    meauring current . . . . . . . . . . . . . . . . 172
    specific problems and remedies . . . . . . . . 173
    using a multimeter . . . . . . . . . . . . . . . 170

uninterruptible power supply (UPS) . . . . . . . . 127
Vmp, defined . . . . . . . . see maximum power point
Voc, defined . . . . . . . . . . . see open circuit voltage
volt, defined . . . . . . . . . . . . . . . . . . . . . . 10
voltage drop index . . . . . . . . . . . . . . . . . . 103
voltage drop, defined . . . . . . . . . . . . . . . . . 96
voltage step down controllers . . . . . . . . . . . . 76
voltage, how to measure . . . . . . . . . . . . . . . 171

water pumping

    developing world applications . . . . . . . . . . 148

    pump curves . . . . . . . . . . . . . . . . . . . . . 150

    pump selection criteria . . . . . . . . . . . . . . 150

    pump terminology . . . . . . . . . . . . . . . 148

    pump types . . . . . . . . . . . . . . . . . . . . . 149

    sample system designs . . . . . . . . . . . . . . 151

    sizing exercise . . . . . . . . . . . . . . . . . . . 150

    storage and delivery . . . . . . . . . . . . . . . 150

    typical load . . . . . . . . . . . . . . . . . . . . . 46


watt, defined . . . . . . . . . . . . . . . . . . . . . . 10

watt-hour, defined . . . . . . . . . . . . . . . . . . . 10

waveform, inverter types . . . . . . . . . . . . . . . 84


wiring

    see also:

        disconnects

        grounding

        National Electical Code

        overcurrent protection


    American Wire Gauge (AWG) . . . . . . . . . . 94

    ampacity of wire . . . . . . . . . . . . . . . . . . 94

    color coding of wires . . . . . . . . . . . . . . . 93

    conductor material . . . . . . . . . . . . . . . . 92

    conduit, and . . . . . . . . . . . . . . . . . . . . 93

    connection materials . . . . . . . . . . . . . . . 160

    connection types . . . . . . . . . . . . . . . . . 160

    continuity, how to check . . . . . . . . . . . . 171

    installation checklist . . . . . . . . . . . 161, 162

    installation considerations . . . . . . . . . . . 160

    insulation . . . . . . . . . . . . . . . . . . . . . . 93

    safety factors . . . . . . . . . . . . . 95, 100, 101

    safety precautions . . . . . . . . . . . . . . . . 170

    safety requirements . . . . . . . . . . . . . . . 181

    sizing equipment grounding conductor . . . 109

    sizing exercises . . . . . . . . . . . . . 95, 96, 100

    sizing grounding electrode conductor . . . . 113

    sizing instructions . . . . . . . . . . . . . . . . 94

    sizing worksheet . . . . . . . . . . . . . . . . . 104

    surge suppression . . . . . . . . . . . . . . . . 113

    troubleshooting . . . . . . . . . . . . . . . . . 170

    types of wire . . . . . . . . . . . . . . . . . . . . 92

    voltage drop . . . . . . . . . . . . . 96, 101, 102

    wire sizing charts . . . . . . . . . . . . 97, 98, 99


worksheets

    . . . . . . . . see List of Tables and Worksheets pg. xvi

## SOLAR ENERGY INTERNATIONAL (SEI)

PO Box 715
Carbondale, Colorado 81623-0715
Phone: 970-963-8855
Fax: 970-963-8866
Email: sei@solarenergy.org
Website: www.solarenergy.org

Solar Energy International (SEI) is a non-profit educational organization whose mission is to help others use renewable energy and environmental building technologies through education and technical assistance. SEI was founded in 1992 in the belief that renewable energy resources of sun, wind and water can improve the quality of life and promote a sustainable future for people throughout the world. SEI educates decision-makers, technicians and users of renewable energy systems. Decision-makers gain the information they need to choose renewable energies with confidence. Technicians and users learn the practical skills they need to implement renewable energy technologies sustainably.

SEI's programs include the Renewable Energy Education Program (REEP), International Training courses, Solar in the Schools, and International Volunteers in Environmentally Sustainable Technologies (INVEST).

SEI's main adult educational program is REEP. Each year hundreds of people from around the world attend *hands-on* workshops seeking practical experience and skills to use renewable energy resources and technologies. Classroom and laboratory work are complemented by case studies, field tours and professional installations with commercial equipment in practical applications. Instruction is provided by industry experts in the following workshops:

- Photovoltaic (PV) Design & Installation
- Advanced Photovoltaics
- Utility-Interactive Photovoltiacs
- Women's PV Design & Installation
- PV Lab Week
- PV Safety
- Residencial Wind Power
- Homebuilt Wind Generators
- Micro-Hydro Power
- Introduction to Renewable Energy
- Sustainable Home Design
- Designing and Building Natural Homes
- Straw-Bale Design and Construction
- Plaster for Natural Homes
- Solar Water Pumping
- Solar and Radiant Heating
- Solar Hot Water
- Renewable Energy for the Developing World
- Appropriate Technology for the Developing World
- Biodiesel Fuel
- Successful Solar Businesses
- Photovoltaic Design Online
- Advanced Photovoltaics Online
- Sustainable Home Design Online

SEI personnel travel domestically and internationally training people in renewable energy technologies. SEI has taught thousands of students how to install renewable energy systems in the United States, Latin America, Asia, Africa, the Pacific and the Caribbean. SEI's international programs stress in-country training of trainers, decision makers, technicians, and end-users. Developing local and regional capabilities at each of these levels is critical to successful renewable energy utilization. Standard training packages and custom programs are available to meet particular program needs.

SEI's Solar In the Schools program targets grade school youth and focuses on experiential learning of energy concepts and issues. The goal is to have students understand energy as it relates to all living things on the planet. SEI's INVEST program (International Volunteers in Environmentally Sustainable Technologies) offers alumni of our workshops an opportunity to volunteer overseas with one of our partner organizations to help bring renewable energy technologies to communities in the developing world.

If you have enjoyed *Photovoltaics: Design and Installation Manual* you might also enjoy other

# BOOKS TO BUILD A NEW SOCIETY

Our books provide positive solutions for people who want to make a difference. We specialize in:

**Environment and Justice • Conscientious Commerce
Sustainable Living • Ecological Design and Planning
Natural Building & Appropriate Technology • New Forestry
Educational and Parenting Resources • Nonviolence
Progressive Leadership • Resistance and Community**

## New Society Publishers

### ENVIRONMENTAL BENEFITS STATEMENT

New Society Publishers has chosen to produce this book on Enviro 100, recycled paper made with **100% post consumer waste**, processed chlorine free, and old growth free.

For every 5,000 books printed, New Society saves the following resources:[1]

| | |
|---:|---|
| 33 | Trees |
| 3,011 | Pounds of Solid Waste |
| 3,313 | Gallons of Water |
| 4,321 | Kilowatt Hours of Electricity |
| 5,474 | Pounds of Greenhouse Gases |
| 24 | Pounds of HAPs, VOCs, and AOX Combined |
| 8 | Cubic Yards of Landfill Space |

[1]Environmental benefits are calculated based on research done by the Environmental Defense Fund and other members of the Paper Task Force who study the environmental impacts of the paper industry.

*For a full list of NSP's titles, please call* **1-800-567-6772** *or check out our website at:*

**www.newsociety.com**

NEW SOCIETY PUBLISHERS